TROPICAL FOREST ECOLOGY

TROPICAL FOREST ECOLOGY

A VIEW FROM BARRO COLORADO ISLAND

EGBERT GILES LEIGH, JR.

New York   Oxford

Oxford University Press

1999

Oxford University Press

Oxford New York
Athens Auckland Bangkok Bogotá Buenos Aires Calcutta
Cape Town Chennai Dar es Salaam Delhi Florence Hong Kong Istanbul
Karachi Kuala Lumpur Madrid Melbourne Mexico City Mumbai
Nairobi Paris São Paulo Singapore Taipei Tokyo Toronto Warsaw

and associated companies in
Berlin Ibadan

Copyright © 1999 by Oxford University Press, Inc.

Published by Oxford University Press, Inc.
198 Madison Avenue, New York, New York 10016

Oxford is a registered trademark of Oxford University Press

All rights reserved. No part of this publication may be reproduced,
stored in a retrieval system, or transmitted, in any form or by any means,
electronic, mechanical, photocopying, recording, or otherwise,
without the prior permission of Oxford University Press.

Library of Congress Cataloging-in-Publication Data
Leigh, Egbert Giles.
Tropical forest ecology : a view from
Barro Colorado Island / by
Egbert Giles Leigh, Jr.
p. cm.
Includes bibliographical references and index.
ISBN 0-19-509602-9; 0-19-509603-7 (pbk.)
1. Forest ecology—Panama—Barro Colorado Island. 2. Barro
Colorado Island (Panama) I. Title.
QH108.P3L45 1999
577.34'097287'5—dc21   97-13770

1 3 5 7 9 8 6 4 2

Printed in the United States of America
on acid-free paper

This book is dedicated to the memory of Martin Moynihan (1928–1996), the founding director of the Smithsonian Tropical Research Institute, in thanks for giving me the opportunity to become a tropical biologist, arranging my acquaintance with a splendid selection of rainforests, and providing me with such a superb setting in which to practice my trade.

Preface

Tropical forest has provided me with a large share of that experience of beauty without which any life would be sadly incomplete. Indeed, I confess to being a creature of my memories of these experiences—the late afternoon sun lending color to the tree trunks of Barro Colorado during walks early in the dry season; sunrise over the misty forest on that island after a night of rain, waiting for the howler monkeys to howl and be counted; the eerie, rather musical whoops of babakotos, Madagascar's largest living lemurs, as they jumped gracefully from tree to tree in the rainforest near Perinet; "conversations" with lemurs near a ridgetop camp overlooking what was then unbroken rainforest sloping to the sea in northeastern Madagascar; the pastel clouds shifting about over the Bay of Antongil at sunset as I was watching from a cloud forest mountaintop on this same Malagasy peninsula; the redwood-like grandeur of the enormous dipterocarps, *Shorea curtisii*, with elegant crowns full of small, silvery leaves, crowning the rainforest on the flank of a ridge at Ulu Gombak, northeast of Kuala Lumpur; eating my first durian in the heath forest atop Gunong Ulu Kali, less than an hour's drive from Kuala Lumpur; and many others.

This book is not a record of my memories—after all, I have not Proust's skill. Rather, it is an attempt to make sense of them. It is an account of my understanding of tropical forest, as it has developed through my own experience, studies of many of my friends on Barro Colorado and elsewhere, and my reading. Thus this book is inevitably a very personal one. I will not try to review the literature about tropical forest. Still less do I claim to present the current consensus among tropical biologists about the organizing principles of tropical forest—I don't think there is any such consensus, and if there were, my experience of scientific fads and bandwagons would lead me to avoid it like the plague. I have no desire to lull my readers to sleep with the deadly ether of soi-disant impersonal objectivity: Polanyi (1958) has shown how dangerous it is to try to cut the person out of the scientist. Instead, I wish to interest readers of goodwill in tropical forests and what can be learned from them. I find this task rather urgent, as I am convinced that the road to a proper understanding of ecology and evolution begins and ends in the tropics.

In 1967, the founding director of the Smithsonian Tropical Research Institute (STRI) in Panama, Martin Moynihan, was encouraging studies to learn how representative the ecology of Barro Colorado's forest and the behavior of its animals were of tropical forests in general. As a young theoretical biologist, I felt that comparing ecological communities in different biogeographic realms, with different evolutionary histories, would be a good way to acquaint myself with the level of detail theoretical prediction ought to strive for. Moynihan engaged me to compare the structure and physiognomy of tropical forests around the world and he sent me and my wife, the former Elizabeth Hodgson, on two trips around the world, in 1968 and 1970, to do this. He also appointed me to the STRI staff in 1969, thus sheltering me from the transformations of American universities, formerly citadels of thought, into organizations where, nowadays, no one has a moment in which to stop and think. The problem of the similarities and differences among tropical forests of different regions continues to preoccupy me. This book is a belated attempt to answer Moynihan's question. As I owe my career, my involvement in tropical biology, the intellectual atmosphere which prevails on Barro Colorado and elsewhere at STRI, and perhaps even my marriage, as well as this book's central question, to Martin Moynihan, I dedicate this book to him.

This book owes much to many others. A. G. Fischer introduced me to the fossil record and to ecology as a tool for making sense of that record. Robert Stallard introduced me to the processes of weathering and soil formation. Robert MacArthur and James Crow introduced me to Ronald

Fisher's (1930) *Genetical Theory of Natural Selection*. MacArthur and Crow helped show me the relevance of the common interest among a genome's genes in distinguishing between the selfish advantage of individual alleles and the fitness of genotypes. MacArthur also taught me how to think about biotic diversity. This book also owes an enormous debt to E. J. H. Corner. I never met him, but his papers, particularly his essay (Corner 1954) on the evolution of tropical forest, his book (Corner 1964), *The Life of Plants*, and his monograph (Corner 1967) on the figs of the Solomon Islands, have given me abundant food for thought. Corner brings out, as no one else could, the idea of tropical forest as a mutualism; moreover, no one else can communicate so effectively the beauty and mystery of tropical forest. Finally, Russell Lande, Montgomery Slatkin, and David Sloan Wilson, among others, showed me how to reconcile Fisher's approach with Corner's.

My wife, Elizabeth, helped me with most of the fieldwork reported in this book. Robin Foster introduced me to the plants of Barro Colorado and showed me how populations of vertebrate herbivores were limited by seasonal shortage of fruit and new leaves. Marcel and Annette Hladik, who were predoctoral fellows on Barro Colorado when I first went there, gave me a 12-year subscription to *La Terre et la Vie*, and otherwise kept me in touch with French work in Africa and South America, a very precious gift. In 1991, they involved me in a UNESCO conference on rainforest peoples, food, and nutrition (Hladik et al. 1993), which showed me the many lessons to be learned from these people, not least the elementary lesson, forgotten by the intellectual elites of the west, that we do not live by bread alone. The Hladiks also provided me the opportunity to assemble, summarize, and publish my thinking on the role of mutualism in evolution: this paper (Leigh and Rowell 1995) was in many ways a trial run for chapter 9. Allen Herre (1996) has shown how a single mutualism, that between fig trees and their pollinating wasps, can shape not only the biology of fig trees, but the ecological characteristics of a tropical forest.

Indeed, this story spreads far beyond Barro Colorado. I am indebted to Benjamin Stone, formerly of the University of Malaya, for introducing me to durians and dipterocarps, cloud forests and *Leptospermum* heaths, in the Malay Peninsula. His recent death was almost as severe a blow to tropical botany as the death of Al Gentry. Alison Jolly, Vololoniaina Jeannoda, Rakotonirina Benja, and Jean Prosper Abraham have all helped acquaint me with the marvelous diversity of plants and habitats in Madagascar. The Smithsonian Foreign Currency Program, which financed our initial trips around the world, made possible subsequent visits to India. Twice, in 1985 and 1995, Madhav Gadgil, Raghavendra Gadagkar, and R. Sukumar invited me to the Centre for Ecological Science at the Indian Institute of Science in Bangalore to lecture, first on why there are so many kinds of tropical trees, and later on the ecology of Barro Colorado Island. These lectures were marvelous opportunities to collect and organize my thoughts for this book; these trips also allowed me to see several of the many different types of forest in south India.

Many people have supplied me with reprints and unpublished data. I must single out Robin Foster and Richard Condit, and their data analysts, Una Smith and Suzanne Loo de Lao, for the abundant data they have provided me on the 50-ha Forest Dynamics Plot on Barro Colorado and similar plots maintained by the Center for Tropical Forest Science elsewhere in the world. Similarly, I am grateful to Donald Windsor, S. Joseph Wright, and Steven Paton for supplying data and analyses from monitoring records of the Smithsonian's Environmental Sciences Program.

S. J. Wright and G. J. Vermeij read the whole manuscript with care and called my attention to errors, omissions, and obscurely written passages. David King and Hubert Herz read chapters 5 and 6, and Richard Condit read chapter 8; they all warned me of other ditches I was falling into. The errors that remain, of course, are my own, but, without their help, there would have been many more.

I must also thank a splendidly various company of artists for their efforts to capture one or another aspect of tropical nature, among them, Alex Murawski, Daniel Glanz, George Angehr, Judith Gradwohl, Arlee Montalvo, Lynn Siri Kimsey, Scott Gross, Gerardo Ravassa, Karen Kraeger, Roxanne Trapp, Marshall Hasbrouck, Hayro Cunampio, Francesco Gattesco, Donna Conlon, Dorsett W. Trapnell, and Deborah Miriam Kaspari. Steven Paton, Robert Stallard, and David Kinner supplied the maps of Barro Colorado Island and the Barro Colorado Nature Monument.

I also thank a host of librarians, especially Carol Jopling, Sylvia Churgin, Tina Lesnick, Vielka Chang-yau, and Angel Aguirre, for procuring copies of all sorts of obscure papers. I thank the American Philosophical Society and Princeton University for help toward my first trip around the world, and the Smithsonian Research Opportunities Fund, its predecessor, and the Smithsonian Tropical Research Institute for other financial support.

I also wish to express my gratitude to the Panamanian government for its hospitality to the Smithsonian Tropical Research Institute, and therefore, to me. In an age where many countries view scientific research as another form of exploitation, to be strictly regulated and controlled, it is a pleasure to acknowledge that in Panama the honor of knowledge for the sake of knowing, so characteristic of the Enlightenment, is still very much alive.

Finally, I thank the current director of the Smithsonian Tropical Research Institute, Ira Rubinoff, for his continued support and encouragement. My style of work is out of fashion, and I bring no money to the institute, yet he has left me the freedom to do my work in peace, with ample encouragement and the necessary financial support. Meanwhile, he has managed to guide this institute through an endless series of political minefields without endangering its commitment to basic research. He has diversified its activities; and thanks to him, I have the opportunity to hobnob with and learn from plant physiologists, anthropologists, archaeologists, palynologists, and even, horror of horrors, the occasional molecular biologist, as well as the ethologists, ecologists, and naturalists who first established the patterns of basic research at STRI. This book has benefitted from all of them.

Barro Colorado Island E. G. L., Jr.
Feast of the Conversion of St. Paul
1997

References

Corner, E. J. H. 1954. The evolution of tropical forest, pp. 34–46. In J. Huxley, A. C. Hardy, and E. B. Ford, eds., *Evolution as a Process*. George Allen and Unwin, London.

Corner, E. J. H. 1964. *The Life of Plants*. World Publishing, Cleveland, OH.

Corner, E. J. H. 1967. *Ficus* in the Solomon Islands and its bearing on the post-Jurassic history of Melanesia. *Philosophical Transactions of the Royal Society of London*, B 253: 23–159.

Fisher, R. A. 1930. *The Genetical Theory of Natural Selection*. Clarendon Press, Oxford.

Herre, E. A. 1996. An overview of studies on a community of Panamanian figs. *Journal of Biogeography* 23: 593–607.

Hladik, C. M., A. Hladik, O. F. Linares, H. Pagezy, A. Semple, and M. Hadley (eds). 1993. *Tropical Forests, People and Food*. UNESCO, Paris, and Parthenon Publishing, Park Ridge, NJ.

Leigh, E. G. Jr., and T. E. Rowell. 1995. The evolution of mutualism and other forms of harmony at various levels of biological organization. *Écologie* 26: 131–158.

Polanyi, M. 1958. *Personal Knowledge*. University of Chicago Press, Chicago.

Contents

Introduction xiii
   References xvi

ONE
Barro Colorado Island: The Background 3
   The Study Area 3
   The History of Barro Colorado's Biota 3
   Concluding Remarks 12
   References 12

TWO
Dramatis Personae 15
   Principal Plants of Barro Colorado 15
   Principal Animals of Barro Colorado 27
   Concluding Remarks 39
   References 39

THREE
Tropical Climates 46
   Climate and Vegetation Type 46
   Rainfall 46
   Factors Affecting Evapotranspiration 53
   Evapotranspiration in Montane Forests 59
   Concluding Remarks 59
   Appendix 3.1. Estimating Evapotranspiration 59
   References 63

FOUR
Runoff, Erosion, and Soil Formation 67
   Modes of Runoff 67
   Runoff and Weathering 70
   Soil and Vegetation 75
   Concluding Remarks 77
   References 77

FIVE
Telling the Trees from the Forest:
Tree Shapes and Leaf Arrangement 79
   Different Architectures of Tropical Trees 80
   Designing Trees: Goals and Constraints 85
   Comparing Leaf Arrangements, Tree Architectures,
     and Forest Structure 94
   Concluding Remarks 112
   Appendix 5.1: How Tall Can a Tree Be? 113
   Appendix 5.2: Collecting the Data for
     Tables 5.10–5.12 113
   References 116

SIX
Biomass and Productivity of Tropical Forest 120
   The "Constants of Tropical Forest" 120
   Leaf Area Index and Leaf Fall in Different
     Forest Types 122
   The Gross Production (Total Photosynthesis)
     of Tropical Forest 125
   How is a Forest's Energy Allocated? 130
   Concluding Remarks 139
   Appendix 6.1: Basal Area and Litterfall in
     Different Forests 139
   References 142

SEVEN
The Seasonal Rhythms of Fruiting and Leaf Flush
and the Regulation of Animal Populations 149
   The Seasonal Rhythm of the Forest's
     Food Production 149
   The Timing of Leaf Flush, Flowering,
     and Fruiting 152
   What the Forest Makes and Who Eats It 156

Regulation of Populations of Vertebrate Folivores
  and Frugivores   156
Insects of Tree Crowns   163
Concluding Remarks   169
Appendix 7.1: Number of Species of Plants of
  Different Categories Dropping Fruit Into
  Traps Each Week on the Central Plateau of
  Barro Colorado, 1992–1995   170
Appendix 7.2: The Mammals of BCI and Manú   172
References   173

EIGHT

Tropical Diversity   179
Documenting Tree Diversity   179
Necessary Conditions for Tropical Diversity   184
Explaining the Diversity of Tropical Trees   185
Consequences of Tree Diversity   195
Concluding Remarks   197
Appendix 8.1: Diversity in Different
  Forest Plots   198
Appendix 8.2: The Distribution of
  Tree Abundances   204
References   205

NINE

The Role of Mutualism in Tropical Forest   211
Mutualism in a Competitive World   211
Ecosystems as Commonwealths   219
Concluding Remarks   223
Appendix 9.1: The Varieties of Group Selection   223
References   226

TEN

The Rainforest Endangered   232
References   234

Name Index   237

Subject Index   239

Introduction

This book is about the ecology of tropical forest and the interrelationships among its plants and animals. Its starting point is the tropical forest on Barro Colorado Island, Panama, because that is the forest I know best. Its fundamental purpose, however, is to learn what Barro Colorado can tell us about other tropical forests—and what other tropical forests can tell us about Barro Colorado.

In this introduction, I describe some of the fundamental assumptions underpinning this book's approach and outline the argument of the book.

FUNDAMENTAL PRESUPPOSITIONS

The Prevalence of Adaptation

This book presupposes the prevalence of adaptation among organisms and their interrelationships. Aristotle recognized that organisms are like human artifacts that are built for a purpose, insofar as one understands neither artifact nor organism nor any part of an organism unless one understands the function for which it was designed (*Parts of Animals* 639b14–20; Barnes 1984: p. 995). Moreover, an organism whose structure or behavior is abnormal for its species is usually dysfunctional (*Physics* 199a33–b4; Barnes 1984: p. 340), as if *normal* organisms are adapted to their ways of life. For those who lack time to search through Aristotle's scattered references on the subject, Nussbaum (1978: pp. 59–106) gives a good account of Aristotle's attitude and its relevance today.

Aristotle recognized the purposiveness of organisms and emphasized that they cannot be understood without resorting to functional explanation, but he was unable to explain the source of this purposiveness (Gilson 1971: p. 20, 24). Darwin (1859: p. 3) dismissed as worthless theories of evolution that failed to explain adaptation. For Darwin, as for Aristotle, adaptation was the central feature of biological organization. No one who has spent a lifetime studying tropical nature could disagree with them.

Natural Selection, Competition, and Cooperation

Darwin (1859) realized that the competitive *mechanism* of natural selection—the process whereby individuals which survive better and reproduce more effectively contribute more genes to future generations—serves as the source of design in organisms (Dennett 1995). In ecology as well, competition (in the broad sense, including predation) is a primary factor shaping the organization of communities and ecosystems (Hutchinson 1959, Connell 1961, Paine 1966, MacArthur 1972).

Biology cannot be understood without reference to competition, but reckoning with mutualism and interdependence is equally essential to biological understanding. A central problem of animal behavior concerns animals that live in groups. Members of a group associate because they depend on each other in some way, but they are also each other's closest competitors for food, mates, and other limiting resources. What circumstances prevent competition among group members from destroying their common interest in their group's welfare, the common good created by their cooperation? This problem has counterparts at all levels of biology. Cancer, for example, shows that conflicts can arise between an individual and certain of its cells. How are such conflicts normally avoided or suppressed? Indeed, the ethologist's problem, with its many and varied solutions, provides one of the grand unifying themes of biology. The key to this problem's solution is Aristotle's concept of the common good, the foundation on which he constructed his *Politics*. To understand the evolution of cooperation, we must show what circumstances make it profitable for *all* the members of the group to serve, or the overwhelming majority of them to enforce, the common good of their group, for therein lies the link between competition and cooperation.

Competition and Interdependence in Ecology

The most bitterly contested of the many issues among ecologists is the extent, if any, to which ecosystems are designed for the common good. Hurst et al. (1996: p. 318) open their paper with the superb remark that "In the social sciences, it is common to distinguish between two competing traditions. One tradition concentrates on the *function* of different social institutions, while the other prefers to view society as an arena for social *conflicts*." The same is true in ecology. Nowadays, the view of ecosystems as competitive arenas is dominant, yet the very real concern with the disruptive effects of human activities on the ecology of our planet suggests the extent of the interdependence among species, a "common good" which is endangered by a humanity run wild. In this account, I will endeavor to do justice to both the competition and the interdependence among the denizens of tropical forest.

THE ARGUMENT OF THE BOOK

Some background information on Barro Colorado is needed to set this book's argument in perspective. What factors have shaped the island's biota and its ecological organization? The history of the institution which now administers Barro Colorado, the Smithsonian Tropical Research Institute (STRI), will be mentioned only insofar as it bears on this question. How the policies of this institute have influenced our understanding of Barro Colorado is a story worth telling (Hagen 1990, Rubinoff and Leigh 1990), but it is a *different* story. It is a big enough job to convey an understanding of the forest's plants and animals, and their interrelationships, along with enough detail on how this understanding was attained to allow the reader to judge the validity of my interpretation.

The dazzling diversity of tropical organisms and the unquestionable delicacy and precision of many of their adaptations is often ascribed to the immunity of tropical settings from the climatic revolutions that sent glaciers marching across the temperate zone and the perfection during ages without end of relationships among a stable set of partners. Nevertheless, the tropical rainforest climates we know today date only from the beginning of the Cenozoic about 65 million years ago. In the late Cretaceous, rainfall within the tropics averaged about 1200 mm/year (Wolfe and Upchurch 1987). Tropical Latin America has suffered major changes of climate over the ages, as well as an extinction of all its largest mammals about 12,000 years ago. Barro Colorado's current population of plants and animals stems from the mixture of two very different biotas, occasioned by the appearance of a land bridge between the Americas a few million years ago. Barro Colorado has had its current biota for less than 10,000 years. Fifteen thousand years ago, the composition of its forest was probably quite unlike that of any forest on the earth today. The prevalence of mutualism in tropical forest cannot be ascribed to the ageless propinquity of species we see growing together today. To drive home this point, we must consider how the biotas mixed, the changes of climate affecting this biota in the past, and the human impacts, past and prospective, impinging upon this biota.

To complete the reader's stock of background information on Barro Colorado requires the introduction of some of the dramatis personae. This consists of outlines of the ecology and natural history of selected species of vertebrates (mostly mammals), insects (mostly social), and plants (all angiosperms, mostly trees and shrubs, with two epiphytes). These species are all well enough known to be worth talking about, and their stories all shed light on an interaction, relationship, or process that is important to understanding the forest. The dramatis personae recapitulates the intellectual history of Barro Colorado Island in this sense: our understanding of this island is built on a series of careful studies of different kinds of organism in their ecological and evolutionary context. The dramatis personae also illustrate the biases of, and suggest some of the lacunae in, work on this island.

The main body of the book is in two parts. The first part is comparative. At first glance, lowland rainforests in different climates look surprisingly alike. They are much more similar to each other than is summer-green deciduous forest on the shores of the Chesapeake at 39° N to the summer-dry redwood forests with their immensely tall trees at the same latitude, or to the various *Eucalyptus* forests at 37° S. How similar are the climate, structure, physiognomy, and productivity of tropical forest on Barro Colorado Island to those of lowland and montane tropical forests elsewhere in the world? Annual evapotranspiration is remarkably similar, roughly 1400 mm/year, in lowland moist and wet tropical forests around the world, as are average annual temperature, about 26°C, average diurnal temperature range, usually about 9°C, and average annual solar radiation (about 180 W/m^2). On the other hand, the frequency of potentially disruptive events, such as thunderstorms and hurricanes, differ markedly from place to place.

Second only to climate, soil is a decisive influence on vegetation. How does soil form, how rapidly does it erode away, and how does tropical forest influence and protect its soil? Is there a "standard tropical soil" on which different soil types converge?

Having considered climate and soil, we can at last turn to the forests themselves. Tropical trees come in a variety of shapes and sizes. What can principles of tree engineering and tree design and comparisons of the spectrum of "tree architectures" among different strata of a forest and among different forests tell us about the meaning of a tree's leaf arrangement and branching pattern? Despite the remarkable diversity of tree shapes, lowland tropical forests share a variety of features in common. Most reasonably mature lowland forests have a basal area (total cross-sectional area of tree-trunks ≥ 10 cm in diameter) near 30 m²/ha, carry about 7 or 8 ha of leaves/ha of ground, and drop about 7 tons dry weight of leaves/ha · year. Finally, the death rate of trees ≥ 10 cm in trunk diameter is between 1% and 2% per year. When one considers all species conjointly, death rate varies remarkably little with diameter class. Why these similarities? To answer, I resort first to physiological enquiry: how much does a forest photosynthesize, and how does it apportion the energy thus stored? Then I ask what can be learned by comparing different forests. Why is a mangrove forest, which is periodically flooded by salt water, more similar to a lowland rainforest in structure and productivity, but with so many fewer tree species, than a windblown, stunted elfin woodland on a misty, cloudbound tropical mountaintop? Our comparisons prompt several questions which occupy the last part of this book:

1. What role does seasonal shortage of fruit and/or new leaves play in limiting populations of vertebrate herbivores? Do the forest's plants conspire to limit their herbivores by seasonal shortage of suitable food? If so, how do different species of plants "know" when to flower, fruit, or flush new leaves? Does the seasonal shortage of suitable food suffice in itself to limit the forest's herbivores, or does the forest need the help of predaceous animals to control its herbivore populations?

To answer these questions, I first document the seasonal rhythm of leaf flush and fruit fall on Barro Colorado and ask how different plants "know" when to flush new leaves, flower, and fruit. I examine the hydraulic architecture of plants for clues concerning the factors that trigger flowering or leaf flush, and discuss a grand experiment where two 2.25-ha plots on Barro Colorado were irrigated for five successive dry seasons to learn the relative roles of changes in atmospheric conditions (humidity, sunlight, etc.) versus moisture content of the soil in governing the timing of flowering, fruiting, and leaf flush in different species of tree. Then I turn to the impact of seasonal shortage of fruit and new leaves on animal populations. I review the "trophic dynamics" of Barro Colorado, and the Parque Manú, to assess whether enough fruit falls in the season of shortage to feed the forest's frugivorous vertebrates, and whether predators eat enough to limit herbivore populations. I then compare the impact on animal populations of the contrast between neotropical and paleotropical rhythms of fruit production and leaf flush.

2. Why are there so many kinds of tropical trees? How can 90 species of trees be represented on a single hectare of Barro Colorado by trees with trunk diameter over 10 cm, not to speak of the 300 species represented in a single hectare of lowland forest in Ecuador? This question is one of the greatest of the many mysteries of tropical biology. Exploring this mystery involves a survey of the empirical correlates of tree diversity, a review of the evidence that individual species of tropical tree are kept rare enough by their pests to allow other species to coexist with them, and a theoretical enquiry of the potential contribution to tree diversity of "temporal niche differentiation," whereby different tree species recruit more successfully in different years.

3. What role does mutualism play in the ecological organization of tropical forest? From the standpoint of conservation planning, this is the most important question in the book. Unfortunately, this question presupposes both an interest in natural history, currently an unfashionable subject, and a willingness to consider relationships between very different kinds of organisms: it has accordingly been dangerously neglected. Yet we cannot prescribe for the preservation of a tree species unless we know what animals it needs to pollinate its flowers or disperse its seeds, and, in turn, what other needs these animals have; not only what other foods they require, but whether they must migrate to other areas at certain seasons, and the like. Tropical forest is riddled with unexpected interdependencies: to preserve it, we must familiarize ourselves with these interdependencies.

More generally, tropical forest reminds us that intense competition favors mutualism: one cannot compete effectively unaided. The ecological organization of tropical forest involves the niche differentiation that allows different species to coexist and the predators or diseases that prevent one species from competitively excluding another. Equally important, however, are the mutualists that permit a plant species to be rare enough to survive in equilibrium with its pests, or that permit its seeds to escape insect attack. The ability a plant's pollinators and seed dispersers confer on a plant species to persist when rare permits plants of that species to divert resources from defense to fast growth. A forest of such fast-growing species is better for the soil, for such plants need not poison their soil with the persistent pesticides abundant species need to keep their leaves from being eaten. The prevalence of mutualism changes the complexion of tropical forest and may greatly enhance its productivity.

Finally, in the last chapter, I reflect on the future of tropical forest, how essential tropical forest is for the welfare of the people around it, and the human attitudes that must change if tropical forest is to survive. Tropical forest is at risk because those who decide its fate are submitting to an economic determinism more rigid than that of Marx, a determinism where the only recognized goods are the increased generation and appropriate distribution of merchandise. In this scheme, beauty is a merchandise like any other, so that, by definition, peoples of the rainforest, who have no money, cannot afford to keep their own home, even if it is their very life. We confront a series of related paradoxes. The tropical forest will survive only if we recognize its value over and above its usefulness to ourselves, just as scientific research will continue to bear useful technological fruit only if some research remains motivated by an interest in nature unrelated to its usefulness to ourselves, and just as humanity and civilization will survive only if its leaders recognize goods higher than the survival of humanity and civilization.

References

Barnes, J. (ed). 1984. *The Complete Works of Aristotle.* Princeton University Press, Princeton, NJ.

Connell, J. H. 1961. Effect of competition, predation by *Thais lapillus*, and other factors on natural populations of the barnacle *Balanus balanoides. Ecological Monographs* 31: 61–104.

Darwin, C. R. 1859. *On the Origin of Species by Means of Natural Selection.* John Murray, London.

Dennett, D. C. 1995. *Darwin's Dangerous Idea.* Simon and Schuster, New York.

Gilson, É. 1971. *D'Aristote au Darwin et retour.* J. Vrin, Paris.

Hagen, J. B. 1990. Problems in the institutionalization of tropical biology: the case of the Barro Colorado Island biological laboratory. *History and Philosophy of Life Science* 12: 225–247.

Hurst, L. D., A. Atlan, and B. O. Bengtsson. 1996. Genetic conflicts. *Quarterly Review of Biology* 71: 317–364.

Hutchinson, G. E. 1959. Homage to Santa Rosalia, or why are there so many kinds of animals? *American Naturalist* 93: 145–159.

MacArthur, R. H. 1972. *Geographical Ecology.* Harper and Row, New York.

Nussbaum, M. C. 1978. Aristotle's *De Motu Animalium.* Princeton University Press, Princeton, NJ.

Paine, R. T. 1966. Food web complexity and species diversity. *American Naturalist* 100: 65–75.

Rubinoff, I., and E. G. Leigh Jr. 1990. Dealing with diversity: the Smithsonian Tropical Research Institute and tropical biology. *Trends in Ecology and Evolution* 5: 115–118.

Wolfe, J. A., and G. R. Upchurch Jr. 1987. North American nonmarine climates and vegetation during the late Cretaceous. *Palaeogeography, Palaeoclimatology, Palaeoecology* 61: 33–77.

TROPICAL FOREST ECOLOGY

ONE

Barro Colorado Island

The Background

THE STUDY AREA

Our story centers on Barro Colorado Island (9°9' N, 79°51' W), a 1500-ha island isolated from the surrounding mainland between 1910 and 1914, when the Rio Chagres was dammed to form the central part of the Panama Canal (Fig. 1.1). The island suffers a severe dry season, which usually starts some time in December, and ends in April or early May: the median rainfall for the year's first 91 days is less than 100 mm (Leigh and Wright 1990). Otherwise, the climate is what one would expect for a lowland, moist forest, as will be discussed in detail in the next chapter. Barro Colorado lies at the midpoint of a gradient between the deciduous dry forests of the Pacific shore, with an annual rainfall of 1800 mm, and the rainforests of the Caribbean side, with an annual rainfall of well over 3000 mm/year (Windsor 1990).

The island is completely forested (Fig. 1.2), save for a few small man-made clearings, of which the largest is the laboratory clearing. Twenty-five of the 211 tree species on Barro Colorado that can attain a height > 10 m lose all or nearly all of their leaves for at least part of the dry season (Croat 1978: p. 30). Other canopy trees, and nearly all the midstory and understory plants, are evergreen. The forest on the island's northeast half is recovering from widespread cutting and clearing late in the nineteenth century, during the French attempt to build a canal: a few small farms still remained in 1923. The other half of the island has been little disturbed since the cruelties and plagues brought on by the Spanish conquest emptied Panama of its human populations (Foster and Brokaw 1982). Parts of the old forest were little disturbed for at least the 1500 years preceding the conquest (Piperno 1990).

Barro Colorado Island (BCI) was declared a reserve in 1923. It has been administered by the Smithsonian Institution since 1946. When the Smithsonian Tropical Research Institute (STRI) was established in 1965, Barro Colorado Island became one of its research sites. In 1979, the Panama Canal treaties granted STRI custodianship of the Barro Colorado Nature Monument, which was established under the Western Hemisphere Convention on Nature Protection and Wildlife Preservation of 1940. In addition to BCI, this Nature Monument now includes five nearby mainland peninsulas whose total area (excluding BCI) is 4000 ha. The Nature Monument shares a border with Panama's 12,000-ha Parque Nacional Soberania.

BCI has been attracting biologists since 1916. A laboratory and dormitory were constructed there in 1924. Since 1957, the island has been rendered readily accessible from Panama City, and the pace of study there has accelerated remarkably. Now it is one of the best-studied tracts of tropical forest in the world. Leigh et al. (1982) provide an overview of the ecological work done there during the 12 years ending in 1980; Leigh and Wright (1990) and Leigh (1996) briefly review later work; and Leigh (1990) gives a somewhat more detailed history of BCI's field station and its development.

THE HISTORY OF BARRO COLORADO'S BIOTA

Tropical rainforests date farther back than their counterparts of the temperate zone, but they are not ageless. During the late Cretaceous, tropical vegetation was typically an open woodland of stout, evergreen trees, spaced widely apart. Their seeds were small, for abundant light was available for germination (Wolfe and Upchurch 1987). The vegetation suggests a relatively aseasonal rainfall of about 1200 mm/year (Wolfe and Upchurch 1987) and tropical temperatures that are 5–7°C cooler than than today's (Wolfe 1990). Only when a bolide smashed into the Yucatan shore, extinguishing the dinosaurs and marking the end of the Cre-

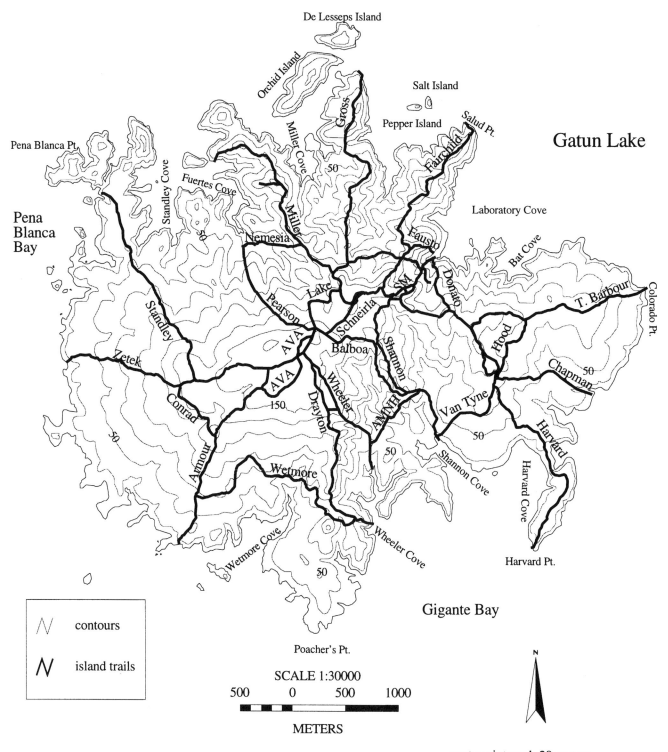

Figure 1.1. Barro Colorado Island and its surroundings. Left. Map of Barro Colorado Island. Wibke Thies surveyed the trails; David Kinner prepared the topography from the 1957 1:25,000 maps of the relevant portion of the Panama Canal; Robert Stallard adjusted the plateau topography; David Clark established the Global Positioning System fixes; and Robert Stallard reconciled the entire data set. Right. Map of Barro Colorado Nature Monument (Barro Colorado Island and the adjoining peninsulae of Gigante, Pena Blanca, Bohio, Buena Vista, and Frijoles) and the situation of this Monument in the Republic of Panama. Maps prepared by S. Paton.

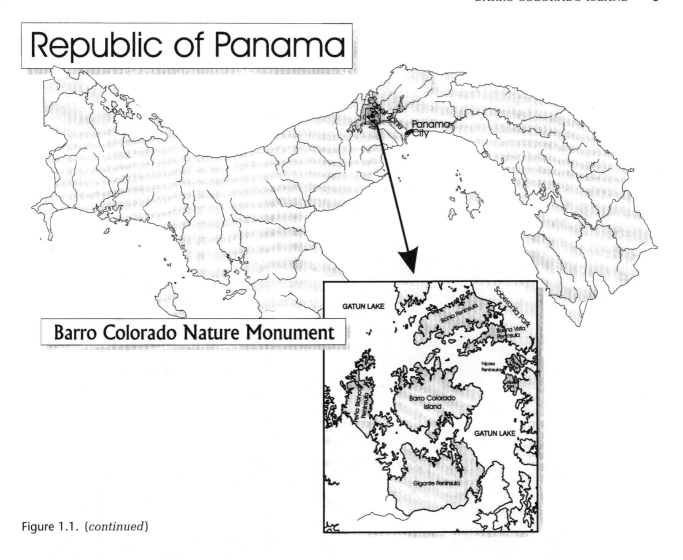

Figure 1.1. (*continued*)

taceous with scenes of unspeakable catastrophe (Wolfe 1991), did humid climates return to tropical settings, enabling the great expansion, if not the origin, of rainforest (Wolfe and Upchurch 1986, 1987). As rainforest appeared, so it continues, buffeted by disruptions. What revolutions gave rise to Barro Colorado's biota, and what changes have affected it since?

The Isthmus of Panama and the Great Biotic Interchange

Panama proclaims itself the "puente del mundo," the world's bridge, because the Panama Canal crosses its territory. Three million years ago (Coates and Obando 1996), Panama first became a bridge between two worlds, worlds as different as those now separated by Wallace's famous line. Of the two continents joined by this bridge, North America was once part of Laurasia, South America part of Gondwana. These two great continents were formed 140 million years ago when the Tethys, a circumtropical sea of which the Mediterranean is the largest surviving remnant, split Pangaea, the single great landmass of Permo-Triassic time (Donnelly 1985, Pitman et al. 1993).

As an erstwhile fragment of Laurasia, North America was frequently connected to Asia, and occasionally, as in the Eocene, accessible from Europe. Consequently, North America was often invaded by mammals and birds, not to mention other animals and plants, from Eurasia. Humans were the most spectacular and the most ominous of these invaders. These invasions enhanced the competitiveness of North America's mammals and birds. Laurasia had also endowed North America with a rich tropical forest flora, attested by the early Oligocene Brandon lignite (Tiffney 1980). This forest, which resembled the one preserved in the London Clay, was widespread in the Eocene. Its flora resembled that of today's Malesian flora more closely than that of the modern Neotropics (Gentry 1982). After the Eocene, however, increasing cold or aridity extirpated nearly all of North America's tropical forest (Wolfe 1975, Gentry 1982), including invaders from Gondwana such as *Bombax* or *Ceiba* (Wolfe 1975) and stingless bees of the genus *Trigona* (Michener and

Figure 1.2. Forest on Barro Colorado. (Above). View across ravine of old forest from R. C. Shannon Trail, near marker 2. Drawing by Francesco Gattesco. (Facing, Top). View across another ravine from near Shannon 1. Drawing by Daniel Glanz. (Facing, Bottom). View of forest from dock at laboratory cove, Barro Colorado. Drawing by Deborah Miriam Kaspari.

Grimaldi 1988) that had reached North America in the Cretaceous.

Gondwana originally included Africa, Australia, India, Madagascar, and Antarctica as well as South America (Raven and Axelrod 1974). South America, however, was separate from Africa by 90 million years ago (Kimsey 1992, Pitman et al. 1993). South America's communication with Australia via Antarctica was disrupted 53 million years ago, and South America was already separated from Antarctica 30 million years ago (Crook 1981, Pitman et al. 1993). The Americas approached each other during the late Cretaceous. Between 80 and 49 million years ago, lines of islands, which may sometimes have coalesced into a real corridor, connected the Americas (Donnelly 1985, Pitman et al. 1993). Nevertheless, the isthmus of Panama seems to have been the first real land bridge between the Americas since the Tethys sundered Pangaea during the Jurassic. Before this isthmus formed, South America had been evolving in splendid isolation (Simpson 1980), with a flora (Gentry 1982) and insect fauna (Kimsey 1992) deriving overwhelmingly from Gondwana. Tropical forest flourished in South America, thus "preadapting" South America's biota for Panama.

The biota of Barro Colorado Island is a mixture of elements from different sources. Its plants derive primarily from South America. Its trees and lianes belong mostly to species centered in Amazonia, while its shrubs and epiphytes derive mostly from rapidly speciating lineages of the northern Andes (Gentry 1982). Most of BCI's insects also seem to derive from South America, and probably reached BCI with the plants on which they depend. An instructive exception are the insect commensals, which have come with their mammal hosts from North America (Kimsey 1992).

The mammals of Barro Colorado are a more complicated mixture. The provenance of BCI's roughly 40 species of terrestrial mammals is well understood, thanks to the paleontological researches of Simpson (1980) and many others. Squirrels, rice rats (*Oryzomys*), peccaries, deer, tapir, and carnivores came from North America; marsupials, sloths, monkeys, and caviomorph rodents such as agoutis, pacas, and spiny rats came from the south.

Birds, bats, amphibians, and reptiles have much less adequate fossil records. South American camelids, such as guanacos and llamas, whose ancestors walked in through Panama, but which today live only in southern South America, where they are rather diverse, warn us of the danger of inferring past events from present distributions. Nevertheless, most of Barro Colorado's insectivorous birds—antbirds, flycatchers, cotingas, and the like—are suboscines. This group was once widespread (there are a few relict suboscines in Madagascar and elsewhere in the Old World tropics), but suboscines are now almost entirely Neotropical. Here, the suboscines have radiated spectacularly. These suboscines have spread into Panama along with South America's tropical forest (Feduccia 1980). Hummingbirds also appear to have evolved in South America (Mayr 1964). Except for tinamous, and perhaps toucans, which hail from South America, and guans, which came from the north, we do not know where most of BCI's frugivorous birds evolved (Mayr 1964). The story is remarkably similar for bats. Thirty-nine (Kalko et al. 1996) of BCI's 70 (E. K. V. Kalko, p. c.) species of bats belong to the family Phyllostomatidae, a family endemic to the Neotropics and firmly centered in South America (Koopman 1982). This family includes BCI's largest bat, the 175-g carnivore *Vampyrum spectrum*; its guild of fig-eating fruit bats (*Artibeus* spp., *Vampyrodes*, *Uroderma*, and *Chiroderma*), its guild of *Piper*-eating understory fruit bats (*Carollia* spp.), blood-sucking vampires (*Desmodus rotundus*), frog-eaters (*Trachops cirrhosus*), bats which glean katydids and other large insects from foliage (*Tonatia* spp., *Micronycteris* spp.), and nectar-eating bats (*Glossophaga*). We do not know where BCI's other bats evolved. The ultimate origins of BCI's reptiles and amphibians is an even more complicated question. Much of this herpetofauna, however, appears to have radiated in middle America (Rand and Myers 1990).

What are the implications of the Great Interchange for BCI's biology today? Smythe (1989) notes that fruits of the spiny palm (*Astrocaryum*) are an important source of food for both agoutis and squirrels. Agoutis, which came with the palm from South America, play an essential role in dispersing its seeds, while the squirrels, recent invaders, are merely seed predators. Monkeys, caviomorph rodents, and phyllostomid bats seem to be BCI's most important seed dispersers, and they arrived with BCI's plants from South America. Has the long association between these plants and animals favored coevolution? To what extent have the irruptions from North America disrupted the harmony of Neotropical forest communities? These are questions that we may not yet be able to answer, but which should be kept in mind.

The Cyclic Revolutions of the Pleistocene

After the isthmus of Panama was completed, did Panama's biota evolve in a stable setting, or has it been subject to violent changes in climate or to devastating human activities? This question is only beginning to receive clear answers.

Webb (1978) observed that the animals that first crossed the isthmus of Panama were types that would have required a corridor of savanna or thorn scrub through the isthmus. When glaciers are widespread in the temperate zone and sea levels are accordingly low, tropical climates are cooler and drier. Savanna habitats are accordingly most widespread, in the isthmus and elsewhere, at times of low sea level. In fact, sea levels were particularly low 3 million years ago, 2 million years ago, and 1.4 million years ago, while the major episodes of exchange of mammals between the two Americas were 2.8–2.5 million years ago, 2.0–1.9 million years ago, and 1.4 million years ago: apparently, drier (savanna?) conditions greatly facilitated the exchange of mammals (Marshall 1988).

During the past million years, polar glaciers have expanded and receded according to a 100,000-year rhythm dictated by cyclic changes in the eccentricity of the earth's orbit around the sun (Hays et al. 1976). How were tropical climates affected during these interludes? When glaciers marched across North America, tree lines on high mountains in east Africa and northern South America dropped as much as 1500 m in altitude, and lakes there shrank or dried up altogether; warmer, wetter conditions returned just 10,000 to 13,000 years ago (Livingstone and van der Hammen 1978, Bradbury et al. 1981, Leyden 1985). Thirty-two thousand years ago, alders and meadowlands of a type now characteristic of subparamo settings at 2600 m appear to have been the dominant vegetation at 1100 m in Amazonian Ecuador, suggesting that temperatures there were 7.5°C lower than they are today (Bush et al. 1990). In Guatemala, land that is now tropical, moist forest was savanna and juniper scrub 13,000 years ago (Leyden 1984).

Such facts led biologists to conclude that, when glaciers marched across Europe and North America, tropical climates grew drier and forests receded from large expanses of South America, returning when glaciers in northern lands retreated (Prance 1982). Colinvaux (1993) observed that the Galapagos were clearly far drier during glacial times, and that in monsoon climates of northern South America and the Caribbean, dry seasons were longer during this period. The llanos of Colombia and Venezuela, now under savanna, were largely covered by sand dunes during the late Pleistocene (van der Hammen and Absy 1994: p. 248). Was Amazonia also drier then? How much so?

Colinvaux et al. (1996a,b) found that in Amazonia and the great Andean wall separating Amazonia from the Pacific Ocean, temperature dropped by at least 5°C during glacial times. Before 1990, it was thought that tropical sea surface temperatures were no lower during glacial times than now. Why should the land cool, but not the sea? More reliable methods of inferring temperature from fossil corals, which are not affected by the influence of glaciation on $^{16}O/^{18}O$ ratios have since been discovered. Beck et al. (1992) found at 19° S in Vanuatu, in the southwest Pacific, and Guilderson et al. (1994) found at 13° N in Barbados, that sea surface temperatures were indeed 5°C cooler during glacial times than they are now. The cooling of the sea had almost no effect on the marine biota: fossil near-shore mollusk faunas from what was Panama's Caribbean coast show no trace of cooler seas in Pleistocene times (Jackson et al. 1993). But less water evaporates from cooler seas, leading to lower rainfall on land (Piperno 1996).

At least during the "last glacial maximum" 22,000–13,000 years ago, Amazonia was also drier than today. During this period, phytoliths of C_4 savanna grasses, charcoal fragments, and other evidence of fire were far more common in the "Amazon fan," riverine sediments flowing hundreds of kilometers out to sea before settling to the bottom, than in sediments deposited a few thousand years ago in the Amazon delta (Piperno 1997), even though agriculture was widespread and intense along Amazonian rivers during the last few thousand years before Columbus (Roosevelt 1989). Cores from glaciers at 6000 m in the Peruvian Andes show far less dust, and rather more NO_3^-, in the ice during the last few thousand years than at the end of the last glacial. Since the wind there blows from the Atlantic, these changes suggest that forest cover in Amazonia was much sparser during the latest part of the last glacial (Thompson et al. 1995).

Just how much drier was Amazonia 15,000 years ago, during the last glacial maximum? In western Amazonia, near Uaupés, Brazil (0° N 67° W), annual rainfall is now 2900 mm/year, the driest month averages 160 mm, and the vegetation is now rainforest. This area was forested during all the last glacial maximum (Colinvaux et al. 1996b). Along the middle Caquetá in Colombian Amazonia, just north of the equator at about 73° W, rainfall is now near 3000 mm/year, yet during the last glacial maximum, there were patches of savanna among the forest (van der Hammen and Absy 1994). At 9° S, 63° W in the Amazonian state of Rondônia, rainfall now averages 2300 mm/year, with a definite 3-month dry season, and the vegetation is semi-evergreen forest. During the last glacial maximum this area was savanna, with almost no trees (van der Hammen and Absy 1994). Piperno (1997) concluded that rainfall was 25% lower 15,000 years ago and that the Amazonian forest block was separated into a larger western half and a smaller eastern portion by a great band of savanna passing through east-central Amazonia. Van der Hammen and Absy (1994) concluded that the rainfall was 40% lower 15,000 years ago, leaving a large block of forest in western Amazonia, and scattered patches in the east, as well as "galleries" along the rivers.

Earlier during the last glacial, at least parts of Amazonia were quite wet (van der Hammen and Absy 1994, Piperno 1996); the relation between northern glaciation and Amazonian drought is not a lockstep. But there are some puzzling features: there was savanna along the Napo more than 60,000 years ago, in what is now a very wet area (Piperno 1997). Just what happened to Amazonia during the Pleistocene is a puzzle.

How did these vicissitudes affect Panama? Bartlett and Barghoorn (1973) report a 12,000-year pollen record from near what is now the Caribbean end of Gatun Lake, 11–16 km northwest of Barro Colorado. Piperno et al. (1992: pp. 114–115) note that the presence of *Myrica* pollen from 11,300 to 10,000 years ago, and its disappearance thereafter save for a mid-Holocene dry snap, coupled with the absence of pollen from *Faramea* and sundry Myrtaceae before 9600 years ago, suggest that conditions in the lower Chagres Valley were decidedly drier before 9600 BP. Phytoliths (silica inclusions occurring in the leaves and stems of plants, often with features peculiar to a given species, genus or family; Piperno 1988) of many species now common on Barro Colorado first appear only 9600 years ago, while the older phytoliths more nearly resemble those now found in soils beneath deciduous "dry forest" in Guanacaste, Costa Rica (Piperno et al. 1992). The Pleistocene vegetation of the lower Chagres Valley, like that of many other Pleistocene sites, both in Panama (Bush et al. 1992) and elsewhere (Guthrie 1984, Bush et al. 1990, Colinvaux et al. 1996a), contained a mix of species that do not grow together today. Indeed, this was true in some more recent settings (MacPhee et al. 1985). It does appear, nonetheless, that conditions in this valley were much drier 10,000 years ago than they are now (Piperno et al. 1992: pp. 114f).

Cores from lake sediments at 500 m in El Valle, 75 km south-southwest of BCI and 24 km north of Panama's Pacific coast, and at 650 m at La Yeguada in Veraguas, 140 km southwest of BCI, 15 km north of the Pacific coastal plain, and 20 km south of the continental divide, both suggest that the temperature was 5°C lower 11,000 years ago than it is now: the vegetation represented in these cores is closest in composition to communities now confined to altitudes higher than 1500 m (Piperno et al. 1991a, Bush et al. 1992). Lake levels were lower near the end of the Pleistocene than in the early Holocene, when the vegetation was what one would expect today in the absence of agriculture, suggesting that conditions were drier in the Pleistocene than in the early Holocene, with perhaps 30% less rainfall per annum than today (Piperno et al. 1991a). Finally, phytoliths in sediments from a lake at Monte Oscuro, 20 m above sea level and 50 km west of Panama City, suggest that between 20,000 and 11,000 years ago, the surrounding vegetation was a mixture of sedges, C_4 grasses, and a savanna tree, *Curatella*, a vegetation very different from the deciduous forest that would grow there today. Rainfall there, which today is nearly 2 m/year, was at least 35% lower during this period (D. R. Piperno, p.c.). During the last glacial maximum, Panama's Pacific coastal plain was covered by a mixture of savanna and thorn scrub (Piperno et al. 1991a, D. R. Piperno, p. c.).

How did Barro Colorado's climate change during this period? The pollen and phytolith record from Gatun Lake clearly indicate that during the Pleistocene, Barro Colorado's climate was too dry for its present vegetation. Indeed, smaller changes have had dramatic effects. From 1926 through 1985, rainfall on Barro Colorado declined, on average, by 8 mm/year, a drop of 17% in 60 years (Windsor 1990). As a result, populations of at least 16 species of shrub and treelet with affinities for moister habitats have declined sharply: on Barro Colorado's 50-ha forest dynamics plot, numbers of stems ≥ 1 cm in diameter for each of these species have declined by more than 33% between 1982 and 1990 (Condit et al. 1996). These 16 species include *Acalypha diversifolia*, *Erythrina costaricensis*, three species of *Piper*, and *Olmedia aspera*; trees such as *Poulsenia armata*, *Ocotea whitei*, and *Virola surinamensis* are also threatened. Nevertheless, between 1982 and 1985, a period which includes the savage El Niño drought of 1982–1983, the common species on the island's 50-ha forest dynamics plot grew more common: it was the rare species, by and large, which declined in abundance (Hubbell and Foster 1990). Many species fruit more abundantly after a severe dry season (Foster 1982a,b). The annual mortality in a surprising number of species was not particularly higher from 1982 to 1985 than from 1985 to 1990, which included no unusual droughts (Condit et al. 1995). It almost seems that Barro Colorado's vegetation is dominated by species that would be favored by conditions somewhat drier than today's. During the last few thousand years, sharp changes in climate appear to have been a fact of life on Barro Colorado, and, indeed, also in Amazonia (Piperno and Becker 1996).

The Effects of Human Irruptions

The most recent, and potentially most devastating, threat to Barro Colorado's biota is the successive waves of human invasion and colonization. Humans first left traces in Panama 11,000 years ago. Around La Yeguada, they first started burning the forest and favoring second-growth vegetation about this time (Piperno et al. 1991a). How might these invaders have affected the biota?

The first effect of their appearance seems to have been the extinction 11,000 years ago of America's Pleistocene "megafauna," gomphotheres, ground sloths, glyptodonts, and the like (Martin 1973), as well as some 25-kg monkeys (Cartelle and Hartwig 1996, Hartwig and Cartelle 1996). Martin (1973, 1984) has convincingly attributed the disappearance of these large mammals to an irruption of human hunters about 11,000 years ago. Janzen and Martin (1982) argued that this megafauna must have played an important role in dispersing seeds of the vegetation where they lived, and that it included the primary seed dispersers for many plants still present in the forests of Central America. Jackson and Budd (1996: p. 5) insist that the megafauna maintained a much more open vegetation even in climates which today would support closed rainforest. The first proposition must be true; the others are more doubtful. Dispersers are sometimes essential to the persistence of the plant populations they serve (Alexandre 1978, Smythe 1989, Leigh et al. 1993). Did the extinction of the megafauna cause extinctions among tropical trees, and did closed forest form only after this extinction? Or was the neotropical megafauna concentrated primarily in savanna and thorn scrub, as Africa's is today (Gruhn and Bryan 1984), and as Piperno et al. (1991a) implicitly assume was true in Panama? Webb and Rancy (1996) classify nearly all the megafauna of South America as savanna animals, but they mention that these animals were quite abundant in southwest Amazonia, which was covered by savanna in the late Pleistocene, and also along the Napo, which is now a very wet area. As yet, we have precious little evidence by which to judge how the disappearance of the megafauna affected tropical forests.

Second, human invaders began clearing large tracts of forest soon after they arrived. Cultivation of root crops near La Yeguada apparently began about 7000 years ago. Maize cultivation was in full swing there about 4000 years ago, by which time La Yeguada's catchment was completely deforested. Agricultural pressure on La Yeguada began to relax about 2000 years ago, as if cultivation had worn out the land (Piperno et al. 1991b). This pattern of deforestation seems quite representative of the Pacific foothills and plains of Panama (Piperno et al. 1991b). Forest animals such as agoutis and pacas disappeared with their habitat from much of the Pacific side of Panama (Cooke and Ranere 1992). However, plenty of rainforest was left on the Caribbean slopes of Bocas del Toro and Veraguas (Linares 1976, Cooke and Ranere 1992).

How did pre-Columbian agriculture affect Barro Colorado's biota? Piperno (1990) searched 50 ha of old forest on Barro Colorado's central plateau for traces of human occupation, taking surface samples and digging occasional pits. She found evidence of human settlement on this 50-ha plot, but almost no evidence of agriculture. Some tracts in this plot were not cleared at any time during the last 2000 years. Agriculture may well have been practiced on the surrounding slopes, especially those facing the Chagres, where soils are more fertile (Leigh and Wright 1990) and yield water more readily late in the dry season (Becker et al. 1988). In sum, however, pre-Columbian agriculture does not seem to have disrupted the integrity of Barro Colorado's vegetation.

Third, the onslaught of the industrial age, which began with the building of the Panama Railroad and was accentuated by the digging of the Panama Canal, has accelerated sharply in the last 40 years. The Republic of Panama has more people, and its land is being cleared more rapidly, than ever before, to the point that one wonders how any forest can possibly be spared. This onslaught has affected Barro Colorado in many ways, and is still doing so. I have already remarked that when Barro Colorado was declared a reserve in 1923, there were a few active farms there, and half the island was in young second growth, the result of cutting and clearing during the late 1800s (Foster and Brokaw 1982). By 1923, moreover, spider monkeys and jaguars were already absent from Barro Colorado (Enders 1935), although a jaguar was seen there in January 1983, and another in January 1993.

The isolation, completed by 1914, of Barro Colorado Island from mainland Panama by the waters of Gatun Lake, accelerated extinction on this island because populations threatened by local extinction were now much less likely to be reinforced or reestablished by colonists from elsewhere. A recrudescence of hunting on Barro Colorado, beginning in 1932, drove its number of ocelots down from 12 to 4 and dangerously depleted its population of pumas during the next 5 years (Enders 1939). In 1958, Martin Moynihan stumbled over the last puma seen on Barro Colorado. White-lipped peccaries, and various smaller mammals, birds, amphibians, and reptiles have disappeared from Barro Colorado during the last 60 years (Willis 1974, Glanz 1982, Rand and Myers 1990). Poachers still roamed the more distant portions of Barro Colorado almost at will in the early 1970s, but poaching on the island was almost entirely suppressed by 1985. The extent to which extinctions of mammals from Barro Colorado have affected the ecological organization of the forest community is a much disputed topic (Terborgh and Winter 1980, Terborgh 1988, Glanz 1990), to which I return in chapter 7. The isolation of Barro Colorado and the destruction of much of the forest on the surrounding mainland may have eliminated or seriously diminished populations of some animals that migrate seasonally within Panama, such as the day-flying moth *Urania fulgens* (Smith 1990). For the moment, we cannot assess the impact of this process.

Human disturbance is also altering BCI's climate. Human activity, especially in industrialized nations, is responsible for global trends that may affect the future climate of BCI, such as the increase in the atmosphere's content of carbon dioxide and methane: how these trends will affect tropical forest has been the subject of much speculation (Hogan et al. 1991). "Global change" could also cause temporary or permanent changes of climatic "state." Chapter 3 discusses El Niño and its connection with the "Southern Oscillation." El Niño refers to a warm current that flows south along the Peruvian coast, bringing what are sometimes catastrophic floods to that desert coast and snuffing out the coastal upwellings that nourish a fishery on which humans and myriads of guano-producing seabirds depend (Hutchinson 1950: pp. 13–14). When this current flows, the dry season in Panama is especially long and severe (Windsor 1990). The Southern Oscillation refers to an atmospheric "see-saw" in the southern Pacific and Indian Oceans. When atmospheric pressure is low over Tahiti and the equatorial eastern Pacific warms, bringing the El Niño to life, atmospheric pressure is high over Darwin in northern Australia, and drought strikes tropical Australia and Indonesia. When, on the other hand, atmospheric pressure is low over Darwin, and Indonesia and tropical Australia are receiving abundant rain, atmospheric pressure over Tahiti is high, and upwellings off Peru deliver their normal nourishment to hosts of seabirds, marine mammals, and fishermen (Diaz and Kiladis 1992). Nowadays, El Niño is an episodic event, which rarely lasts two years. According to legend, one came about every seven years, but it is a much less regular event, and also far less clearcut, for El Niños come in all degrees from the barely detectable to the devastating (Hutchinson 1950: p. 55). When a strong El Niño occurs, the water level in Lake Titicaca drops, rainfall over Amazonia diminishes, and the direction of transport of sand along the Brazilian coast south of Salvador reverses. Geological traces of these phenomena show that El Niño conditions lasting several decades at a time occurred seven times between 5100 and 3900 years ago, never between 3600 and 2800 years ago, and thrice during the last 2500 years—once 2200 years ago, once 1300 years ago, and once rather recently (Martin et al. 1993: p. 342). Might human-induced changes bring on more of these super El Niños?

CONCLUDING REMARKS

Barro Colorado is a well-studied tract of tropical forest. An enormous stock of background information is available for this forest. Students from many countries and a variety of backgrounds have exchanged ideas and pooled techniques to understand this forest and its animals. Their results would fill a book much bigger than this.

Nevertheless, the message of this chapter is that Barro Colorado's biota is in no sense "pristine." As in any Neotropical site, Barro Colorado's biota represents a mingling which began in earnest only three million years ago when the Isthmus of Panama connected the Americas. Like all forests, tropical or temperate, Barro Colorado's has suffered many changes of climate. Thirteen thousand years ago, that site's vegetation was very different, and no place in Panama had a vegetation equivalent to Barro Colorado's present-day forest. Even over eight years, some plant populations change remarkably in response to shocks of climate. Finally, the invasion of humans around 14,000 years ago greatly altered the Neotropical fauna. Now, the human impact on the forest is accelerating in diverse ways. In sum, this forest's biota did not coevolve as a unit, and this forest's environment has not always been what it is today. What degree of interdependence can evolve among the forest's plants and animals against such a background of change?

References

Alexandre, D.-Y. 1978. Le rôle disséminateur des éléphants en forêt de Taï, Côte d'Ivoire. *La Terre et la Vie* 32: 47–72.

Bartlett, A. S., and E. S. Barghoorn. 1973. Phytogeographic history of the Isthmus of Panama during the past 12,000 years (a history of vegetation, climate and sea-level change), pp. 203–209. In A. Graham, ed., *Vegetation and Vegetational History of Northern Latin America*. Elsevier, New York.

Beck, J. W., R. L. Edwards, E. Ito, F. W. Taylor, J. Recy, F. Rougerie, P. Joannot, and C. Henin. 1992. Sea-surface temperature from coral skeletal strontium/calcium ratios. *Science* 257: 644–647.

Becker, P., P. E. Rabenold, J. R. Idol, and A. P. Smith. 1988. Water potential gradients for gaps and slopes in a Panamanian tropical moist forest's dry season. *Journal of Tropical Ecology* 4: 173–184.

Bradbury, J. P., B. Leyden, M. Salgado-Labouriau, W. M. Lewis Jr., C. Schubert, M. W. Binford, D. G. Frey, D. R. Whitehead, and F. H. Weibezahn. 1981. Late Quaternary environmental history of Lake Valencia, Venezuela. *Science* 214: 1299–1305.

Bush, M. B., and P. A. Colinvaux. 1990. A pollen record of a complete glacial cycle from lowland Panama. *Journal of Vegetation Science* 1: 105–118.

Bush, M. B., P. A. Colinvaux, M. C. Wiemann, D. R. Piperno, and K.-B. Liu. 1990. Late Pleistocene temperature depression and vegetation change in Ecuadorian Amazonia. *Quaternary Research* 34: 330–345.

Bush, M. B., D. R. Piperno, P. A. Colinvaux, P. E. de Oliveira, L. A. Krissek, M. C. Miller, and W. E. Rowe. 1992. A 14,300-yr paleoecological profile of a lowland tropical lake in Panama. *Ecological Monographs* 62: 251–275.

Cartelle, C., and W. C. Hartwig. 1996. A new extinct primate among the Pleistocene megafauna of Bahia, Brazil. *Proceedings of the National Academy of Sciences, USA* 93: 6405–6409.

Coates, A. G., and J. A. Obando. 1996. The geologic evolution of the Central American isthmus, pp. 21–56. In J. B. C. Jackson, A. F. Budd, and A. G. Coates, eds., *Evolution and Environment in Tropical America*. University of Chicago Press, Chicago.

Colinvaux, P. 1993. Pleistocene biogeography and diversity in tropical forests of South America, pp. 473–499. In P. Goldblatt, ed., *Biological Relationships between Africa and South America*. Yale University Press, New Haven, CT.

Colinvaux, P. A., K-B. Liu, P. de Oliveira, M. B. Bush, M. C. Miller, and M. Steinitz Kannan. 1996a. Temperature depression in the lowland tropics in glacial times. *Climatic Change* 32: 19–33.

Colinvaux, P. A., P. E. de Oliveira, J. E. Moreno, M. C. Miller, and M. B. Bush. 1996. A long pollen record from lowland Amazonia: forest and cooling in glacial times. *Science* 274: 85–88.

Condit, R., S. P. Hubbell, and R. B. Foster. 1995. Mortality rates of 205 neotropical tree and shrub species and the impact of a severe drought. *Ecological Monographs* 65: 419–439.

Condit, R., S. P. Hubbell, and R. B. Foster. 1996. Changes in tree species abundance in a neotropical forest over eight years: impact of climate change. *Journal of Tropical Ecology* 12: 231–256.

Cooke, R. G., and A. J. Ranere. 1992. Precolumbian influences on the zoogeography of Panama: an update based on archaeofaunal and document data. *Tulane Studies in Zoology and Botany* (supplement) 1: 21–58.

Croat, T. B. 1978. *Flora of Barro Colorado Island*. Stanford University Press, Stanford, CA.

Crook, K. W. A. 1981. The break-up of the Australian-Antarctic segment of Gondwanaland, pp. 1–14. In A. Keast, ed., *Ecological Biogeography of Australia*. W. Junk, The Hague.

Diaz, H. F., and G. N. Kiladis. 1992. Atmospheric teleconnections associated with the extreme phases of the Southern Oscillation, pp. 7–28. In H. F. Diaz and V. Markgraf, eds., *El Niño*. Cambridge University Press, Cambridge.

Donnelly, T. W. 1985. Mesozoic and Cenozoic plate evolution of the Caribbean region, pp. 89–121. In F. G. Stehli and S. D. Webb, eds., *The Great American Biotic Interchange*. Plenum Press, New York.

Enders, R. K. 1935. Mammalian life histories from Barro Colorado Island, Panama. *Bulletin of the Museum of Comparative Zoology* 78: 385–502.

Enders, R. K. 1939. Changes observed in the mammal fauna of Barro Colorado Island, 1929–1937. *Ecology* 20: 104–106.

Feduccia, A. 1980. *The Age of Birds*. Harvard University Press, Cambridge, MA.

Foster, R. B. 1982a. Famine on Barro Colorado Island, pp. 201–212. In E. G. Leigh, Jr., A. S. Rand, and D. M. Windsor, eds., *The Ecology of a Tropical Forest*. Smithsonian Institution Press, Washington, DC.

Foster, R. B. 1982b. The seasonal rhythm of fruitfall on Barro Colorado Island, pp. 151–172. In E. G. Leigh, Jr., A. S. Rand, and D. M. Windsor, eds., *The Ecology of a Tropical Forest*. Smithsonian Institution Press, Washington, DC.

Foster, R. B., and N. V. L. Brokaw. 1982. Structure and history of the vegetation of Barro Colorado Island, pp. 67–81. In E. G. Leigh, Jr., A. S. Rand, and D. M. Windsor, eds., *The Ecology of a Tropical Forest*. Smithsonian Institution Press, Washington, DC.

Gentry, A. H. 1982. Neotropical floristic diversity: phytogeographical connections between Central and South America, Pleistocene climatic fluctuations, or an accident of the Andean orogeny? *Annals of the Missouri Botanical Garden* 69: 557–593.

Glanz, W. E. 1982. The terrestrial mammal fauna of Barro Colorado Island: censuses and long-term changes, pp. 455–468. In E. G. Leigh Jr., A. S. Rand, and D. M. Windsor, eds., *The Ecology of a Tropical Forest*. Smithsonian Institution Press, Washington, DC,

Glanz, W. E. 1990. Neotropical mammal densities: how unusual is the community on Barro Colorado Island, Panama? pp. 287–311. In A. H. Gentry, ed., *Four Neotropical Rainforests*. Yale University Press, New Haven, CT.

Gruhn, R., and A. L. Bryan. 1984. The record of Pleistocene megafaunal extinctions at Taima-taima, northern Venezuela, pp. 113–127. In P. S. Martin and R. G. Klein, eds., *Quaternary Extinctions, a Prehistoric Revolution*. University of Arizona Press, Tucson, AZ.

Guilderson, T. P., R. G. Fairbanks, and J. L. Rubenstone. 1994. Tropical temperature variations since 20,000 years ago: modulating interhemispheric climate change. *Science* 263: 663–665.

Guthrie, R. D. 1984. Mosaics, allelochemics, and nutrients: an ecological theory of late Pleistocene megafaunal extinctions, pp. 259–298. In P. S. Martin and R. G. Klein, eds., *Quaternary Extinctions, a Prehistoric Revolution*. University of Arizona Press, Tucson, AZ.

Hartwig, W. C., and C. Cartelle. 1996. A complete skeleton of the giant South American primate *Protopithecus*. *Nature* 381: 307–311.

Hays, J. D., J. Imbrie, and N. J. Shackleton. 1976. Variations in the earth's orbit: pacemaker of the ice ages. *Science* 194: 1121–1132.

Hogan, K. P., A. P. Smith, and L. H. Ziska. 1991. Potential effects of elevated CO_2 and changes in temperature on tropical plants. *Plant, Cell and Environment* 14: 763–778.

Hubbell, S. P., and R. B. Foster. 1990. Structure, dynamics and equilibrium status of old-growth forest on Barro Colorado Island, pp. 522–541. In A. H. Gentry, ed., *Four Neotropical Rainforests*. Yale University Press, New Haven, CT.

Hutchinson, G. E. 1950. Survey of contemporary knowledge of biogeochemistry. 3. The biogeochemistry of vertebrate excretion. *Bulletin of the American Museum of Natural History* 96: 1–554 + 16 plates.

Jackson, J. B. C., and A. F. Budd. 1996. Evolution and environment: introduction and overview, pp. 1–20. In J. B. C. Jackson, A. F. Budd, and A. G. Coates, eds., *Evolution and Environment in Tropical America*. University of Chicago Press, Chicago.

Jackson, J. B. C., P. Jung, A. G. Coates, and L. S. Collins. 1993. Diversity and extinction of tropical American mollusks and emergence of the Isthmus of Panama. *Science* 260: 1624–1626.

Janzen, D. H., and P. S. Martin. 1982. Neotropical anachronisms: the fruits the gomphotheres ate. *Science* 215: 19–27.

Kalko, E. K. V., C. O. Handley Jr., and D. Handley. 1996. Organization, diversity, and long-term dynamics of a Neotropical bat community, pp. 503–553. In M. Cody and J. Smallwood, eds., *Long-term Studies of Vertebrate Communities*. Academic Press, San Diego, CA.

Kimsey, L. S. 1992. Biogeography of the Panamanian region, from an insect perspective, pp. 14–24. In D. Quintero A. and A. Aiello, eds., *Insects of Panama and MesoAmerica: Selected Studies*. Oxford University Press, Oxford.

Koopman, K. F. 1982. Biogeography of bats of South America, pp. 273–302. In A. Mares and H. H. Genoways, eds., *Mammalian Biology in South America*. Special Publication Series 6. Pymatuning Laboratory of Ecology, University of Pittsburgh, Pittsburgh, PA.

Leigh, E. G. Jr. 1990. Introducción, pp. 15–23. In E. G. Leigh, Jr., A. S. Rand, and D. M. Windsor, eds., *Ecología de un Bosque Tropical: Ciclos estacionales y cambios a largo plazo*. Smithsonian Tropical Research Institute, Balboa, Panama.

Leigh, E. G. Jr. 1996. Epilogue: research on Barro Colorado Island, 1980–94, pp. 469–503. In Leigh, E. G. Jr., A. S. Rand, and D. M. Windsor. 1996, *The Ecology of a Tropical Forest* (2nd ed). Smithsonian Institution Press, Washington, DC.

Leigh, E. G. Jr., A. S. Rand, and D. M. Windsor (eds). 1982. *Ecology of a Tropical Forest*. Smithsonian Institution Press, Washington, DC.

Leigh, E. G. Jr., and S. J. Wright. 1990. Barro Colorado Island and tropical biology, pp. 28–47. In A. H. Gentry, ed., *Four Neotropical Rainforests*. Yale University Press, New Haven, CT.

Leigh, E. G. Jr., S. J. Wright, F. E. Putz, and E. A. Herre. 1993. The decline of tree diversity on newly isolated tropical islands: a test of a null hypothesis and some implications. *Evolutionary Ecology* 7: 76–102.

Leyden, B. W. 1984. Guatemalan forest synthesis after Pleistocene aridity. *Proceedings of the National Academy of Sciences, USA* 81: 4856–4859.

Leyden, B. W. 1985. Late Quaternary aridity and Holocene moisture fluctuations in the Lake Valencia basin, Venezuela. *Ecology* 66: 1279–1295.

Linares, O. F. 1976. "Garden hunting" in the American tropics. *Human Ecology* 4: 331–349.

Livingstone, D. A., and T. van der Hammen. 1978. Paleogeography and paleoclimatology, pp. 61–90. In *Tropical Forest Ecosystems*. UNESCO, Paris.

MacPhee, R. D. E., D. A. Burney, and N. A. Wells. 1985. Early Holocene chronology and environment of Ampasambazimba, a Malagasy subfossil lemur site. *International Journal of Primatology* 6: 463–489.

Marshall, L. G. 1988. Land mammals and the Great American Interchange. *American Scientist* 76: 380–388.

Martin, L., M. Fournier, P. Mourguiart, A. Sifeddine, B. Turcq, M. L. Absy, and J.-M. Flexor. 1993. Southern Oscillation signal in South American paleoclimatic data of the last 7000 years. *Quaternary Research* 39: 338–346.

Martin, P. S. 1973. The discovery of America. *Science* 179: 969–974.

Martin, P. S. 1984. Prehistoric overkill: the global model. pp. 354–403. In P. S. Martin and R. G. Klein, eds., *Quaternary Extinctions, a Prehistoric Revolution*. University of Arizona Press, Tucson, AZ.

Mayr, E. 1964. Inferences concerning the Tertiary American bird faunas. *Proceedings of the National Academy of Sciences, USA* 51: 280–288.

Michener, C. D., and D. A. Grimaldi. 1988. The oldest fossil bee: Apoid history, evolutionary stasis, and antiquity of social behavior. *Proceedings of the National Academy of Sciences, USA* 85: 6424–6426.

Piperno, D. R. 1988. *Phytolith Analysis, An Archaeological and Geological Perspective*. Academic Press, San Diego, CA.

Piperno, D. R. 1990. Fitolitos, arqueología y cambios prehistóricos de la vegetación en un lote de cincuenta hectáreas de la isla de Barro Colorado, pp. 153–156. In E. G. Leigh, Jr., A. S. Rand, and D. M. Windsor, eds., *Ecología de un Bosque Tropical*. Smithsonian Tropical Research Institute, Balboa, Panama.

Piperno, D. R. 1997. Phytoliths and microscopic charcoal from Leg 155: a vegetational and fire history of the Amazon Basin during the last 75,000 years. *Proceedings of the Ocean Drilling Program* 155: 411–418.

Piperno, D. R., and P. Becker. 1996. Vegetational history of a site in the central Amazon Basin derived from phytolith and charcoal records from natural soils. *Quaternary Research* 45: 202–209.

Piperno, D. R., M. B. Bush, and P. A. Colinvaux. 1991a. Paleoecological perspectives on human adaptation in central Panama. I. The Pleistocene. *Geoarchaeology* 6: 201–226.

Piperno, D. R., M. B. Bush, and P. A. Colinvaux. 1991b. Paleoecological perspectives on human adaptation in Central Panama. II. The Holocene. *Geoarchaeology* 6: 227–250.

Piperno, D. R., M. B. Bush, and P. A. Colinvaux. 1992. Patterns of articulation of culture and the plant world in prehistoric Panama: 11,500 B. P.–3000 B. P, pp. 109–127. In O. R. Ortiz-Troncoso and T. van der Hammen, eds., *Archaeology and Environment.* Instituut voor pre- en protohistorische archaeologie Albert Egges van Griffen (IPP), Universiteit van Amsterdam, Amsterdam.

Pitman, W. C. III, S. Cande, J. LaBrecque, and J. Pindell. 1993. Fragmentation of Gondwana: the separation of Africa from South America, pp. 15–34. In P. Goldblatt, ed., *Biological Relationships between Africa and South America.* Yale University Press, New Haven, CT.

Prance, G. T. 1982. *Biological Diversification in the Tropics.* Columbia University Press, New York.

Rand, A. S., and C. W. Myers. 1990. The herpetofauna of Barro Colorado Island, Panama: an ecological summary, pp. 386–409. In A. H. Gentry, ed., *Four Neotropical Rainforests.* Yale University Press, New Haven, CT.

Raven, P. H., and D. I. Axelrod. 1974. Angiosperm biogeography and past continental movements. *Annals of the Missouri Botanical Garden* 61: 539–673.

Roosevelt, A. C. 1989. Resource management in Amazonia before the Conquest: beyond ethnographic projection. *Advances in Economic Botany* 7: 30–62.

Simpson, G. G. 1980. *Splendid Isolation. The Curious History of South American Mammals.* Yale University Press, New Haven, CT.

Smith, N. G. 1990. El porqué de la migración del lepidóptero diurno *Urania fulgens* (Uraniidae: Geometroidea), pp. 415–431. In E. G. Leigh, Jr., A. S. Rand, and D. M. Windsor, eds., *Ecología de un Bosque Tropical: Ciclos Estacionales y Cambios a Largo Plazo.* Smithsonian Tropical Research Institute, Balboa, Panama.

Smythe, N. 1989. Seed dispersal in the palm *Astrocaryum standleyanum*: evidence for dependence upon its seed dispersers. *Biotropica* 21: 50–56.

Terborgh, J. 1988. The big things that run the world—a sequel to E. O. Wilson. *Conservation Biology* 2: 402–403.

Terborgh, J., and B. Winter. 1980. Some causes of extinction, pp. 119–133. In M. E. Soulé and B. Wilcox, eds., *Conservation Biology: An Ecological-Evolutionary Perspective.* Sinauer Associates, Sunderland, MA.

Thompson, L. G., E. Mosley-Thompson, M. E. Davis, P.-N. Lin, K. A. Henderson, J. Cole-Dai, J. F. Bolzan, and K.-B. Liu. 1995. Late glacial stage and Holocene tropical ice-core records from Huascarán, Peru. *Science* 269: 46–50.

Tiffney, B. H. 1980. Fruits and seeds of the Brandon lignite. V. Rutaceae. *Journal of the Arnold Arboretum* 61: 1–40.

van der Hammen, T., and M. L. Absy. 1994. Amazonia during the last glacial. *Palaeogeography, Palaeoclimatology, Palaeoecology* 109: 247–261.

Webb, S. D. 1978. A history of savanna vertebrates in the new world. Part II. South America and the Great Interchange. *Annual Review of Ecology and Systematics* 9: 393–426.

Webb, S. D., and A. Rancy. 1996. Late Cenozoic evolution of the Neotropical mammal fauna, pp. 335–358. In J. B. C. Jackson, A. F. Budd, and A. G. Coates, eds., *Evolution and Environment in Tropical America.* University of Chicago Press, Chicago.

Willis, E. O. 1974. Populations and local extinctions of birds on Barro Colorado Island, Panama. *Ecological Monographs* 44: 153–169.

Windsor, D. M. 1990. Climate and moisture variability in a tropical forest: long-term records from Barro Colorado Island, Panama. *Smithsonian Contributions to the Earth Sciences* 29: 1–145.

Wolfe, J. A. 1975. Some aspects of plant geography of the northern hemisphere during the late Cretaceous and Tertiary. *Annals of the Missouri Botanical Garden* 62: 264–279.

Wolfe, J. A. 1990. Palaeobotanical evidence for a marked temperature increase following the Cretaceous/Tertiary boundary. *Nature* 343: 153–156.

Wolfe, J. A. 1991. Palaeobotanical evidence for a June 'impact winter' at the Cretaceous/Tertiary boundary. *Nature* 352: 420–423.

Wolfe, J. A., and G. R. Upchurch Jr. 1986. Vegetation, climatic and floral changes at the Cretaceous-Tertiary boundary. *Nature* 324: 148–152.

Wolfe, J. A., and G. R. Upchurch Jr. 1987. North American nonmarine climates and vegetation during the late Cretaceous. *Palaeogeography, Palaeoclimatology, Palaeoecology* 61: 33–77.

T W O

Dramatis Personae

Enumerating the principal actors in Barro Colorado's ecological theater and characterizing them by suitable thumbnail sketches would be a book in itself—quite a substantial one, like Janzen's (1983) *Costa Rican Natural History*.

One could easily center an account of the ecology of the rocky weather coasts of Tatoosh Island (Paine and Levin 1981) in the northeastern Pacific on the dynamics and interactions of the principal space occupiers and their consumers. The story of the factors which "zone" the intertidal, that is to say, which restrict the vertical range of the few principal space occupiers, would provide us with a synoptic account of the rocky intertidal at Tatoosh. Such a story would tell us what limits the kelp *Laminaria* to the lowermost intertidal and how the starfish *Pisaster* eats the mussels it can reach, restricting the mussel zone to the upper intertidal (Paine 1974) at all but the most wave-beaten sites, thereby making room for a complex of kelps and coralline algae (Paine 1992) between the mussels and the *Laminaria*. It would tell us how the crashing waves rolling in from vast expanses of open ocean allow the shrubby, narrow-leaved *Lessoniopsis* to grow between *Laminaria* and the mussels at the most exposed sites and how these same waves tear great gaps in exposed angles of the mussel bed, immobilize herbivores, and enable sea palms, *Postelsia palmaeformis* (Paine 1979) to use the incoming light with extraordinary effectiveness (Leigh et al. 1987, Wing and Patterson 1993), so that these gaps fill with sea palms. It would tell us what governs the succession of organisms that colonize gaps of different sizes torn by winter waves from other sections of the mussel zone (Paine and Levin 1981). Finally, it would tell us what limits the zone of springy, inch-high golden-green turf of *Iridaea cornucopiae* above the mussels, tenacious but slow to colonize, and the barnacle zone of the uppermost intertidal. This account would necessarily assess with some completeness the processes that shape the ecological organization of the community and would very likely represent a triumph of ecological reductionism.

Alas, there are too many actors involved in tropical forest to approach its ecology in this manner. Here, I can only introduce representatives of various categories of actors. Doing so, however, serves several purposes. To begin with, familiarity with some of the actors is better than not knowing any of them at all. Second, research traditions on Barro Colorado were shaped largely by the ethologist Martin Moynihan, founder of the Smithsonian Tropical Research Institute, and by the botanist Robin Foster. Both these biologists were interested in specific features of specific organisms and had no use for approaches that treated organisms as abstract lines in an energy flow diagram. More recently, Edward Allen Herre has awakened interest in the far-reaching implications of specific coevolutionary processes. As a result, our understanding of the ecology of Barro Colorado is based on case histories which illustrate a variety of specific processes. Here, I describe a sample of organisms and the ecological processes they exemplify.

PRINCIPAL PLANTS OF BARRO COLORADO

Dipteryx panamensis (Leguminosae, Papilionoideae)

The almendro, *Dipteryx panamensis*, is a striking canopy tree with smooth, pale pinkish-salmon bark whose trunk repeatedly forks to produce a graceful, somewhat hemispherical crown of compound leaves spiralled around its twigs (Fig. 2.1). There is one adult ≥ 30 cm dbh (diameter at breast height; 1.3 m above the ground) per hectare on Barro Colorado's central plateau (R. Condit, S. P. Hubbell, and R. B. Foster, unpublished data): *Dipteryx* are more common along the island's northeast shore.

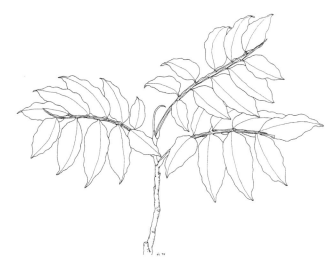

Figure 2.1. *Dipteryx oleifera* (= *panamensis*): seedling with crown 50 cm wide. Drawing by Donna Conlon.

In June and July *Dipteryx* produce bunches of pink flowers at the ends of their twigs, and the forest canopy is speckled with the purple crowns of flowering *Dipteryx*. These trees apparently flower in response to the onset of the rainy season in late April or early May. In 1970 and 1993, the transition from dry to rainy season was indistinct, and *Dipteryx* flowering and fruiting was much reduced. *Dipteryx* fruit ripens between December and April, alleviating Barro Colorado's most prolonged and severe fruit shortage of the year (Smythe 1970, Foster 1982a). Individual trees tend to alternate good and poor years, a heavy fruit crop one year seeming to leave little energy for the next (Bonaccorso et al. 1980). Combined with the likelihood of fruit failure when the onset of rainy season is too indistinct, this variation in fruit crop makes *Dipteryx* an unreliable palliative to the fruit shortage of the late rainy season.

The fruit is a single-seeded pod: the seed is a sort of kidney bean, and the pod is a thick, hard wooden casing, covered by a thin layer of sweet green pulp. Fruits weigh 18–26 g (Bonaccorso et al. 1980), and trees can produce 20 or more fruits per square meter of crown in a good year (Giacalone et al. 1990). When a fruit crop ripens, swarms of animals flock to the feast: kinkajous and bats, monkeys, coatis, and squirrels. Agoutis and peccaries, pacas, spiny rats, and the occasional tapir seek out the fruits that fall to the forest floor. Most of these animals simply eat the sweet pulp covering the fruit: only peccaries, squirrels, agoutis, and spiny rats gnaw through the wooden casing to reach the kidney bean inside (Bonaccorso et al. 1980).

A *Dipteryx* fruit must be carried far from its parent tree, and buried, if it is to have any chance of becoming another tree (De Steven and Putz 1984, Forget 1993). Although monkeys leaving *Dipteryx* trees sometimes carry fruit with them, the 70-g *Artibeus lituratus*, Barro Colorado's largest fruit-eating bat, disperses by far the largest number of *Dipteryx* fruits (Bonaccorso et al. 1980). These bats carry their fruits off to a feeding roost where they can chew off the pulp in peace, far from those predators that lurk in fruiting trees to snare unwary fruit-eaters (Howe 1979).

Fruits dropped by bats far from a fruiting *Dipteryx* usually escape the attentions of squirrels (Forget 1993), which eat many seeds, but never "plant" any (Smythe 1986). Agoutis carry off fruit the bats have dropped and bury some of them: agoutis are less likely to bury *Dipteryx* fruit found near their parent trees. Normally, agoutis dig up most of these seeds or eat their seedlings when they germinate. In a year of abundant *Dipteryx* fruit, however, seeds buried later in the fruiting season will often germinate and grow in peace if, in the meantime, trees of other species begin to drop fruit nearer the agouti's den.

Thus *Dipteryx* may need two animals, the bat *Artibeus lituratus* and the agouti, to give its seeds the opportunity to become new trees. Even so, on Barro Colorado *Dipteryx panamensis* does not seem to be replacing itself: in most years, nearly every seedling is eaten within its first year of life (De Steven and Putz 1984). Are agoutis overabundant because BCI lacks big cats? After all, *Dipteryx* seedlings and saplings were common enough on mainland peninsulas near Barro Colorado when hunters were keeping their mammal populations low (De Steven and Putz 1984). Or is *Dipteryx* a tree of early succession, which can outgrow its enemies if situated in a big enough gap (De Steven 1988)?

Platypodium elegans (Leguminosae, Papilionoideae)

Platypodium elegans is a canopy tree with a characteristically "braided" trunk and pinnately compound leaves whose leaflets are roughly 30 × 8 mm (Fig. 2.2). There are about two *Platypodium elegans* > 30 cm dbh/3 ha on Barro Colorado's central plateau (R. Condit, S. P. Hubbell, and R. B. Foster, unpublished data). The tops of small saplings lean over to cast horizontal sprays of foliage, whereupon a new branch sprouts up from the bend, to lean over in its own turn. This seems to be a cheap but reasonably effective way to build a sapling (King 1994).

Platypodium trees are pollinated by bees and almost entirely outcrossed (Hamrick and Murawski 1990). In 1997 an average of 22.4 seedlings were genotyped per maternal tree, and the average number of pollen parents found per maternal tree was 15.7; 1 year later, 20.1 seedlings were genotyped per maternal tree, and the number of pollen parents found per maternal tree was 8.1. The mean distance from which the pollen came was at least 368 m (Hamrick and Murawski 1990); long-distance pollination seems to be usual in this species. *Platypodium* fruits look like giant maple-wings, 10 × 3 cm (Augspurger 1986: Fig. 1). These "samaras" are attached to their twig by the tip of their wing. A samara has a dry weight of about 2 g, to which the seed contributes 0.4 g (Augspurger 1988). Unripe samaras are green and remain this way for many months. They dry out

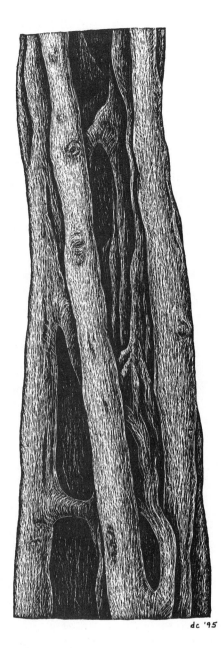
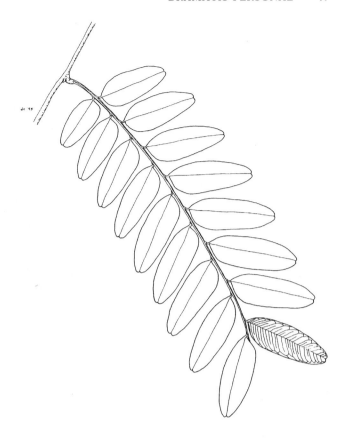

Figure 2.2. *Platypodium elegans.* L. Trunk 30 cm wide. R. Compound leaf 15 cm long. Drawings by Donna Conlon.

after the dry season arrives and are dispersed by the dry season winds. Seeds often are blown 50 m or more. Shaded seedlings within 20 or 30 m of the parent tree tend to die quickly, many from a fungal pathogen related to the Irish potato blight, but the fungus does not attack seedlings in light gaps (Augspurger 1983b). Seed dispersal serves two functions: it conveys seeds away from the parent and from the diseases that lurk near it, and it increases the probability that some seedlings will encounter a light gap.

Astrocaryum standleyanum (Palmae)

The black palm, *Astrocaryum standleyanum* (Fig. 2.3), normally has a trunk diameter of 14–18 cm. Reproductive individuals are 7–15 m tall and carry about 15 full-grown fronds, forming a crown up to 8 m in diameter (Hladik and Hladik 1969, De Steven et al. 1987). There are 4 adults/ha on Barro Colorado's central plateau (R. Condit, S. P. Hubbell, and R. B. Foster, unpublished data), about 10/ha in young forest near the laboratory clearing (Thorington 1975), and 40/ha in even younger forest in the northeast of the island, toward Barbour Point (Lang and Knight 1983).

These palms replace three to five (average 3.8) fronds per year, so fronds live about 4.5 years after they have expanded to full size. The trunk is ringed by the scars of old fronds. These rings are spaced 17–30 cm apart near the base and < 5 cm apart near the top of mature palms. One 7-m palm on Barro Colorado had 11 leaf-scar rings on the bottom 2 m of its trunk, and about 70 more on its trunk's remaining 5 m. These rings leave a fascinating record of the

Figure 2.3. *Astrocaryum standleyanum* 12 m tall. Drawing by Hayro Cunampio.

palm's growth rate. Does their growth slow because the canopy is closing, because taller stems are costlier to maintain, or because taller palms devote more of their resources to reproduction?

Between leaf scars, the trunk is covered by black, downward-pointing spines 5–15 m long. These spines seem "rough-sharpened." They are covered by an irritating principle which causes the cuts they make to fester. The bases and rachides of their fronds and the stalks (peduncles) supporting the great pendant clusters (spadices) of fruit are also covered by black spines.

For its size, the black palm is one of the most bounteous producers of fruit in the forest. This palm bears its fruit on great pendant spadices. In a good year, a palm with a 50-m² crown can ripen 6 spadices, each with 500 25-g fruits apiece, which amounts to 1.5 kg fresh weight of fruit/m² of crown, but, when conditions are less favorable, a spadix may carry 300 (a usual number), sometimes only 100, fruits (Hladik and Hladik 1969). Black palms flower in July and August (De Steven et al. 1987), apparently in delayed response to the onset of the rainy season. Their fruit begins to develop in September and ripens between March and June, helping to end the October–March fruit shortage, the worst of Barro Colorado's yearly cycle. Like *Dipteryx panamensis*, *Astrocaryum* failed to fruit in 1971, following the very wet dry season of 1970 (Foster 1982a: p. 207), and its fruit crop was very sparse in 1994, following the indistinct onset of rainy season in 1993.

During the October–March food shortage, *Astrocaryum*'s ripening fruit depends on its abundance of unpleasant spines for protection against hungry animals. Sometimes, however, squirrels and capuchins (white-faced monkeys) can cross from adjacent trees to get at the ripening fruit (Smythe 1989). When ripe, the fruits are bright orange. The 2 × 3-cm seed is covered by a tough wooden endocarp 1.5–3 mm thick, forming a nut or stone covered in turn by a layer of sweet pulp 4–5 mm thick (Smythe 1989). When these fruit ripen and drop to the floor, a variety of animals—opossums, pacas, coatis, and the like—come to eat the sweet pulp, sometimes moving the seeds about in the process. Only peccaries, agoutis, squirrels, and some smaller rodents, however, gnaw through the endocarp to reach the seed (Smythe 1989).

Black palms were the first species whose reproductive success was shown to depend on agoutis. Bruchid beetles lay eggs on black palm fruit, and seem to find just about every one. Unless agoutis peel the orange flesh from the palm nuts before the newly hatched bruchid larvae have reached the nut and bury the seed, thereby preventing other bruchids from laying eggs on it, the seed appears to have no chance of becoming a successful seedling. Agoutis benefit from peeling away the flesh before burying the seed, because seeds thus peeled have some chance of remaining sound and edible until they germinate (Smythe 1989). *Astrocaryum* seeds can remain in the soil, dormant but viable, for over 30 months, thus providing a reliable resource for agoutis (N. Smythe, p. c.).

Scheelea zonensis (Palmae)

The palma real, *Scheelea zonensis*, is the stoutest palm in the forest, and it lacks spines. It is another copious producer of fruit. *Scheelea*'s trunk diameter often exceeds 30 cm, and reproductive individuals are normally 6–15 m tall. There are up to 20 adult *Scheelea* per hectare in young forest (Thorington et al. 1982, Lang and Knight 1983), but fewer than 2 per hectare on Barro Colorado's central plateau (R. Condit, S. P. Hubbell, and R. B. Foster, unpublished data).

An adult *Scheelea* carries about 21 mature fronds and replaces an average of 5 fronds per year (De Steven et al. 1987). The foliage of adult *Scheelea* is rather multilayered; these palms need well-lit sites (Hogan 1986).

The crowns of some large, healthy *Scheelea* are 11 m in diameter. Wright (1990) records that a fruiting *Scheelea* produces one or two spadices carrying several hundred to a thousand 30-g fruits each, while De Steven et al. (1987: Table 8) record that flowering *Scheelea* produce 1–7 inflorescences (which become fruit-bearing spadices), averaging 4.5—compared to *Astrocaryum*'s 1–6, averaging 2.8.

Perhaps, in a good year, a *Scheelea* with a 100-m² crown produces 3 pendant spadices with 1000 fruits each, 90 kg fresh weight of fruit, or 900 g/m² of crown. The hard-shelled palm nut is surrounded by a layer of bright yellow, fibrous, fatty flesh, of which capuchins are fond (Hladik and Hladik 1969). *Scheelea* palm nuts are a crucial resource for squirrels (Glanz et al. 1982) and much sought after by agoutis (Forget et al. 1994).

Ripe *Scheelea* fruit is normally available somewhere on Barro Colorado from May through November, but an individual *Scheelea* palm bears ripe fruit for only a month and a half (Wright 1990). Nearly all fruit dropped from May through July that is not stripped of its pulp and buried by agoutis falls prey to bruchids (Wright 1983, 1990). By September, however, bruchids are no longer laying eggs on freshly fallen fruit (Wright 1990). From July on, progressively more of the fallen fruit is taken by agoutis and squirrels, as fruit becomes progressively scarcer in the forest as a whole. On Barro Colorado, a bruchid egg laid on a *Scheelea* fruit in September would have a small chance of completing its development before a hungry mammal ate the seed it was in (Forget et al. 1994).

Virola spp. (Myristicaceae)

Barro Colorado has three species in the nutmeg family Myristicaceae, all in the genus *Virola* (Fig. 2.4). In all three species, the sexes are separate. Saplings of all three species (like those of every species in the family) have vertical stems and horizontal branches in bunches along the stem. Leaves are distichous (arranged along the sides of horizontal twigs); thus saplings have discrete horizontal layers of foliage. *Virola sebifera* is a midstorey species, whose leaves are widest no more than halfway from base to tip. Sapling leaves have a soft, velvety texture on the underside. There are about five *Virola sebifera* ≥ 20 cm dbh/ha both on the central plateau (R. Condit, S. P. Hubbell, and R. B. Foster, unpublished data) and in young forest near the laboratory (Thorington et al. 1982). *Virola nobilis* (which figures in much of the literature as *Virola surinamensis*) is a common canopy tree (about 1.6 stems ≥ 30 cm dbh/ha on Barro Colorado's central plateau: R. Condit, S. P. Hubbell, and R. B. Foster, unpublished data, and 7/ha in young forest near the laboratory clearing: Thorington 1975) whose leaves are smooth and widest near the tip. The third species, code named *Virola* "bozo," is not yet described: it is a canopy tree whose leaves are smaller, not more than 10 cm long and narrower than the others. *Virola* "bozo" is concentrated in old forest on the side of Barro Colorado opposite the laboratory clearing.

The typical growth form of the family is well illustrated by saplings of *Virola sebifera*, which are quite common on Barro Colorado. In these saplings, the layering of the foliage is particularly clear, and small, shaded saplings arrange leaves in their top layer so as to catch as much as possible of the incoming light without shading each other. The distinction between vertical trunk and horizontal branches normally lasts all the life of a *Virola nobilis* (it is a bit less rigid in mature *Virola sebifera*). In large, old *Virola nobilis*, however, high, large horizontal branches turn up to form (or sprout near their tips) vertical shoots which "reiterate" (Hallé et al. 1978) the architecture of *Virola* saplings. The *Virola nobilis* just behind Lutz catchment's meteorological tower, however, has lost its "apical dominance," apparently because its crown was thoroughly infested by lianes. The trunk of this tree forks several times, and a vertical shoot reiterating the architecture of a *Virola* sapling sprouted from near the base of a horizontal branch.

The two canopy species, *Virola nobilis* and *Virola* "bozo" are significantly more common on the slopes surrounding Barro Colorado's central andesitic cap than on the cap itself: at the height of the dry season, soil moisture is more readily available on these slopes than atop the plateau (Condit et al. 1995). On these wetter slopes, more seedlings survived artificial defoliation than on the drier andesitic cap (Howe 1990b). Moreover, the total numbers of stems ≥ 1 cm dbh of these two species declined markedly between 1982 and 1990 (Condit et al. 1996). It appears that these two *Virola* prefer moister conditions.

A *Virola* fruit consists of a round seed 1 cm or more in diameter, covered by a waxy red aril, and enclosed in a thick orange casing. *Virola* are therefore convenient subjects for studying the production and removal of fruits: the number of casings under the tree denote the number of fruits the tree has ripened, and the number of casings less the number of seeds under that tree denotes the number of seeds carried away (most of which are dispersed). A *Virola sebifera* seed has a fresh weight of 0.7 g, while its aril weighs 0.23 g (Howe 1981); a *Virola nobilis* seed has a fresh weight of 3.1 g, while its aril weighs about 1.7 g (Howe and Vande Kerckhove 1981). In a good year, *Virola nobilis* produce about 70 fruits/m² of crown (Hladik and Hladik 1969). Arils are a rich food: half of their dry matter is lipid. In both *Virola sebifera* and *Virola nobilis*, more than half the seeds are removed—a remarkably high proportion (Howe 1981, 1986). *Virola* attracts a limited array of frugivores. Of Barro Colorado's 62 species of fruit-eating birds and 18 species of fruit-eating mammals, *Virola nobilis* attracts 7 species of birds, kinkajous (*Potos flavus*), and 2 species of monkeys, while *Virola sebifera* attracts 7 species of birds, 6 of which visit *Virola nobilis*.

Are *Virola* "choosing" a restricted coterie of seed dispersers? It would seem so. *Virola* seeds survive only if dispersed: seeds dropped 45 m from the parent crown have 44 times the chance of becoming 12-week-old seedlings that seeds under the crown do (Howe et al. 1985). Indeed, *Virola nobilis* is a textbook example of a tree whose seedlings can only survive far enough away that there is room between a parent and its surviving offspring for trees of other species (Janzen 1970).

In both *Virola nobilis* and *Virola sebifera*, over a third of the seeds removed are dispersed by chestnut-mandibled

Figure 2.4. *Virola sebifera* (L) (= *surinamensis*) and *Virola nobilis* (R), canopy trees. Drawing by Judith Gradwohl.

toucans, *Ramphastos swainsonii* (Howe 1981, 1986). The habits of these toucans are eminently convenient for *Virola*. They appear at dawn, remove some seeds to a convenient perch tens of meters away, where they mouth off the aril, drop the seeds, and fly back for more. It is easy to imagine how the *Virola* retain the toucans' loyalty: they fruit faithfully and predictably once each year, and their fruit crops vary much less in size from year to year than, for example, those of *Dipteryx*. *Virola* have protracted fruiting seasons: *Virola nobilis* fruit from March to September, and *Virola sebifera* from September to February (Howe and Vande Kerckhove 1981), as if to assure the toucans a reliable source of food. It is less clear how the *Virola* "choose" their dispersers: are "less desirable" frugivores repelled by tannins or other compounds in the aril?

Nonetheless, *Virola nobilis* may also need agoutis as well as toucans, for weevils destroy most of the seeds of *Virola nobilis* that agoutis do not bury (Forget and Milleron 1991).

Tetragastris panamensis (Burseraceae)

Tetragastris is a common canopy tree, which can bear fruit when only 11 cm in trunk diameter. *Tetragastris* trees can attain a trunk diameter of 60 cm and a height of 38 m (Howe 1980). There are 118 *Tetragastris* ≥ 30 cm dbh on Barro Colorado's 50-ha forest dynamics plot (R. Condit, S. P. Hubbell, and R. B. Foster, unpublished data); it is much rarer near the laboratory (Thorington et al. 1982). The fruits contain one to six (usually five) seeds, arranged radially, like the segments of an orange: the seeds average 0.5 g (wet) apiece (Howe 1982, De Steven 1994). Each seed is surrounded by a sugary white aril, and the group of seeds is enclosed by a purple husk. When ripe, the husk opens, dangling the white fruit from its red container. *Tetragastris panamensis* normally flowers in early July and bears fruit from January to June of the following year (Howe 1980, 1990a, Foster 1982b). The fall of fruit husks under the crown measures fruit production; the number of seeds removed is five times the number of husks under the crown, less the number of seeds there (Howe 1982).

Tetragastris fruits much less reliably than *Virola*. In the year following a wet dry season, or a dry season whose end is uncommonly indistinct, *Tetragastris* bears no fruit. *Tetragastris* did not fruit in 1971, following the very wet dry season of 1970. This species also failed to fruit in the years following the relatively weak dry seasons of 1972 and 1978 (Howe 1980, Foster 1982a). On the other hand, *Tetragastris* fruited abundantly in the year following the 6-month dry season of 1976–1977, and in the years following the El Niño drought of 1982–1983 (Howe 1990a) and 1992 (S. Paton, p. c.).

Indeed, unlike *Virola*, *Tetragastris* seems to thrive on dry weather. The number of *Tetragastris* stems ≥ 1 cm dbh on Barro Colorado's 50-ha forest dynamics plot increased markedly from 1982 to 1985, despite the savage El Niño drought of 1982–1983.

During the bumper year of 1978, *Tetragastris* fruit was consumed by at least 8 species of mammals and 14 species of birds. Despite this diversity of visitors, only 28% of the seeds produced by *Tetragastris* that year were removed or eaten that year, compared to 54% for *Virola* between 1979 and 1983 (Howe 1980, 1990a). Again, in contrast with *Virola*, where the proportion of a tree's seeds removed did not depend on the size of its fruit crop (Howe and Vande Kerckhove 1981), the proportion of seeds removed from *Tetragastris* trees with small fruit crops was quite low. Most *Tetragastris* trees producing fewer than 11,000 fruits apiece had less than 15% of their seeds removed, while those trees producing between 12,000 and 80,000 fruits apiece had about 30% of their seeds removed (Howe 1980: Fig. 8). It appears that *Tetragastris* requires a fairly large fruit crop to attract substantial attention from frugivores.

Young *Tetragastris* seedlings survive much better under their parent trees than do their *Virola* counterparts. One month after the end of the bumper season of fruit fall in 1978, Howe (1980) found an average of 55 *Tetragastris* seedlings/m² of ground. About 15 seeds fall/m² under the crowns of *Virola nobilis*, and only 11.5% of the seeds placed under *Virola* crowns survived 4 weeks to become germinated seedlings (Howe et al. 1985). The contrast between the lawn of *Tetragastris* seedlings under parent trees and the near absence of *Virola* seedlings under their parents is quite striking (Howe 1989, 1990a). *Tetragastris* seems much less dependent on its seed dispersers than does *Virola*.

Faramea occidentalis (Rubiaceae)

Faramea occidentalis is an understory tree which begins to fruit at a trunk diameter of 3.5 cm. Occasional individuals attain 20 cm dbh. Young plants have a characteristic growth form: horizontal branches, decussate on an erect main stem. Leaves look as if they were meant to be decussate around their twigs, but were instead flattened out to form a horizontal spray of foliage (Fig. 2.5). A twig ends in a pair of leaves, between whose bases is a pair of slender, 5-mm long "cross-billed" stipules.

Faramea is the second most common stem ≥ 1 cm dbh on Barro Colorado's 50-ha plot. In 1982 the plot contained 23,465 *Faramea* among its 236,000 stems ≥ 1 cm dbh; in 1990 there were 26,912 *Faramea* among its 244,000 stems ≥ 1 cm dbh (R. Condit, S. P. Hubbell, and R. Foster, unpublished data). *Faramea* is more common atop the flat surfaces of the central plateau than on the surrounding slopes (Hubbell and Foster 1986), even though, at the height of the dry season, moisture is more readily available on the slopes (Becker et al. 1988).

Faramea flowers early in the rainy season, in May or June, bearing corollas like white trumpets < 2 cm long flaring into slender lobes, heavy with the scent of jasmine (Croat 1978). Its fruit, 1-cm berries, ripens between October and February, turning nearly black: these fruits are

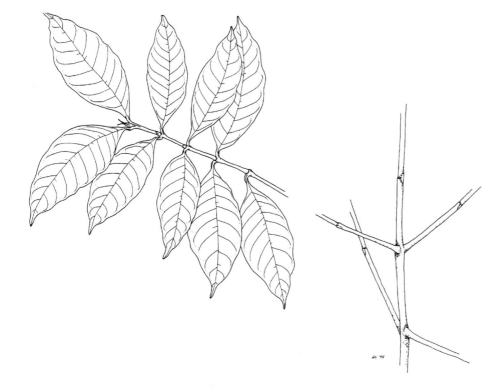

Figure 2.5. *Faramea occidentalis*: sprig of foliage 20 cm long and portion of main stem showing decussate branching. Drawing by Donna Conlon.

eaten by monkeys and guans. In the 1980s, *Faramea* was reputed to fruit in synchrony every other year—it fruited at the end of 1982, 1983, 1985, and 1987 and failed to fruit in 1981, 1984, and 1986 (Schupp 1990). This tidy biennial rhythm, however, broke down after 1987 (S. Paton, p. c.). The wet dry seasons of 1970 (Foster 1982a) and 1981 (Schupp 1990), or an indistinct onset of rainy season, like 1993 (S. Paton, p. c.) were followed by fruit failure, and a severe dry season such as the El Niño dry seasons of 1982–1983 (Schupp 1990) and 1992 (S. Paton, p. c.) were followed by bumper crops even though *Faramea* had fruited during the preceding years. Curiously, another rubiaceous treelet, *Coussarea curvigemmia*, which is one-tenth as abundant as *Faramea*, behaves in an opposite way. *Coussarea* fruited plentifully after the wet dry season of 1970, when *Faramea* failed (Foster 1982a), and after the indistinct onset of rainy season in 1993 (S. Paton, p. c.). *Coussarea* and *Faramea* never fruit heavily during the same year.

When *Faramea*'s seeds fall to the ground, they normally remain dormant until well into the following rainy season: their germination peaks in July. A heavy dry season rain, however, can cause early germination, in which case the whole fruit crop can die of desiccation, as happened to the 1983 seed crop during the 1984 dry season. Mortality among seeds is heaviest when fruit is in shortest supply in the forest as a whole and animals are accordingly hungriest (Schupp 1990).

Seeds and seedlings survive better in the shade than in gaps: *Faramea* is not a gap-dependent species (Schupp 1990). As do many shade-tolerant species, *Faramea* normally appear to require more than a year to accumulate the resources required to flower and fruit. This appears to be why they normally fruit every 2 years. Their synchrony is apparently maintained by the similarity of their responses to unusually dry and unusually wet dry seasons (Schupp 1990). There is no evidence that seeds have better prospects in years of heavy fruiting.

Faramea attracts dispersers by enclosing its seeds in a fleshy berry. Normally, seeds survive at least three times better 5 m from conspecific adults than under conspecific crowns (Schupp 1988). Yet where *Faramea* is most dense (> 300 adults/ha), their seeds survive better under their parents' crown than farther away (Schupp 1992). High seedling mortality where adults are most dense limits the density of *Faramea* on BCI (see chapter 8).

Hybanthus prunifolius (Violaceae)

Hybanthus, a shrub, is the most common stem ≥ 1 cm dbh on Barro Colorado's 50-ha plot. In 1982, 40,000 of this plot's 236,000 stems ≥ 1 cm dbh were *Hybanthus*. A *Hybanthus* stem leans over to cast a horizontal spray of foliage: the shrub continues its growth by sprouting a branch from the bend, which repeats the process.

Hybanthus flowers in response to the first dry season rain exceeding 12 mm that occurs after the soil has dried sufficiently (Augspurger 1982). Flowering begins a little more than a week after the rain and lasts only a few days. If the rain is heavy enough and follows a sufficiently protracted and severe dry spell, the mass flowering it causes is quite spectacular, with 2.5-cm white, snapdragon-like flowers hanging in dense rows from the curving branches (Fig. 2.6). Fruits develop in a few weeks, dry out, and disperse their small seeds explosively. If the rain that stimulates flowering is late enough that the rainy season follows soon afterward, *Hybanthus* seedlings germinate when rainy season is just beginning at which time soil nutrients are most abundant. If, however, the dry season rain comes too early, the fruit crop can be destroyed by drought (Augspurger 1982).

Watering the base of a *Hybanthus* shrub brings it into flower if a sufficiently long and intense drought precedes watering (Augspurger 1982). One must water a large plant, or a clump of plants, if their flowers are to attract pollinating bees down from the canopy. The "point" of populationwide synchrony in flowering, however, is that synchronous *fruiting* satiates caterpillars of the species of microlepidopteran that destroy nearly all the seeds of those *Hybanthus* that fruit "out of turn" (Augspurger 1982).

Flowering is followed by a leaf flush which accounts for at least 80% of a plant's annual leaf production (Aide 1992). Both leafing in synchrony with conspecifics and flushing leaves late in the dry season when insect populations are low should reduce herbivory. Is this true? To find out, during the dry season of 1987 Aide (1992) placed plastic sheeting over the soil, extending at least 1 m out from the stems of experimental plants that had not flowered, in order to keep rainwater from reaching their roots. He measured the percentage of leaf surface that was damaged during the first month after leaf flush in these experimental plants and in control plants whose soil was left exposed. In 1987, the rainy season began on 25 April (Windsor 1990: p. 57). Until the plastic was removed, experimental plants did not flush leaves. The control plants that flushed leaves

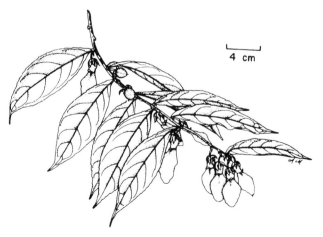

Figure 2.6. *Hybanthus prunifolius*: flowering branch. Drawing by Arlee Montalvo.

after a rain on 11 February lost a median of 6% of their leaf area, other controls flushing after a rain on 7–8 April lost a median of 4%, plants that flushed leaves after their plastic was removed on 4 May lost a median of 9%, and those whose plastic was removed on 20 May lost a median of 27% of their leaf area during the first month after flushing (Aide 1992). Leafing out of turn is dangerous, especially if the out-of-turn flush occurs well after rainy season begins, when herbivore populations have had a chance to build up.

Psychotria spp. (Rubiaceae)

Psychotria is a circumtropical genus of understory shrubs. There are about 20 species of *Psychotria* on Barro Colorado, some of which are quite common or conspicuous. In 1982, the 50-ha plot contained 6168 *P. horizontalis* and 592 *P. marginata* among its 236,000 stems ≥ 1 cm dbh (Condit et al. 1995). Many of these species germinate, grow, and reproduce in the shade (Wright et al. 1992). Their fruits are small berries, which are an important resource for manakins during the rainy season (Worthington 1990).

Shrubs are liable to receive crushing blows from falling branches. If a fallen piece of wood forces a branch of *Psychotria acuminata* into the dirt, that branch will take root. If one "plants" a green leaf of *Psychotria horizontalis* in the soil, that leaf will sprout roots and grow to form an independent plant (Sagers 1993).

Psychotria horizontalis grows much faster and flowers sooner if protected from insect pests (Sagers and Coley 1995). Among cuttings of similar initial size from different plants, which are stuck in a "common garden" in the forest, exposed to insect pests, those with the most tannin-rich leaves suffered the least loss of foliage to herbivores during the year after planting. Nonetheless, there was no correlation between a plant's weight gain during the year after planting and the tannin content of its leaves (Sagers and Coley 1995). Does this mean that defensive compounds are useless? By no means! In other cuttings from the same parent plants, grown at the same sites but screened from insects, plants with more tannin-rich leaves grew more slowly. It therefore costs these *Psychotria* something to make tannins. Under natural conditions, the decrease in herbivory caused by higher tannin content just balances the cost of the extra tannins (Sagers and Coley 1995), as if these plants had optimized the tannin content of their leaves.

Psychotria marginata has domatia—minute mite-houses—sandwiched between the midrib and the primary veins on the underside of the leaf. In other shrubs these domatia greatly increase the abundance of both predatory and fungus-eating mites, which presumably reduce the abundance of arthropod pests and fungal pathogens on the leaf (Walter and O'Dowd 1992, O'Dowd 1994).

Many *Psychotria* flower at the beginning of the rainy season, or in response to heavy rains late in the dry season (Augspurger 1983a, Wright 1991). Flowering is much less conspicuous, and lasts much longer than in *Hybanthus*. Flowers and fruit are usually produced at the tips of the twigs. In *Psychotria acuminata*, twigs that have flowered fork symmetrically before the next round of flowering (Fig. 5.4), a process that leads to an umbrella crown of thin 10 × 5 cm leaves, decussate around the ends of upright twigs (Hamilton 1985), in accord with Leeuwenberg's model (Hallé et al. 1978).

Other *Psychotria* flower at their branch tips, after which these tips fork (Hamilton 1985). The main stem of *P. deflexa* forks asymmetrically into one nearly upright and one nearly horizontal branch (Fig. 5.6). Horizontal branches will sometimes fork symmetrically in a horizontal plane, but they are short: *P. deflexa* tend to be far taller than wide, with a single main stem and horizontal branches spaced alternately and evenly along it, bearing opposite leaves. In *P. furcata*, the main stem's forking is less obviously asymmetric, but the twig that becomes a branch bifurcates repeatedly to cast a leaning spray of 6 × 3 cm leaves (Fig. 2.7). The plant elaborates a series of partially overlapping sprays of such foliage. *P. chagrensis* bears a few discrete layers of leaves. In each, a horizontal twig may bifurcate in a horizontal plane or fork asymmetrically into a short, inclined twig, which will eventually bear a quasi-rosette of decussate leaves and a longer horizontal shoot (Fig. 2.7). A horizontal spray of *P. chagrensis* leaves looks a bit like a *Terminalia* branch with its horizontal spray of rosettes.

Different species of *Psychotria* also have contrasting physiologies. *P. limonensis*, which has a tap root 1 m deep, produces leaves, flowers, and fruit all year long (Wright 1991). Its big leaves (20 × 10 cm) live 4 years. *P. marginata*, with a rooting depth of only 38 cm, flushes leaves at the beginnings of both the dry and the rainy seasons. Dry season leaves are designed to tolerate drought. They are heavier per unit area and have higher water use efficiency than their rainy season counterparts (Mulkey et al. 1992). *P. furcata*, whose rooting depth is only 29 cm, flushes leaves only at the beginning of the rains, and these leaves have little capacity to reduce water loss during the dry season (Wright 1991, Wright et al. 1992). *P. chagrensis* has a rooting depth of 51 cm, but it is more drought sensitive than the others mentioned: it wilts more readily as dry season advances, even though it is restricted to the island's wetter sites.

All these species are adapted to cope with drought. Even when irrigated, *Psychotria limonensis*, *P. chagrensis*, *P. marginata*, and *P. horizontalis* increase the osmotic tension (lower the osmotic potential) of their leaf tissue and increase the rigidity of the cell walls in their leaves during the dry season, changes which reduce the likelihood of wilting (Wright et al. 1992).

Rooting depth in *Psychotria horizontalis* is only 33 cm (Wright et al. 1992). The top 35 cm of soil on Barro Colorado becomes quite dry during a severe dry season: at the end of 1992's dry season, predawn water potential in

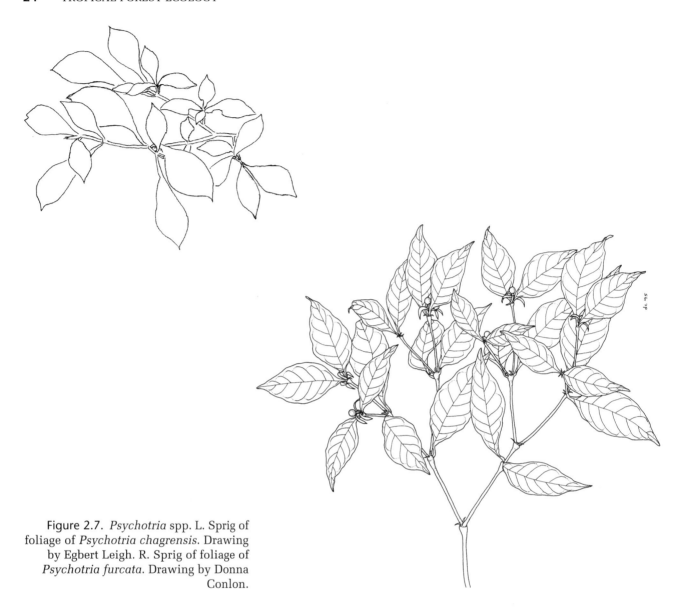

Figure 2.7. *Psychotria* spp. L. Sprig of foliage of *Psychotria chagrensis*. Drawing by Egbert Leigh. R. Sprig of foliage of *Psychotria furcata*. Drawing by Donna Conlon.

Psychotria horizontalis was −3.5 MPa, −35 atmospheres (Kursar 1997). Accordingly, xylem vessels of this species are so strongly built that they can withstand water tensions of 50 atmospheres (Zotz et al. 1994, Table 3). These adaptations work. On the 50-ha plot, *Psychotria horizontalis* and *P. marginata* suffer a death rate of 4–5% per year, twice as high as the average mortality rate for trees ≥ 10 cm dbh, but mortality rate was lower from 1982 to 1985, a period including the El Niño drought of 1982–1983, than from 1985 to 1990 (Condit et al. 1995). Irrigation during dry season affects the timing, but not the total number, of leaves and inflorescences produced per year by *P. horizontalis*, *P. furcata*, *P. marginata* and *P. limonensis*.

The most striking effect on *Psychotria furcata*, *P. horizontalis*, and *P. marginata* of irrigating their surroundings through four successive dry seasons is to render the seasonal peaks of leafing and flowering in these species progressively less sharp in successive years. Irrigated *P. furcata* also leafed somewhat earlier every year, as if leaf production were timed by a circannual clock which was normally reset every year by the transition from dry to rainy season (Wright 1991).

Epiphyllum phyllanthus (Cactaceae)

The epiphytic cactus *Epiphyllum phyllanthus* grows on the trunks and primary branches of canopy trees, especially those whose trunk diameter is 50 cm or more. A plant consists of a clump of pendant, straplike stems, often 30 or more. Each stem is 2–3 cm wide, about 2 mm thick, and up to 1 m long (Andrade and Nobel 1996). One expects cacti in deserts: how can they succeed in tropical forest?

The primary problems facing epiphytes growing on the trunks or branches of trees are reliable access to nutrients and water. Trees and lianes obtain nutrients and water from soil on the forest floor, which epiphytes cannot reach. Epiphytic strapferns such as *Polypodium crassifolium* grow root baskets to trap dust and litter, which form a nutrient-rich, water-holding soil. Root baskets allow these strapferns to grow on a variety of tree species and on many parts of any given tree, but the presence of root baskets means that *Polypodium crassifolium* has an average of 1.86 g dry weight of roots/g dry weight of shoot (Andrade and Nobel 1997). *Epiphyllum* usually grows from a crotch or crevice in the bark of the upper trunk of a canopy tree or from a hole in the upper trunk or primary branch of a canopy tree, where their roots can tap humic soil and decaying organic matter (Andrade and Nobel 1996a). Some *Epiphyllum* grow, along with other epiphytes, on ant nests (Croat 1978), forming "ant gardens": wastes that ants leave in the nest "fertilize" these ant gardens (Davidson 1988). *Epiphyllum* does not gather its own soil, and it has only 0.22 g dry weight of root/g dry weight of shoot (Andrade and Nobel 1997), a root/shoot ratio typical for cacti.

Epiphyllum pays a double price, however, for skimping on roots. First, its choice of habitats is restricted. Barro Colorado's 50-ha forest dynamics plot has about seven *Epiphyllum* plants per hectare. More than half of these are found on two species of host tree, *Platypodium elegans* and *Tabebuia guayacan*. *Epiphyllum* may favor *Platypodium* because its "braided" trunk (Fig. 2.2) provides ready-made cavities where dust and litter may be trapped, while the uncommonly deep fissures in the bark of *Tabebuia guayacan* provide the same service. Second, *Epiphyllum* must conserve water. These succulent plants contain 150–250 mg water/cm stem length. In 4 weeks of drought, *Epiphyllum phyllanthus* loses 45% of its water, while *Polypodium crassifolium* loses 98% (Andrade and Nobel 1997). *Epiphyllum* also saves water by employing crassulacean acid metabolism. Their stomates open at night when far less water is lost per unit CO_2 obtained than during daytime transpiration, and the CO_2 thus obtained is stored in malic acid molecules until its release for photosynthesis the next day (Winter and Smith 1996). Photosynthesis is accordingly restricted. At maximum, an *Epiphyllum* takes up 1.4 $\mu mol\ CO_2/m^2$ stem per second (Andrade and Nobel 1996), while the maximum uptake of a square meter of sun-leaves of the average canopy tree is about 12.6 $\mu mol\ CO_2/sec$ (Bossel and Krieger 1991, Table 1).

Like bromeliads, cacti live in both deserts and the crowns of rainforest trees. Presumably, adaptations for desert life allowed both cacti and bromeliads to become epiphytes. Apparently, trapping their own soil has enabled bromeliads to diversify far more extensively in tropical forest than cacti have (Pittendrigh 1948).

Catasetum viridiflavum (Orchidaceae)

The epiphytic orchid *Catasetum viridiflavum* grows from 10 to 30 m above the ground, on both live branches and dead wood. It bears leaves on the single shoot from its most recent "pseudobulb": these leaves are shed in December when the dry season starts. A new shoot begins to form in mid-February, which produces 6–12 rather thin, plicate leaves and a new pseudobulb. Flowering can occur at any time from April to December, but, in any given year, only 40% of the plants flower (Zimmerman 1991). Walking along the trails of Barro Colorado, Zimmerman (1991) found about one *Catasetum viridiflavum* per 100 m of trail. About 30% of the *Catasetum* he censused from forest trails were growing on dead trees or branches; the rest were seated on live branches (Zimmerman 1991: Table 2).

In addition to the leaf-bearing bulb, a *Catasetum viridiflavum* plant has up to 10 "back bulbs": the oldest bulbs eventually rot away and disappear (Zimmerman 1990). These bulbs serve to some extent as storehouses of carbohydrates, nutrients, and water. In small plants, pseudobulbs provide resources for both leaf growth and flowering. In large plants with many "back bulbs," severing connections between the preceding year's pseudobulb and older ones has no effect on leaf growth but reduces the number of flowers produced (Zimmerman 1990).

Catasetum has no capacity for crassulacean acid metabolism. Over its lifetime, a *Catasetum viridiflavum* leaf fixes 16 g carbon for every gram of carbon in its leaf: the corresponding figure for the thicker fronds of *Polypodium crassifolium* is 5.3. Curiously, both species fix about 1 kg CO_2/m^2 of leaf per year, compared to 2.6 kg/m^2 for *Ceiba pentandra* sun-leaves (Zotz and Winter 1994a,b).

The epiphytic orchid *Dimerandra emarginata*, which also produces one new shoot each year but has a capacity for crassulacean acid metabolism (Zotz and Tyree 1996), apparently takes 30 years to grow to full size, at which point its current shoot is 30 cm long (Zotz 1995). *Catasetum viridiflavum* may grow faster, at least when its roots have access to rotting wood, but its demography is not yet known. Nonetheless, *Catasetum* is no exception to the rule that the world of epiphytes is a world of scanty resources. How does this scarcity affects its reproductive behavior?

For *Catasetum*, as for other epiphytic orchids (Zimmerman and Aide 1989) and plants of the shaded understory (Clark and Clark 1988), fruit-bearing is a costly business, which reduces the next year's leaf production. Even the well-lit *Catasetum* on the dead trees of Gatun Lake, which flower more often than do *Catasetum* in the forest, have lower leaf production in the year following fruit-bearing (Zimmerman 1991). Plants in marginal settings, moreover, can mobilize enough resources to produce pollen long before they can afford to bear fruit. Thus, like the cycads that live on the forest floor at La Selva, Costa Rica (Clark and Clark 1987), the smallest *Catasetum* in flower bear male flowers. In general, those *Catasetum* receiving more light are more likely to flower as females. In short, *Catasetum* (like the La Selva cycads) start life as males, where small size is less costly. Soon, however, light level becomes the most decisive influence on a *Catasetum*'s sex. A female *Catasetum* will become male if shaded enough to cut off 65% of its light (Zimmerman 1991).

Scarcity of resources places these *Catasetum* in an awkward quandary. *Catasetum viridiflavum* is pollinated by the euglossine bee *Eulaema cingulata*, which weighs 500–600 mg (Romero and Nelson 1986, Courtney Murran, p. c.). Male bees are attracted to these orchids by the prospect of gathering fragrances which attract female bees (Schemske and Lande 1984). *Eulaema* are usually most abundant in April or May, when flowers are most abundant in the forest as a whole. The few *Catasetum* that can mobilize the resources to flower in April or May, at the beginning of their growing season, have high pollination success (50–80%), but most *Catasetum* flower between August and December, and their pollination success hovers around 5%. The fruit capsules dehisce in the following dry season, releasing great quantities of minute seed. In some years, more than half the fruit on early-flowering plants rots before dispersing seed: even so, early-flowering plants have higher reproductive success (Zimmerman et al. 1989). Most *Catasetum*, however, are limited by shortage of both pollinators and resources for fruit-bearing, a predicament that appears typical of epiphytic orchids (Zimmerman and Aide 1989).

Male *Catasetum* flowers have a mechanism described by Darwin (1877) that flings a large pollinarium (packet of pollen) onto the back of a visiting bee. *Catasetum viridiflavum* imposes a 40 mg pollinarium on its 500-mg pollinator (Courtney Murran, p. c.). A bee finds this experience sufficiently off-putting that it tends thereafter to avoid male *Catasetum* of the species that victimized it, a reaction which benefits the victimizer, as imposing a second pollinarium on the bee often disables the first one. Female *Catasetum* are pollinated only because their flowers differ enough from male flowers not to trigger the bee's aversion to male flowers (Romero and Nelson 1986). *Catasetum*'s pollination strategy presumably only works insofar as *Catasetum* flowers are too rare to play a large part in the economy of any given bee. For obvious reasons, this strategy also favors rapid divergence in appearance of male flowers of coexisting species of *Catasetum* (Romero and Nelson 1986).

Ficus spp. (Moraceae)

There are two conspicuous categories of fig tree on Barro Colorado. First are the free-standing fig trees, among which the most prominent are the two pioneer trees, *Ficus insipida* and *Ficus yoponensis*, with pale, smooth trunks and wide, flat crowns of erect or inclined leaves spiralled about the tips of their twigs. On Barro Colorado, where these two species colonized abandoned fields, there were five canopy trees of these species per hectare in 1976, with an average crown area of > 300 m^2 apiece, in the 25 ha of young forest surrounding the laboratory clearing (Morrison 1978). These trees are rapidly dying of disease as the forest ages (C. O. Handley Jr., p. c.). Before humans were clearing fields, *Ficus insipida* were colonizing accreting riverbanks; around Peru's Rio Manu, *Ficus insipida* are dominant trees on land formed 50–160 years ago (Foster et al. 1986, Foster 1990).

The second category is "strangler figs," each of which begins as an epiphyte on another tree, drops roots to the ground which rapidly enclose its host's trunk in a strangling latticework, causing the host to die and rot away, leaving a magnificent giant fig with a hollow trunk (Corner 1940, 1988). On Barro Colorado, strangler figs are most common in mature forest: mature forest on Barro Colorado's central plateau contains one canopy strangler fig/2 ha (R. Condit, S. P. Hubbell, and R. B. Foster, unpublished data).

Fig trees have acquired remarkably effective pollinators, which in turn have had a fundamental impact on both the fig trees and the forest as a whole. Corner (1940, 1988) described fig fruits and their pollinators in some detail. Each species of fig tree has its own species of pollinating wasp. A fig fruit starts as a flower head turned outside in—the flowers line the inside of a ball with a hole, an "ostiole" at one end. When the flowers are ready, one or more fertilized female wasps enter the fruit, pollinate the flowers, and lay eggs in some of them. Each wasp larva feeds within, and the resulting adult hatches out from, a single fig seed. When a fruit's wasps hatch, they mate among themselves, and the fertilized females fly off in search of other fruits to pollinate. As a tree's fruits ripen in synchrony (in most species: Kalko et al. 1996), it has no fruits ready for pollination when its fertilized female wasps are emerging, so these wasps must find fruits on other trees to pollinate. Thus, whatever the time of year, some trees of each fig species are bearing fruit ready to pollinate (Milton et al. 1982, Windsor et al. 1989).

A fig's wasps can carry its pollen surprisingly far. J. Nason and E. A. Herre collected 28 fruits one day from a single strangler fig tree, *Ficus obtusifolia*: each of these fruits was pollinated by just one wasp. They planted the seeds from each fruit, and genotyped the seedlings by starch gel electrophoresis. They identified both the maternal genotype and the genotype of each fruit's pollen parent. They found that the 28 wasps that had pollinated the collected fruits came from 22 different parent trees (Nason et al. 1996). These wasps probably live less than 4 days after emerging from their natal seeds: they appear not to eat as adults. Barro Colorado's forest has only 1 adult *Ficus obtusifolia*/14 ha. A *Ficus obtusifolia* ripens no more than two fig crops per year: as each fig crop ripens, pollen-bearing wasps are being released over a period of no more than 5 days. Thus about 1 of every 26 of the mature *Ficus obtusifolia* would have been supplying pollen-bearing wasps when the 28 collected fruits needed them (Nason et al. 1996). Thus the 28 pollinating wasps must have come from at least 14 × 22 × 26 or about 8000 ha of forest—an area much larger than Barro Colorado. Despite the relative rarity of *Ficus obtusifolia*, the proportion of loci in this species that are heterozygous is higher than the mean for common trees studied on Barro Colorado: other species of strangler figs also have a high proportion of heterozygous loci (Hamrick and Murawski 1991).

"By leaf, fruit, and easily rotted wood fig-plants supply an abundance of surplus produce" (Corner 1967: p. 24). *Ficus insipida* has the highest photosynthetic rate of any full-grown tree yet measured: these trees can fix up to 33 μmol CO_2/m^2 of leaf per second, a rate more typical of crop plants than of forest trees. Both in the rainy season and in the dry season, these trees fix an average of 0.5 mol CO_2/m^2 of sun-leaf per day (Zotz et al. 1995). To make sense of these numbers, 0.5 mol CO_2 represents the CO_2 content of 3.5 m^3 of air. Moreover, "fixing" 0.5 mol CO_2 adds 227 kilojoules (kJ) to a plant's energy reserves (Nobel 1983: p. 241), while "fixing" this much CO_2 per day represents a power input, averaged over 24 hr, of 2.6 W/m^2 of leaf. Photosynthesizing flat out, fig leaves attain a power input of 10 W/m^2 of leaf (an 8-kg howler monkey's energy usage averages 33 watts).

Such copious photosynthesis supports rapid growth and abundant fruiting. Moreover, such high photosynthetic rates presuppose an abundance of nitrogen in the leaves (Zotz and Winter 1994a). Thus fig leaves are an important resource for Barro Colorado's howler monkeys (Milton 1980). Fig fruit is also an important resource. When the pollen-bearing wasps leave a fig tree, its fruit crop ripens quickly, providing a rich feast which often attracts a host of mammals and birds. A *Ficus insipida* often produces > 100 9-g figs/m^2 of crown—40,000 9-g fruits in a single crop (Morrison 1978). Moreover, the maintenance of the pollinating wasps implies that, at any time of year, there will be some trees in fruit in each species of fig (Milton et al. 1982, Windsor et al. 1989). This steady food supply is a keystone resource (Terborgh 1986) for Barro Colorado's animals during the season of food shortage (Foster 1982a). Moreover, on Barro Colorado, this steady supply of fig fruit is the primary resource for a guild of seven relatively common species of fruit-eating bat (Bonaccorso 1979). Finally, rotting fig logs attract an extraordinary community of decomposers (Zeh and Zeh 1992, 1994).

PRINCIPAL ANIMALS OF BARRO COLORADO

Dasyprocta punctata (Dasyproctidae)

Agoutis, *Dasyprocta punctata*, are terrestrial rodents (Fig. 2.8), weighing about 3 kg when adults. They resemble small forest duikers of Africa and the mouse deer of Southeast Asian forests (Smythe 1978). Their closest ecological counterpart is the 5-kg blue duiker, *Cephalophus monticola*, of African rainforests (Dubost 1980): like agoutis, blue duikers are diurnal, frugivorous, monogamous, and territorial. Near the laboratory clearing, there is a pair of adult agoutis/2 ha (Smythe et al. 1982). Agoutis are the mammals most often seen by visitors on forest walks.

Figure 2.8. *Dasyprocta punctata*. Drawing by Dorsett Trapnell.

Agoutis belong to the guild of terrestrial frugivores (Smythe 1978, 1986), which also includes spiny rats (*Proechimys semispinosus*), pacas (*Agouti paca*), coatis (*Nasua narica*), common opossums (*Didelphis marsupialis*), and collared peccaries (*Tayassu tayacu*). Study of these frugivores of the forest floor provided the first evidence that Barro Colorado's populations of vertebrate frugivores are limited by seasonal shortage of food. During the season of food shortage from November through January, and especially in December, too little fruit falls to feed the frugivores (Smythe 1970, Smythe et al. 1982). When fruit is in short supply, each species resorts to a fallback specialty: agoutis live on previously buried seeds, pacas browse seedlings, coatis search for arthropods and other small animals in the leaf litter, peccaries browse and root more. Yet these animals are clearly hungrier during the season of fruit shortage. At this season, they enter baited traps more readily, they spend more time looking for food and range farther in search of it, and they defend their territories more vigorously. In addition, young are forced off their parents' territory more quickly and grow more slowly, if it all, during the season of fruit shortage (Smythe 1978, Smythe et al. 1982).

Agoutis have between two and three litters per year. Two-thirds of these contain a single young, the other third, twins (Smythe et al. 1982). Although agoutis are born at all times of year, the birth rate is highest from March through July, toward the beginning of the season of fruit abundance. Only agoutis born during this favorable period have a substantial chance of surviving to independence (Smythe 1978: p. 38). Agoutis can mature within 6 months, but they must secure a territory in order to reproduce. Blue duikers populations are less dynamic: a female blue duiker bears an average of one young every 13.3 months (Dubost 1980: p. 249), and they become capable of reproducing only when 20 months old (Dubost 1980: p. 208, 248).

Agoutis bury seeds, scattered over their territory, against future need. Perhaps to eliminate odors that might attract robbers, agoutis peel flesh or pulp off the seeds before burying them, a habit which has the gratifying consequence that eggs or larvae of seed predators are stripped away before the seed is buried (Smythe 1978, 1989). In the dry forest of Santa Rosa, Costa Rica, the larger a seed, the more likely an agouti is to bury it rather than to eat it immediately, and agoutis are also likely to carry larger seeds farther before burying them (Hallwachs 1994). Buried seeds are an essential resource for agoutis when food is short. We have already mentioned several plants, however, whose seeds must be buried to escape attack from insects and other consumers: on Barro Colorado, agoutis are the only animals that reliably bury seeds in a manner which enhances the future prospects of unretrieved seeds. Indeed, agoutis may be keystone animals for preserving the diversity of Neotropical trees (Forget and Poletto-Forget 1992, also see chapter 8).

Agouti paca (Dasyproctidae)

Pacas are large rodents, whose adults average 9 kg, and they live on fruit that has fallen to the forest floor. In Lutz catchment, near the laboratory clearing, there are two adult pairs of pacas per 3 ha (Smythe et al. 1982). Pacas look like giant agoutis, colored chestnut brown with rows of cream-colored spots like those on a fawn or a baby tapir. Nonetheless, their habits are quite different. Pacas are nocturnal, agoutis diurnal. Pacas store food as fat on their bodies, while agoutis store food reserves as buried seeds scattered over their territories (Smythe 1983). Pacas cannot chew through the wooden shell of a *Dipteryx* fruit to extract the seed because, unlike agoutis, they lack sufficient dexterity and strength in their forepaws to hold it still while chewing it at a fixed point (D. McClearn, p. c.). Finally, a diet of fruit, monkey chow, and meat on which agoutis thrive is too rich in protein for pacas: such a diet gives pacas kidney failure, and drains them of calcium (D. McClearn, p. c.).

In the wild, pacas, like agoutis, are monogamous and territorial. This behavior, however, is not genetically "fixed": it is imprinted during the first 30 days of life by the example of its parents (Smythe 1991). Smythe created a social strain of pacas as follows:

1. For the first 30 days after birth, Smythe separated young of wild-caught pacas from their mothers at night, when pacas are active. These baby pacas were bottle fed, handled, preferably with young of other mothers, for 2 hr per night, left in a cage with other baby pacas for the rest of the night, and returned to their mothers during the day.
2. When over a month old, young pacas were no longer returned to their mother during the day. Rather, they were kept with other young of similar age in a group, which was joined by younger animals each night.
3. Several females and a male were then placed in a group in a larger cage, with no hiding places.

Young of such a group grew to be social animals with no further interference from the experimenter. They not only slept in groups; they nursed each other's young. Thus a territorial animal was transformed into a social one in a single generation, without the aid of genetic change, raising the question, just what is the relation between an animal's behavior and its genes?

Proechimys semispinosus (Echimyidae)

Spiny rats, *Proechimys semispinosus*, are rodents weighing up to 500 g which eat fruits and seeds that have fallen to the forest floor. In Lutz catchment, a spiny rat's territory is three-fifths as wide as an agouti's (Smythe et al. 1982). This catchment probably has three adult female spiny rats per hectare, and fewer adult males.

Until recently, the most thorough study of spiny rats was that of Gliwicz (1973, 1984) on Orchid Island, a 16-ha island lying just off Barro Colorado. Orchid Island has a resident troop of howler monkeys, but relatively few terrestrial frugivores other than spiny rats (Smythe et al. 1982). On Orchid Island, fruit was most abundant and varied from April or May to November; spiny rat densities were lowest at the end of April (7.2/ha) and highest at the end of October (9.7/ha). Judging by the proportion of young in the traps, birth rate was lowest during the season of food shortage (December through February), although reproduction never ceased entirely.

A study of spiny rats on 8 (now 12) 2-ha islands near Barro Colorado in Gatun Lake is providing more decisive evidence of the role of seasonal fruit shortage in the limitation of *Proechimys* populations. These islands have no resident mammals other than *Proechimys*. The species composition of trees on these islets is so varied (Leigh et al. 1993) that different islands have very different seasons of fruit shortage. Adding the population numbers on all eight islands, the average density of spiny rats on these islands is > 20/ha. The seasonal cycle resembles that on Orchid Island: density is lowest in March (19/ha), and highest in September (25/ha); birth rate is lowest from October through January, and highest in April (Adler 1994). Summing the densities on all eight islands, however, masks great differences from one island to the next. Each island's *Proechimys* cycle is different, reflecting the seasonal alternation of food abundance and food shortage on that island (Adler 1994).

The evidence for population limitation by seasonal shortage of fruit is quite strong for spiny rats. We have seen how reproductive rate is lowest in the season of fruit shortage and how *Proechimys* populations are higher where there are fewer competing frugivores. Moreover, *Proechimys* populations on small islands cycle in apparent accord with the seasonal rhythm of the food supply on their particular island.

Comparing different islands, and following year-to-year changes on a given island also suggests that spiny rat populations are food limited and that this food limitation is reflected in density-dependent population change. Adler and Beatty (1997) followed changes in population density of spiny rats and recorded each new young rat as it reached trappable age, inferring its birth date from its size, on four islands where rats were neither added nor removed. During the 4 reported years of their study, the higher the average density during a breeding season of trees and lianas bearing ripe fruit on an island, the higher that island's density of spiny rats at that time, suggesting food limitation (Adler and Beatty 1997). Moreover, this food limitation is reflected by density-dependent reproductive effort. If population change is density dependent then, as a population grows, its per capita reproductive effort should diminish. Reproductive effort should begin increasing again only after the population begins to decline, eventually causing the population to increase again and the cycle to repeat. Thus, plotting an island's density of spiny rats during a breeding season on the x axis and the number of births per adult female during that season on the y, the point x, y should cycle clockwise: this happened on all four islands (Adler and Beatty 1997).

Nonetheless, this evidence is only correlative. Adler is now testing this proposition experimentally. In two successive years, half the adult male *Proechimys* are being removed from four 2-ha islands, half the adult females from four others, while the four already discussed are being left as controls. Then, in the third year, adult males will be added to four islands and adult females to four others. These manipulations may reveal more about the mode of population regulation.

Nasua narica (Procyonidae)

Coatis are the "trademark" of Barro Colorado. The Canal Zone stamp issued on the 25th anniversary of Barro Colorado's establishment as a reserve depicted a coati. They are rather like raccoons, only slenderer, with more streamlined noses (Fig. 2.9); like ring-tailed lemurs, they tend to carry their tails erect. Coatis are diurnal. Adult female coatis average 3.8 kg, adult males, 4.8 kg. Coatis are usually found on the ground, searching the leaf litter for arthropods, worms and small vertebrates, but they sleep in trees, and one can see them scrambling in canopy crowns, far above the ground, looking for fruit (Kaufmann 1962, Russell 1982, McClearn 1992). Coatis have a legendary ability to get into things, including "coati-proof" garbage cans, and to recover fruit from difficult places (Chapman 1935). Coatis have been credited with the rarity on Barro Colorado of poisonous snakes such as fer-de-lances and are also blamed for the absence of various ground-nesting birds whose eggs they supposedly eat.

Coatis have a striking social structure. Adult females live in bands with subadult and juvenile offspring of both sexes; adult males are normally solitary. Mating occurs in late January or early February, occasioning damaging fights among the males (Chapman 1935). Life in bands is advantageous for several reasons. A band of females can oust solitary males from fruiting trees so their young can forage there (Gompper 1996). A band has several animals to share the watch for predators and the defense of their young. Band members also groom each other for ectoparasites, and sometimes nurse each others' young (Gompper 1994). Why do band members cooperate? Females normally live all their lives in the band in which they were born. Consequently, a band's females are usually rather closely related, which increases the advantage of cooperating. If a band dwindles, however, it will seek to join another. Even though those joining are treated as "second-class citizens" and are groomed less and suffer more aggression, it is important for them to belong to a band "big enough to look out for itself" (Gompper 1994, 1996, Gompper et al. 1997).

Adult female coatis leave their band in April or May to nest in trees and bear young (Russell 1982). Thus coatis come in distinguishable age classes, so it is easy to assess year-to-year differences in reproductive rate. Indeed, for this reason, coatis provided some of the first evidence for questioning the idea that all tropical populations are stable (Kaufmann 1962: p. 180). In normal years, average band size drops from 11–15 in June to 7–9 the following March (Russell 1982, Gompper 1994, 1997). A band's home range is about 33 ha. Barro Colorado normally has about 50 coati bands and 100 adult males, about 800 animals in all (see Gompper 1994, 1997). Low availability of important fruits between September and March, however, can reduce or annihilate coati reproduction the following May. In early 1959, *Dipteryx* and *Astrocaryum* did not fruit on Barro Colorado, and coatis did not reproduce (Kaufmann 1962: p. 182). On Barro Colorado, coatis also failed to reproduce in 1972, and reproduction was very low in 1976, presumably due to the few *Dipteryx* which were fruiting at mating time (Russell 1982).

Alouatta palliatta (Cebidae)

Howler monkeys are black monkeys which weigh an average of 5.5 kg (Eisenberg and Thorington 1973); large males

Figure 2.9. *Nasua narica*. Drawing by Dorsett Trapnell.

weigh 8 kg (Milton 1982). They have prehensile tails. Howler monkeys are entirely herbivorous: on Barro Colorado, they live primarily on equal parts of fruit and new leaves (Hladik and Hladik 1969, Milton 1980). They are normally slow moving and spend an inordinate amount of time resting. Barro Colorado's average howler troop contains about 20 animals, among which are 3–4 adult males, 7–9 adult females, and 8–9 young, of which 5 or more are infants less than 14 months old; a female gives birth to a single young once every 17 months, on average (Milton 1982).

At first it seemed puzzling why, when the forest was always full of leaves, and when monkeys appeared so wasteful of fruit, Barro Colorado had so few howlers. It since became evident that, like African buffalo (Sinclair 1975), howlers are limited by seasonal shortage of suitable food. Howlers depend on finding leaves with a reasonably high ratio of protein to fiber content, thus their preference for young leaves (Milton 1979). They need fruit as a source of calories (Milton 1982). Thus, howlers are limited by seasonal shortage of fruit and/or new leaves late in the rainy season or early in the dry season. Adult howlers die most frequently between August and October (Milton 1990), and birth rate is lower late in the rainy season than at other times. The population is probably regulated by mortality among infants and juveniles (Milton 1982). Screwworm larvae, which infest botfly wounds, are the agents which kill the monkeys, but poor condition of the monkeys brought on by shortage of food enables them to do so (Milton 1982, 1996, cf. Sinclair and Arcese 1995).

Howler monkeys were the first animals on Barro Colorado to be accurately censused and the first to have their food intake accurately estimated. Until 1957, howlers on Barro Colorado were subject to occasional yellow fever epidemics, and their numbers fluctuated violently: since then, their numbers have slowly stabilized at 1200 to 1300 animals (Milton 1982: Table 1; Milton 1996).

Nowadays, howlers are counted in two stages. First, Milton counts the number of monkeys per troop for a good many troops to find the average troop size. Then about 35 people (nearly everyone on Barro Colorado) are placed in an even grid over the island at 5 A.M. on two successive mornings, with watches synchronized and compasses in hand. They record the time and relative loudness of the howls heard and the direction from which the howls come as the sun rises above the misty forest and the howlers sound off at dawn, as is their custom, to tell neighboring troops to keep their distance (Chivers 1969). From such data, Milton (1982) plotted the location of each of BCI's howler troops and counted the number of troops on the island (about 65).

To estimate feeding rate, Nagy and Milton (1979) first used doubly-labeled water containing a known proportion of tritium among its hydrogen atoms and a known proportion of ^{18}O among its oxygen atoms, to measure the metabolic rate of free-living monkeys. One injects an animal with this doubly-labeled water, and, when the isotopes have had a chance to mix evenly, measures the proportion of tritium and ^{18}O in blood withdrawn from the animal. The animal is released, recaptured after a week or so of normal life in the forest, and the content of tritium and ^{18}O in its blood measured again. The decline in tritium content reflects the rate of water loss, while the decline in ^{18}O content reflects the rate of water loss plus the rate of CO_2 loss; the difference between the rate of ^{18}O loss and the rate of tritium loss gives the respiration rate as expressed by CO_2 loss (Nagy 1987).

If 21.8 kJ is metabolized for every liter of CO_2 exhaled, a howler metabolizes 355 kJ/kg body weight per day (Nagy and Milton 1979). This is a metabolic rate of 4.1 W/kg body weight, about 23 watts/monkey: it is sobering to think that the life of so marvellous an animal can be supported by an energy expenditure so much less than that of a light bulb. Howlers eat leaves, fruit, and some flowers; all have an energy content of about 19.1 kJ/g dry weight. Howlers metabolize about 35% of the energy in the leaves and fruit they eat, suggesting that they eat 355/19(0.35) = 54 g dry matter/kg body weight per day (Nagy and Milton 1979). Forty percent of this dry matter is leaves (Hladik and Hladik 1969). A 5.5-kg howler thus eats 44 kg dry weight of leaves per year; howlers eat 35 kg dry weight of leaves, and 90 kg dry weight of food of all kinds, per hectare per year.

Cebus capucinus (Cebidae)

White-faced monkeys, *Cebus capucinus*, are black monkeys with white heads and shoulders, averaging 2.6 kg (Eisenberg and Thorington 1973). A normal *Cebus* troop averages about 21 members: 2 adult males, 7 adult females, 2 subadult males, 7 juveniles, and 3 infants (Mitchell 1989). Mitchell also observed a "runt" troop with only 5 members. A typical troop's home range is about 90 ha (Hladik and Hladik 1969, Mitchell 1989). In 1988 there were 16–18 troops on the island (Mitchell 1989); if there were 14 normal troops and 3 "runt" troops, Barro Colorado then had 310 *Cebus* monkeys, about one/5 ha. Adult *Cebus* monkeys have a low mortality rate (Oppenheimer 1982, Mitchell 1989), and females normally produce a single young every 2 years. Accordingly, the *Cebus* population on Barro Colorado appears to have been stable in recent years. Oppenheimer (1982) thought there were 18 troops on the island in 1968.

Cebus differ markedly from howlers. Over the year, only 15% of the diet of *Cebus* is green matter—young leaves, tender shoots, the pith of stems—they eat more greenery in the rainy season and less during the dry. Sixty-five percent of their diet is fruit, and 20% is insects and other small animals (Hladik and Hladik 1969). Unlike howlers, which will stay for hours in one tree, *Cebus* are always on the move, stopping only briefly in any one fruiting tree. They spend many hours searching foliage or probing "wormy" fruit for insects, larval and adult; they spend much more time looking for insects than in traveling to fruit trees or eating fruit.

In the old forest of Barro Colorado's central plateau, *Cebus* respond differently from most of the island's other animals to the seasonal rhythm of fruit availability. From mid-December through mid-August, *Cebus* take most of their fruit from a few tall, large-crowned trees: most of these fruit, like that of *Dipteryx, Astrocaryum,* and *Scheelea,* are large and difficult for young *Cebus* to handle. From mid-August through September, fruits *Cebus* eat are relatively more abundant. These fruits are of intermediate size and difficulty of handling, but again, most of them are provided by relatively few large-crowned trees, especially *Quararibea*. From October through mid-December, fruit eaten by *Cebus* is very abundant and scattered over a large number of small-crowned treelets and shrubs such as *Faramea occidentalis* and *Psychotria horizontalis*. The fruits of these species are small and easy for baby monkeys to handle. During the season of abundant small fruits, aggressive encounters within *Cebus* troops are least frequent, although this is also the season when *Cebus* spend the least time resting. Most *Cebus* are born between February and May; the season of birth seems adjusted to ensure that infant monkeys are weaned in October, at the beginning of the season when fruits infants can easily reach and eat are most abundant (Mitchell 1989). The season most favorable for newly weaned *Cebus*, however, overlaps almost entirely with the season when other frugivores and herbivores are suffering most severely from shortage of suitable food (Leigh and Windsor 1982).

Bradypus variegatus (Bradypodidae)

Three-toed sloths, *Bradypus variegatus*, live nearly entirely on leaves. Adults weigh an average of 4 kg. Sloths, more nearly than howlers, are the ecological counterparts of the leaf-eating *Colobus* of Africa and *Presbytis* of Southeast Asia. In behavior, home range size, and social system, however, sloths are much more like the leaf-eating koalas of Australia (Lee and Martin 1988).

Sloths seem to organize their lives so as to minimize energy expenditure. They change trees only once every 1.5 days (Montgomery 1983). For sloths, as for coatis, not all tree crowns are equally accessible: sloths move from one tree crown to another by connecting lianes, or occasionally, by reaching from one tree crown to another (Montgomery 1983). Although sloths engage to some extent in behavioral thermoregulation, the diurnal temperature range in their body temperature is about equal to the diurnal temperature range in their habitat (Montgomery and Sunquist 1978). Sloths process their food carefully. Their gut contents are 18% of their body weight, and they are digested for a very long time: sloths defecate and urinate only once a week, climbing down to the base of their tree to do so (Montgomery and Sunquist 1975).

A three-toed sloth has a home range of about 1.5 ha. Although neighboring home ranges overlap greatly, two adult sloths are seldom found in a single tree. A female sloth bears a single young once each year; when weaned, the young sloth remains in a part of its mother's home range for a year before dispersing (Montgomery 1983). The young inherits its mother's dietary preferences, perhaps because it inherits its mother's gut flora. Both sexes take 3 years to attain sexual maturity. Death rate among adult sloths may be less than 10% per year.

Nagy and Montgomery (1980) used doubly-labeled water to learn just how slow, indeed, how reptilian, the metabolism of sloths is. Three-toed sloths metabolize 147 kJ/kg body weight per day. A 4-kg sloth thus metabolizes 588 kJ/day, a metabolic rate of 6.8 W. These sloths assimilate 61% of the energy in their food. If they metabolize 90% of what they assimilate, and if a gram dry weight of what they assimilate contains 18 kJ, sloths consume 147/(0.549)(18) = 15 g dry weight of leaves/kg body weight per day. Thus, a 4-kg three-toed sloth eats 60 kg dry weight of leaves/day (Nagy and Montgomery 1980), or 22 kg of leaves/year.

It is an open question how many sloths there are on Barro Colorado. Glanz (1982) estimated that BCI has 8000 three-toed sloths, but his Table 2 suggests that his censuses found only 1 of about every 200 sloths he thought were there. In second growth near Cayenne, Guyane Française, Charles-Dominique et al. (1981) estimated a population density of 3.5–7 sloths per hectare, but found only 0.8 per hectare. Why do people think there are so many more sloths than they see?

Montgomery and Sunquist (1975) once found 3 two-toed sloths, *Choloepus hoffmanni*, and 10 three-toed sloths in 1.3 ha of young forest on Barro Colorado, suggesting a density of 10 sloths per hectare, 77% of which were three-toed. Montgomery and Sunquist (1975) cleared 0.324 ha of ground in Barro Colorado's young forest and found that 14 piles of sloth fecal pellets accumulated over the next 35 days. If sloths defecate once a week, it would appear that about three sloths occupied this area, suggesting a density of nine sloths per hectare, most of which are probably three-toed. Thus there is some reason to think that Barro Colorado has an abundance of sloths, but it is hard to judge just how many. If BCI has 27.7 kg of three-toed sloths/ha, these sloths might eat 152 kg dry weight of leaves/ha · year, about 2% of the island's leaf production.

Artibeus jamaicensis (Phyllostomidae)

The most common bat on Barro Colorado is the Jamaican fruit bat, *Artibeus jamaicensis*, whose adults average 50 g. *A. jamaicensis*, and nine other species of phyllostomid bats live mainly on figs (Kalko et al. 1996). These bats spend only 15–35 min digesting a food item before defecating it, so trees using bats to disperse their fruits must present a readily digestible reward. Reproductive females roost during the day in groups of 3–14, with their nursing or recently weaned young, in tree hollows. Each group has a single adult male (Morrison and Morrison 1981). Solitary males and subadults roost in foliage.

Females spend an average of 1.4 hr per night flying to their food—usually figs—and looking for other trees with ripening fruit (Morrison and Morrison 1981). A female flies, perhaps as much as 1–2 km, to a fruiting fig tree, plucks a fig, and carries it 25–200 m away to eat it in peace, away from the predators that lurk in fruit-bearing trees. The bat keeps returning to that tree, or shifts to one or two others, until it has taken at least 70 g fresh weight (13 g dry weight) of fruit. Of the 13 g dry weight taken, the bat spits out 60% as pellets, but digests 64% of what it actually swallows (Morrison 1980). The energetic content of a 10-g *Ficus insipida* fruit is 33 kJ, of which the bat uses 8 kJ. On an average night, an adult female bat derives at least 56 kJ from the fruit that it eats. Nagy (1987) predicted that a 50-g bird eats 8.3 g dry weight of food per day; these bats appear rather economical.

Flying costs a 45-g bat 5 W (300 J/min); resting costs 0.43 W (25 J/min). Thus a female which flies 1.4 hr per night and rests 22.6 hr of the 24 expends 59 kJ/day, or 0.68 W, roughly equal to its intake.

A roost's male energetically defends his females from intruding males. To keep watch, he returns to the roost to eat any fruit he takes, even if this fruit is from a tree > 1 km away. At night he ceases feeding and patrols or watches over the roost, when a female returns to rest, avoid bright moonlight, or nurse her young. How much does this vigilance cost? A roost's male spends an average of 3.1 hr per night flying, so that his total energy expenditure is 87 kJ/day. A roost's male feeding from a *Ficus insipida* for seven successive nights ate an average of 10.9 figs per night, thus consuming 85 kJ/night, about 1 W, which was enough to pay for the 2.9 hr per night he spent flying during this period.

Artibeus jamaicensis is the only tropical bat whose population density has been estimated with reasonable accuracy. The data for this estimate came from marking and recapturing bats during an intense 5-year program of bat-netting over the whole of Barro Colorado and at nearby mainland sites (Handley et al. 1991), and from an earlier mark–recapture program in 1973 (Bonaccorso 1979). The *Artibeus jamaicensis* population on Barro Colorado can be treated as if it were nearly isolated.

In March or April, and again in July or August, during Barro Colorado's season of fruit abundance (Foster 1982b), an average of 90% of Barro Colorado's adult female *Artibeus jamaicensis* give birth, each to a single young. Immediately after giving birth, the females mate again, but the fetus formed after the second birth only begins to develop at the normal rate sometime in November, as if to avoid births during the early dry season food shortage (Fleming 1971).

Young females suffer only slightly higher mortality than adults, and the progressive rate of dilution of bats marked by Bonaccorso among those caught by Handley's program after September 1976 suggests that the survival rate of adult females is about 79% per 6 months.

Knowing the life table provides a way to estimate the number of adult female bats on Barro Colorado (Leigh and Handley 1991). Despite their striking tendency to avoid nets after being caught once, several years after Handley et al.'s (1991) program began, most adult bats were already marked. The total number $N_{TF}(t)$ of marked adult female *Artibeus jamaicensis* alive at time t can be estimated by.

1. Recording the number $N_{MF}(t-s)$ of females marked at each netting time $t-s$, which would now be adult if still alive,
2. Multiplying this number by $p(s)$, the proportion of these bats still living after the lapse of s time units, and.
3. Summing these products over all times $t-s$ when female *Artibeus jamaicensis* were marked on Barro Colorado, to find

$$N_{TF}(t) = \sum_s N_{MF}(t-s)\, p(s) \qquad (2.1)$$

The number of adult females plateaued at about 1500 after 3 years of netting. The total number of adult males can be estimated by the method used for adult females, but separately, because they suffer higher mortality: the number of adult males plateaued at about 800. The number of juveniles can easily be estimated from the number of adult females and the time of year. The average number of *Artibeus jamaicensis* on Barro Colorado must be about 4000, slightly over two per hectare.

Mixed Antwren Flocks (Formicariidae)

When walking through the forest, one may hear a number of small birds twittering in the understory, flitting from branch to branch. On Barro Colorado, the core of such a flock would be a pair of 11-g checker-throated antwrens, *Myrmotherula fulviventris*, and a family group (pair and attendant young) of 8-g dot-winged antwrens, *Microrhapias quixensis*. A single pair of each species jointly defend a territory of about 1.3 ha, which usually contains two conspicuous vine tangles (Gradwohl and Greenberg 1980, Greenberg and Gradwohl 1986). During the northern winter, their territory is also occupied by a migrant 9-g chestnut-sided warbler, *Dendroica pensylvanica*, which spends much of its time foraging with the flock. Birds with smaller territories join the flock when it crosses their land: conspicuous among these is the 23-g slaty antshrike, *Thamnophilus punctata*, a pair of which occupies a territory averaging 0.75 ha, centered on a single vine tangle. Some birds with larger territories, such as the 8-g whiteflanked antwren, *Myrmotherula axillaris*, a pair of which occupies an average of 2.3 ha, move from one flock to another (Greenberg and Gradwohl 1986). The flock moves ahead at an average pace of 0.88 m/min (Jones 1977)—fast enough to cover its territory several times per day.

Why do these birds flock together? They are all insectivores; indeed, these understory flocks include some of Barro Colorado's most abundant insectivorous birds (Greenberg and Gradwohl 1986). Dot-winged antwrens, white-flanked antwrens, and chestnut-sided warblers all glean insects from the undersides of leaves (Greenberg and Gradwohl 1980). Yet, within a flock, there is a clear niche differentiation. Checker-throated antwrens probe for arthropods in dead curled leaves caught in vine tangles. White-flanked antwrens forage in the low or middle understory, averaging 5 m above the ground, preferably in rather open areas, while dot-winged antwrens forage higher up, preferably in vine tangles. Slaty antshrikes forage at the same level as white-flanked antwrens, but they use a greater variety of foraging techniques, and they prefer vine tangles (Greenberg and Gradwohl 1985). Chestnut-sided warblers also glean from the undersides of leaves, but 78% of their larger prey are insect larvae such as beetle grubs, whereas more than half the prey of foliage-gleaning antwrens are Orthoptera and less than 15% are grubs and caterpillars (Greenberg 1981). Apparently, the primary advantage of flocking together is to avoid predators: there are more animals to share the watch, some predators can be dealt with by mobbing, and it may be harder for a predator to home in on its prey when there are so many other birds about (Willis 1972: pp. 135–147, Moynihan 1979, Powell 1985).

Predation does influence antwren dynamics. There are many fewer insects on understory leaves during the dry season than during the rains, so most antwrens in understory flocks only begin nesting in April, when the rains are approaching. The earliest successful nests, however, are in June: eggs laid earlier are apparently eaten by snakes seeking protein to fuel their own early-rainy season burst of reproductive activity (Gradwohl and Greenberg 1982). Mortality among checker-throated and white-flanked antwrens and slaty antshrikes averages about 40% per year, much higher than among manakins of similar size, but the territory boundaries of checker-throated antwrens and slaty antshrikes (unlike those of white-flanked antwrens) remained unchanged over 7 years as new owners replaced old (Greenberg and Gradwohl 1986).

In the Parque Nacional Manú of Peru's Amazonia, understory flocks are much larger (Munn and Terborgh 1979) and have larger territories, averaging 7 ha apiece (Munn 1985). A flock's territory is defended jointly by one pair each of two species of antshrike, *Thamnomanes* spp., and three species of antwren, *Myrmotherula* spp., along with a single foliage-gleaner, *Philydor*. There is a corresponding diversity of less regular visitors who join as the flock moves through their smaller territories, or which have large territories within which they move from flock to flock (Munn 1985). Unlike Panama, the core species of the Manú flocks are not dependent on vine tangles. Does the larger flock size betoken a more intense predation pressure?

Iguana iguana (Iguanidae)

On Barro Colorado, green iguanas (*Iguana iguana*) are the principal reptilian herbivores. Mature females average 2 kg (Rand and Bock 1992). The largest males, which are dull gray, gold, or tan, weigh twice as much (Dugan 1982). Captive female iguanas lay eggs when 3 years old. On Barro Colorado, wild females lay eggs for the first time when between 2 and 8 years old—the variation is striking (Dugan 1982, Zug and Rand 1987). Courtship begins in November, with large males displaying vigorously; females only copulate after several weeks of courtship, and may mate with several males (Dugan 1982). In January and February, female iguanas converge from up to 3 km away on one of a few nesting sites surrounding the island (Montgomery et al. 1973, Bock et al. 1985). Most nesting sites are offshore islets with an expanse of bare earth exposed to the sun for at least part of the day (Bock et al. 1985). Each female digs a burrow and lays an average of 41 eggs in it (Rand 1984). Hatchling lizards emerge in early May and migrate in groups from their nesting site to Barro Colorado (Burghardt et al. 1977). The hatchlings actively associate with adults, from which they presumably obtain the gut flora they need to break down the walls of plant cells (Troyer 1982).

Judging by regressions from Nagy (1982), a 2-kg iguana metabolizes 100 kJ per day: in other words, it has a power demand of 1.13 W. To maintain this metabolism, the iguana must eat 10.5 g dry weight of food per day, 3.8 kg per year. A 2-kg mammalian herbivore would eat 145 g dry weight of food per day, 53 kg per year (Nagy 1987), to meet a power demand of 17 watts. An iguana matures in 3 years; a chicken matures in 4 months, but it takes about the same amount of feed to produce 1 kg of each animal (Werner 1991).

Bock et al. (1985) identified four major sites around Barro Colorado where iguanas nest. Fifty to 70 were nesting at Slothia, an 0.3-ha islet in the laboratory cove, between 1980 and 1983, and perhaps 150 or more between 1973 and 1976 (Burghardt et al. 1977). There are as many adult males as adult females on Flamenco, an island on the Pacific side of Panama (Dugan 1982). If this is also true on Barro Colorado, it may have about 1000 adult iguanas. Most of these iguanas live near shore, where one rarely sees more than one adult iguana per 50 m of shore (A. S. Rand, p. c.). This observation suggests that Barro Colorado has slightly less than 1000 iguanas. The dry forest on Flamenco has more than 10 adult iguanas per hectare (Dugan 1982), which presumably eat as much foliage per hectare as Barro Colorado's howler monkeys.

The demography of Barro Colorado's iguanas is somewhat puzzling. On Slothia, 22% of the eggs laid fail to hatch

(Rand and Dugan 1980). Moreover, no more than half the iguanas that nest there one year do so the next, and very few of those that miss one year return the year after (Rand and Bock 1992). If we let p be the probability that a female hatchling survives to reproduce at least once, then a hatchling female's prospective output of hatchling female offspring is

$$p(0.78)(20 + 10 + 5 + 2.5 + 1.25 + 0.625 + \ldots) = 31p$$

The population is in balance when $p = 0.032$. Yet, only 7 of 700 hatchlings marked at Slothia were ever seen to nest there. A crocodile also nests at Slothia, and in the course of defending her nest from the iguanas eats some of them and disturbs some of their nests (Dugan et al. 1981). Does this circumstance prevent Slothia's iguanas from replacing themselves? Or are many of the hatchlings nesting somewhere else?

Anolis limifrons (Iguanidae)

Anolis limifrons is a small insectivorous lizard which is frequently seen in the understory at Barro Colorado and other Central American forests. Adults weigh 2 g or less; a 1.6-g adult's daily intake of food has an energetic content averaging 0.4 kJ. Their home ranges are restricted: adult females captured four or more times moved a median distance of 4 m between successive recaptures (Andrews and Rand 1982).

Populations of this lizard fluctuate markedly from one year to the next. Between 1970 and 1989, the number of *Anolis limifrons* on an 890-m^2 plot in Lutz catchment varied from a maximum of 1/9 m^2 in 1978–1979 to a minimum near 1/1000 m^2 after the El Niño dry season of 1982–1983 (Andrews and Rand 1982, Andrews 1991). Population fluctuations occurred in rough synchrony over the island as a whole. These fluctuations seemed to be dictated by some aspect of the seasonal distribution of rainfall, but it was not clear which aspect was decisive (Andrews and Rand 1982, 1990, Andrews 1991). From 1971 through 1976, survival rate per 30 days of *Anolis limifrons* was 77%, and from 1976 through 1979 it was 67%. Within each period, survival rate was independent of season and was the same for males and females, adults and young (Andrews and Nichols 1990). These lizards lay few eggs in the dry season, but early in the rainy season a female lays one egg approximately every 9 days (Andrews and Rand 1982).

To learn why the populations of this lizard fluctuate so much, *Anolis limifrons* were censused from 1986 through 1988 on plots supplied by sprinklers with 6 mm of water per working day during each dry season and on nearby unwatered control plots (Andrews and Wright 1994). Irrigation affected neither food supply nor adult survival. In 1986, there was a population explosion of *Anolis limifrons* on the irrigated plots, apparently because watering greatly increased the survival of eggs during the dry season (Andrews 1988, Andrews and Wright 1994). In 1987 and 1988, *Anolis* populations on irrigated plots fell back to levels typical of the unwatered plots, perhaps because sustained watering increased populations of egg-eating ants or egg-destroying pathogens (Andrews and Wright 1994). Experiments are now in progress to assess the effects of different levels of moisture on consumption of eggs by ants.

Physalaemus pustulosus (Leptodactylidae)

The tungara frog, *Physalaemus pustulosus* is a small, nocturnal frog, 25–35 mm long. Adults live in the leaf litter of the forest floor and also in gardens, fields, and scrub pasture (Rand 1983). At a mainland site near Barro Colorado, these frogs eat mostly ants and termites (Ryan 1985: p. 155). Males of this species aggregate at night in temporary pools and call to attract mates. The call consists of a "whine," a clear, high-pitched, descending call lasting a third of a second, which is one of the most characteristic nocturnal sounds of Barro Colorado's laboratory clearing. If other males are calling, each whine is followed by one or more short "chucks." After a female has chosen to mate with one of the calling males, she lays eggs while the male in amplexus elaborates a foam nest for them at the side of the pool. Under laboratory conditions, the (aquatic) tadpoles take about 4 weeks to metamorphose and another few months to attain sexual maturity (Rand 1983).

The mechanics of sexual selection in this frog have been worked out in exceptional detail. Females prefer males whose whines are followed by chucks, but the chucks also allow frog-eating bats to find the calling male more easily (Tuttle and Ryan 1981). Larger males attract more mates. If a female mates with a male much smaller than herself, some of her eggs will not be fertilized (Ryan 1985). Larger frogs tend to make deeper-toned (lower frequency) chucks, and, in "speaker-choice" experiments, female *Physalaemus* prefer calls with lower-frequency chucks (Ryan 1980).

What drives the attraction of female *Physalaemus* for calls with chucks? Calls of male *Physalaemus coloradorum* always lack chucks, but in speaker-choice experiments, female *coloradorum* prefer synthetic calls with *pustulosus* chucks following the normal *coloradorum* call to their own males' normal call (Ryan and Rand 1990); however, female *coloradorum* prefer unaltered *coloradorum* to unaltered *pustulosus* calls. Ryan and Rand inferred that the evolution of chucks in *Physalaemus pustulosus* exploited a preexisting female preference shared by *coloradorum* females. Ryan and Rand (1990) also found that several different varieties of synthetic chuck were as attractive to *pustulosus* females as the natural variety. It seems that the female sensory system constrains and biases the evolution of the males' calls, but there are a variety of ways to exploit the female's sensory preferences. This story has a parallel among fiddler crabs, *Uca beebei*, where some males build miniature pillars of sand by their burrows. These pillars exploit a preexisting tendency, shared by other species of

Uca, to run when frightened by predators toward structures behind which they can hide (Christy 1995). Not all *Uca beebei* males build pillars. A pillar (like a deep-toned chuck) apparently also serves as a useful indicator of the builder's quality as a mate (Backwell et al. 1995).

Physalaemus pustulosus belongs to a group of six closely related frog species that are considered to have evolved in strict isolation from each other (Ryan and Rand 1993). Speaker-choice experiments show that females vastly prefer the calls of males of their own species to those of any of the other five. In this group, reproductive isolation evolved in the absence of selection to suppress hybridizing (Ryan and Rand 1993). Was this simply a consequence of sexual selection?

Eberhard (1985, 1996) has documented an adversarial aspect of mating. A female tries to suspend response to a courting male, yielding fully only if the male truly proves his worth, while not only does the male try to persuade the female to mate, even when mating, he tries to stimulate her more fully, so that she will let him fertilize more of her eggs. There is an analogous contest between a human fetus, whose hormones cause the mother to supply more nourishment, and maternal countermeasures meant to keep this supply within limits consistent with the mother's welfare (Haig 1993). Sexual selection thus drives a steady flux of new male signals and female responses or habituation, which continues whether or not related species are present (Eberhard 1993). Moynihan characterized the dynamics of this process as follows:

> the probable evolution of displays is closely comparable to what is known to happen to certain human signals. A particularly "strong" word or phrase, e. g. an obscenity or striking new technical term, is apt to be very impressive and effective when it first begins to be used. Simply because it is effective, it tends to be used more and more frequently. And, unless it is constantly reinforced by new variations or further elaborations, it eventually becomes essentially meaningless. (1970: p. 103).

The direction of this coevolution between male trait and female response is hardly more constrained than fashions of dress. In isolated populations, this coevolution takes divergent directions, until members of neither population recognize those of the other as suitable mates.

How long does it take isolated populations to become new species? Roughly 3 million years ago, the isthmus of Panama sundered the Atlantic from the Pacific (Coates et al. 1992). *Bathygobius* (fish) populations separated 3 million years ago (Gorman et al. 1976) can still produce viable, fertile hybrids (Rubinoff and Rubinoff 1971). In three pairs of snapping shrimp (*Alpheus* spp.) divided by the Isthmus of Panama, individuals from the two oceans often interbreed and sometimes produce fertile clutches (Knowlton et al. 1993). Three million years of isolation are not enough to *assure* reproductive isolation [a conclusion supported by the work of Coyne and Orr (1989) on *Drosophila*]. On the other hand, members of a pair of snapping shrimp populations divided 8 million years ago by a strong current can no longer interbreed (Knowlton et al. 1993).

Eciton burchelli (Formicidae)

Army ants, *Eciton burchelli*, create one of the more remarkable spectacles of Neotropical forest. A swarm of roughly 100,000 ants moves over a front averaging 6 m wide, which advances at an average pace of 15 m/hr (Willis 1967, Franks 1982a). These ants dismember any insect they catch, while ant-following antbirds hang about the swarm, snatching insects flushed by the ants (Willis 1967). The life of an *Eciton burchelli* colony is remarkably cyclical. For roughly 13 days, they raid each day, reaching an average distance of 109 m from their bivouac, and returning to it with their booty. Each of these 13 nights, the whole colony moves, carrying their larvae, to a new bivouac site an average of 81 m from its predecessor. At the end of this "nomadic phase" the larvae weave cocoons, and the colony enters "statary phase": the bivouac does not move again until the pupae hatch 21 days later. On 13 of these 21 days, ants swarm out from the bivouac to raid. Raids are least frequent during the middle of statary phase, when the queen is laying a new batch of eggs (Willis 1967).

Eciton burchelli was the first insect species on Barro Colorado whose demography, population density, and feeding rate were assessed. Willis (1967: p. 7) found 12 statary bivouacs during an 11-month search of a 40-ha plot in 1961 and 1962. He inferred that the plot contained an average of 12(35)/335 colonies, and that Barro Colorado thus contained slightly more than 3 colonies/km². Franks (1982b) crossed 13 *Eciton burchelli* raids in 83.4 km of late afternoon walks on Barro Colorado in 1978 and 1979. He assumed that each raid was a line of random direction positioned randomly on Barro Colorado, independently of the others. Then the total length, S, of *Eciton burchelli* raids on Barro Colorado is $\pi NA/2L$, where $N = 13$ is the number of raids crossed, $A = 15.6$ km² is the island's area, and $L = 83.4$ km is the distance he walked. At 4 P.M., the total length of all *Eciton burchelli* raids on Barro Colorado averaged 3.8 km. If the average raid length is 91 m, and if only 76% of the colonies raid on a given day, then Barro Colorado had 55 colonies, 3.6/km², near the beginning of 1979 (Franks 1982b).

An average *Eciton burchelli* colony has a single queen and 400,000 workers. Its total dry weight is about 1 kg (Franks 1982a). Workers come in four castes: minims, which stay in the bivouac and nurse the brood; media, jacks of all trades which supply the "troops" for swarm raids; submajors, specialized to carry booty back to the bivouac, and majors, which defend the colony (Franks 1989a). The queen lays 55,000 eggs every 35 days (Franks 1982b); average worker lifetime is thus $35(400,000/55,000) \approx 250$ days, nearly 8 times that of a honeybee worker (Seeley 1985: p. 31). A colony splits in half every 3 years (Franks and

Hölldobler 1987: p. 233). Thus, if the number of *Eciton burchelli* colonies is stable, a third of the colonies must die every year.

During raids, *Eciton burchelli* workers bring an average of 10 food items per 12.4 sec back to their bivouac. The dry weight of these items averages 2.14 mg apiece. Taking into account times when raiding does not occur, *Eciton burchelli* bivouacs receive an average of 42 g dry weight of animal matter per day (Franks 1982a). For comparison, a 3-kg coati eats about 200 g dry weight of food per day, judging by the regression for herbivorous eutherians in Nagy (1987).

How do *Eciton burchelli* affect their food supply? *Eciton burchelli* sweep 1/400 of Barro Colorado's area per day (Franks and Bossert 1983). A swarm of these ants reduces the abundance of crickets, roaches, and other large mobile insects by 50%, but immigrants replace these losses within a week. The raid greatly reduces the abundance of their favorite ant prey: it takes 100 days for their prey to recover to 50% of pre-raid levels (Franks 1982a). New colonies of favored prey species establish more frequently where *Eciton burchelli* have recently raided, as if these raids have reduced the intensity of established competition (Franks 1982c).

Before 1979, *Eciton burchelli* were absent from Orchid Island, an 18-ha island just off Barro Colorado. Where *Eciton burchelli* is absent, competition is more intense, enforcing more even spacing: colonies of favored prey species were more evenly dispersed on Orchid than on Barro Colorado. When, in 1979, an *Eciton burchelli* colony was introduced to Orchid Island, 67% of its prey was social insect larvae, compared to 39% for their Barro Colorado counterparts (Franks 1982c), suggesting that *Eciton burchelli* reduces the average abundance of their favored prey.

Eciton burchelli workers serve their colony's good without any explicit central direction. The queen does not physically enforce a division of labor—how could she control so many workers? She need not do so: since her workers cannot reproduce on their own, they can only contribute to the spread of their genes by helping the queen (usually their mother) reproduce. These colonies are marvels of self-organization (Franks 1989a). Ant metabolism generates more than enough heat to warm their bivouac, which is made of the bodies of living ants: they adjust their positions to allow enough heat to escape to keep their bivouac's core temperature within a degree of 28.5°C, day in and day out (Franks 1989b). *Eciton burchelli* colonies also follow foraging rules that avoid return to recently raided areas (Franks and Fletcher 1983). The colony's self-organized function is an expression of its workers' common interest in its welfare. This community of interest can express itself in striking ways. When a colony splits, workers in both halves choose both their queen and her mate, sometimes rejecting the existing queen (Franks and Hölldobler 1987).

Atta colombica (Formicidae)

Near Barro Colorado's laboratory clearing, one often sees lines of leaf-cutter ants, *Atta colombica*, carrying fragments of leaves back to their nest. The nest's site is revealed by great piles of reddish earth, surrounded by an area of sparse understory whose few stems are largely stripped of leaves. Pathways 10 cm or more wide snake across the forest floor and disappear down tunnel entrances into the nest. Downslope from the nest is a "rubbish tip" of orange fragments, which is diagnostic of the species: other species of leaf-cutter ants bury their wastes underground (Haines 1978). If this rubbish is spread around the base of a favored plant, leaf-cutter ants will not cross the barrier until it is washed away or loses its odor. Leaf-cutter nests are quite abundant near the laboratory clearing: the 80 ha nearest the laboratory contained 41 major colonies of *Atta colombica* and 6 major colonies of *Atta cephalotes*: 1 leaf-cutter colony/1.70 ha (H. Herz, p. c.). More than 700 m from the laboratory clearing, however, leaf cutters are rare. Leaf-cutter ant colonies are quite common on forested islets of 2 ha or less near Barro Colorado in Gatun Lake, but very rare in the young forest of Gigante, a mainland area facing Barro Colorado's south side. A 28-ha tract of 42-year-old forest on a nearby mainland site, 6 km northwest of Gamboa, contained 21 active *Atta colombica*, and two active *Atta cephalotes* nests, 1 nest/1.22 ha (Haines 1978).

Leaf-cutter ants take their leaf fragments underground to caves where the fragments nourish a fungus on which the leaf cutters feed. Leaf cutters ensure that selection on the fungus renders it more serviceable by (1) feeding on what would otherwise become the reproductive organs of the fungus and (2) spreading fungicidal compounds about the nest to prevent spores of foreign fungi from germinating. Thus the fungus can only reproduce when a newly mated leaf-cutter queen takes a mouthful of the fungus to start a new colony (Hölldobler and Wilson 1990). Using a fungus as a digestive organ is not unique to leaf-cutter ants. A whole subfamily of termites, the Macrotermitinae, appear to do the same (Darlington 1994). In Malaysia, the termite *Macrotermes carbonarius* nourishes the fungus it eats on litter leaves collected from the forest floor and carried into growth chambers inside the nest (Matsumoto 1978).

Near the laboratory clearing, leaf-cutter ants are the principal herbivore. Wirth et al. (1997) measured the intake of a leaf-cutter ant colony in Lutz catchment over a year. Their nest received an average of 134,000 leaf fragments per day, of average area 0.79 cm^2 and average weight 5.51 mg apiece—a total daily intake of 737 g dry weight of leaves. For comparison, a nest of *Atta colombica* in lowland rainforest on the Osa Peninsula in Costa Rica, studied for 50 hr in the rainy season, received 142,000 leaf fragments per day, averaging 8 mg dry weight and 1.06 cm^2 apiece—an average daily intake of 1136 g dry weight of leaves (Lugo et al. 1973: Table 10. I have corrected decimal points misplaced in that paper). The leaf-cutter colony in Lutz catchment also harvested 304 g dry weight per day of other vegetable matter from the forest floor—flowers, stipules, and fallen, yellowed leaves. This colony drew its resources from about 1 ha of forest (R. Wirth, p. c.), whereas the colony in Costa Rica drew its resources from 1.4 ha (Lugo et al. 1973: Table 7). This hectare supplied the leaf cutters with 270 kg dry weight (386 m^2) of leaves per year (R. Wirth, ms). This is 4% of the hectare's total annual weight, but more than 5% of its total area of leaf litter fall (Leigh and Windsor 1982).

Leaf-cutter ant colonies are territorial, and in many cases their colonies are dispersed more evenly than would occur by chance (Rockwood 1973). Leaf-cutter queens disperse from the parental nest near the beginning of the rainy season, when there is a maximum of palatable foliage in the forest, but most of the leaf-cutter queens that survive the gauntlet of waiting predators are killed by workers from nearby established colonies (Fowler et al. 1986).

Leaf-cutter ants feed from a variety of trees. Many have tried to sort out the criteria by which leaf cutters choose their leaves (Cherrett 1968, 1972, Rockwood 1976, Howard 1990). Leaf cutters often forage in a tree one day and abandon it the next, or defoliate one branch and leave the next (Howard 1990). Leaf chemistry and protein content are both demonstrably relevant to leaf-cutter choices (Hubbell et al. 1983, Howard 1990). In Lutz catchment, the average dry weight of a square meter of leaf cut by leaf cutters is 70 g (Wirth et al. 1997), while the average dry weight of a square meter of fallen leaves is nearly 100 g (Leigh and Windsor 1982). More generally, leaf-cutter ants prefer young leaves (Rockwood 1976, de Vasconcelos 1990) or thin leaves designed for quick replacement, which have relatively little fiber per unit protein (Cherrett 1972). In agricultural areas, leaf-cutter ants can be devastating. A Brazilian politician once declared that either Brazil would kill the leaf cutters, or the leaf cutters would kill Brazil.

Even more than army ants, leaf-cutter ants are marvels of self-organization (Wilson 1980a,b, 1983). *Atta sexdens* workers come in four castes: cultivators of fungus (gardeners)/nurses of brood, jacks of all trades, foragers/excavators, and defenders. Members of the three smaller castes change to a second group of tasks as they age (Wilson 1980a). The cutting of leaf fragments, their transformation into "fungus food," and their inoculation with fungal hyphae involves a series of tasks, each successively executed by smaller workers (Wilson 1980a). Workers can have surprising duties: some minims, which are normally gardeners or nurses, ride leaf fragments carried by other leaf cutters to fend off parasitic flies (Feener and Moss 1990). In the variety of species they do eat, and the greater variety they avoid, leaf-cutter ants face foraging decisions remarkably analogous to, and as complex as, those of howler monkeys (Rockwood and Glander 1979). The complexity of choices faced by monkeys that live on fruit and young leaves have been considered a crucial stimulus for the evolution of primate intelligence (Milton 1981)—what are we to say of the leaf cutters?

Stingless bees (Meliponinae)

Stingless bees pollinate many Neotropical trees and rob nectar from others (Roubik 1989). Mayans cultivated stingless bees for their honey. Recently, however, honeybees (*Apis mellifera*) introduced from tropical Africa have spread rapidly throughout the Neotropics (Roubik 1991). "Africanized" honeybees have not had an obviously detectable impact on the stingless bees of dense tropical forests (Roubik 1991), but when these honeybees colonized Central America, yields from Mayan stingless bee colonies were much reduced (D. Roubik, p. c.).

Honeybee queens mate with many males. Thus, most of a honeybee worker's companions are half sisters. A honeybee worker is less related to a half sister's son than to a queen's son, so workers eat eggs laid by half sisters, ensuring that nearly all the hive's males are sons of the queen (Ratnieks 1993). Such mutual policing creates a common interest among workers in helping their queen reproduce (Ratnieks and Visscher 1989). The queens of the stingless bee colonies so far studied, however, mate with only one male. Thus it is more profitable for a stingless bee worker to raise a sister's son than a son of the queen. In some stingless bee colonies a substantial proportion of the males are sons of workers (Machado et al. 1984). Yet, in other respects, stingless bee colonies are highly organized. In some species, such as *Melipona panamica*, foragers appear to be able to communicate the three-dimensional location of food sources by means as complex and specific as the famous dances of honeybees (Nieh and Roubik 1995).

Counting stingless bees and assessing their energetic impact would provide a minimum estimate of what the forest pays for pollination services (Roubik 1993). In the dry forest of Guanacaste, Costa Rica, there are four stingless bee colonies per hectare (one can count them when the forest is leafless). When 5 ha of tropical forest were cut down on the Caribbean slope of Panama, 30 stingless bee nests were found in the fallen vegetation. Roubik believes that there are 6 colonies of stingless bee, averaging 6000 0.01 g bees each (of which a third are foragers), and 8300 other bees, averaging 0.02 g each, per hectare on Barro Colorado, for a total of 44,300 bees, weighing 526 g in all. A stingless bee colony raises an average of 8 broods per year, with 21,000 immatures per brood. As it takes 38.1 kJ, or about 2.3 g of sugar, to make a gram of adult bee, stingless bee reproduction consumes 380,000 kJ, or 23 kg worth of sugar, per hectare per year. Other bees average 2.5 broods per year, raise 4 immatures per adult, and thereby consume 45,000 kJ, or about 1.2 kg worth of sugar, per hectare per year.

An adult stingless bee consumes 0.016 g of nectar, containing 0.0064 g of sugar, per day, so 36,000 bees (a hectare's worth) consume the equivalent of 84 kg sugar per hectare per year. Other bees consume .032 g of nectar per day each, or 0.013 g worth of sugar each. Thus 8300 of them (a hectare's worth) consume the equivalent of 39 kg sugar per hectare per year.

Finally, the average stingless bee colony needs 1.4 million 0.016 g loads of resin a year to maintain its nest: six such colonies use about 134 kg of resin, the energetic equivalent of 270 kg sugar, per hectare per year. The average stingless bee colony also stores 2 l of honey and 1 l of pollen per year, the former the equivalent of 2 kg of sugar, and the latter the equivalent of 1.6 kg, which consumes the equivalent of another 22 kg of sugar per hectare per year. Roubik (1993) sums these terms to conclude that the bees consume the energetic equivalent of 440 kg sugar per hectare per year. One might argue that, on occasion, the same puppy has been bought twice: when bees eat stored honey they do not forage, and foraging must supply the energy to raise young, etc. However, it does appear that BCI's bees account for the equivalent of nearly 400 kg sugar per hectare per year and collect twice their dry weight in forest products each day. As sugar is about as energy-rich as dried leaves, bees "cost" the forest the energetic equivalent of about 6% of its leaf litter fall. If we take into account the costs of flower production and the energy demands of other pollinators, will we find that the forest expends as much energy in getting its flowers pollinated as in getting its seeds dispersed? For it would appear that Barro Colorado produces about a ton dry weight of fruit per hectare per year (Hladik et al. 1993).

Pseudostigmatidae

Walking through a treefall gap in the rainy season, one's eye may be caught by a fluttering giant damselfly (Pseudostigmatidae), with a slender abdomen up to 10 cm long and a wingspan of 17 cm. They pluck spiders 3–6 mm in body length from their webs (Fincke 1992a: p. 107). The most conspicuous species is *Megaloprepus coerulatus*. This species has a bluish-black band covering all but the very tip of the distal third of each wing. Wings of male *Megaloprepus* have large white patches on their leading edges, forward of the dark band; females have white spots distal to the dark wing bands (Fincke 1992a). In a tree fall gap, a territorial male *Megaloprepus* resembles a "pulsating blue and white beacon" when he is hovering, beating his wings in synchrony, a few strokes per second (Fincke 1992a: 109). Barro Colorado also has three other species of giant damselfly: two species of *Mecistogaster* and a rare species of *Pseudostigma* (Fincke 1984, 1992a).

These giant damselflies all lay eggs in water-filled tree holes, especially those in tree fall gaps. These tree holes contain between 0.1 and 32 l of water. Tree holes are clearly a limiting resource: late in the rainy season, 99% of the tree holes containing more than a liter of water shelter at least one odonate larva (Fincke 1992c: p. 93). Moreover, these larvae depress resource levels in these holes: "a reliable indication that a tree hole was occupied by an odonate larva was the conspicuous paucity of mosquito larvae, the most ubiquitous and abundant prey" (Fincke 1992c: 87). Larger larvae reduce future competition for food by killing the smaller larvae they encounter, reducing the abundance of odonate larvae in their tree hole to less than one per liter (Fincke 1992c, 1994). Different sizes of tree holes contribute disproportionately to the recruitment of different species of damselfly (and dragonfly). *Mecistogaster* spp. lay eggs in tree holes once they fill with water at the beginning of rainy season. *Megaloprepus* begin laying eggs later, but they keep laying eggs all through the rainy season (Fincke 1992a). In small tree holes containing less than 1 l of water, *Mecistogaster* can usually find and kill later-hatching *Megaloprepus* larvae before the latter outgrow them, so *Mecistogaster* usually emerge from such holes. In larger tree holes, later-hatching *Megaloprepus* often outgrow and kill *Mecistogaster* larvae in their hole. In the largest tree holes, larvae of the aeshnid dragonfly *Gynacantha membranalis* outgrow and eat the larvae they encounter of all the species of giant damselfly, although in such large holes *Megaloprepus* larvae sometimes escape detection and manage to mature even when *Gynacanthus* are present. Thus holes with a few liters of water contribute most to *Megaloprepus* recruitment, while *Gynacantha* recruit mostly from the largest holes (Fincke 1992a,c).

Judging by Figure 1 in Fincke (1992b), there is about one water-filled tree hole containing giant damselfly larvae within reach of a ground-based observer per hectare. About one site per 12 ha was defended by an adult male *Megaloprepus*. If 1 of 3 adult *Megaloprepus* males is territorial for at least part of its life (Fincke 1992b: p. 454), Barro Colorado has somewhat more than 1 adult male per 4 ha during the mating season, or 800 adult *Megaloprepus* of both sexes in all. This calculation is infinitely crude, but the number of adult *Megaloprepus* on Barro Colorado is undoubtedly closer to 1000 than to 100 or 10,000.

Adult male *Megaloprepus* defend holes which are larger, and more often in or near a tree fall gap, than the average hole that shelters *Megaloprepus* larvae. Large holes support two or three generations of *Megaloprepus* larvae per year, whereas holes small enough to permit *Mecistogaster* to develop safely allow only one generation of larvae per year, thanks to the restricted stock of prey (Fincke 1992a).

The ecological characteristics of larval giant damselflies dictate the characteristics of their adults' social behavior. Larger tree holes allow *Megaloprepus* larvae to mature at a larger size. In turn, larger *Megaloprepus* males mate more often because they can keep smaller males away from tree holes containing > 1 l of water, where females prefer to lay eggs. A large male benefits from holding a territory because he can keep a female from laying eggs in his hole unless she mates with him first (Fincke 1984, 1992a–c). *Megaloprepus* males defend holes big enough to produce several larvae: their choice of territory depends not only on the availability of visiting females, but on the suitability of the defended tree hole for larvae (Fincke 1992b). It does not pay *Mecistogaster* males to defend large tree holes because they cannot prevent female *Megaloprepus* from laying eggs there. *Mecistogaster linearis* males establish reproductive territories in tree fall gaps, where they mate with females foraging there. On the other hand, the mating season of *Mecistogaster ornata* is apparently too short for males to profit by establishing reproductive territories (Fincke 1992c).

CONCLUDING REMARKS

Sketches of a selection of plants and animals on Barro Colorado have been presented. The selection, with its emphasis on trees, mammals, and social insects, reflects the work that has been done there. Certain fundamental themes emerge. To learn how Barro Colorado can support so many species, we must become acquainted with the species themselves. What problems do they face? How do they find nourishment and mates, and how do they avoid being eaten? How does the success of their efforts depend on other members of their own, or of other, species?

An ethologist sees this problem presented in miniature by a group of social animals, whose members may be competing with each other for food, mates, and other resources but which nonetheless depend on each other for help in procuring food, caring for young, coping with ectoparasites, or escaping predators, as the case may be. How are the potential conflicts among group members suppressed or resolved to a degree sufficient to allow these individuals to function effectively as a group?

The ethologist's conundrum applies equally to ecological communities. Here, one issue is how all these species can coexist without competition leading to the ouster of some by others. Another issue, equally urgent, is to what degree, and how stringently, do some species depend on the services of others? Competition and interdependence are both essential parts of one story. At bottom, the fundamental question is, how can relationships of interdependence, cooperation, and other forms of harmony evolve when the motor of evolution, natural selection, is driven more truly by individual self-interest than any free-market economy can ever hope to be?

Before we enquire further into this play, with its intricate interweaving of competition and interdependence, more needs to be said about the stage upon which it is acted. To this topic we now turn.

References

Adler, G. H. 1994. Tropical forest fragmentation and isolation promote asynchrony among populations of a frugivorous rodent. *Journal of Animal Ecology* 63: 903–911.

Adler, G. H., and R. P. Beatty. 1997. Changing reproductive rates in a Neotropical forest rodent, *Proechimys semispinosus*. *Journal of Animal Ecology*, 66: 472–480.

Aide, T. M. 1992. Dry season leaf production: an escape from herbivory. *Biotropica* 24: 532–537.

Andrade, J. L., and P. S. Nobel. 1996. Habitat, CO_2 uptake and growth for the CAM epiphytic cactus *Epiphyllum phyllanthus* in a Panamanian tropical forest. *Journal of Tropical Ecology* 12: 291–306.

Andrade, J. L., and P. S. Nobel. 1997. Microhabitats and water relations of epiphytic cacti and ferns in a lowland neotropical forest. *Biotropica*, 29: 261–270.

Andrews, R. M. 1988. Demographic correlates of variable egg survival for a tropical lizard. *Oecologia* 76: 376–382.

Andrews, R. M. 1991. Population stability of a tropical lizard. *Ecology* 72: 1204–1217.

Andrews, R. M., and J. D. Nichols. 1990. Temporal and spatial variation in survival rates of the tropical lizard *Anolis limifrons*. *Oikos* 57: 215–221.

Andrews, R. M., and A. S. Rand. 1982. Seasonal breeding and long-term population fluctuations in the lizard *Anolis limifrons*, pp. 405–412. In E. G. Leigh, Jr., A. S. Rand, and D. M. Windsor, eds., *Ecology of a Tropical Forest*. Smithsonian Institution Press, Washington, DC.

Andrews, R. M., and A. S. Rand. 1990. Adición: nuevas percepciones derivadas de la continuación de un estudio a largo plazo de la lagartija *Anolis limifrons*, pp. 477–479. In E. G. Leigh, Jr., A. S. Rand, and D. M. Windsor, eds., *Ecología de un Bosque Tropical*. Smithsonian Tropical Research Institute, Balboa, Panama.

Andrews, R. M., and S. J. Wright. 1994. Long-term population fluctuations of a tropical lizard: a test of causality, pp. 267–285. In L. J. Vitt and E. R. Pianka, eds., *Lizard Ecology: Historical and Experimental Perspectives*. Princeton University Press, Princeton, NJ.

Augspurger, C. K. 1982. A cue for synchronous flowering, pp. 133–150. In E. G. Leigh, Jr., A. S. Rand, and D. M. Windsor, eds., *Ecology of a Tropical Forest*. Smithsonian Institution Press, Washington, DC.

Augspurger, C. K. 1983a. Phenology, flowering synchrony and fruit set of six Neotropical shrubs. *Biotropica* 15: 257–267.

Augspurger, C. K. 1983b. Seed dispersal of the tropical tree, *Platypodium elegans*, and the escape of its seedlings from fungal pathogens. *Journal of Ecology* 71: 759–771.

Augspurger, C. K. 1986. Morphology and dispersal potential of wind-dispersed diaspores of neotropical trees. *American Journal of Botany* 73: 353–363.

Augspurger, C. K. 1988. Mass allocation, moisture content, and dispersal capacity of wind-dispersed tropical diaspores. *New Phytologist* 108: 357–368.

Backwell, P. R. Y., M. D. Jennions, J. H. Christy, and U. Schober. 1995. Pillar building in the fiddler crab *Uca beebei*: evidence for a condition-dependent ornament. *Behavioral Ecology and Sociobiology* 36: 185–192.

Becker, P., P. E. Rabenold, J. R. Idol, and A. P. Smith. 1988. Water potential gradient for gaps and slopes in a Panamanian tropical moist forest's dry season. *Journal of Tropical Ecology* 4: 173–184.

Bock, B. C., A. S. Rand, and G. M. Burghardt. 1985. Seasonal migration and nesting site fidelity in the green iguana, pp. 435–443. In M. A. Rankin, ed., *Migration: Mechanisms and Adaptative Significance*. University of Texas Marine Science Institute, Port Aransas, TX.

Bonaccorso, F. J. 1979. Foraging and reproductive ecology in a Panamanian bat community. *Bulletin of the Florida State Museum, Biological Sciences* 24: 359–408.

Bonaccorso, F. J., W. E. Glanz, and C. M. Sandford. 1980. Feeding assemblages of mammals at fruiting *Dipteryx panamensis* (Papilionaceae) trees in Panama: seed predation, dispersal, and parasitism. *Revista de Biologia Tropical* 28: 61–72.

Bossel, H., and H. Krieger. 1991. Simulation model of natural tropical forest dynamics. *Ecological Modelling* 59: 37–71.

Burghardt, G. M., H. W. Greene, and A. S. Rand. 1977. Social behavior in hatchling green iguanas: life at a reptile rookery. *Science* 195: 689–691.

Chapman, F. M. 1935. José: two months from the life of a Barro Colorado coati. *Natural History* 35: 299–309.

Charles-Dominique, P., M. Atramentowicz, M. Charles-Dominique, H. Gérard, A. Hladik, C. M. Hladik, and F. Prévost. 1981. Les mammifères frugivores arboricoles nocturnes d'une forêt guyanaise: inter-relations plantes-animaux. *Revue d'Écologie (La Terre et la Vie)* 35: 341–435.

Cherrett, J. M. 1968. The foraging behaviour of *Atta cephalotes* L. (Hymenoptera, Formicidae) I. Foraging pattern and plant species attacked in tropical forest. *Journal of Animal Ecology* 37: 387–403.

Cherrett, J. M. 1972. Some factors involved in the selection of vegetable substrate by *Atta cephalotes* (L.)(Hymenoptera: Formicidae) in tropical forest. *Journal of Animal Ecology* 41: 647–660.

Chivers, D. J. 1969. On the daily behavior and spacing of free-ranging howler monkey groups. *Folia Primatologica* 10: 48–103.

Christy, J. H. 1995. Mimicry, mate choice, and the sensory trap hypothesis. *American Naturalist* 146: 171–181.

Clark, D. A., and D. B. Clark. 1987. Temporal and environmental patterns of reproduction in *Zamia skinneri*, a tropical rain forest cycad. *Journal of Ecology* 75: 135–149.

Clark, D. B., and D. A. Clark. 1988. Leaf production and the cost of reproduction in the neotropical rain forest cycad, *Zamia skinneri*. *Journal of Ecology* 76: 1153–1163.

Coates, A. G., J. B. C. Jackson, L. S. Collins, T. M. Cronin, H. J. Dowsett, L. M. Bybell, P. Jung, and J. A. Obando. 1992. Closure of the Isthmus of Panama: the near-shore marine record of Costa Rica and western Panama. *Geological Society of America Bulletin* 104: 814–828.

Condit, R., S. P. Hubbell, and R. B. Foster. 1995. Mortality rates of 205 Neotropical tree and shrub species and the impact of a severe drought. *Ecological Monographs* 65: 419–439.

Condit, R., S. P. Hubbell, and R. B. Foster. 1996. Changes in tree species abundance in a Neotropical forest: impact of climate change. *Journal of Tropical Ecology* 12: 231–256.

Corner, E. J. H. 1940. *Wayside Trees of Malaya*. Government Printer, Singapore.

Corner, E. J. H. 1967. Ficus in the Solomon Islands and its bearing on the post-Jurassic history of Melanesia. *Philosophical Transactions of the Royal Society of London* B 253: 23–159.

Corner, E. J. H. 1988. *Wayside Trees of Malaya* (3rd ed). Malayan Nature Society, Kuala Lumpur.

Coyne, J. A., and H. A. Orr. 1989. Patterns of speciation in *Drosophila*. *Evolution* 43: 362–381.

Croat, T. B. 1978. *Flora of Barro Colorado Island*. Stanford University Press, Stanford, CA.

Darlington, J. P. E. C. 1994. Nutrition and evolution in fungus-growing termites, pp. 105–130. In J. H. Hunt and C. A. Nalepa, eds., *Nourishment and Evolution in Insect Societies*. Westview Press, Boulder, CO.

Darwin, C. 1877. *The Various Contrivances by which Orchids are Fertilized by Insects*. John Murray, London (reprinted by University of Chicago Press).

Davidson, D. W. 1988. Ecological studies of neotropical ant gardens. *Ecology* 69: 1138–1152.

De Steven, D. 1988. Light gaps and long-term seedling performance of a Neotropical canopy tree (*Dipteryx panamensis*, Leguminosae). *Journal of Tropical Ecology* 4: 407–411.

De Steven, D. 1994. Tropical tree seedling dynamics: recruitment patterns and their population consequences for three canopy species in Panama. *Journal of Tropical Ecology* 10: 369–383.

De Steven, D., and F. E. Putz. 1984. Impact of mammals on early recruitment of a tropical canopy tree, *Dipteryx panamensis*, in Panama. *Oikos* 43: 207–216.

De Steven, D., D. M. Windsor, F. E. Putz, and B. de León. 1987. Vegetative and reproductive phenologies of a palm assemblage in Panama. *Biotropica* 19: 342–356.

de Vasconcelos, H. L. 1990. Foraging activity of two species of leaf-cutting ants (*Atta*) in a primary forest of the central Amazon. *Insectes Sociaux* 37: 131–145.

Dubost, G. 1980. L'écologie et la vie sociale du Céphalophe bleu (*Cephalophus monticola* Thunberg), petit ruminant forestier africain. *Zeitschrift für Tierpsychologie* 54: 205–266.

Dugan, B. 1982. The mating behavior of the green iguana, *Iguana iguana*, pp. 320–341. In G. M. Burghardt and A. S. Rand, eds., *Iguanas of the World*. Noyes Publications, Park Ridge, NJ.

Dugan, B. A., A. S. Rand, G. M. Burghardt, and B. C. Bock. 1981. Interactions between nesting crocodiles and iguanas. *Journal of Herpetology* 15: 409–414.

Eberhard, W. G. 1985. *Sexual Selection and Animal Genitalia*. Harvard University Press, Cambridge, MA.

Eberhard, W. G. 1993. Evaluating models of sexual selection: genitalia as a test case. *American Naturalist* 142: 564–571.

Eberhard, W. G. 1996. *Female Control: Sexual Selection by Cryptic Female Choice*. Princeton University Press, Princeton, NJ.

Eisenberg, J. F., and R. W. Thorington, Jr. A preliminary analysis of a neotropical mammal fauna. *Biotropica* 5: 150–161.

Feener, D. H. Jr., and K. A. G. Moss. 1990. Defense against parasites by hitchhikers in leaf-cutting ants: a quantitative assessment. *Behavioral Ecology and Sociobiology* 26: 17–29.

Fincke, O. M. 1984. Giant damselflies in a tropical forest: reproductive biology of *Megaloprepus coerulatus* with notes on *Mecistogaster* (Zygoptera: Pseudostigmatidae). *Advances in Odonatology* 2: 13–27.

Fincke, O. M. 1992a. Behavioural ecology of the giant damselflies of Barro Colorado Island, Panama (Odonata: Zygoptera: Pseudostigmatidae), pp. 102–113. In D. Quintero and A. Aiello, eds., *Insects of Panama and Mesoamerica: Selected Studies*. Oxford University Press, Oxford.

Fincke, O. M. 1992b. Consequences of larval ecology for territoriality and reproductive success of a Neotropical damselfly. *Ecology* 73: 449–462.

Fincke, O. M. 1992c. Interspecific competition for tree holes: consequences for mating systems and coexistence in Neotropical damselflies. *American Naturalist* 139: 80–101.

Fincke, O. M. 1994. Population regulation of a tropical damselfly in the larval stage by food limitation, cannibalism, intraguild predation and habitat drying. *Oecologia* 100: 118–127.

Fleming, T. H. 1971. *Artibeus jamaicensis*: delayed embryonic development in a Neotropical bat. *Science* 171: 402–404.

Forget, P.-M. 1993. Post-dispersal seed predation and scatter-hoarding of *Dipteryx panamensis* (Papilionaceae) seeds by rodents in Panama. *Oecologia* 94: 255–261.

Forget, P.-M., and T. Milleron. 1991. Evidence for secondary seed dispersal by rodents in Panama. *Oecologia* 87: 596–599.

Forget, P.-M., E. Munoz, and E. G. Leigh Jr. 1994. Predation by rodents and bruchid beetles on seeds of *Scheelea* palms on Barro Colorado Island, Panama. *Biotropica* 25: 420–426.

Forget, P.-M., and C. Poletto-Forget. 1992. L'agouti, sauveur des forêts. *Science et Vie* 892: 66–67.

Foster, R. B. 1982a. Famine on Barro Colorado Island, pp. 201–212. In E. G. Leigh, Jr., A. S. Rand, and D. M. Windsor, eds., *The Ecology of a Tropical Forest*. Smithsonian Institution Press, Washington, DC.

Foster, R. B. 1982b. The seasonal rhythm of fruitfall on Barro Colorado Island, pp. 151–172. In E. G. Leigh Jr., A. S. Rand,

and D. M. Windsor, eds., *The Ecology of a Tropical Forest*. Smithsonian Institution Press, Washington, DC.

Foster, R. B. 1990. Long-term change in the successional forest community of the Rio Manu floodplain, pp. 565–572. In A. H. Gentry, ed., *Four Neotropical Forests*. Yale University Press, New Haven, CT.

Foster, R. B., J. Arce B., and T. S. Wachter. 1986. Dispersal and the sequential plant communities in Amazonian Peru floodplain, pp. 357–370. In A. Estrada and T. H. Fleming, eds., *Frugivores and Seed Dispersal*. W. Junk, Dordrecht.

Fowler, H. G., V. Pereira-da-Silva, L. C. Forti, and N. B. Saes. 1986. Population dynamics of leaf-cutting ants: a brief review, pp. 123–145. In C. S. Lofgren and R. K. Vander Meer, eds., *Fire Ants and Leaf-Cutting Ants*. Westview Press, Boulder, CO.

Franks, N. 1982a. Ecology and population regulation in the army ant *Eciton burchelli*, pp. 389–395. In E. G. Leigh, Jr., A. S. Rand, and D. M. Windsor, eds., *The Ecology of a Tropical Forest*. Smithsonian Institution Press, Washington, DC.

Franks, N. R. 1982b. A new method for censusing animal populations: the number of *Eciton burchelli* army ant colonies on Barro Colorado Island, Panama. *Oecologia* 52: 266–268.

Franks, N. R. 1982c. Social insects in the aftermath of swarm raids of the army ant *Eciton burchelli*, pp. 275–279. In M. D. Breed, C. D. Michener, and H. E. Evans, eds., *The Biology of Social Insects*. Westview Press, Boulder, CO.

Franks, N. R. 1989a. Army ants: a collective intelligence. *American Scientist* 77: 138–145.

Franks, N. R. 1989b. Thermoregulation in army ant bivouacs. *Physiological Entomology* 14: 397–404.

Franks, N. R., and W. H. Bossert. 1983. The influence of swarm raiding army ants on the patchiness and diversity of a tropical leaf litter ant community, pp. 151–163. In S. L. Sutton, T. C. Whitmore, and A. C. Chadwick, eds., *Tropical Rain Forest: Ecology and Management*. Blackwell Scientific Publications, Oxford.

Franks, N. R., and C. R. Fletcher. 1983. Spatial patterns in army ant foraging and migration: *Eciton burchelli* on Barro Colorado Island, Panama. *Behavioral Ecology and Sociobiology* 12: 261–270.

Franks, N. R., and B. Hölldobler. 1987. Sexual competition during colony reproduction in army ants. *Biological Journal of the Linnean Society* 30: 229–243.

Giacalone, J., W. E. Glanz, and E. G. Leigh, Jr. 1990. Adición: fluctuaciones poblacionales a largo plazo de *Sciurus granatensis* en relación con la disponibilidad de frutos, pp. 331–335. In E. G. Leigh, Jr., A. Stanley Rand, and D. M. Windsor, eds., *Ecología de un Bosque Tropical*. Smithsonian Tropical Research Institute, Balboa, Panama.

Glanz, W. E. 1982. The terrestrial mammal fauna of Barro Colorado Island: censuses and long-term changes, pp. 455–468. In E. G. Leigh, Jr., A. S. Rand, and D. M. Windsor, eds., *The Ecology of a Tropical Forest*. Smithsonian Institution Press, Washington, DC.

Glanz, W. E., R. W. Thorington, Jr., J. Giacalone-Madden, and L. R. Heaney. 1982. Seasonal food use and demographic trends in *Sciurus granatensis*, pp. 239–252. In E. G. Leigh, Jr., A. S. Rand, and D. M. Windsor, eds., *The Ecology of a Tropical Forest*. Smithsonian Institution Press, Washington, DC.

Gliwicz, J. 1973. A short characteristics of a population of *Proechimys semispinosus* (Tomes, 1860)—a rodent species of the tropical rain forest. *Bulletin de l'Academie Polonaise des Sciences, Série des sciences biologiques*, Cl. II. 21: 413–418.

Gliwicz, J. 1984. Population dynamics of the spiny rat *Proechimys semispinosus* on Orchid Island (Panama). *Biotropica* 16: 73–78.

Gompper, M. E. 1994. The importance of ecology, behavior, and genetics in the maintenance of coati (*Nasua narica*) social structure. PhD dissertation, University of Tennessee, Knoxville.

Gompper, M. E. 1996. Sociality and asociality in white-nosed coatis (*Nasua narica*): foraging costs and benefits. *Behavioral Ecology* 7: 254–263.

Gompper, M. E. 1997. Population ecology of the white-nosed coatis (*Nasua narica*) on Barro Colorado Island, Panama. *Journal of Zoology* 241: 441–455.

Gompper, M. E., J. L. Gittleman, and R. K. Wayne. 1997. Genetic relatedness, coalitions and social behaviour of white-nosed coatis, *Nasua narica*. *Animal Behaviour* 53: 781–797.

Gorman, G. C., Y. J. Kim, and R. Rubinoff. 1976. Genetic relationships of three species of *Bathygobius* from the Atlantic and Pacific sides of Panama. *Copeia* 1976: 361–364.

Gradwohl, J., and R. Greenberg. 1980. The formation of antwren flocks on Barro Colorado Island, Panamá. *Auk* 97: 385–395.

Gradwohl, J., and R. Greenberg. 1982. The breeding season of antwrens on Barro Colorado Island, pp. 345–351. In E. G. Leigh, Jr., A. S. Rand, and D. M. Windsor, eds., *The Ecology of a Tropical Forest*. Smithsonian Institution Press, Washington, DC.

Greenberg, R. 1981. Dissimilar bill shapes in New World tropical versus temperate forest foliage-gleaning birds. *Oecologia* 49: 143–147.

Greenberg, R., and J. Gradwohl. 1980. Leaf surface specializations of birds and arthropods in a Panamanian forest. *Oecologia* 46: 115–124.

Greenberg, R., and J. Gradwohl. 1985. A comparative study of the social organization of antwrens on Barro Colorado Island, Panama, pp. 845–855. In P. A. Buckley, M. S. Foster, E. S. Morton, R. S. Ridgely, and F. G. Buckley, eds., *Neotropical Ornithology*. Ornithological Monographs no. 36. American Ornithologists' Union, Washington, DC.

Greenberg, R., and J. Gradwohl. 1986. Constant density and stable territoriality in some tropical insectivorous birds. *Oecologia* 69: 618–625.

Haig, D. 1993. Genetic conflicts in human pregnancy. *Quarterly Review of Biology* 68: 495–532.

Haines, B. L. 1978. Element and energy flows through colonies of the leaf-cutting ant, *Atta colombica*, in Panama. *Biotropica* 10: 270–277.

Hallé, F., R. A. A. Oldeman, and P. B. Tomlinson. 1978. *Tropical Trees and Forests*. Springer-Verlag, Berlin.

Hallwachs, W. 1994. The clumsy dance between agoutis and plants: scatterhoarding by Costa Rican dry forest agoutis (*Dasyprocta punctata*: Dasyproctidae: Rodentia). PhD dissertation, Cornell University, Ithaca, NY.

Hamilton, C. W. 1985. Architecture in Neotropical *Psychotria* L. (Rubiaceae): dynamics of branching and its taxonomic significance. *American Journal of Botany* 72: 1081–1088.

Hamrick, J. L., and D. A. Murawski. 1990. The breeding structure of tropical tree populations. *Plant Species Biology* 5: 157–165.

Hamrick, J. L., and D. A. Murawski. 1991. Levels of allozyme diversity in populations of uncommon Neotropical tree species. *Journal of Tropical Ecology* 7: 395–399.

Handley, C. O. Jr., D. E. Wilson, and A. L. Gardner (eds). 1991. *Demography and Natural History of the Common Fruit Bat, Artibeus jamaicensis, on Barro Colorado Island, Panamá*. Smithsonian Institution Press, Washington, DC.

Hladik, A., and C. M. Hladik. 1969. Rapports trophiques entre végétation et primates dans la forêt de Barro Colorado (Panama). *La Terre et la Vie* 26: 149–215.

Hladik, A., E. G. Leigh, Jr., and F. Bourlière. 1993. Food production and nutritional value of wild and semi-domesticated species—background, pp. 127–138. In C. M. Hladik et al., eds.,

Tropical Forests, People and Food. UNESCO, Paris, and Parthenon Publishing, Park Ridge, NJ.

Hogan, K. P. 1986. Plant architecture and population ecology in the palms Socratea durissima and Scheelea zonensis on Barro Colorado Island, Panama. *Principes* 30: 105–107.

Hölldobler, B., and E. O. Wilson. 1990. *The Ants*. Harvard University Press, Cambridge, MA.

Howard, J. J. 1990. Infidelity of leaf-cutting ants to host plants: resource heterogeneity or defense induction? *Oecologia* 82: 394–401.

Howe, H. F. 1979. Fear and frugivory. *American Naturalist* 114: 925–931.

Howe, H. F. 1980. Monkey dispersal and waste of a neotropical fruit. *Ecology* 61: 944–959.

Howe, H. F. 1981. Dispersal of a neotropical nutmeg (*Virola sebifera*) by birds. *Auk* 98: 88–98.

Howe, H. F. 1982. Fruit production and animal activity in two tropical trees, pp. 189–199. In E. G. Leigh, Jr., A. S. Rand, and D. M. Windsor, eds., *The Ecology of a Tropical Forest*. Smithsonian Institution Press, Washington, DC.

Howe, H. F. 1986. Consequences of seed dispersal by birds: a case study from Central America. *Journal of the Bombay Natural History Society* 83 (supplement): 19–42.

Howe, H. F. 1989. Scatter- and clump-dispersal and seedling demography: hypothesis and implications. *Oecologia* 79: 417–426.

Howe, H. F. 1990a. Seed dispersal by birds and mammals: implications for seedling demography, pp. 191–218. In K. S. Bawa and M. Hadley, eds., *Reproductive Ecology of Tropical Forest Plants*. UNESCO/Parthenon Publishing, Park Ridge, NJ.

Howe, H. F. 1990b. Survival and growth of juvenile *Virola surinamensis* in Panama: effects of herbivory and canopy closure. *Journal of Tropical Ecology* 6: 259–280.

Howe, H. F., E. W. Schupp, and L. C. Westley. 1985. Early consequences of seed dispersal for a neotropical tree (*Virola surinamensis*). *Ecology* 66: 781–791.

Howe, H. F., and G. A. Vande Kerckhove. 1981. Removal of wild nutmeg (*Virola surinamensis*) by birds. *Ecology* 62: 1093–1106.

Hubbell, S. P., and R. B. Foster. 1986. Commonness and rarity in a Neotropical forest: implications for tropical tree conservation, pp. 205–231. In M. E. Soulé, ed., *Conservation Biology*. Sinauer Associates, Sunderland, MA.

Hubbell, S. P., D. F. Wiemer, and A. Adejare. 1983. An antifungal terpenoid defends a neotropical tree (*Hymenaea*) against attack by fungus-growing ants (*Atta*). *Oecologia* 60: 321–327.

Janzen, D. H. 1970. Herbivores and the number of tree species in tropical forests. *American Naturalist* 104: 501–528.

Janzen, D. H. (ed). 1983. *Costa Rican Natural History*. University of Chicago Press, Chicago.

Jones, S. E., 1977. Coexistence in mixed species antwren flocks. *Oikos* 29: 366–375.

Kalko, E. K. V., E. A. Herre, and C. O. Handley Jr. 1996. Relation of fig fruit characteristics to fruit-eating bats in the New and Old World tropics. *Journal of Biogeography* 23: 565–576.

Kaufmann, J. H. 1962. Ecology and social behavior of the coati, *Nasua narica*, on Barro Colorado Island, Panama. *University of California Publications in Zoology* 60: 95–222.

King, D. A. 1994. Influence of light level on the growth and morphology of saplings in a Panamanian forest. *American Journal of Botany* 81: 948–957.

Knowlton, N., L. A. Weigt, L. A. Solórzano, D. K. Mills, and E. Bermingham. 1993. Divergence in proteins, mitochondrial DNA, and reproductive compatibility across the Isthmus of Panama. *Science* 260: 1629–1632.

Kursar, T. A. 1997. Relating tree physiology to past and future changes in tropical rainforest tree communities. *Climatic Change*, in press.

Lang, G. E., and D. H. Knight. 1983. Tree growth, mortality, recruitment, and canopy gap formation during a 10-year period in a tropical moist forest. *Ecology* 64: 1075–1080.

Lee, A., and R. Martin. 1988. *The Koala: a Natural History*. New South Wales University Press, Kensington, Australia.

Leigh, E. G. Jr., and C. O. Handley, Jr. 1991. Population estimates, pp. 77–87. In C. O. Handley, Jr., D. E. Wilson, and A. L. Gardner, eds., *Demography and Natural History of the Common Fruit Bat*, Artibeus jamaicensis, *on Barro Colorado Island, Panamá*. Smithsonian Institution Press, Washington, DC.

Leigh, E. G. Jr., R. T. Paine, J. F. Quinn, and T. H. Suchanek. 1987. Wave energy and intertidal productivity. *Proceedings of the National Academy of Sciences, USA* 84: 1314–1318.

Leigh, E. G. Jr., and D. M. Windsor. 1982. Forest production and regulation of primary consumers on Barro Colorado Island, pp. 111–122. In E. G. Leigh, Jr., A. S. Rand, and D. M. Windsor, eds., *The Ecology of a Tropical Forest*. Smithsonian Institution Press, Washington, DC.

Leigh, E. G. Jr., S. J. Wright, F. E. Putz, and E. A. Herre. 1993. The decline of tree diversity on newly isolated tropical islands: a test of a null hypothesis and some implications. *Evolutionary Ecology* 7: 76–102.

Lugo, A. E., E. G. Farnsworth, D. Pool, P. Jerez, and G. Kaufman. 1973. The impact of the leaf cutter ant *Atta colombica* on the energy flow of a tropical wet forest. *Ecology* 54: 1292–1301.

McClearn, D. 1992. Locomotion, posture, and feeding behavior of kinkajous, coatis, and raccoons. *Journal of Mammalogy* 73: 245–261.

Machado, M. F. P. S., E. P. B. Contel, and W. E. Kerr. 1984. Proportion of males sons-of-the-queen and sons-of-workers in *Plebeia droryana* (Hymenoptera, Apidae) estimated from data of an MDH isozymic polymorphic system. *Genetica* 65: 193–198.

Matsumoto, T. 1978. Population density, biomass, nitrogen and carbon content, energy value and respiration rate of four species of termites in Pasoh Forest Reserve. *Malayan Nature Journal* 30: 335–351.

Milton, K. 1979. Factors influencing leaf choice by howler monkeys: a test of some hypotheses of food selection by generalist herbivores. *American Naturalist* 114: 362–378.

Milton, K. 1980. *The Foraging Strategy of Howler Monkeys*. Columbia University Press, New York.

Milton, K. 1981. Distribution patterns of tropical plant foods as an evolutionary stimulus to primate mental development. *American Anthropologist* 83: 534–548.

Milton, K. 1982. Dietary quality and demographic regulation in a howler monkey population, pp. 273–289. In E. G. Leigh, Jr., A. S. Rand, and D. M. Windsor, eds., *The Ecology of a Tropical Forest*. Smithsonian Institution Press, Washington, DC.

Milton, K. 1990. Annual mortality patterns of a mammal community in central Panama. *Journal of Tropical Ecology* 6: 493–499.

Milton, K. 1996. Effects of bot fly (*Alouattamyia*) parasitism on a free-ranging howler monkey (*Alouatta palliata*) population in Panama. *Journal of Zoology* 239: 39–63.

Milton, K., D. M. Windsor, D. W. Morrison, and M. A. Estribi. 1982. Fruiting phenologies of two neotropical *Ficus* species. *Ecology* 63: 752–762.

Mitchell, B. J. 1989. Resources, group behavior, and infant development in white-faced capuchin monkeys, *Cebus capucinus*. PhD Dissertation, Department of Zoology, University of California, Berkeley.

Montgomery, G. G. 1983. *Bradypus variegatus* (Perezoso de tres dedos, three-toed sloth), pp. 453–456. In D. H. Janzen, ed., *Costa Rican Natural History*. University of Chicago Press, Chicago, IL.

Montgomery, G. G., A. S. Rand, and M. E. Sunquist. 1973. Post-nesting movements of iguanas from a nesting aggregation. *Copeia* 1973: 620–622.

Montgomery, G. G., and M. E. Sunquist. 1975. Impact of sloths on neotropical forest energy flow and nutrient cycling, pp. 69–98. In F. B. Golley and E. Medina, eds., *Tropical Ecological Systems: Trends in Terrestrial and Aquatic Research*. Springer-Verlag, New York.

Montgomery, G. G., and M. E. Sunquist. 1978. Habitat selection and use by two-toed and three-toed sloths, pp. 329–359. In G. G. Montgomery, ed., *The Ecology of Arboreal Folivores*. Smithsonian Institution Press, Washington, DC.

Morrison, D. W. 1978. Foraging ecology and energetics of the frugivorous bat *Artibeus jamaicensis*. *Ecology* 59: 716–723.

Morrison, D. W. 1980. Efficiency of food utilization by fruit bats. *Oecologia* 45: 270–273.

Morrison, D. W., and S. H. Morrison. 1981. Economics of harem maintenance by a Neotropical bat. *Ecology* 62: 864–866.

Moynihan, M. H. 1970. Control, suppression, decay, disappearance and replacement of displays. *Journal of Theoretical Biology* 29: 85–112.

Moynihan, M. 1979. *Geographic Variation in Social Behavior and in Adaptations to Competition among Andean Birds*. Nuttall Ornithological Club, Cambridge, MA.

Mulkey, S. S., A. P. Smith, S. J. Wright, J. L. Machado, and R. Dudley. 1992. Contrasting leaf phenotypes control seasonal variation in water loss in a tropical forest shrub. *Proceedings of the National Academy of Sciences, USA* 89: 9084–9088.

Munn, C. A. 1985. Permanent canopy and understory flocks in Amazonia: species composition and population density, pp. 683–712. In P. A. Buckley, M. S. Foster, E. S. Morton, R. S. Ridgely, and F. G. Buckley, eds., *Neotropical Ornithology*. Ornithological Monographs no. 36. American Ornithologists' Union, Washington, DC.

Munn, C. A., and J. W. Terborgh. 1979. Multi-species territoriality in Neotropical foraging flocks. *Condor* 81: 338–347.

Nagy, K. A. 1982. Energy requirements of free-living iguanid lizards, pp. 49–59. In G. M. Burghardt and A. S. Rand, eds., *Iguanas of the World*. Noyes Publications, Park Ridge, NJ.

Nagy, K. A. 1987. Field metabolic rate and food requirement scaling in mammals and birds. *Ecological Monographs* 57: 111–128.

Nagy, K. A., and K. Milton. 1979. Energy metabolism and food consumption by wild howler monkeys (*Alouatta palliatta*). *Ecology* 60: 475–480.

Nagy, K. A., and G. G. Montgomery. 1980. Field metabolic rate, water flux and food consumption in three-toed sloths (*Bradypus variegatus*). *Journal of Mammalogy* 61: 465–472.

Nason, J. D., E. A. Herre, and J. L. Hamrick. 1996. Paternity analysis of the breeding structure of strangler fig populations: evidence for substantial long-distance wasp dispersal. *Journal of Biogeography* 23: 501–512.

Nieh, J. C., and D. W. Roubik. 1995. A stingless bee (*Melipona panamica*) indicates food location without using a scent trail. *Behavioral Ecology and Sociobiology* 37: 63–70.

Nobel, P. S. 1983. *Biophysical Plant Physiology and Ecology*. W. H. Freeman, San Francisco.

O'Dowd, D. J. 1994. Mite association with the leaf domatia of coffee (*Coffea arabica*) in North Queensland, Australia. *Bulletin of Entomological Research* 84: 361–366.

Oppenheimer, J. R. 1982. *Cebus capucinus*: home range, population dynamics, and interspecific relationships, pp. 253–272. In E. G. Leigh, Jr., A. S. Rand, and D. M. Windsor, eds., *Ecology of a Tropical Forest*. Smithsonian Institution Press, Washington, DC.

Paine, R. T. 1974. Intertidal community structure: experimental studies on the relationship between a dominant competitor and its principal predator. *Oecologia* 15: 93–120.

Paine, R. T. 1979. Disaster, catastrophe, and local persistence of the sea palm *Postelsia palmaeformis*. *Science* 205: 685–687.

Paine, R. T. 1992. Food-web analysis through field measurement of per capita interaction strength. *Nature* 355: 73–75.

Paine, R. T., and S. A. Levin. 1981. Intertidal landscapes: disturbance and the dynamics of pattern. *Ecological Monographs* 51: 145–178.

Pittendrigh, C. S. 1948. The bromeliad-Anopheles-malaria complex in Trinidad. I-The bromeliad flora. *Evolution* 2: 58–89.

Powell, G. V. N. 1985. Sociobiology and adaptive significance of interspecific foraging flocks in the Neotropics, pp. 713–732. In P. A. Buckley, M. S. Foster, E. S. Morton, R. S. Ridgely, and F. G. Buckley, eds., *Neotropical Ornithology*. Ornithological Monographs no. 36. American Ornithologists' Union, Washington, DC.

Rand, A. S. 1983. *Physalaemus pustulosus* (Rana, sapito túngara, foam toad, mudpuddle frog), pp. 412–415. In D. H. Janzen, ed., *Costa Rican Natural History*. University of Chicago Press, Chicago.

Rand, A. S. 1984. Clutch size in *Iguana iguana* in central Panama, pp. 115–122. In R. A. Seigel et al., eds., *Vertebrate Ecology and Systematics: A Tribute to Henry S. Fitch*. University of Kansas Museum of Natural History Special Publication no. 10. Lawrence, KS.

Rand, A. S., and B. C. Bock. 1992. Size variation, growth and survivorship in nesting green iguanas (*Iguana iguana*) in Panama. *Amphibia-Reptilia* 13: 147–156.

Rand, A. S., and B. Dugan. 1980. Iguana egg mortality within the nest. *Copeia* 1980: 531–534.

Ratnieks, F. L. W. 1993. Egg-laying, egg-removal and ovary development by workers in queenright honey bee colonies. *Behavioral Ecology and Sociobiology* 32: 191–198.

Ratnieks, F. L. W., and P. K. Visscher. 1989. Worker policing in the honeybee. *Nature* 342: 796–797.

Rockwood, L. L. 1973. Distribution, density and dispersion of two species of *Atta* (Hymenoptera: Formicidae) in Guanacaste Province, Costa Rica. *Journal of Animal Ecology* 42: 803–817.

Rockwood, L. L. 1976. Plant selection and foraging patterns in two species of leaf-cutting ants (*Atta*). *Ecology* 57: 48–61.

Rockwood, L. L., and K. E. Glander. 1979. Howling monkeys and leaf-cutting ants: comparative foraging in a tropical deciduous forest. *Biotropica* 11: 1–10.

Romero, G. A., and C. E. Nelson. 1986. Sexual dimorphism in *Catasetum* orchids: forcible pollen emplacement and male flower competition. *Science* 232: 1538–1540.

Roubik, D. W. 1989. *Ecology and Natural History of Tropical Bees*. Cambridge University Press, New York.

Roubik, D. W. 1991. Aspects of Africanized honey bee ecology in tropical America, pp. 259–281. In M. Spivak, D. J. C. Fletcher, and M. D. Breed, eds., *The "African" Honey Bee*. Westview Press, Boulder, CO.

Roubik, D. W. 1993. Direct costs of forest reproduction, bee-cycling and the efficiency of pollination modes. *Journal of Bioscience* 18: 537–552.

Rubinoff, R. W., and I. Rubinoff. 1971. Geographic and reproductive isolation in Atlantic and Pacific populations of *Bathygobius*. *Evolution* 25: 88–97.

Russell, J. K. 1982. Timing of reproduction by coatis (*Nasua narica*) in relation to fluctuations in food resources, pp. 413–431. In E. G. Leigh Jr., A. S. Rand, and D. M. Windsor, eds., *Ecology of a Tropical Forest*. Smithsonian Institution Press, Washington, DC.

Ryan, M. J. 1980. Female mate choice in a Neotropical frog. *Science* 209: 523–525.

Ryan, M. J. 1985. *The Túngara Frog*. University of Chicago Press, Chicago.

Ryan, M. J., and A. S. Rand. 1990. The sensory basis of sexual selection for complex calls in the túngara frog, *Physalaemus pustulosus* (sexual selection for sensory exploitation). *Evolution* 44: 305–314.

Ryan, M. J., and A. S. Rand. 1993. Species recognition and sexual selection as a unitary problem in animal communication. *Evolution* 47: 647–657.

Sagers, C. L. 1993. Reproduction in Neotropical shrubs: the occurrence and some mechanisms of asexuality. *Ecology* 74: 615–618.

Sagers, C. L., and P. D. Coley. 1995. Benefits and costs of defense in a Neotropical shrub. *Ecology* 76: 1835–1843.

Schemske, D. W., and R. Lande. 1984. Fragrance collection and territorial display by male orchid bees. *Animal Behaviour* 32: 935–937.

Schupp, E. W. 1988. Seed and early seedling predation in the forest understory and in treefall gaps. *Oikos* 51: 71–78.

Schupp, E. W. 1990. Annual variation in seedfall, postdispersal predation, and recruitment of a neotropical tree. *Ecology* 71: 504–515.

Schupp, E. W. 1992. The Janzen-Connell model for tropical tree diversity: population implications and the importance of spatial scale. *American Naturalist* 140: 526–530.

Seeley, T. D. 1985. *Honeybee Ecology*. Princeton University Press, Princeton, NJ.

Sinclair, A. R. E. 1975. The resource limitations of trophic levels in tropical grassland ecosystems. *Journal of Animal Ecology* 44: 497–520.

Sinclair, A. R. E., and P. Arcese. 1995. Population consequences of predation-sensitive foraging: the Serengeti wildebeest. *Ecology* 76: 882–891.

Smythe, N. 1970. Relationships between fruiting seasons and seed dispersal methods in a Neotropical forest. *American Naturalist* 104: 25–35.

Smythe, N. 1978. The natural history of the Central American agouti (*Dasyprocta punctata*). *Smithsonian Contributions to Zoology* 257: 1–52.

Smythe, N. 1983. *Dasyprocta punctata* and *Agouti paca* (Guatusa, Cherenga, Agouti, Tepezcuintle, Paca), pp. 463–465. In D. H. Janzen, ed., *Costa Rican Natural History*. University of Chicago Press, Chicago.

Smythe, N. 1986. Competition and resource partitioning in the guild of Neotropical terrestrial frugivorous mammals. *Annual Reviews of Ecology and Systematics* 17: 169–188.

Smythe, N. 1989. Seed survival in the palm *Astrocaryum standleyanum*: evidence for dependence upon its seed dispersers. *Biotropica* 21: 50–56.

Smythe, N. 1991. Steps toward domesticating the Paca (*Agouti = Cuniculus paca*) and prospects for the future, pp. 202–216. In J. G. Robinson and K. H. Redford, eds., *Neotropical Wildlife Use and Conservation*. University of Chicago Press, Chicago.

Smythe, N., W. E. Glanz, and E. G. Leigh, Jr. 1982. Population regulation in some terrestrial frugivores, pp. 227–238. In E. G. Leigh, Jr., A. S. Rand, and D. M. Windsor, eds., *Ecology of a Tropical Forest*. Smithsonian Institution Press, Washington, DC.

Terborgh, J. 1986. Keystone plant resources in the tropical forest, pp. 330–344. In M. E. Soulé, ed., *Conservation Biology*. Sinauer Associates, Sunderland, MA.

Thorington, R. W. Jr. 1975. Tree mapping program, pp. 192–222. In D. M. Windsor, ed., *1974 Environmental Monitoring and Baseline Data Compiled under the Smithsonian Institution Environmental Monitoring Program, Tropical Studies*. Unpublished report, Smithsonian Institution, Washington, DC.

Thorington, R. W. Jr., B. Tannenbaum, A. Tarak and R. Rudran. 1982. Distribution of trees on Barro Colorado Island: a five-hectare sample, pp. 83–94. In E. G. Leigh, Jr., A. S. Rand, and D. M. Windsor, eds., *Ecology of a Tropical Forest*. Smithsonian Institution Press, Washington, DC.

Troyer, K. 1982. Transfer of fermentative microbes between generations in a herbivorous lizard. *Science* 216: 540–542.

Tuttle, M. D., and M. J. Ryan. 1981. Bat predation and the evolution of frog vocalizations in the Neotropics. *Science* 214: 677–678.

Walter, D. E., and D. J. O'Dowd. 1992. Leaves with domatia have more mites. *Ecology* 73: 1514–1518.

Werner, D. I. 1991. The rational use of green iguanas, pp. 181–201. In J. G. Robinson and K. H. Redford, eds., *Neotropical Wildlife Use and Conservation*. University of Chicago Press, Chicago.

Willis, E. O. 1967. The behavior of bicolored antbirds. *University of California Publications in Zoology* 79: 1–132.

Willis, E. O. 1972. *The Behavior of Spotted Antbirds*. Ornithological Monographs no. 10. American Ornithologists' Union, Washington, DC.

Wilson, E. O. 1980a. Caste and division of labor in leaf-cutter ants (Hymenoptera: Formicidae: *Atta*) I. The overall pattern in *A. sexdens*. *Behavioral Ecology and Sociobiology* 7: 143–156.

Wilson, E. O. 1980b. Caste and division of labor in leaf-cutter ants (Hymenoptera: Formicidae: *Atta*). II. The ergonomic optimization of leaf cutting. *Behavioral Ecology and Sociobiology* 7: 157–165.

Wilson, E. O. 1983. Caste and division of labor in leaf-cutter ants (Hymenoptera: Formicidae: *Atta*) IV. Colony ontogeny of *A. cephalotes*. *Behavioral Ecology and Sociobiology* 14: 55–60.

Windsor, D. M. 1990. Climate and moisture variability in a tropical forest: long-term records from Barro Colorado Island, Panamá. *Smithsonian Contributions to the Earth Sciences* 29: 1–145.

Windsor, D. M., D. W. Morrison, M. A. Estribi, and B. de Leon. 1989. Phenology of fruit and leaf production by 'strangler' figs on Barro Colorado Island, Panamá. *Experientia* 45: 647–653.

Wing, S. R., and M. R. Patterson. 1993. Effects of wave-induced light-flecks in the intertidal zone on photosynthesis in the macroalgae *Postelsia palmaeformis* and *Hedophyllum sessile* (Phaeophyceae). *Marine Biology* 116: 519–525.

Winter, K., and J. A. C. Smith. 1996. An introduction to crassulacean acid metabolism. Biochemical principles and ecological diversity, pp. 1–13. In K. Winter and J. A. C. Smith, eds., *Crassulacean Acid Metabolism: Biochemistry, Ecophysiology and Evolution*. Springer, Berlin.

Wirth, R., W. Beyschlag, R. J. Ryel, and B. Hölldobler. 1997. Annual foraging of the leaf-cutting ant *Atta colombica* in a semideciduous rainforest in Panama. *Journal of Tropical Ecology* 13: 741–757.

Worthington, A. H. 1990. Comportamiento de forrajeo de dos especies de saltarines en respuesta a la escasez de frutos, pp. 285–304. In E. G. Leigh, Jr., A. S. Rand, and D. M. Windsor, eds., *Ecología de un Bosque Tropical*. Smithsonian Tropical Research Institute, Balboa, Panamá.

Wright, S. J. 1983. The dispersion of eggs by a bruchid beetle among *Scheelea* palm seeds and the effect of distance to the parent palm. *Ecology* 64: 1016–1021.

Wright, S. J. 1990. Cumulative satiation of a seed predator over the fruiting season of its host. *Oikos* 58: 272–276.

Wright, S. J. 1991. Seasonal drought and the phenology of understory shrubs in a tropical moist forest. *Ecology* 72: 1643–1657.

Wright, S. J., J. L. Machado, S. S. Mulkey, and A. P. Smith. 1992. Drought acclimation among tropical forest shrubs (*Psychotria*, Rubiaceae). *Oecologia* 89: 457–463.

Zeh, D. W., and J. A. Zeh. 1992. Emergence of a giant fly triggers phoretic dispersal in the neotropical pseudoscorpion, *Semeiochernes armiger* (Balzan) (Pseudoscorpionida: Chernetidae). *Bulletin of the British Arachnological Society* 9: 43–46.

Zeh, J. A., and D. W. Zeh. 1994. Tropical liaisons on a beetle's back. *Natural History* 103(3): 36–43.

Zimmerman, J. K. 1990. Role of pseudobulbs in growth and flowering of *Catasetum viridiflavum* (Orchidaceae). *American Journal of Botany* 77: 533–542.

Zimmerman, J. K. 1991. Ecological correlates of labile sex expression in the orchid *Catasetum viridiflavum*. *Ecology* 72: 597–608.

Zimmerman, J. K., and T. M. Aide. 1989. Patterns of fruit production in a neotropical orchid: pollinator vs. resource limitation. *American Journal of Botany* 76: 67–73.

Zimmerman, J. K., D. W. Roubik, and J. D. Ackerman. 1989. Asynchronous phenologies of a neotropical orchid and its euglossine bee pollinator. *Ecology* 70: 1192–1195.

Zotz, G. 1995. How fast does an epiphyte grow? *Selbyana* 16: 150–154.

Zotz, G., G. Harris, M. Königer, and K. Winter. 1995. High rates of photosynthesis in the tropical pioneer tree, *Ficus insipida* Willd. *Flora* 190: 265–272.

Zotz, G., and M. T. Tyree. 1996. Water stress in the epiphytic orchid, *Dimerandra emarginata* (G. Meyer) Hoehne. *Oecologia* 107: 151–159.

Zotz, G., M. T. Tyree, and H. Cochard. 1994. Hydraulic architecture, water relations and vulnerability to cavitation of *Clusia uvitana*: a C_3-CAM tropical hemiepiphyte. *New Phytologist* 127: 287–295.

Zotz, G., and K. Winter. 1994a. Annual carbon balance and nitrogen-use efficiency in tropical C_3 and CAM epiphytes. *New Phytologist* 126: 481–492.

Zotz, G., and K. Winter. 1994b. Photosynthesis of a tropical canopy tree, *Ceiba pentandra*, in a lowland forest in Panama. *Tree Physiology* 14: 1291–1301.

Zug, G. R., and A. S. Rand. 1987. Estimation of age in nesting female *Iguana iguana*: testing skeletochronology in a tropical lizard. *Amphibia-Reptilia* 8: 237–250.

THREE

Tropical Climates

CLIMATE AND VEGETATION TYPE

The latitudinal line passing through New York City crosses a great variety of climates, each with its characteristic vegetation (Schimper 1903: p. 543). Near New York, summers are wet enough to support mesic deciduous forest which, mature and fully leafed, bears some striking structural resemblances to rainforest. West of the Mississippi, the native vegetation of the Great Plains is a spacious grassland subject to sporadic fires which exclude trees. Between the Rockies and the Cascades is a dry zone, a desert world of saltbush and sagebrush. Near the Pacific Coast, summers are dry, and winters are wet and not very harsh. Here the native vegetation is an evergreen, coniferous forest of giant trees, much larger, older, thicker, and more widely spaced than the trees of any tropical forest.

The lowlands traversed by the equator have no such variety of climate and vegetation. In nearly all the equatorial lowlands, east Africa alone excepted, the native vegetation is rainforest: broad-leaved evergreen forest. Lowland rainforests in Latin America, Africa, Madagascar, New Guinea, and Australia resemble each other much more closely than any of them resembles the coniferous rainforest of western Washington or a cove forest in the Great Smokies (Richards 1952, Leigh 1975). Why is the vegetation of the lowland tropics so much more uniform than that of temperate-zone lowlands?

A preliminary answer begins with the observation that the diversity of temperate-zone vegetation types arises from the diverse ways the seasonal rhythm of temperature can be related to the seasonal rhythm of rainfall: summer-dry forests are very different from their summer-wet counterparts. In the tropics, seasonal variation in temperature is nearly nil, and rainfall is the primary aspect of climate that varies seasonally (Schimper 1903: p. 211).

However, there is more to the similarity of lowland tropical climates. If one compares differences between annual rainfall and annual runoff in different equatorial catchments, selecting only those catchments "watertight" enough that water flows out only above ground, in the streams draining them, we find that this difference is similar in different equatorial forests (Bruijnzeel 1989). The annual difference between the amount of water entering such a watertight catchment as rain and the amount leaving it as runoff in the stream which drains it is the amount of water used by the forest that year. A forest's annual "water use," its annual evapotranspiration, is the water transpired plus the amount evaporated from wet surfaces. Annual evapotranspiration is a good predictor of a community's productivity (Rosenzweig 1968). Might the similarity of annual evapotranspiration in different lowland tropical settings explain the similarity in their vegetation?

RAINFALL

The very term "rainforest" reflects the importance of abundant water to this forest. Barro Colorado averages about 2600 mm of rain per year. Between 1925 and 1986, the wettest year was 1981, with 4133 mm, and the driest was 1976, with 1679 mm. More than 3130 mm of rain fell in 6 of these years, > 2900 mm in 15 years, < 2250 mm in 15 years, and < 2050 mm in 6 years (Windsor 1990: Table A4).

The Seasonal Rhythm

The alternation of dry and rainy seasons is crucial to the biology of Barro Colorado Island. The dry season, which is quite severe, usually begins some time in December and ends in April or early May. The median and average rainfall for the year's first 3 months are 96 and 125 mm, respectively. Between 1925 and 1986, the least rain for this period was 16 mm, in 1959, and the most, 480 mm, in 1981.

Five years had ≤ 26 mm, 15 years had ≤ 66 mm, 15 years had ≥ 165 mm, and 5 years had ≥ 267 mm rain during their first 3 months (Windsor 1990: Table A3).

This seasonal rhythm is thought to be driven in large part by changes in the sun's zenith (Windsor 1990). The sun heats most effectively when at its zenith, but heat continues increasing after the sun has passed its zenith. This heat builds up a loose band of low pressure and cloudiness, the Intertropical Convergence Zone, where vapor rises to condense as rain in convective storms. The risen air, cold and deprived of its moisture, diverges to either side of this zone, sinks, and forms drying trade winds bringing air to replace that which has risen (Lauer 1989). The Intertropical Convergence Zone follows the sun's zenith with a lag of about 5–10 weeks (see Fig. 16 in Hallé 1993). Its winter shift northward brings Barro Colorado its dry season. Its summer shift to the south of BCI sometimes brings the island, and more often brings Panama's north shore, spells of dry weather and trade winds in July (Cubit et al. 1988, Windsor et al. 1990).

Perturbations of the Rhythm

The severity of Barro Colorado's dry season differs from year to year, sometimes quite markedly. When overheating of equatorial waters in the eastern Pacific pushes a particularly strong El Niño current of warm water south along the coast of Peru, as happened in 1982–1983, the dry season in Panama is uncommonly long or severe (Windsor 1990). Strong El Niños reflect a worldwide climatic disturbance, involving the "Southern Oscillation," with low barometric pressure over Tahiti, high pressure over north Australia, severe drought in eastern Borneo, the Malay Peninsula, north Australia, southern Africa and Amazonia, high sea levels and violent storms on the west coast of North America, floods in coastal Peru, diminished Nile floods in Egypt, and weak summer monsoons in India (Rasmussen and Wallace 1983, Paine 1986, Glynn 1988, 1990, Diaz and Kiladis 1992, Quinn 1992).

An El Niño/Southern Oscillation event can be devastating. The 1982–1983 El Niño/Southern Oscillation so dried out the forest in Borneo that fires burned millions of hectares there (Leighton and Wirawan 1986). This same disturbance inflicted devastating mortality on the corals of Panama's Pacific coast (Glynn 1988). The correlation between the intensities of El Niño droughts in Panama and Southern Oscillation droughts in Borneo, however, is not perfect. Although the 1982 drought in Borneo and the early 1983 drought in Panama were considered once-a-century events (Leigh et al. 1990), an even worse drought struck Borneo in 1992 (Salafsky 1994), while Panama's 1992 El Niño drought, however severe while it lasted, was not unusually prolonged and was far less devastating than the 1983 drought. In Borneo, the severity (but not the frequency) of Southern Oscillation droughts has increased over the last 25 years or so (Salafsky 1994); this does not appear to be true in Panama. Although El Niños and Southern Oscillations are clearly related, they do not march in exact lockstep (Diaz and Pulwarty 1992, Cole et al. 1993).

Curiously, though the largest and/or most severe dry seasons on Barro Colorado occur in El Niño years, the primary effect on Central America of El Niños is to reduce rainfall from July through October. Ropelewski and Halpert (1987) found no significant effect of El Niños on dry season rainfall in Central America.

The animals of Barro Colorado suffer more from a very wet dry season, or from a dry season that grades indistinctly and indecisively into the following rainy season, than from prolonged or severe El Niño dry seasons (Foster 1982a, Leigh et al. 1990). Indistinct dry seasons, however, also seem to be connected with El Niños. On Barro Colorado, severe famines have followed the wet dry seasons of 1931, 1958, 1970, and 1993 (Foster 1982, Wright et al. 1996), which in turn followed the medium-severity El Niños of 1931 and 1969 and the strong ones of 1957 and 1992 (Quinn 1992, Wright et al. 1996).

Rainfall and Transpiration

There is a sharp rainfall gradient across the Isthmus of Panama from Cristobal and Colon on the Caribbean coast, which average 3300 mm/year, to Panama City and Balboa on the Pacific coast, which average 1800 mm/year (Table 3.1). Rainfall on Panama's Caribbean coast is mainly orographic, that is to say, most of it results from air being forced to rise when it strikes land. As a result, rainfall is more or less equally likely, and more or less equally intense, at any hour of the day (Table 3.2; also see p. 13 and fig. 11 of Windsor 1990). On Barro Colorado, however, rain falls much more often between 2 and 6 P.M., when convective thunderstorms occur. In Amazonia, convective thunderstorms "recycle" water transpired by local vegetation (Salati and Vose 1984, Lauer 1989). More generally, the Intertropical Convergence Zone is wider over land than over the oceans (Lauer 1989), as if, by transpiring, "tropical forest makes its own storms" (Corner 1964). Is this true for Panama?

Between 1925 and 1986, the regression of Barro Colorado's annual rainfall on the calendar year when it fell showed a decrease of 8 mm/year ($0.01 < p < 0.02$). Similar declining trends occurred at other inland stations, but not on the coasts. We all thought that deforestation was reducing convective rainfall in central Panama by reducing evapotranspiration or increasing the reflectivity of the land surface (Windsor 1990, Windsor et al. 1990). Between 1972 and 1994, however, the regression of Barro Colorado's rainfall on calendar year showed an (statistically insignificant) *increase* of 5.7 mm/year. The increase would have to continue for 30 more years to be statistically detectable. Meanwhile, we are left to wonder whether the effect of deforestation on rainfall is temporarily being obscured by

Table 3.1 Average annual rainfall in successive 10-year periods at various sites along the Panama Canal (data from Table A4 of Windsor 1990)

Period	Site				
	Cristobal	Monte Lirio	Barro Colorado	Gamboa	Balboa
1925–34	3361	3036	2720	2112	1799
1935–44	3627	3098	2854	2286	1833
1945–54	3248	2643	2586	1961	1715
1955–66*	3207	2711	2625	1993	1888
1967–76	3059	2522	2331	2133	1884
1977–86	3023	2442	2443	2070	1686

*Omitting 1960 and 1964, for which Monte Lirio lacks records.

other factors, or whether, in Panama at least, deforestation does not affect rainfall (perhaps because Panama's rain originates from vapor from the ocean).

Duration of Rainfall Events

Rainstorms on Barro Colorado (and elsewhere in the tropics) are said to be much briefer than temperate-zone rains (MacArthur 1972). In 1967, 77 out of 336 rains recorded at Trenton, New Jersey, lasted 10 hr or more, compared to 5 out of 456 rains recorded on Barro Colorado (MacArthur 1972: pp. 17–18) MacArthur considered the brevity of tropical rainstorms to be an essential aspect of the stability of tropical environments because in the tropics, rains are much less likely to prevent birds, bats, and insects from feeding for days at a time than in the temperate zone.

The spectrum of rainfall durations at Barro Colorado Island, Panama and Barro Branco, in the Ducke Reserve near Manaus, Brazil, are rather similar: if anything, rains lasting more than 3 hr are rarer at Barro Branco (Table 3.3). On Barro Colorado, a rainstorm is defined as a series of consecutive hours with recorded rainfall of 0.01 inch or more. This definition is a bit arbitrary: there are days where 0.01 inch is recorded every 2 or 3 hr due to a continual drizzle rather than a series of brief showers. On the other hand, a rainfall of 0.01 inch (0.254 mm) per hour, which could be constituted by 135 raindrops 1 mm in diameter, or 70 1.5-mm drops, or 30 2-mm drops, or 5 3-mm drops, or 2 4-mm drops, per square meter per second, is very close to what Barat (1957) considered the minimum detectable rainfall. Unfortunately, Franken and Leopoldo (1984) did not give their criteria for delineating rainstorms. Some of the differences recorded in Table 3.3 between Barro Colorado and Barro Branco, perhaps including the larger amount of rain falling at Barro Branco in storms of less than 1 hr, could be artifacts.

Rainfall Intensity

There are two aspects to rainfall intensity. First, how much rain falls in a given period of time? World record rainfalls, in millimeters, for different durations, D, in hours, can be expressed as $422D^{0.475}$ (Paulhus 1965): 422 mm in 1 hr, 1.9 m in 24 hr, or 3.8 m in 100 hr. For intervals less than 1 hr, the increase of maximum rainfall with duration is steeper—$17d^{0.784}$—where d is duration in minutes. The maximum recorded rainfall for 1 min is 17 mm, and for 5 min is 63 mm (Jennings 1950). Windsor (1990: Fig. 12) noted several instances where a tipping-bucket rain gauge recorded rainfalls between 25 and 91 mm in 1 min, but tipping-bucket rain gauges "chatter" when rainfall is intense, and Windsor's figures are almost unquestionably artifacts (R. F. Stallard, p. c.). World records are truly extraordinary events, as can be seen by comparing them with the maximum rainfalls for different time intervals at Barro Colorado and Ducke Reserve in central Amazonia (Table 3.4).

Table 3.2 Average amounts of rainfall and average hours of rain, per year, in successive two-hour periods of the day at Galeta Marine Laboratory (Caribbean Coast) and on Barro Colorado Island, 1980–84 (data from Table 4 of Windsor et al. 1990)

	Morning			Afternoon			Night					
	6–7	8–9	10–11	12–1	2–3	4–5	6–7	8–9	10–11	12–1	2–3	4–5
Rainfall, Galeta												
mm	248	274	194	258	263	297	193	162	194	207	279	191
hours	54	50	41	52	55	68	52	39	41	42	51	47
Rainfall, Barro Colorado												
mm	140	144	168	252	569	461	269	159	86	109	126	120
hours	18	21	30	50	75	60	36	22	14	16	19	19

Table 3.3 Numbers of rainstorms of different duration and size (total amount of rainfall in each category, mm, in parentheses) in different sample years at Barro Colorado, Panama, and Barro Branco, Brazil (data from Smythe 1975, 1976, Franken & Leopoldo 1984)

	Barro Colorado		Barro Branco
Duration	1974	1975	23 Sept 1976–22 Sept 1977
< 1 hr, 0.25 mm	161 (41)	131 (33)	
< 1 hr, > 0.3 mm	62 (156)	78 (197)	(tot <1 hr) 310 (685)
1–< 2 hr	72 (358)	119 (517)	51 (360)
2–< 3 hr	29 (223)	42 (357)	18 (195)
3–< 4 hr	30 (500)	22 (257)	(tot ≥3 hr) 37 (773)
4–< 6 hr	20 (341)	24 (524)	
6–<10 hr	10 (410)	8 (371)	
≥10 hr	0 (0)	2 (44)	
Rainstorm Size			
0.25 mm	161 (41)	131 (33)	
0.3–< 5 mm	132 (216)	190 (342)	(tot <5 mm) 328 (392)
5.0–<10 mm	31 (236)	42 (298)	49 (352)
10–< 20 mm	26 (384)	27 (378)	40 (578)
20–< 30 mm	20 (500)	15 (363)	10 (241)
30–< 40 mm	6 (207)	4 (136)	2 (68)
40–< 50 mm	6 (265)	7 (303)	3 (131)
≥50 mm	2 (177)	7 (438)	5 (309)

Second, what is the energetic impact of rainfall? This depends on the numbers, sizes, and speed of falling raindrops. In Madagascar, Barat (1957) measured the numbers of raindrops of different sizes falling per unit area per second by exposing a square of absorbent paper, whose dye changed color when wet, to the rain for a fixed, brief interval, a technique also employed by Mosley (1982) and Vis (1986). By letting drops of known diameter fall on his paper, he established a "normal curve" relating a raindrop's diameter to the diameter of the spot it leaves. He assumed that the spot's diameter was independent of the drop's speed.

Table 3.4 Maximum rainfall, in mm, during selected time intervals recorded in selected sample years, Panama and Brazil (Data from Panama Canal Commission and Marques Filho et al. 1981)

	5 min	10 min	15 min	60 min	2 hr	year total
Barro Colorado, Panama						
1957	13	23	32	74	80	2553
1958	18	29	38	67	94	2545
1959	15	25	37	94	161	2403
1960	19	29	42	72	73	3475
1961	18	29	42	72	80	2545
Max 1925–1971	33	42	—	104	—	3642
Ducke Reserve, Brazil						
1967	12	20	26	51	60	2174
1968	12	22	32	89	119	2904
1969	13	22	28	75	101	2341
1970	20	28	35	61	84	2330
1971	10	17	24	62	74	2744
Max 1967–1980	20	28	35	98	180	2904

He found that, for a steady, homogeneous rain, drop size is often fairly uniform: 1–2 mm in diameter. In a thunderstorm, some raindrops are much larger: twice in his career, he recorded storms with some 7-mm drops. Moreover, the spectrum of drop sizes usually changes radically during a thunderstorm. A normal thunderstorm releases 1300–2700 drops/m^2 sec, amounting to 0.6—1 mm of rain/min: the diameter of its raindrops usually range from 1 to 4 mm. The average intensity of the thousand thunderstorms measured by Barat (1957: p. 56) was 0.83 mm/min, or 2600 drops/m^2 · sec (Table 3.5). In the most intense storm Barat observed (Table 3.5), a squall at Madagascar's Montagne d'Ambre which dropped 318 mm of rain in 70 mins, rainfall intensity hovered around 4.8 mm/min, or 4200 drops/m^2 · sec, sustaining Barro Branco's maximum 5-min rainfall rate for over an hour.

Barat (1957) then calculated the erosive power of a rainstorm from the numbers and sizes of its raindrops. He related a raindrop's volume, terminal velocity, and kinetic energy to its diameter. Barat's velocity estimates came from a physicist who measured the terminal velocity of a falling raindrop by placing it in a wind tunnel and measuring the speed of the updraft required to keep the drop still. These terminal velocities accord reasonably well with those published elsewhere (cf. Reynolds et al. 1989).

What, precisely, is the energetic impact of a rainstorm, and how does this energy influence erosion? The impact power of the falling rain Barat calculated for his supersquall on the Montagne d'Ambre averaged 2.4 W/m^2. This is trivial compared to the 500 W/m^2 of full sunlight. Like good generals, however, raindrops do not distribute their energy evenly over the surface available: they focus it, as

Table 3.5 Volume (mm³), maximum and average flow rates* (drops/m²·sec), terminal velocity (m/sec), kinetic energy (10^{-9} kg·m²/sec² = 10^{-9} joules) and impact power** (W/mm²) of individual raindrops with different diameters (from Barat 1957)

Diameter	1.0	1.5	2.0	2.5	3.0	3.5	4.0	4.5	5.0	5.5	6.0	7.0
Volume	.5	1.7	4.2	8.1	14.0	22.3	33.3	47.4	65.0	86.3	112	178
Rate of fall												
Avg.		960	600	850	125	47						
Max.	860		600	1000		750	750			185		60
Velocity	4.4	5.2	5.9	6.5	6.9	7.3	7.7	7.9	8.0	8.1	8.2	8.4
Velocity†	3.8		6.5		7.8		8.5					
Energy	5.0	23	72	177	334	591	986	1478	2080	2831	3776	6293
Impact power**	0.028	0.040	0.068	0.094	0.109	0.128	0.151	0.163	0.169	0.175	0.183	0.196

*Maximum "observed" fall is flow rate, drops/m² per second, for given drop size in the most intense rain observed by Barat (squall of 27 February 1939 on the Montagne d'Ambre, Madagascar).
**Impact power of a drop is the energy of a drop, divided by its cross-sectional area, and the time elapsed from the beginning to the end of its impact on the ground, which is assumed equal to its diameter divided by its terminal velocity.
†From Reynolds et al. (1989).

if to overwhelm the enemy at one specific point by concentrating their force there. The energy of a 4-mm raindrop, concentrated on the 12.6 mm² of soil under its silhouette during the 4/7700 sec of its impact, supplies an average of 150,000 W/m² to this impact area during the brief duration of its impact. When a 3-m (10 ft.) wave crashes upon a rocky shelf or bench, the speed at impact of its fastest moving water, the water of its crest, is 8 m/sec (see Denny 1985: Eq. 7). Who could stand against such a wave? Yet, when a 4-mm raindrop hits the earth, it is moving at this same speed of 8 m/sec (Table 3.5): for the brief instant of its impact, then, the soil struck by this raindrop feels all the force of a breaking wave. If the drop strikes a "well-integrated" material like wood, stone, or even human skin, the bonds of this material share the impact over a wide enough area that the drop does no damage. If this drop strikes fresh soil, whose particles are not closely linked to each other, it exerts a readily detectable effect.

Barat (1957), like many others (see Franken and Leopoldo 1984: p. 508; Sioli 1984), emphasized the importance of forest cover in protecting the soil from erosion. Barat argued that forest cover breaks larger drops into smaller ones before they hit the soil. He had observed empirically that drops 0.001–0.1 mm in diameter, droplets suspended as aerosol, fall very slowly and do not change shape when they strike the surface but rest on it until they evaporate. Larger drops, 0.1–2.5 mm in diameter, spread without bursting when they strike the soil and are immediately absorbed. Drops ≥ 3 mm in diameter burst upon impact, leaving a "crater" in the soil. Moreover, some of their water may flow overland, provoking erosion. Barat (1957: pp. 65, 66) offers a sample comparison of the spectrum of drop sizes falling simultaneously in the open and on the forest floor, showing that the forest cuts the rain's impact power by 40%, from 0.37 to 0.21 W/m² (Table 3.6). In the open, more than half the total impact power was supplied by drops ≥ 4.5 mm in diameter, while in the forest understory there were no drops > 3.5 mm in diameter, and drops > 2.5 mm in diameter, drops big enough to crater soft, bare soil, were responsible for only one-eighth of the rain's total power.

The impact energy of "throughfall" striking the forest floor is now believed to be higher than that of rainfall striking open ground, because the foliage causes rain to coalesce into larger drops (Bruijnzeel 1990). In a typical rainstorm in New Zealand, where 51 mm of rain fell in 36 hr, kinetic energy of throughfall averaged 1.5 times that of rainfall. The kinetic energy of rainfall striking open ground increased more rapidly with rainfall intensity than that of throughfall for rainfall of 0–8 mm/hr (Mosley 1982); extrapolated, the regressions actually cross at 40 mm/hr, 0.7 mm/min, the intensity of a normal tropical thunderstorm. In montane forest of Colombia, the kinetic energy of throughfall is said to be 4–30% higher than rainfall on open ground from the same storm (Vis 1986: Table IV). Vis sampled throughfall thoroughly, being well aware of its heterogeneity (as Barat may not have been). Vis (1986: Fig. 5, Table IV) thought, however, that the distribution of a rainfall's

Table 3.6 Comparison of numbers (N), volume (V, mm²), and total energy (E, joules) of raindrops of different diameter (D, mm) falling per square meter per second on open ground and on the forest floor (data from Barat 1957: pp. 65, 66)

	Clearing			Forest Floor		
D	N	V	E	N	V	E
5.0	44	2860	0.100	0	0	0
4.5	99	4738	0.150	0	0	0
3.5	99	2229	0.060	9	222	0.006
3.0	55	772	0.021	61	842	0.020
2.5	121	975	0.020	600	4872	0.106
2.0	0	0	0	900	4492	0.065
1.5	600	1026	0.015	733	1253	0.015
1.0	0	0	0	1000	500	0.005
Totals	1298	12600	0.366	3303	12181	0.216

*Barat gives this value as 0.0106, which must be an error: why would five times as many drops have only half as much power as in the open?

drop size was independent both of its intensity and the altitude where it occurred, and that all rain consisted nearly entirely of drops between 0.7 and 3.2 mm in diameter, so that 1 mm of rain delivers 19 J/m^2 ground (a reasonable number: 1 mm of rain in Barat's "average" storm delivers 21 J/m^2 ground). Rainfall energy, however, does vary with time and place (Barat 1957). The question of how forest cover affects the energy of throughfall striking the forest floor is still unsettled, although Bruijnzeel (1990) must be right to conclude that fallen leaves covering the forest floor play a central role in breaking the force of rainfall or throughfall.

Tropical forest does seem to protect its soil. The Amazon catchment's 5.8 × 10^6 km^2 exports an annual average of 9 × 10^8 tons of sediment (Sioli 1984), roughly 1.1 × 10^9 m^3 of soil, equivalent to a layer of soil 0.2 mm thick over the whole catchment. Presumably, much more sediment must be transported from one place to another within the catchment. A 9.7-ha catchment on Barro Colorado Island exports an annual average of 4.5 tons of sediment/ha (R. F. Stallard. p. c.), equivalent to a layer of soil 0.56 mm deep (Dietrich et al. 1982: p. 44), or, assuming that 65% of the soil's volume is air or water (Dietrich et al. 1982: p. 36), a layer of bedrock 0.19 mm deep. The largely deforested Gatun Lake catchment contributes four times as much sediment per hectare to Gatun and Madden Lakes as this 9.7-ha catchment does to Lutz creek, occasioning warnings about the dangers deforestation poses to the Panama Canal (Isaza 1986). The 1.3 × 10^6 km^2 of China's Huang Ho catchment exports more sediment than does the Amazon (Sioli 1984). Rainfall within this catchment averages less than 500 mm/year, compared to the Amazon catchment's 2300 mm (Salati and Marques 1984), suggesting that the Amazon catchment is (5.8/1.3)(23/5) ≈ 20 times more effective than the Huang Ho's at protecting its soil. In the next chapter I say much more about the role of rainforest in protecting its soil and nutrients.

Where Does the Rainwater Go?

What is the fate of a forest's rain? The best of the earlier studies, such as those by Manokaran (1979), Leopoldo et al. (1982), and Franken et al. (1982a), suggest that only 75–80% of falling rain reaches the forest floor (Table 3.7). Lloyd and Marques Filho (1988), using 36 127-mm diameter rain gauges rotated weekly at random over 505 sites in a 4 × 100 m plot in the understory of the Ducke Reserve, found that 90.9 ± 2.2% (mean ± SEM) of the incident rainfall dropped to the forest floor. The distribution of this throughfall over the forest floor is heterogeneous and quite skewed, with 1.6% of the sites registering throughfall over twice the rainfall outside the forest. If there are not sufficient collectors to sample this heterogeneity, throughfall will be underestimated. Franken et al. (1982a) sampled throughfall at Barro Branco with 20 gauges that were not moved about, and found that throughfall was 80 ± 12% of total rainfall (Lloyd and Marques Filho 1988).

The forest intercepts a higher proportion of rainfall from storms of less than 10 mm; indeed, Jordan and Heuveldop (1981) assert that rainfalls of less than 5 mm are entirely intercepted. Otherwise, there seems to be little relationship between a storm's size and the proportion of its rain which is intercepted (Table 3.8).

The earlier studies of Manokaran (1979) and Franken et al. (1982b) suggested that stemflow, rainwater oozing as a film over the stems of trees and saplings, accounts for less than 1% of total rainfall. Jordan and Heuveldop (1981) found that 8% of the rainfall reached the ground as stemflow in the forest at San Carlos de Rio Negro, of which 89% was moving down stems < 10 cm in diameter at breast height. Manokaran did not measure stemflow on trees < 10 cm in diameter, and Franken et al. (1982b) measured stemflow on trees of diameters ranging from 68 to 516 mm. Lloyd and Marques Filho (1988) measured stemflow on 18 trees ranging from 3.8 to 52 cm in diameter, and found it to be 3.3% of total rainfall: 56% of this stemflow was accounted for by a single palm. Accurate estimates of forest-wide stemflow do not yet exist, and I will not try to estimate how much rainfall reaches the ground as stemflow.

Nonetheless, stemflow is a fascinating, and sometimes important, topic. In Australian rainforest, trees with steeply ascending branches serve as stemflow guides that can concentrate water flow to the trunk's base a hundredfold, if one "credits" the stemflow to an area equal to the tree-

Table 3.7 Total rainfall, throughfall, stemflow, and water intercepted by vegetation (mm) in selected forests

Site and Time	Total Rain	Throughflow	Stemflow	Interception	Sampling regime
Bacia Modelo, April 1980–May 1981	1705	1337	6	362	30 100-cm^2 gauges, not moved about
Ducke Reserve, Sept. 1976–Sept. 1977	2076	1688		388	20 100-cm^2 gauges, not moved about
Ducke Reserve, Nov. 1976–Dec. 1977	2570	2062		509	20 100-cm^2 gauges, not moved about
Ducke Reserve, Jul. 1984–Aug. 1985	2721	2480	90	151	36 127-cm^2 gauges, moved weekly, 505 sites
San Carlos de Rio Negro, annual	3664	3188	292*	184	4 8750-cm^2 gauges, 20 others, not moved
Pasoh, Malaysia, 1973	2381	1847	15	519	13 640-cm^2 gauges in 100 m^2, not moved

*Of this stemflow, 32 mm contributed by trees ≥ 10 cm in diameter at breast height.

Data for Bacia Modelo from Franken et al. (1982b). Data for Ducke Reserve, Sept. 1976–Sept. 1977 from Leopoldo et al. 1982b; Nov. 1976–Dec. 1977 from Franken et al. (1982a); and Jul. 1984–Aug. 1985 from Lloyd and Marques Filho (1988). Data for San Carlos de Rio Negro from Jordan and Heuveldop (1981), and data for Pasoh from Manokaran (1979).

Table 3.8 Percentage of rainfall intercepted from storms of different sizes (data from Franken et al. 1982a,b)

Size of Storm (mm)	Bacia Modelo		Ducke Reserve	
	N	% Intercepted	N	% Intercepted
< 10	2	34.9	4	33.9
10–20	8	27.4	7	22.1
20–30	3	27.6	9	18.7
30–40	4	21.6	4	8.5
40–50	7	24.5	2	18.1
50–60	5	22.5	3	15.2
60–70	6	25.2	7	16.8
> 70	3	12.6	13	18.0

trunk's cross-section (Herwitz 1986). This concentration can cause overland flow. Here, buttresses play an important role in dividing the flow of water down the trunk, so it hits the ground as several streams, which infiltrate the soil more easily than a single, undivided flow (Herwitz 1988). In montane rainforest of Australia, one tree species has adventitious roots on its stems and branches, solely to scavenge nutrients from stemflow (Herwitz 1991).

Rainfall Minus Runoff: Evapotranspiration?

How much of a forest's rain leaves the catchment as runoff and flows out to sea? Between 1976 and 1994, the difference between annual rainfall and annual runoff on a wholly forested 9.7-ha catchment near Barro Colorado Island's laboratory clearing varied from 1335 mm in 1978–1979 to 1915 mm in 1983–1984. The coefficient of variation of rainfall less runoff was less than half that in rainfall, which in turn was less than half that in runoff (Table 3.9). The correlation of rainfall with rainfall less runoff is 0.5: annual rainfall accounts for 25% of the variation in rainfall minus runoff over these 18 years. Barro Colorado may be a leaky catchment: in a watertight catchment at Hubbard Brook, New Hampshire, correlation between rainfall and rainfall less runoff is only 0.24 (Table 3.9). For lowland rainforest catchments the world around, rainfall less runoff is nearly independent of annual rainfall (Table 3.10).

If rainwater only leaves a catchment by being evaporated or transpired as water vapor into the atmosphere, or by flowing over the recording weir—that is to say, if water cannot flow underground from (or to) the catchment—then, over a period of time so long that changes in the amount of water stored in the soil are trivial compared to rainfall and runoff, the difference between rainfall and runoff is the forest's "evapotranspiration," the rainwater evaporated from a forest's leaves before it reaches the ground, plus the water absorbed from the ground and transpired through the forest's stomates. Because runoff is what is left over after evapotranspiration has taken its nearly constant "draft," runoff varies much more from year to year than does total rainfall or evapotranspiration (Table 3.9).

In fact, some of the water falling into small catchments with sandy soils, such as Bacia Modelo and Barro Branco

Table 3.9 Rainfall, P; runoff, R; and rainfall minus runoff, $P - R$ (mm) in successive water years

Year	P	R	$P - R$	Year	P	R	$P - R$	Year	P	R	$P - R$
A. Lutz Catchment, Barro Colorado Island*											
1976	1890	282	1608	1982	1824	360	1464	1988	2665	1132	1533
1977	2698	1224	1474	1983	2849	906	1943	1989	2139	567	1572
1978	2093	734	1359	1984	2646	1124	1522	1990	2899	1214	1685
1979	2901	1119	1782	1985	2273	729	1544	1991	2471	899	1572
1980	2620	877	1743	1986	2232	562	1670	1992	3204	1420	1784
1981	4165	2354	1811	1987	2927	1168	1759	1993	2666	1186	1480
Mean	2620	992	1628	S.D.	543	466	152	C.V.	0.207	0.470	0.093
B. Forested Catchment, Hubbard Brook, New Hampshire†											
1956	1220	700	520	1962	1240	800	440	1968	1280	860	420
1957	1440	950	490	1963	1160	680	480	1969	1300	840	460
1958	1030	560	470	1964	950	500	450	1970	1260	770	490
1959	1620	1100	520	1965	1240	730	510	1971	1230	720	510
1960	1130	590	540	1966	1320	800	520	1972	1510	980	530
1961	1050	540	510	1967	1420	910	510	1973	1850	1360	490
Mean	1295	801	494	S.D.	216	212	34	C.V.	0.167	0.265	0.069

At the bottom of each table are given means for the 18 years of data, their standard deviations, S. D., and their coefficients of variation, C. V., which is S. D. divided by the mean.
*Data from Windsor (1990: Table G2) and unpublished records of the Environmental Sciences Program. Windsor's figures for runoff have been divided by 0.96 in accord with the revised estimate of the area of Lutz catchment. Water year x extends from April 1 of year x to March 31 of the following year.
  At Lutz, the regression of evapotranspiration, $P - R$, on rainfall, P, is $1181 + 0.1705P$, $r^2 = 0.37$; the regression of runoff, R, on rainfall, P, is $R = 0.8295P - 1181$, $r^2 = 0.93$.
†Data from Likens et al. (1977: Fig. 3). Water year x extends from 1 June of year x through 31 May of following year.
  At Hubbard Brook, the regression of evapotranspiration, $P - R$, on rainfall, P, is $P - R = 0.038P + 445$. $r^2 = 0.06$; the regression of runoff, R, on rainfall, P, is $R = 0.96P - 445$, $r^2 = 0.99$.

in Amazonia or Tonka in Surinam, flows out underground. At 1000 m in montane forest near Perinet, Madagascar, the difference between rainfall and runoff is lower for a 100-ha catchment than for the subbasins within it; water that leaves the small basins underground flows out from the larger basin over the weir (Bailly et al. 1974: pp. 30–33). Similarly, rainfall less runoff is greater in small catchments of Amazonia or the Guayana Shield than for the Amazonian basin as a whole, or for a large blackwater catchment in Venezuela (Table 3.10; see Bruijnzeel 1991). Nevertheless, measuring evapotranspiration as rainfall minus runoff is vastly more accurate than estimating evapotranspiration by subtracting the amount of water moving downward through the soil (as estimated by lysimeters or the like) from rainfall (Russell and Ewel 1985, Bruijnzeel 1991). The most (perhaps the only) plausible estimate of evapotranspiration by the rate of depletion of water from the soil was that of Nepstad et al. (1994) in evergreen forest at Paragominas, south of Belem in eastern Amazonia (3° S latitude). There, 95 mm of rain fell during the 5.5-month dry season beginning June 1992. During this period, the loss of water from the top 2 m of soil was 130 mm, and the net loss of water from soil between 2 and 8 m deep was 380 mm, suggesting an evapotranspiration rate of 605 mm/5.5 months, or 1320 mm/year. Thanks to their deep roots, this forest transpired at a normal rate during the El Niño dry season of 1992. Tracking the depletion of water deep in the soil was essential to an accurate estimate of evapotranspiration (Nepstad et al. 1994).

FACTORS AFFECTING EVAPOTRANSPIRATION

Table 3.10 suggests that annual evapotranspiration is rather similar in different lowland rainforests but lower in the temperate zone. What shapes this similarity among tropical forests? To begin with, rainforests, by definition, receive as much water as they need, at least for most of the year. How long a forest's dry season can be without imperiling its status as a rainforest is still an open question, but differences in rainfall should not account for much variance among forests in evapotranspiration.

Given that there is enough water, what factors influence a forest's evapotranspiration? Penman (1948) derived a

Table 3.10 Rainfall, P; runoff, R; evapotranspiration, $P - R$; and transpiration, T (evapotranspiration less intercepted rainfall) in catchments of different area (A, km²)

	P	R	$P-R$	T	A
Tropical catchments					
Amazon Basin (average)	2300	950	1350		5.8×10^6
Bacia Modelo, Amazonia, 1980–81	2089	541	1548	1014	23.5
Barro Branco, Amazonia, 1976–77	2075	400	1675	1287	1.3
Barro Branco, 1982	2372	909	1463		1.3
Barro Branco, 1983	1906	482	1424		1.3
Grégoire I, Fr. Guiana, 8 years	3676	2148	1528		
Grégoire II, Fr. Guiana, 8 years	3697	2260	1437		
Grégoire III, Fr. Guiana, 8 years	3751	2307	1444		
Tonka, Surinam, 1979–84	2145	515	1630		1.4
Rio Caura, Venezuela, 1982–84	3850	2425	1425		47500.00
Barro Colorado, Panamá, 1974–75	2570	904	1666		0.10
Guma, Sierra Leone	5795	4578	1217		8.7
Congo Basin, Average	1510	337	1173		3.8×10^6
Sungei Tekam, Malaysia, 6 years	1727	229	1498		
Sungei Lui I, Malaysia, 3 years	2410	894	1516		
Ulu Langat, Malaysia, 1 year	2482	1215	1267		
Ei Creek, Papua, 1972–73	2700	1480	1220		16.3
Lowland Queensland, Australia	4037	2624	1413		0.26
Temperate-zone Catchments					
NW of Baltimore, MD, 3 years	997	159	807		0.38
Hubbard Brook, NH, 18 years	1295	801	494		Not given
Coweeta, NC, Watershed 18, 30 years	1813	955	858		0.12

Rainfall for Amazon Basin from Salati and Marques (1984); runoff from Sioli (1984) and Leopoldo et al. (1987). Bacia Modelo data from Leopoldo et al. (1982a), Barro Branco 1976–77 data from Franken and Leopoldo (1984); Barro Branco 1982, 1983, data from Franken and Leopoldo (1987); BCI data from Windsor (1990); Congo data from Clark and Cooper (1987); data for Queensland and Sierra Leone quoted from Bruijnzeel (1982) by Walter and Breckle (1986); data for R. Caura, Tonka, Ei Creek, and Bukit Berembun are from Bruijnzeel (1991); data for other sites from Bruijnzeel (1989).

Data for NW of Baltimore from Cleaves et al. (1970); for Hubbard Brook from Likens et al. (1977); for Coweeta from Johnson and Swank (1973).

formula that relates evapotranspiration to air temperature, relative humidity, incident radiation, and wind speed above the canopy. More precisely, he calculated "potential evapotranspiration" from a vegetated surface, the evapotranspiration which would occur were sufficient water available. We shall now outline the derivation of this widely used formula to identify the climatic variates most relevant to transpiration.

Penman (1948) started from four basic assumptions:

1. The flow of heat from the foliage to the air above, in joules per square centimeter of canopy per second, is proportional to the difference between the concentration of heat (J/cm³) in air close enough to the leaves to be at the leaf temperature, T_1, and heat concentration in open air of temperature, T_a, above the "boundary layer" of air warmed and humidified by proximity to the canopy. This difference is the heat required to warm 1 cm³ of air from T_a to T_1, which is the heat required to warm 1 g of air by 1°C at constant pressure, c_p (the specific heat of air at constant pressure, J/g · °C), times the density ρ' of air (cm³/g), times $T_1 - T_a$ (this air must be warmed to T_1, not $T_a + 1$). Thus the heat flow, H, from the canopy (J/cm² · sec), is

$$H = K_H \rho' c_p (T_1 - T_a) \quad (3.1)$$

(Marshall and Holmes 1988: p. 319), where the atmospheric heat conductance, K_H, has units (J/cm² · sec)/(J/cm³), or cm/sec. The canopy cools faster when the wind blows: K_H increases with wind speed, as will be described later.

2. Evapotranspiration, the flow, E, of water vapor to the atmosphere per unit area of canopy (g/cm² · sec), is proportional to the difference between the vapor concentration (absolute humidity, g water/cm³ air) in canopy leaves, which is assumed to be the absolute humidity at saturation for the temperature T_1 of canopy leaves, $\chi_s(T_1)$ (air in the leaf is presumed to be in equilibrium with the free water drawn up from the xylem) and the absolute humidity, χ_a, above the canopy's boundary layer. The absolute humidity, χ, is $\rho'q$, where ρ' is the density of air and q is the specific humidity, g vapor/g air (Monteith and Unsworth 1990: p. 12). Thus the evapotranspiration, E, is

$$E = K_H [\chi_s(T_1) - \chi_a] = K_H \rho'[q_s(T_1) - q_a]. \quad (3.2)$$

Here, $q_s(T_1)$ is specific humidity at saturation at the temperature T_1. The conductance, K_H, is assumed to be the same as for heat (Marshall and Holmes 1988: p. 319). In fact, water vapor faces an additional resistance, that from the leaves' stomates, which Penman ignored (Shuttleworth 1988: p. 328).

3. Evaporating a gram of water from a surface at temperature T requires $\lambda = 2500.25 - 2.365T$ joules of heat; λ is the latent heat of vaporization.

4. Net radiation, R_n, incident upon the forest canopy is precisely balanced by the flux of heat and evapotranspiration of water from the canopy foliage: none of this radiation heats the ground or the understory air. Thus

$$R_n = \lambda E + H = \lambda E + K_H \rho' c_p (T_1 - T_a). \quad (3.3)$$

To express E in terms of measurable aspects of climate, T_1 must be eliminated. This takes two steps. First, set

$$q_s(T_1) = q_s(T_a) + \Delta' (T_1 - T_a), \Delta' = \frac{dq_s}{dT}\bigg|_{T=T_a} \quad (3.4)$$

$$E = K_H \rho'[\Delta' (T_1 - T_a) + q_s(T_a) - q_a]. \quad (3.5)$$

Next, use equation 3.3 to replace $T_1 - T_a$ in equation 3.5 by $(R_n - \lambda E)/K_H \rho' c_p$, and solve for E. We obtain

$$E + \lambda E \frac{\Delta'}{c_p} = \frac{\Delta' R_n}{c_p} + K_H \rho' [q_s(T_a) - q_a]. \quad (3.6)$$

Now let $\gamma' = c_p/\lambda$. Like $\Delta' = dq/dT$, γ' has units 1/°C. At 25.5°C, $\gamma' = 0.000414$. Then evapotranspiration, E, is

$$E = \frac{\Delta'(R_n/\lambda) + \gamma' K_H \rho'[q_s(T_a) - q_a]}{\Delta' + \gamma'}. \quad (3.7)$$

If we let $K_H \rho' [q_s(T_a) - q_a] = E_a = D_a f(u)$, where E_a represents evaporation by wind, u is windspeed, and D_a is a measure of vapor pressure deficit (deficit relative to saturation), we obtain the familiar Penman equation (Dunne and Leopold 1978: p. 114, eq. 4.22, Ribeiro and Villa Nova 1979, Dietrich et al. 1982, Marshall and Holmes 1988: p. 323, eq. 12.23, Monteith and Unsworth 1990: p. 185, eq. 11.26). The units, however, are different, and quantities like K_H and R_n must be specified more precisely if E is to be expressed in terms of climatic variates.

If net radiation, R_n, is expressed as W/m² = J/m² · sec, then latent heat of vaporization must be expressed as J/kg, K_H as m/sec, and ρ' as kg/m³ to give E in kg/m² · sec. Only theoreticians, however, talk about grams of vapor per gram of air. Vapor content is usually measured as a pressure, in pascals (newtons/m²: 101,325 Pa = 1 atmosphere), millibars (1 mb = 10 Pa), or mm Hg (760 mm Hg = 1 atmosphere). Monteith and Unsworth (1990) show quite elegantly how to convert from specific humidity to vapor pressure, as well as how to calculate saturation vapor pressure $e_s(T)$ as a function of T. It is remarkably pleasant to see utterly abstract principles of thermodynamics contribute to a concrete understanding of evapotranspiration. Alas, there is only space to repeat their conclusions. Specific humidity, q, is related to vapor pressure, e, by the equation $q = \varepsilon e/p$, where p is atmospheric pressure, e is measured in the same units as p, and ε is the molecular weight of water (18) divided by the molecular weight of air (29), or 0.622 (Monteith and Unsworth 1990: p. 12). Setting $e = pq/\varepsilon$ in equation 3.7 requires replacing $\gamma' = c_p/\lambda$ by the "psychometric constant" $c_p p/\lambda\varepsilon$ (Monteith and Unsworth 1990), which can be expressed as $0.00066p(1 + 0.00115T_a)$ mb/°C, where p is atmospheric pressure in millibars, and T_a is air temperature.

Finally, how may we express R_n and K_H? Net radiation, R_n, can be expressed as

$$R_n = Q_s(1 - \alpha) - Q_{1w},\qquad(3.8)$$

(Dietrich et al., 1982), where Q_s is the incident solar radiation, α is the albedo (the proportion of this radiation reflected back to space, which lies between 0.12 and 0.15 for tropical forest) (Salati and Marques 1984, Shuttleworth 1988), and Q_{1w} is the net flux of long-wave radiation from the vegetation back to space. Dunne and Leopold (1978: p. 109) express this long-wave radiation as

$$Q_{1w} = \sigma\,(T_a + 273)^4(0.56 - 0.08\sqrt{e_a})\,(1 - aC).\qquad(3.9)$$

Here σ, the Stefan-Boltzmann constant, is 5.67×10^{-8} W/m^2(°K)^4, C is the average proportion of the sky covered by cloud, a is a "cloud-type constant" varying with the height and thickness of the clouds, here set $= 0.6$, and vapor pressure is e_a, in millibars. Finally, Dietrich et al. (1982) set $E_a = (0.013+0.00016V)(e_s(T_a)-e_a)$ cm/day $\approx (0.013 + 0.00016V)(e_s - e_a)/86400$ g/cm$^2 \cdot$ sec $= (0.013 + 0.00016V)(e_s - e_a)/8640$ kg/m$^2 \cdot$ sec, where V is wind run, in kilometers per day.

Daily or monthly averages of solar radiation, temperature, wind speed, cloudiness, and relative humidity are normally used in formulae such as equation 3.7. As we shall see in chapter 7, most canopy trees on Barro Colorado have year-round access to water, so potential and actual evapotranspiration are nearly the same. On BCI, Dietrich et al. (1982: Table 1) calculated potential evapotranspiration of 1409, 1527, and 1508 mm for the water years 1977, 1978, and 1979: rainfall minus runoff for these water years were 1433, 1335, and 1746 mm, respectively. The agreement is crude, yet remarkable, given the multiplicity of assumptions. The data required to calculate evapotranspiration from monthly averages are tabulated for a variety of stations, tropical and temperate, by Müller (1982).

Penman's derivation, however, is based on principles exact enough to provide a reasonable basis for hourly estimates of evapotranspiration from each of five layers in an Amazonian rainforest (Roberts et al. 1993). Does the similarity of annual evapotranspiration in lowland tropical rainforests reflect similarity in their average temperature, radiation, cloudiness, humidity, and wind speed?

Humidity

On BCI, relative humidity varies from nearly saturated at night and in the early morning to as low as 60% or 50% at 1 P.M. on a particularly sunny day. The air is drier in the dry season, and drier in the open and above the forest canopy than at the forest floor (Tables 3.11, 3.12). Humidity in the clearing and atop a 40-m tower jutting a few meters above the forest canopy match less closely than do temperatures at these two sites (Tables 3.11–3.13). As far as one can judge, the behavior of relative humidity on Barro Colorado is similar to its behavior at other tropical sites (Lauer 1989), but measurements of relative humidity are difficult to automate reliably. Data on the subject are less than complete and not strictly comparable.

Wind

Air movement is important for several reasons. Exchange between air inside or just above the forest and air higher up enhances the availability to leaves of CO_2. In tree fall gaps on Barro Colorado, wind speed is often lower than 25 cm/sec and rarely exceeds 50 cm/sec. Thus the boundary layer of air saturated with vapor and depleted of CO_2 that surrounds the foliage is so thick that opening stomates 10% wider than average increases transpiration only by 2.5% (Meinzer et al. 1995). In the northeastern Pacific, the kelps of still water have ruffles and other features to increase turbulence around their fronds, enhancing nutrient uptake by mixing richer water from farther away into the boundary layer of water their fronds have depleted of nutrients and CO_2 (Koehl and Alberte 1988). Are there ways trees can construct or arrange their leaves so as to enhance air movement (by, for example, convective heating of the surrounding air), or to take better advantage of existing air movement? Brunig (1970) considered how "aerodynamic roughness" of the canopy might facilitate gas exchange. Most discussions of this topic, however, have focused on how to minimize disadvantageous effects of air movement (see Leigh 1975).

In tropical forests, convective thunderstorms spawn localized violent winds that fell individual trees and sometimes destroy a hectare or more of forest (Whitmore 1984). Tree growth is paced in large part by tree death: the frequency of windblown tree falls may help pace a forest's growth. Thunderstorms vary in frequency from place to place (Whitmore 1984: p. 81), although the causes of differences in thunderstorm frequency and the role played therein, if any, by the vegetation, are not yet known. On Barro Colorado, most trees die by falling over: wind must trigger most of these deaths (Putz et al. 1983). In Pasoh, Malaysia, nearly half the trees die standing (Putz and Appanah 1987). Differences in the ways most trees die influences a forest's spectrum of sizes of treefall gap and the proportion of its trees belonging to pioneer species (Putz and Appanah 1987).

Too much air movement can be devastating, as hurricanes show. We are only just beginning to understand the causes of hurricanes: curiously, hurricanes seem to be among the better illustrations of the idealized Carnot "heat engines" used to illustrate the principles of thermodynamics in freshman physics (Emanuel 1988). Hurricanes presuppose a considerable depth (60 m or more) of water over 26°C on the sea surface, and, often, a terrestrial source of storm "waves" or "clusters." When storms bring heavy rain to the Sahel, that unstable borderland between the Sahara and the wetter tropics to the south, strong hurricanes strike the eastern United States far more frequently (Gray 1990). Nowadays, hurricanes rarely strike Panama or Malaysia,

Table 3.11 Hourly values of relative humidity (percent), March and October 1986, in the laboratory clearing, Barro Colorado Island (data from Windsor 1990)

	1:00	2:00	3:00	4:00	5:00	6:00	7:00	8:00	9:00	10:00	11:00	12:00
Mar. 1986, A.M.	93	93	94	94	94	94	94	96	95	88	77	72
Oct. 1986, A.M.	94	95	95	95	95	94	95	95	94	89	85	82
Mar. 1986, P.M.	69	70	71	73	76	80	85	88	90	91	92	92
Oct. 1986, P.M.	80	80	82	84	87	90	92	93	94	94	94	94

and they seem to be unheard of in Amazonia—how global warming might change this circumstance, we do not know. Hurricanes, cyclones and typhoons are now the bane of forests between 10° and 20° N and between 10° and 20° S (Whitmore 1984: p. 81). Such storms greatly influence the fates of forest in the Caribbean (Tanner et al. 1991), the Philippines and Australia (Whitmore 1984), and Madagascar (Ganzhorn 1995).

Temperature

The average annual temperature (average of daily maxima and minima) in the laboratory clearing of Barro Colorado Island is 27°C. Temperature behaves similarly in this clearing and above the forest canopy (Table 3.13). The average diurnal temperature range has fallen from 9° to 7°C between 1972 and 1989 (Table 3.14). Windsor (1990) commented that, while average diurnal temperature maxima have shown no consistent trend during the 20 years ending in 1989, average temperature minima have increased with the passage of time. Average daily temperature minima are similar in the laboratory clearing and near the forest floor, but average daily maxima tend to run 3°–4°C lower inside the forest.

In other lowland sites of the wet tropics within 10° latitude of the equator, the average annual temperature is 25°–27°C, with monthly averages varying within a range of only 2°–4°C (Table 3.15; also see Salati and Marques 1984: Table 4; Lauer 1989: Table 2.2). At other tropical sites, as on Barro Colorado, the average diurnal temperature range in the open is 6°–9°C (Table 3.15). For lowland sites throughout peninsular Malaysia, average diurnal temperature range lies between 6° and 11°C (Dale 1963).

In general, temperature varies much less from month to month than do radiation reaching the forest canopy, hours of bright sunshine per day, or even the radiation incident on the upper atmosphere. At Manaus, an annual average of 413 W/m² of solar radiation impinges on the upper edge of the atmosphere, of which 192 W/m² penetrates the atmosphere and its clouds to the forest. The forest reflects 15% of this radiation back to the atmosphere (Salati and Marques 1984). If evapotranspiration is 1300 mm/year, that is to say, 3.56 mm/day, or about 0.041 cm³/m² · sec, and if it takes 2430 J to evaporate 1 cm³ of water from the forest's leaves, then, averaging throughout the year, evapotranspiration accounts for 100 W/m² of this radiation, over half the radiation that reaches the forest. The energy used for photosynthesis is trivial by comparison.

Does the rainforest function as a giant air conditioner that maintains the atmospheric conditions appropriate to the ideal glasshouse, the wet warmth which is most suitable for plant growth? One function of transpiration is to avoid overheating. Do the measures each plant takes to avoid overheating also help to maintain the climate for the forest as a whole? When radiation impinging on the outer atmosphere increases, does this increase bring about increased cloudiness, and vice versa, thus regulating the forest's temperature? Table 3.15 offers a hint that this is indeed the case. Moreover, where tropical forest is cut down, heat becomes harsher, and, very often, rainfall drops.

It appears, however, that had this study been done 16,000 years ago, we might well have concluded that tropical forest set its thermostat to achieve an average temperature of 22°–23°C—at least during the rainy season when it had enough water available to regulate its temperature (Colinvaux et al. 1996)—rather than the 25°–27°C we find today. For the last million years, glaciers have been marching and retreating, paced by a 100,000-year rhythm in the eccentricity of the earth's orbit around the sun (Hays et al. 1976), and when the glaciers spread, tropical rainforests retreated, and their climates became cooler and drier.

Table 3.12 Monthly averages of midday relative humidity (percent) at various levels of the forest on Barro Colorado Island, 1978 (data from Windsor 1990)

	Jan.	Feb.	Mar.	Apr.	May	June	July	Aug.	Sep.	Oct.	Nov.	Dec.
Lutz Tower, 1 m	83	84	79	84	91	95	95	97	92	96	97	91
Lutz Tower, 13 m	81	79	76	78	88	92	91	91	85	92	92	83
Lutz Tower, 26 m	78	77	75	78	86	90	90	89	81	87	87	81
Lutz Tower, 40 m	72	69	69	73	80	88	85	84	76	82	78	74
Laboratory clearing	72	73	71	77	83	90	88	88	83	86	84	81

Table 3.13 Monthly rainfall and average of daily temperature maxima and minima (°C) in the laboratory clearing and 5 m above the forest canopy at Barro Colorado Island, 1982 and 1983 (data from Windsor 1990, Tables A3, B1, B8)

	1982					1983				
	Rain	Clearing		Tower		Rain	Clearing		Tower	
Month	mm	Max	Min	Max	Min	mm	Max	Min	Max	Min
Jan.	114	31.5	23.0	30.4	22.6	23	31.3	24.3	30.1	23.7
Feb.	18	32.3	23.6	30.2	23.2	0	32.4	23.9	30.0	23.3
Mar.	5	32.4	23.6	30.2	23.3	3	33.8	24.7	31.6	24.0
April	94	32.2	24.0	30.4	23.4	94	33.6	25.0	31.7	24.1
May	168	32.4	24.0	31.7	23.2	246	31.9	25.0	31.4	24.2
June	178	32.1	24.2	31.7	23.4	284	32.0	24.6	31.8	23.7
July	170	30.7	24.2	30.5	23.4	305	31.4	24.8	31.5	24.5
Aug.	295	31.8	24.5	31.0	23.5	229	31.3	24.6	31.3	24.3
Sept.	287	31.5	23.8	31.2	22.9	345	30.4	23.9	30.5	23.6
Oct.	290	30.6	23.1	30.4	22.2	269	30.7	23.6	30.8	23.2
Nov.	102	31.5	23.8	31.0	22.8	422	31.2	24.4	31.4	23.6
Dec.	30	31.8	23.9	31.0	23.4	269	30.0	23.4	29.8	23.3
Total/Avg.	1750	31.7	23.8	30.8	23.1	2489	31.7	24.4	31.0	23.8

Chapter 1 had a lot to say about the effects on the earth's climate of "Milankovitch cycles"—cyclic perturbations in features of the earth's orbit, such as its eccentricity, the angle between the earth's axis of rotation and the plane of the earth's orbit, and the direction in which this axis is leaning when the earth is farthest from the sun (the precessional rhythm). Such cycles have often exerted a striking impact on the earth's climate. For 35 million years of the Triassic, the 23,000-year cycle in the precession of the earth's axis of rotation dictated the cyclic drying up and refilling of a series of great rift lakes at 15° N latitude and a correlated desertification and reforestation of their surroundings (Olsen et al. 1978, Olsen 1984). This was not an unusual happening (Fischer 1986). The striking effects of these relatively minor changes in and redistribution of energy reaching the earth's surface suggest that tropical forests have little power to maintain "homeostasis" in their climate.

If we lengthen our sight still farther, however, yet a different picture emerges. During the 3.8 billion years since life began on earth, energy received from the sun has increased by a third or more (Lovelock 1988). This period has witnessed many revolutions. Photosynthetic organisms appeared very early on; they must have consumed much of the abundant CO_2 that had helped to keep the earth warm. Somewhat over 2 billion years ago, the oxygen content of the earth's atmosphere increased sufficiently to drastically reduce the atmosphere's stock of methane (another "greenhouse gas"). The last 600 million years have witnessed two great cycles, each beginning with active vulcanism, splitting of continents, rising midocean ridges, and high sea levels, during which CO_2 was being abun-

Table 3.14 Monthly average daily temperature (degrees C) maxima (ADT max) and minima (ADT min) in the laboratory clearing, and 1 m above the ground, in closed forest on Barro Colorado Island, 1972 and 1989 (Data from Windsor 1990 Tables B1, B3)

	Jan.	Feb.	Mar.	Apr.	May	Jun.	Jul.	Aug.	Sep.	Oct.	Nov.	Dec.	Average
Laboratory clearing, 1972													
ADT max	30.3	31.1	32.0	31.8	32.0	31.7	32.0	30.6	31.1	30.5	31.9	31.9	31.4
ADT min	21.8	22.2	21.8	22.8	23.4	22.8	23.5	22.8	23.0	22.5	22.8	22.8	22.7
Forest floor, 1972													
ADT max	26.7	27.7	28.7	29.3	29.6	28.1	28.8	27.7	27.4	27.1	27.6	27.8	27.8
ADT min	21.4	21.9	21.9	23.3	23.7	23.5	23.8	23.0	22.8	22.1	22.7	23.0	22.7
Laboratory clearing, 1989													
ADT max	30.4	29.9	30.5	32.0	31.6	30.7	30.3	29.6	30.9	29.9	29.6	30.0	30.6
ADT min	23.4	23.1	22.8	24.0	24.1	24.3	23.9	23.8	23.8	23.5	23.8	23.0	23.6
Forest floor, 1989													
ADT max	26.8	26.8	27.3	28.3	28.1	26.8	26.4	25.9	26.5	25.5	25.4	25.7	26.7
ADT min	23.6	23.3	23.2	24.7	24.2	24.7	23.7	23.6	23.7	23.4	23.7	23.3	23.8

Table 3.15 Features of climate at selected tropical sites

	Jan.	Feb.	Mar.	Apr.	May	Jun.	Jul.	Aug.	Sep.	Oct.	Nov.	Dec.	Total/Avg
Barro Colorado (9° N)													
Solar radiation above atmosphere	366	397	421	433	429	421	421	429	426	402	373	354	406
Solar radiation at canopy, 1984–1989	209	218	242	220	185	154	157	158	167	137	153	174	181
Day length (sunrise to sunset)	11.7	11.9	12.1	12.3	12.5	12.6	12.6	12.4	12.2	11.9	11.7	11.6	12.1
Average daily temperature max, 1971–89	31.0	31.3	31.9	32.3	31.8	30.9	30.7	30.7	30.9	30.7	30.5	30.7	31.1
Average daily temperature min, 1971–89	22.8	22.9	23.1	23.5	23.7	23.5	23.4	23.2	23.1	22.8	23.0	22.9	23.2
Monthly rainfall, 1972–1989	71	37	23	106	245	275	237	322	309	364	360	202	2551
Manaus, Amazonia (3° S)													
Solar radiation above atmosphere	428	438	436	415	384	364	374	398	423	433	429	424	413
Solar radiation at canopy, Q, Mar 78–Feb 79													
observed	163	191	148	156	162	209	196	224	235	242	197	176	192
calculated	152	161	162	151	156	199	191	220	215	208	185	162	180
Day length	12.2	12.2	12.1	12.0	11.9	12.0	12.0	12.1	12.1	12.2	12.3	12.3	12.1
Hours sunshine/day, Mar 78–Feb 79	2.3	2.6	2.7	2.5	3.5	6.8	6.0	7.1	6.0	5.4	4.2	3.0	4.3
Hours sunshine/day, average	3.8	3.9	3.6	3.8	5.4	6.9	7.9	8.2	7.5	6.6	5.9	4.9	5.7
Cloudiness (percent cover)	83	84	85	85	79	69	64	61	69	76	78	80	76
Average daily temperature	25.9	25.8	25.8	25.8	26.4	26.6	26.9	27.5	27.9	27.7	27.3	26.7	26.7
Monthly rainfall (40-yr average)	276	277	301	287	193	98	61	41	61	112	165	228	2101
Average atmospheric vapor (rain eq.)	45	47	47	48	45	44	43	41	39	40	45	47	44
Singapore, Singapore (1° N)													
Solar radiation at canopy	222	245	252	240	222	218	219	226	231	220	207	188	224
Hours sunshine/day	5.1	6.3	6.2	5.9	5.9	6.0	6.2	5.9	5.6	5.2	4.6	4.5	5.6
Mean daily maximum temperature	30.0	31.1	31.1	31.1	31.7	31.1	31.1	30.6	30.6	30.6	30.6	30.6	30.6
Mean daily minimum temperature	22.8	22.8	23.9	23.9	23.9	23.9	23.9	23.9	23.9	23.9	23.3	23.3	23.3
Monthly rainfall	251	173	193	188	173	173	170	196	178	208	254	256	2413
Abidjan, Côte d'Ivoire, (6° N)													
Solar radiation at canopy, Adiopodoumé	168	206	227	222	209	156	133	127	152	195	192	180	180
Hours sunshine/day, Adiopodoumé	5.2	6.2	6.3	6.1	5.5	2.8	2.8	2.4	2.8	5.1	6.1	5.4	4.7
Average daily temperature max, Adiopodoumé	31.2	32.1	32.3	31.9	31.1	28.8	27.8	27.4	28.0	29.3	30.6	30.5	30.0
Average daily temperature min, Adiopodoumé	22.0	22.8	23.1	23.2	22.9	22.4	21.7	21.2	21.8	22.5	22.4	22.1	22.3
Monthly rainfall, Banco National Park	41	55	106	138	282	602	267	61	102	191	166	84	2095

Solar radiation is in W/m² (24-hr average); rainfall is in mm.

BCI above-atmosphere solar radiation from Dietrich, Windsor, and Dunne (1982); day length from Balboa/Cristobal Tide Tables published by the Panama Canal Commission for 1977; other BCI data from Windsor (1990). Data for above-atmosphere radiation, observed solar radiation, day length, and observed hours of bright sunshine per day March 1978–February 1979 are from Ribeiro et al. (1982); other Manaus data are from Salati and Marques (1984), except for calculated solar radiation at canopy level, Q, which was obtained from above-atmosphere radiation Q_0 by the formula $Q = Q_0(0.26 + 0.5n/N)$, where n is the number of hours of bright sunshine per day, as given by Ribeiro et al. (1982), and N is average day length. This equation averages Ribeiro et al.'s (1982) $Q = Q_0(0.26 + 0.49n/N)$ and Ribeiro and Villa Nova's (1979) $Q = Q_0(0.26 + 0.51n/N)$. Singapore data are from Müller (1982). Data for the Côte d'Ivoire are from Bernhard-Reversat et al. (1978).

dantly "outgassed" to the atmosphere, while relatively little land surface was available for the weathering processes which consume CO_2, and ending with reduced vulcanism, the rejoining of continents, the subsidence of midocean ridges, and a fall in sea level, with a consequent excess of CO_2 consumption by weathering over its production by vulcanism (Fischer 1984). The resulting variation in the atmosphere's CO_2 content has been substantial. Berner (1993) estimates on geochemical grounds that in the early Devonian, when the earth was in a "greenhouse" state (Fischer 1984), the concentration of CO_2 in the atmosphere was 12 times the current level. The stomatal index—number of stomates/(number of stomates + number of epidermal cells) on a photosynthetic surface—of selected Devonian plants were one-fifth those of their nearest modern equivalents, *Juncus* and *Psilotum* (McElwain and Chaloner 1995): plants need fewer stomates when CO_2 is plentiful. In the late Carboniferous and early Permian, the earth was in icehouse state, and Berner (1993) believes that the CO_2 content of the atmosphere then was similar to today's. One conifer from the late Carboniferous, and another from the early Permian, had stomatal indices 1.5 times that of a modern *Araucaria* (McElwain and Chaloner 1995). In the mid-Cretaceous, CO_2 levels were 4–18 times higher than today's (Arthur et al. 1988, Berner 1993). Yet, during the 3.8 billion years that the earth has harbored life, the earth never froze entirely, nor did it ever heat up beyond the tolerance of living things (Lovelock 1988). Today, were it not for the impact of life on the geochemistry of earth and its atmosphere, the earth's surface temperature would be well over 200°C, far too hot to live on (Lovelock 1988). Living things clearly played an essential role in keeping the earth's temperature within a livable range. Whether this is sheer happenstance or the result of each population evolving in its own interest, which happened also (in this instance) to be the interest of all, or something more mysterious is beyond the scope of this book. In any event, the history of the earth's temperature offers abundant food for thought.

EVAPOTRANSPIRATION IN MONTANE FORESTS

We have seen how the difference between rainfall and runoff in watertight catchments of lowland tropical moist or wet forest tends to approximate 1425 mm/year (Bruijnzeel 1989). In montane forests below the cloud line, ranging in altitude from 1000 to 2500 m, this difference averages 1250 mm/year. As in lowland forests, annual evapotranspiration in montane forests does not increase with rainfall; more surprisingly, it is not correlated with altitude (Table 3.16).

Above the cloud line, forests are wreathed in cloud for a substantial fraction of most days, and rainfall minus runoff is much lower (Table 3.16). Cloud forests do extract water from the clouds which envelop them, which rain gauges never record (Cavelier and Goldstein 1989). Thus evapotranspiration in cloud forests exceeds the difference between rainfall and runoff by several hundred millimeters. Evapotranspiration is still much lower than in montane forests outside the cloud belt (Bruijnzeel and Proctor 1995). Cloud forests receive less radiation, they are more humid, and their average diurnal temperature range is much lower than in lowland tropical forests (Table 3.17). The cloud line is much lower on isolated mountaintops and on mountaintops close to the sea than on large inland mountain ranges (Grubb 1977). Why some cloud forests are so stunted and unproductive is a mystery, to which we will return in chapter 5.

CONCLUDING REMARKS

Lowland equatorial climates are remarkably similar, with average annual temperature near 26°C, average diurnal temperature range near 8°C, and 24-hr solar radiation averaging 180 W/m². Schimper (1903: p. 161) remarked that where human activity has not transformed the landscape, climate is the primary influence on vegetation, and soil usually plays only a secondary role. If he is right, than the similitude in structure and physiognomy of lowland equatorial forests reflects that among their climates.

Perhaps the most decisive similarity among lowland tropical climates is that annual evapotranspiration is near 1400 mm/year. I have discussed the evidence for this similitude and the ways to assess it at length.

As one ascends tropical mountains, temperature drops, but annual evapotranspiration declines only slightly until one reaches the cloud line. Above this line, the forest is wreathed in fog for much of the day (or night), the ground is saturated, and evapotranspiration is much lower. Chapter 5 shows how markedly the structure and physiognomy of tropical forest changes at the cloud line. First, however, let us turn to the relation between runoff, erosion, and soil.

APPENDIX 3.1
ESTIMATING EVAPOTRANSPIRATION

Calculating Evapotranspiration

In suitably nonleaky catchments, annual evapotranspiration can be estimated as the difference between rainfall and runoff. In lowland tropical rainforests, such estimates average 1425 mm/year (Bruijnzeel 1989). For shorter periods, or for leaky catchments, Penman's (1948) formula is often used to calculate evapotranspiration from climatic variates.

Calculating Q_{1w} and E_a provide opportunities for disagreement. Were the forest a "black body" of temperature, T°C, in a vacuum, or in clear, dry air, it would emit radiation through a plane just above the canopy at a rate of $\sigma(273 + T)^4$ W/m². The forest, however, is not a black body. Vari-

Table 3.16 Annual rainfall and evapotranspiration in selected montane rainforests (from Bruijnzeel 1989)

Site	Altitude	Rainfall	Evapotranspiration
Outside the fog belt			
Ciwidey, west Java	1800	3306	1170
Perinet, Madagascar	1010	2081	1295
Mbeya, Tanzania	2500	1924	1381
Average ± standard deviation, seven sites*	1864 ± 522	2277 ± 437	1250 ± 87
Cloud forest			
Sierra Nevada de Santa Marta, Colombia	2100	2316	308
Mt. Data, Philippines	2000	3370	450

*The seven sites in question are Sierra Nevada de Santa Marta, Colombia (1300 m), Ramu, Papua New Guinea (1800 m), Kericho, Kenya (2200 m), Kimakia, Kenya (2440 m) and the three sites tabulated. For these seven sites, correlation between altitude and annual evapotranspiration is 0.073, while correlation between annual rainfall and annual evapotranspiration is −0.564.

ous formulae, somewhat discrepant, have been used to calculate Q_{1w}. Dietrich et al. (1982) use

$$Q_{1w} = \sigma(273 + T)^4 (0.56 - 0.08\sqrt{e}) (1 - aC), \quad (3.10)$$

where $a = 0.6$ and C, the mean proportion of the sky covered by clouds, is given by

$$C = (1.863 - 2.183 Q_s/I)^{1/2} - 0.332, \quad (3.11)$$

where I is the solar radiation reaching the outer edge of the earth's atmosphere, and Q_s is solar radiation reaching the forest canopy. Ribeiro and Villa Nova (1979) use

$$\sigma (273 + t)^4(0.56 - 0.078\sqrt{e}) [0.1 + 0.9(n/N)], \quad (3.12)$$

where n is the number of hours of bright sunshine per day, and N is day length [I have multiplied their "*tensão de vapor saturado*" by 1.33 to convert from millimeters Hg to millibars and have adjusted other units accordingly. The sum $0.1 + 0.9n/N$ varies far more from day to day than does $1 - aC$; even monthly averages of these terms are often very different. Ribeiro and Villa Nova (1979) set $E_a = (0.026 + 0.00016V)(e_s - e_a)$, using 0.026 for the constant term, whereas Dietrich et al. (1982) use 0.013].

Using different expressions for Q_{1w} and E_a in the Penman equation for monthly evapotranspiration in the Ducke Reserve, 1966–1973, does not change estimates of annual evapotranspiration by much more than 10%. If one calculates monthly evapotranspiration for the Ducke Reserve, September 1983–August 1984, from the data of Marques Filho et al. (1987), setting Q_n equal to $Q(1 - 0.12) - Q_{1w}$ and $E_a = (e_{sa} - e)(0.26 + 0.00016V)$, and using Dietrich et al's (1982) expression for Q_{1w}, one attains a total evapotranspiration of 1156 mm for the year. If one substitutes the monthly values of net radiation measured by Shuttleworth (1988) for the calculated Q_n, one finds 1299 mm. Shuttleworth (1988) also replaced E_a by a more elaborate and exact expression, requiring him to follow events from hour to hour, keeping track of the wetness of the canopy, and estimated evapotranspiration for this year to be 1344 mm. If this estimate is "exact," then the Brazilian expression for E_a, $(0.026+0.00016V)(e_{sa} - e)$ cm/day, underestimates annual E_a by 16%, while Dietrich *et al.*'s (1982) expression for E_a, $(0.013+0.00016V)(e_{sa} - e)$ cm/day, underestimates annual E_a by 40%.

On the other hand, Marques Filho et al. (1987), using equations similar in appearance to those of Shuttleworth (1988) and data from at least some of Shuttleworth et al.'s (1984a) instruments, estimated potential evapotranspiration in the Ducke Reserve for this same year to be 1652 mm—a sharp contrast with Shuttleworth's "exact" 1344 mm and my estimate of 1156 mm. Why the estimates of Marques Filho et al. (1987) and Shuttleworth (1988) were so different was not explained by either party.

Table 3.17 Climate in selected cloud forests

		Solar Radiation	Temperature	
Site	Altitude (m)	W/m², 24-hr avg	Annual avg	Diurnal range
Monteverde, Costa Rica	1500	107	17.8	
Pico del Oeste, Puerto Rico	1000	127	18.6	2.5
Macuira, Colombia	675	92	23.8	5.0

Radiation data for Monteverde from Lawton (1990); temperature data for Monteverde from Nadkarni and Matelson (1992); data for Puerto Rico from Baynton (1968); data for Macuira from Cavelier and Mejia (1990).

Table 3.18 Annual evaporation, E, and rainfall, P (mm), from the surfaces of Gatun and Madden Lakes and from nearby class A pans

		1965	1966	1967	1968	1969	1970	1971
Pan evaporation								
Pedro Miguel	E	1410		1375	1435	1439	1160	No data
	P	2100		2023	2015	2079	2487	No data
Madden Dam	E	1290		1269	1244	1204	1117	1283
	P	2223		2613	2739	2262	2859	2604
Gatun	E	1321		1376	1391	1374	1220	1426
	P	3176		3051	2669	3202	3952	2588
Lake evaporation*								
Gatun Lake	E	1276	1298	1332	1324	1265	1204	1351

*Data from Panama Canal Commission. Data for lake evaporation is given as water lost by evaporation, in acre-feet: I divided it by 185.5 square miles, the area of lake surface in the former Canal Zone, times 640 acres per square mile to get evaporation from lake surfaces, in feet, and converted from feet to millimeters.

Evapotranspiration and Evaporation

Is it better to measure potential evapotranspiration directly than to calculate it from formulae like Penman's? Potential evapotranspiration has been measured from "class A" pans at Bacia Modelo near Manaus by Marques Filho and Ribeiro (1987) and in the laboratory clearing at Barro Colorado Island by Windsor (1990). Windsor's pan was 1.22 m in diameter and 25.4 cm deep and was mounted on a platform 10 cm high that allowed air to circulate all around it; the details of how the Bacia Modelo pan was positioned were not described. The pan evaporations were remarkably similar: 128 cm for the year beginning August 1981 at Bacia Modelo (Marques Filho and Ribeiro 1987), 132 cm for both 1982 and 1983 on Barro Colorado Island. These values are strikingly similar both to evaporation from the surfaces of Gatun and Madden Lakes in central Panama and to evaporation from class A pans at sites near these lakes (Table 3.18). Evaporation is similar at these various Panamanian sites despite great differences in rainfall from site to site and from year to year. Moreover, unlike rainfall less runoff on Barro Colorado Island (Table 3.9), evaporation is least in very wet years (although not necessarily so in the wetter places; Table 3.18). Evaporation rates in central Panama also agree astonishingly with the "exact" evapotranspiration of 1344 and 1288 mm in Amazonia calculated by Shuttleworth (1988) for the 2 successive years beginning in September 1983. Are we to conclude that, when soil moisture is in sufficient supply, evaporation from a tropical rainforest is equal to evaporation from class A pans and from lake surfaces? It would be truly extraordinary if transpiration through barely visible pores on the undersides of leaves of water sucked up from the soil, supplemented by episodic evaporation from wet leaf surfaces, yields as much vapor to the atmosphere as does evaporation from a surface of standing water.

On the other hand, annual evaporation from a class A pan at San Carlos de Rio Negro is 1970 mm (Heuveldop 1980). Annual evaporation from pans in Malaysia is also high, even though most of the pans are sunken. The one elevated pan, at Singapore, evaporates 1736 mm/year (Nieuwolt 1965); pans at Kuala Lumpur, near the west coast of Malaya, Kuala Lipis, an inland, lowland station, and Mersing, on the east coast, evaporate an average of 1476, 1356, and 1577 mm/year, respectively. As in Panama, evaporation in Malaysia is lowest at the sites farthest inland. But evaporation from these Malaysian pans is surprisingly high, especially considering the other similarities between

Table 3.19 Calculated evapotranspiration and observed pan evaporation (mm/month), Bacia Modelo, August 1981–July 1982

	Aug.	Sep.	Oct.	Nov.	Dec.	Jan.	Feb.	Mar.	Apr.	May	June	Jul.	Total
Calculated evapotranspiration	112	105	121	111	115	105	95	102	99	102	102	109	1278
Observed pan evaporation	112	108	121	102	115	109	87	99	99	105	108	115	1280

Calculations and data from Marquis Filho and Ribeiro (1987).

Marquis Filho and Ribeiro (1987) calculated potential evapotranspiration from monthly averages of the fraction of daylight hours with bright sunshine, n/N, solar radiation reaching the outer atmosphere, R, measured in mm water evaporated per day, air temperature, T, (° C.), vapor pressure, e, (mb), and wind speed, u, (m/sec), according to the formula $H = (0.26 + 0.49n/N)(1 - \alpha) R - Q_{1w}$, where H is flux of radiation into the forest canopy, measured as mm water evaporated per day, α is the proportion of this radiation reflected back to space, and Q_{1w} is long-wave radiation from the forest skyward; $Q_{1w} = \sigma (273.2 + T)^4(0.56 - 0.08\sqrt{e})(0.1 + 0.9n/N)$, where σ is the Stefan-Boltzmann constant 2.01×10^{-9} mm/day·° C.; $E_a = 0.26(e_s - e)(1 + 0.54u)$, where E_a is evaporation by wind (mm/day) and e_s is the saturation vapor pressure at air temperature T; and $PE = (\Delta H + \gamma E_a)/(\Delta + \gamma)$, where PE is potential evapotranspiration (mm/day), $\Delta = d(e_s)/dT$ (mb/° C.), and γ is the psychrometric constant, 0.66 mb/° C.

Table 3.20 Calculated evapotranspiration, observed pan evaporation, rainfall, runoff, and changes in amount of water stored in the top 1.1m of soil, Barro Colorado Island, April 1982–March 1983 (data from Windsor 1990)

	Apr.	May	Jun.	Jul.	Aug.	Sep.	Oct.	Nov.	Dec.	Jan.	Feb.	Mar.	Total/Avg
Q_o, (W/m²)	433	429	421	421	429	426	402	373	354	366	397	421	406
Q, incident radiation (W/m²)	215	182	171	182	184	154	153	188	198	212	221	219	190
Q_{1w}, net long wave radiation, (W/m²)	32	26	27	29	26	26	31	30	31	37	32	30	30
Relative humidity at noon (1976–77)	72	79	78	80	81	80	76	82	81	71	76	68	77
Temperature (°C)	28.1	28.2	28.2	27.5	28.2	27.7	26.9	27.7	27.9	27.8	28.2	29.3	28.0
Saturated vapor pressure (mb)	38.0	38.3	38.3	36.7	38.3	37.2	35.5	37.2	37.6	37.4	38.3	40.8	37.8
Actual vapor pressure (mb)	32.7	34.3	34.0	33.1	34.6	33.4	31.2	33.8	34.0	32.0	33.7	34.2	33.4
Wind speed (km/day)	163	53	29	53	55	22	19	106	158	185	185	190	102
Evaporation by wind E_a (mm/month)	83	43	40	38	40	34	39	44	57	93	72	115	698
Calculated evapotranspiration (mm/month)	138	114	102	109	115	89	88	112	124	138	131	154	1414
Class A evaporation (mm/month)	132	115	96	99	102	102	115	114	90	102	112	136	1315
Rainfall, P (mm/month)	107	133	248	188	322	327	320	112	38	23	3	4	1825
Runoff, R (mm/month)	3	3	9	15	50	94	115	44	9	3	0	0	346
Soil moisture change (mm/month)	28	41	58	14	16	17	−0.4	−26	−102	49	−17	−37	−56
Catchment evaporation, $P − R − \Delta S$	76	89	181	158	255	216	205	93	131	69	20	40	1534

Q_o is radiation incident on the outer edge of the earth's atmosphere, from Dietrich, Windsor, and Dunne (1982), Q is radiation incident on the canopy at Barro Colorado, as tabulated by Becker (1986); Q_{1w} = net long wave radiation reflected back from the forest to space, calculated according to the formula $Q_{1w} = \sigma T^4(0.56 − 0.08\sqrt{e})(1 − aC)$, where $a = 0.6$ and $C = −.332 + (1.863 − 2.183 Q/Q_o)^{1/2}$, from Dietrich, Windsor, and Dunne (1982); relative humidity is that observed at noon in the laboratory clearing; temperature is the average of the mean daily maximum and mean daily minimum for the month; saturation vapor pressure $e_{sa}(T)$ is obtained from table A.3, p. 263 of Monteith and Unsworth (1990); actual vapor pressure e is saturation vapor pressure times (relative humidity at noon + 100)/200; E_a, evaporation by wind is (0.26 + 0.0016V)(vapor pressure at saturation less actual vapor pressure), mm/day, multiplied by the number of days that month; and soil moisture change is the change in the moisture content of the top 1.1 m of soil in Lutz catchment (see text).

To transform $R_n = Q(1 − \alpha) − Q_{1w}$ from W/m² = J/m²·sec to mm/day, I divided by the latent heat of vaporization of water, $\lambda = 2,432,000$ J/kg = $2,423,000$ J/dm³ to obtain the energetic equivalent in dm³ of water evaporated per square meter of ground per second, which is evaporation, in mm/sec. Then I multiplied by 86,400 to get evaporation in mm/da. Values of γ and Δ needed to calculate total evapotranspiration from R_n and E_a, are tabulated by Monteith and Unsworth (1990: Tables A.3, A.4).

the climates of Malaysia and Panama. Moreover, Nieuwolt (1965) repeats a remark of Rohwer (1931) that evaporation from large reservoirs in the United States is 70% of evaporation from nearby class A pans. Why these differences?

Monthly pan evaporation at Bacia Modelo agrees nicely (Table 3.19) with the Penman calculation of Marques Filho and Ribeiro (1987), but their Penman calculation assumed an albedo of 0.25, which is unreasonably high for tropical rainforest (Shuttleworth et al. 1984b). Pan and lake evaporation are decidedly less than the Penman potential evapotranspiration calculated for Barro Colorado over the year starting in April 1982 by the same formulae which *underestimated* evapotranspiration for the year starting September 1983. Pan evaporation in Panama and Amazonia likewise falls short of Bruijnzeel's (1989) estimate from rainfall minus runoff in various nonleaky catchments that average annual evapotranspiration is 1415 mm. Can forest evapotranspiration be higher than evaporation from a lake surface?

Water Budgets

To ascertain trustworthy evapotranspiration estimates, it would help if we could reconcile estimates of evapotranspiration from climatic variables with estimates based on differences between rainfall and runoff. It is ominous that, in Lutz catchment, the annual difference between rainfall and runoff increases with annual rainfall, while in Gatun Lake, annual evaporation decreases with annual rainfall. Windsor (1990) tried to draw up a water budget for Lutz catchment, based on rainfall, runoff, and the amount of water stored in the top 1.1 m of the soil. Windsor noted from his observations and measurements that soil within 10 cm of the surface weighs 0.7 g/cm^3 when dry, and its moisture content by volume can vary from 27 to 57%; while soil 30–40 cm below the surface weighs 1.1 g/cm^3 when dry, and its moisture content by volume can vary from 43 to 67%. He assumed that soil at 30–40 cm depth was representative of soil as deep as 1.1 m, that soil 10–30 cm below the surface weighed 1 g/cm^3 when dry, and that the moisture content by volume of this soil could vary from 39 to 70% by volume. He assumed that the water content of this soil was that measured for the top 10 cm, plus that measured for soil 30–40 cm deep and extrapolated to soil 30–110 cm deep, plus that for soil 10–30 cm deep, inferred from averaging the moisture content by weight of soil 0–10 cm deep and soil 30–40 cm deep. According to these assumptions, the top 1.1 m of soil has a total moisture content of 313 mm. It does seem that, at the beginning of rainy season, when soil is very dry, it takes a good deal of rain to generate substantial runoff: in 1982, cumulative rainfall had to exceed cumulative evapotranspiration by 120 mm or so to generate substantial runoff (Table 3.20). However, water moves more slowly through the soil than these assumptions suggest, and it seems to require the soaking of far less than the top meter of soil to generate substantial runoff. Moreover, Windsor's estimates of monthly evapotranspiration as rainfall minus runoff minus the increase that month in water stored in the top 1.1 m of soil yields some outlandishly high figures for monthly evapotranspiration; for example, 255 mm for the wet month of August. This raises the question of how deeply water penetrates the soil, given time enough, and how much water flows out underground.

Lettau and Hopkins (1991) attempted to reconcile evapotranspiration as calculated by climatic parameters with evapotranspiration calculated from catchment water balance by modeling stored water in a more realistic fashion and by interposing a variable factor, f, representing the changing resistance of the forest to evapotranspiration. They were able to "predict" runoff rather accurately from rainfall and climatic parameters, but to do so their factor f was highest (double the average) in September, in the middle of the rainy season, which seems most implausible.

Although similarity in evaporative potential may play the decisive role in governing the similarity in structure and production of lowland tropical rainforests, we do not know enough about the "water budgets" of tropical forests and their soils to decide this issue.

References

Arthur, M. A., W. E. Dean, and L. A. Pratt. 1988. Geochemical and climatic effects of increased marine organic carbon burial at the Cenomanian/Turonian boundary. *Nature* 335: 714–717.

Bailly, C., G. Benoit de Coignac, C. Malvos, J. M. Ningre, and J. M. Sarrailh. 1974. Étude de l'influence du couvert naturel et de ses modifications á Madagascar. *Bois et Forêts des Tropiques* (supplement). no. 4: 1–115.

Barat, C. 1957. Pluviologie et aquidimétrie dans la zone intertropicale. *Mémoires de l'Institut Français d'Afrique Noire* 49:1–80.

Baynton, H. W. 1968. The ecology of an elfin forest in Puerto Rico, 2. The microclimate of Pico del Oeste. *Journal of the Arnold Arboretum* 49: 419–430.

Becker, P. 1987. Monthly average solar radiation in Panama—daily and hourly relations between direct and global insolation. *Solar Energy* 39: 445–453.

Berner, R. A. 1993. Paleozoic atmospheric CO_2: importance of solar radiation and plant evolution. *Science* 261: 68–70.

Bernhard-Reversat, F., C. Huttel, and G. Lemée. 1978. Structure and functioning of evergreen rain forest ecosystems of the Ivory Coast, pp. 557–574. In *Tropical Forest Ecosystems*. UNESCO, Paris.

Bruijnzeel, L. A. 1989. Nutrient cycling in moist tropical forests: the hydrological framework, pp. 383–415. In J. Proctor, ed., *Mineral Nutrients in Tropical Forest and Savanna Ecosystems*. Blackwell Scientific Publications, Oxford.

Bruijnzeel, L. A. 1990. *Hydrology of Moist Tropical Forests and Effects of Conversion: A State of Knowledge Review*. Free University, Amsterdam.

Bruijnzeel, L. A. 1991. Nutrient input-output budgets of tropical forest ecosystems: A review. *Journal of Tropical Ecology* 7: 1–24.

Bruijnzeel, L. A., and J. Proctor. 1995. Hydrology and biogeochemistry of tropical montane cloud forests: what do we really know? pp. 38–78. In L. Hamilton, J. O. Juvik, and F. N.

Scatena, eds., *Tropical Montane Cloud Forests*. Springer-Verlag, New York.

Brunig, E. F. 1970. Stand structure, physiognomy and environmental factors in some lowland forests in Sarawak. *Tropical Ecology* 11: 26–43.

Cavelier, J., and G. Goldstein. 1989. Mist and fog interception in elfin cloud forests in Colombia and Venezuela. *Journal of Tropical Ecology* 5: 309–322.

Cavelier, J., and C. A. Mejia. 1990. Climatic factors and tree stature in the elfin cloud forest of Serrania de Macuira, Colombia. *Agricultural and Forest Meteorology* 53: 105–123.

Clarke, R. T., and D. M. Cooper. 1987. Water recycling in tropical forests as a stochastic process. *Acta Amazônica* 16/17: 239–252.

Cleaves, E. T., A. E. Godfrey, and O. P. Bricker. 1970. Geochemical balance of a small watershed and its geomorphic implications. *Geological Society of America Bulletin* 81: 3015–3032.

Cole, J. E., R. G. Fairbanks, and G. T. Shen. 1993. Recent variability in the Southern Oscillation: isotopic results from a Tarawa Atoll coral. *Science* 260: 1790–1793.

Colinvaux, P. A., K.-B. Liu, P. de Oliveira, M. B. Bush, M. C. Miller, and M. S. Kannan. 1996. Temperature depression in the lowland tropics in glacial times. *Climatic Change* 32: 19–33.

Corner, E. J. H. 1964. *The Life of Plants*. World Publishing, Cleveland, OH.

Cubit, J. D., R. C. Thompson, H. M. Caffey, and D. M. Windsor. 1988. Hydrographic and meteorological studies of a Caribbean fringing reef at Punta Galeta, Panama: hourly and daily variations for 1977–1985. *Smithsonian Contributions to the Marine Sciences* 32: 1–220.

Dale, W. L. 1963. Surface temperatures in Malaya. *Journal of Tropical Geography* 17: 57–71.

Denny, M. W. 1985. Wave forces on intertidal organisms: a case study. *Limnology and Oceanography* 30: 1171–1187.

Diaz, H. F., and G. N. Kiladis. 1992. Atmospheric teleconnections associated with the extreme phase of the Southern Oscillation, pp. 7–28. In H. F. Diaz and V. Markgraf, eds., *El Niño: Historical and Paleoclimatic Aspects of the Southern Oscillation*. Cambridge University Press, Cambridge, MA.

Diaz, H. F., and R. S. Pulwarty. 1992. A comparison of Southern Oscillation and El Niño signals in the tropics, pp. 175–192. In H. F. Diaz and V. Markgraf, eds., *El Niño: Historical and Paleoclimatic Aspects of the Southern Oscillation*. Cambridge University Press, Cambridge, MA.

Dietrich, W. E., D. M. Windsor, and T. Dunne. 1982. Geology, climate and hydrology of Barro Colorado Island, pp. 21–46. In E. G. Leigh, Jr., A. S. Rand, and D. M. Windsor, eds., *Ecology of a Tropical Forest*. Smithsonian Institution Press, Washington, DC.

Dunne, T., and L. B. Leopold. 1978. *Water in Environmental Planning*. W. H. Freeman, San Francisco.

Emanuel, K. A. 1988. Toward a general theory of hurricanes. *American Scientist* 76: 370–379.

Fischer, A. G. 1984. The two Phanerozoic supercycles, pp. 129–150. In W. A. Berggren and J. A. Van Couvering, eds., *Catastrophes and Earth History*. Princeton University Press, Princeton, NJ.

Fischer, A. G. 1986. Climatic rhythms recorded in strata. *Annual Review of Earth and Planetary Sciences* 14: 351–376.

Foster, R. B. 1982. Famine on Barro Colorado Island, pp. 201–212. In E. G. Leigh, Jr., A. S. Rand, and D. M. Windsor, eds., *The Ecology of a Tropical Forest*. Smithsonian Institution Press, Washington, DC.

Franken, W., and P. R. Leopoldo. 1984. Hydrology of catchment areas of Central-Amazonian forest streams, pp. 501–519. In H. Sioli, ed., *The Amazon: Limnology and Landscape Ecology of a Mighty Tropical River and Its Basin*. W. Junk, Dordrecht.

Franken, W., and P. R. Leopoldo. 1987. Relações entre fluxos de água subterrânea e superficial em bacia hidrográfica caracterizada por cobertura florestal amazônica. *Acta Amazônica* 16/17: 253–262.

Franken, W., P. R. Leopoldo, E. Matsui, and M. de N. G. Ribeiro. 1982a. Estudo da interceptação da água de chuva em cobertura florestal amazônica do tipo terra firme. *Acta Amazônica* 12: 327–331.

Franken, W., P. R. Leopoldo, E. Matsui, and M. de N. G. Ribeiro. 1982b. Interceptação das precipitações em floresta amazônica de terra firme. *Acta Amazônica* 12 (supplement): 15–22.

Ganzhorn, J. U. 1995. Cyclones over Madagascar: fate or fortune? *Ambio* 24: 124–125.

Glynn, P. W. 1988. El Niño-Southern Oscillation 1982–1983: nearshore population, community and ecosystem responses. *Annual Review of Ecology and Systematics* 19: 309–345.

Glynn, P. W. (ed.) 1990. *Global Ecological Consequences of the 1982–83 El Niño-Southern Oscillation*. Elsevier, Amsterdam.

Gray, W. M. 1990. Strong association between West African rainfall and U. S. landfall of intense hurricanes. *Science* 249: 1251–1256.

Grubb, P. J. 1977. Control of forest growth and distribution on wet tropical mountains with special reference to mineral nutrition. *Annual Review of Ecology and Systematics* 8: 83–107.

Hallé, F. 1993. *Un Monde sans Hiver: Les Tropiques, Nature et Societés*. Éditions du Seuil, Paris.

Hays, J. D., J. Imbrie, and N. J. Shackleton. 1976. Variations in the Earth's orbit: pacemaker of the ice ages. *Science* 194: 1121–1132.

Herwitz, S. R. 1986. Infiltration-excess caused by stemflow in a cyclone-prone tropical rainforest. *Earth Surface Processes and Landforms* 11: 401–412.

Herwitz, S. R. 1988. Buttresses of tropical rainforest trees influence hillslope processes. *Earth Surface Processes and Landforms* 13: 563–567.

Herwitz, S. R. 1991. Aboveground adventitious roots and stemflow chemistry of *Ceratopetalum virchowii* in an Australian montane tropical rain forest. *Biotropica* 23: 210–218.

Heuveldop, J. 1980. Das Bioklima von San Carlos de Rio Negro, Venezuela. *Amazoniana* 7: 7–17.

Isaza, C. 1986. Analisis de los factores que influyen en la erosion de la cuenca, pp. 121–141. In S. Heckadon M., ed., *La Cuenca del Canal de Panama*. Impretex, Panamá.

Jennings, A. H. 1950. World's greatest observed point rainfalls. *Monthly Weather Review* 78: 4–5.

Johnson, P. L., and W. T. Swank. 1973. Studies of cation budgets in the southern Appalachians on four experimental watersheds with contrasting vegetation. *Ecology* 54: 70–80.

Jordan, C. F., and J. Heuveldop. 1981. The water budget of an Amazonian rainforest. *Acta Amazônica* 11: 87–92.

Koehl, M. A. R., and R. S. Alberte. 1988. Flow, flapping, and photosynthesis of *Nereocystis luetkeana*: a functional comparison of undulate and flat blade morphologies. *Marine Biology* 99: 435–444.

Lauer, W. 1989. Climate and weather, pp. 7–53. In H. Lieth and M. J. A. Werger, eds., *Tropical Rain Forest Ecosystems: Biogeographical and Ecological Studies*. Amsterdam: Elsevier.

Lawton, R. O. 1990. Canopy gaps and light penetration into a wind-exposed tropical lower montane rain forest. *Canadian Journal of Forest Research* 20: 659–667.

Leigh, E. G. Jr. 1975. Structure and climate in tropical rain forest. *Annual Review of Ecology and Systematics* 6: 67–86.

Leigh, E. G. Jr., D. M. Windsor, A. S. Rand, and R. B. Foster. 1990. The impact of the "El Niño" drought of 1982–83 on a Panamanian semideciduous forest, pp. 473–486. In P. W. Glynn, ed., *Global Ecological Consequences of the 1982–83 El Niño-Southern Oscillation.* Elsevier, Amsterdam.

Leighton, M., and N. Wirawan. 1986. Catastrophic drought and fire in Borneo tropical rain forest associated with the 1982–1983 El Niño-Southern Oscillation event, pp. 75–102. In G. T. Prance, ed., *Tropical Rain Forests and the World Atmosphere.* Westview Press, Boulder, CO.

Leopoldo, P. R., W. Franken, E. Matsui, and E. Salati. 1982a. Estimativa de evapotranspiração de floresta amazônica de tirra firme. *Acta Amazônica* 12, supplement: 23–28.

Leopoldo, P. R., W. Franken, and E. Salati. 1982b. Balanço hidrico de pequena bacia hidrográfica em floresta amazônica de tirra firme. *Acta Amazônica* 12: 333–337.

Leopoldo, P. R., W. Franken, E. Salati, and M. N. Ribeiro. 1987. Towards a water balance in the Central Amazonian region. *Experientia* 43: 222–233.

Lettau, H. H., and E. J. Hopkins. 1991. Evapoclimatonomy III: The reconciliation of monthly runoff and evaporation in the climatic balance of evaporable water in land areas. *Journal of Applied Meteorology* 30: 776–792.

Likens, G. E., F. H. Bormann, R. S. Pierce, J. S. Eaton, and N. M. Johnson. 1977. *Biogeochemistry of a Forested Ecosystem.* Springer-Verlag, New York.

Lloyd, C. R., and A. de O. Marques Filho. 1988. Spatial variability of throughfall and stemflow measurements in Amazonian rainforest. *Agricultural and Forest Meteorology* 42: 63–73.

Lovelock, J. 1988. *The Ages of Gaia: A Biography of Our Living Earth.* W. W. Norton, New York.

MacArthur, R. H. 1972. *Geographical Ecology.* Harper and Row, New York.

McElwain, J. C., and W. G. Chaloner. 1995. Stomatal density and index of fossil plants track atmospheric carbon dioxide in the Paleozoic. *Annals of Botany* 76: 389–395.

Manokaran, N. 1979. Stemflow, throughfall and rainfall interception in a lowland tropical rain forest in peninsular Malaysia. *Malaysian Forester* 42: 174–201.

Marques Filho, A. de O., and M. de N. G. Ribeiro. 1987. Evaporação do tanque classe A e sua relação com os parâmetros climáticos. *Acta Amazônica* 16/17: 263–276.

Marques Filho, A. de O., M. de N. G. Ribeiro, H. M. dos Santos, and J. M. dos Santos. 1981. Estudos climatológos da Reserva Florestal Ducke-Manaus-AM. IV. Precipitação. *Acta Amazônica* 11: 759–768.

Marques Filho, A. de O., M. de N. G. Ribeiro, A. P. Fattori, G. F. Fisch, and M. Januário. 1987. Evaporação potencial de florestas. *Acta Amazonica* 16/17: 277–292.

Marshall, T. J., and J. W. Holmes. 1988. *Soil Physics* (2nd ed). Cambridge University Press, Cambridge.

Meinzer, F. C., G. Goldstein, P. Jackson, N. M. Holbrook, M. V. Gutiérrez, and J. Cavelier. 1995. Environmental and physiological regulation of transpiration in tropical forest gap species: the influence of boundary layers and hydraulic properties. *Oecologia* 101: 514–522.

Monteith, J. L., and M. H. Unsworth. 1990. *Principles of Environmental Physics.* Edward Arnold, London.

Mosley, M. P. 1982. The effect of a New Zealand beech forest canopy on the kinetic energy of water drops and on surface erosion. *Earth Surface Processes and Landforms* 7: 103–107.

Müller, M. J. 1982. *Selected Climatic Data for a Global Set of Standard Stations for Vegetation Science.* W. Junk, The Hague.

Nadkarni, N. M., and T. J. Matelson. 1992. Biomass and nutrient dynamics of fine litter of terrestrially rooted material in a neotropical montane forest, Costa Rica. *Biotropica* 24: 113–120.

Nepstad, D. C., C. R. de Carvalho, E. A. Davidson, P. H. Jipp, P. A. Lefebvre et al. 1994. The role of deep roots in the hydrological and carbon cycles of Amazonian forests and pastures. *Nature* 372: 666–669.

Nieuwolt, S. 1965. Evaporation and water balances in Malaya. *Journal of Tropical Geography* 20: 34–53.

Olsen, P. E. 1984. Periodicity of lake-level cycles in the late Triassic Lockatong Formation of the Newark Basin (Newark Supergroup), New Jersey and Pennsylvania, pp. 129–146. In A. L. Berger, J. Imbrie, J. Hays, G. Kukla, and B. Saltzman, eds., *Milankovitch and Climate,* Part I. D. Reidel, Dordrecht.

Olsen, P. E., C. L. Remington, B. Cornet, and K. S. Thomson. 1978. Cyclic changes in late Triassic lacustrine communities. *Science* 201: 729–733.

Paine, R. T. 1986. Benthic community-water column coupling during the 1982–83 El Niño. Are community changes at high latitudes attributable to cause or coincidence? *Limnology and Oceanography* 31: 351–360.

Paulhus, J. L. H. 1965. Indian Ocean and Taiwan rainfalls set new records. *Monthly Weather Review* 93: 331–335.

Penman, H. L. 1948. Natural evaporation from open water, bare soil and grass. *Proceedings of the Royal Society,* A 193: 120–145.

Putz, F. E., and S. Appanah. 1987. Buried seeds, newly dispersed seeds, and the dynamics of a lowland forest in Malaysia. *Biotropica* 19: 326–333.

Putz, F. E., P. D. Coley, K. Lu, A. Montalvo, and A. Aiello. 1983. Uprooting and snapping of trees: structural determinants and ecological consequences. *Canadian Journal of Forest Research* 13: 1101–1120.

Quinn, W. H. 1992. A study of Southern Oscillation-related climatic activity for A. D. 622–1990 incorporating Nile River flood data, pp. 119–149. In H. F. Diaz and V. Markgraf, eds., *El Niño: Historical and Paleoclimatic Aspects of the Southern Oscillation.* Cambridge University Press, Cambridge.

Rasmussen, E. M., and J. M. Wallace. 1983. Meteorological aspects of the El Niño/Southern Oscillation. *Science* 222: 1195–1202.

Reynolds, K. M., L. V. Madden, D. L. Reichard, and M. A. Ellis. 1989. Splash dispersal of *Phytophthora cactorum* from infected strawberry fruit by simulated canopy drip. *Phytopathology* 79: 425–432.

Ribeiro, M. de N. G., E. Salati, N. A. Villa Nova, and C. G. B. Demétrio. 1982. Radiação solar disponivel em Manaus (AM) e sua relação com a duração do brilho solar. *Acta Amazônica* 12: 339–346.

Ribeiro, M. de N. G., and N. A. Villa Nova. 1979. Estudos climatológos da Reserva Florestal Ducke, Manaus, AM. III. Evapotranspiração. *Acta Amazônica* 9: 305–309.

Richards, P. W. 1952. *The Tropical Rain Forest.* Cambridge University Press, Cambridge.

Roberts, J., O. M. R. Cabral, G. Fisch, L. C. B. Molion, C. J. Moore, and W. J. Shuttleworth. 1993. Transpiration from an Amazonian rainforest calculated from stomatal conductance measurements. *Agricultural and Forest Meteorology* 65: 175–196.

Ropelewski, C. F., and M. S. Halpert. 1987. Global and regional scale precipitation patterns associated with the El Niño/Southern Oscillation. *Monthly Weather Review* 115: 1606–1626.

Rosenzweig, M. L. 1968. Net primary productivity of terrestrial communities: prediction from climatological data. *American Naturalist* 102: 67–74.

Russell, A. E., and J. J. Ewel. 1985. Leaching from a tropical andept during big storms: a comparison of three methods. *Soil Science* 139: 181–189.

Salafsky, N. 1994. Drought in the rain forest: effects of the 1991 El Niño-Southern Oscillation event on a rural economy in West Kalimantan, Indonesia. *Climatic Change* 27: 373–396.

Salati, E., and J. Marques. 1984. Climatology of the Amazon region, pp. 85–126. In H. Sioli, ed., *The Amazon: Limnology and Landscape Ecology of a Mighty Tropical River.* Dordrecht: W. Junk.

Salati, E., and P. Vose. 1984. Amazon Basin: a system in equilibrium. *Science* 225:129–138.

Schimper, A. F. W. 1903. *Plant-Geography upon a Physiological Basis.* Clarendon Press, Oxford.

Shuttleworth, W. J. 1988. Evaporation from Amazonian rainforest. *Proceedings of the Royal Society of London, B* 233: 321–346.

Shuttleworth, W. J. et al. 1984a. Eddy correlation measurements of energy partition for Amazonian forest. *Quarterly Journal of the Royal Meteorological Society* 110: 1143–1162.

Shuttleworth, W. A. et al. 1984b. Observations of radiation exchange above and below Amazonian forest. *Quarterly Journal of the Royal Meteorological Society* 110: 1163–1169.

Sioli, H. 1984. The Amazon and its main effluents: hydrography, morphology of the river courses, and river types, pp. 127–165. In H. Sioli, ed., *The Amazon: Limnology and Landscape Ecology of a Mighty Tropical River.* Dordrecht: W. Junk.

Smythe, N. 1975. Rainfall, tipping bucket, hourly totals, pp. 50–61. In D. M. Windsor, ed., *1974 Environmental Monitoring and Baseline Data Compiled Under the Smithsonian Institution Environmental Sciences Program. Tropical Studies.* Unpublished mimeographed report. Smithsonian Institution, Washington, DC.

Smythe, N. 1976. Rainfall, tipping bucket, hourly totals, pp. 48–59. In D. M. Windsor, ed., *1975 Environmental Monitoring and Baseline Data Compiled under the Smithsonian Institution Environmental Sciences Program. Tropical Studies.* Unpublished mimeographed report. Smithsonian Institution, Washington, DC.

Tanner, E. V. J., V. Kapos, and J. R. Healey. 1991. Synthesis: hurricane effects on forest ecosystems in the Caribbean. *Biotropica* 23: 513–521.

Vis, M. 1986. Interception, drop size distributions and rainfall kinetic energy in four Colombian forest ecosystems. *Earth Surface Processes and Landforms* 11: 591–603.

Walter, H., and S.-W. Breckle. 1986. *Ecological Systems of the Geobiosphere 2. Tropical and Subtropical Zonobiomes.* Berlin: Springer-Verlag.

Whitmore, T. C. 1984. *Tropical Rain Forests of the Far East.* Oxford University Press, Oxford.

Windsor, D. M. 1990. Climate and moisture variability in a tropical forest: long-term records from Barro Colorado Island, Panama. *Smithsonian Contributions to the Earth Sciences* 29:1–145.

Windsor, D. M., A. S. Rand, and W. M. Rand. 1990. Caracteristicas de la precipitación en la isla de Barro Colorado, pp. 53–71. In E. G. Leigh, Jr., A. S. Rand, and D. M. Windsor, eds., *Ecología de un Bosque Tropical.* Balboa, Panama: Smithsonian Tropical Research Institute.

Wright, S. J., C. Carrasco, and O. Calderon. 1996. Community-level fruit failure and famine among mammalian frugivores in a tropical rain forest. *Bulletin of the Ecological Society of America* 77, supplement: 492.

F O U R

Runoff, Erosion, and Soil Formation

How much of the rain falling into a catchment or basin leaves it as runoff? How is the amount and the seasonal variation of the runoff affected by the catchment's vegetation?

Runoff removes soil from a catchment as matter suspended in the moving water (suspended solid load) or pushed along the bottom of the stream (bed load), and as ions dissolved from the soil or bedrock (solute load). How does topography, vegetation, and land use affect erosion rate? These questions have acquired an urgent practical importance. Exploitation of tropical forest settings by inappropriate techniques can lead to landslides and loss of soil from the catchment, ruining the land and causing flooding and siltation of reservoirs downstream.

MODES OF RUNOFF

Runoff is the water left over after evapotranspiration takes its "cut" from rainfall. On Barro Colorado, runoff is measured at two sites: Lutz Stream, near the laboratory clearing, and Conrad Stream, just upstream from Conrad trail 1.9. Conrad Stream drains part of the island's andesite-capped (Johnsson and Stallard 1989) central plateau. At Conrad, water level is measured in a stilling pond behind a 90° V-notch weir, which collects runoff from a 40-ha catchment, much of which lies within the 50-ha forest dynamics plot (fig. 4.1). Reliable monitoring of the water level only began in May 1996. At Lutz, water level is monitored in a stilling pond collecting runoff from a catchment (fig. 4.2), long thought to be 10 ha, but whose area is closer to 9.7 ha (R. F. Stallard, p. c.). As yet, runoff data are available only for Lutz, so this catchment is discussed in more detail.

At Lutz, water flows from the stilling pond over a 120° V-notch weir. Since 1971, water level, as measured by the position of a float, has been tracked by a Stevens A-71 strip chart water level recorder. Beginning in 1993, measurements were recorded every 5 min from an ISCO "bubble gauge" that measures the water level by the pressure from the water above it. The rate dV/dt at which water flows over the V notch, in cubic centimeters per second, is calculated from the height, H, in millimeters, of the water surface in the stilling pond above the bottom of the V notch according to the formula $dV/dt = 0.103534 H^{2.449}$. This formula is transformed from $F = 4.43 h^{2.449}$, given by Hertzler (1938), where F is flow in cubic feet per second, and h is height of the water surface above the bottom of the V notch, in feet.

There are discrepancies between runoff inferred from the chart records and from the bubble gauge (Table 4.1). Calculating outflow is a tricky business, because water flow increases most rapidly with water level when the stream is in spate and water level is changing most rapidly. The bubble gauge is thought to respond more rapidly and more accurately than the chart recorder to swiftly changing water levels.

Monthly rainfall and runoff records from Lutz catchment (Tables 4.1, 4.2) suggest that runoff quickly drops to almost zero during the dry season. When rainy season begins, the thirsty soil soaks up nearly all the water, leaving little to run off. Only after the soil is sufficiently saturated—often 6 weeks or more into the rainy season—does runoff become substantial. Because runoff is what is left over after evaporation has subtracted its nearly constant demand from an already variable rainfall, runoff is particularly changeable from year to year (Table 4.2) and especially from storm to storm. On the other hand, in a 130-ha catchment, Barro Branco, near Manaus (like the 40-ha Conrad catchment on Barro Colorado), runoff seems to respond much more slowly to changes in rainfall—it seems steadier (Table 4.2; see also Leopoldo et al. 1984: fig. 6). At Conrad weir, water level takes perhaps 3 hr to increase after the onset of heavy rain, while at Lutz water level can begin to

Conrad Trail Stream and the 50 Ha Plot

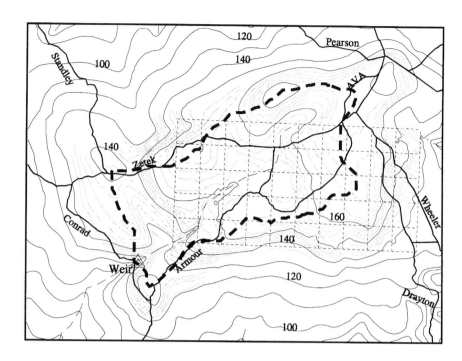

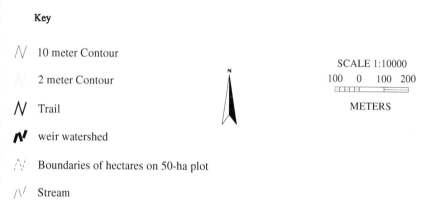

Figure 4.1. A preliminary map of Conrad catchment and the 50-ha Forest Dynamics Plot. The trails were surveyed by Wibke Thies. Topography of the 50-ha plot was provided by S. P. Hubbell and R. B. Foster (p. c.); remaining topography was prepared from the 1957 1:25,000 maps of the Panama Canal by David Kinner. Robert Stallard adjusted the plateau topography. Global Positioning System fixes were supplied by David Clark. All these data were reconciled by Robert Stallard.

increase in a few minutes (R. F. Stallard, p. c.). It is as if, at Conrad, overland flow contributed a far smaller proportion to total runoff.

By plotting runoff as the rate, Q, at which water leaves the catchment via the weir (in liters per second per square kilometer of catchment) as a function of time, we may distinguish between baseflow and delayed flow, on the one hand, and quickflow, on the other. Quickflow is largely supplied by water that runs overland without ever sinking into the soil and which therefore has the greatest potential for erosion: the baseflow and delayed flow have percolated through the soil and subsoil. Quickflow is revealed by sudden rises in water level associated with storms. At the beginning of such a rise, extend a line from the point on the preexisting baseflow, right at the base of the rise, and let this line delineate an imaginary rise in Q of 0.55 every hour, to where it strikes the descending limb of the "hydrograph," that is to say, the declining portion of the graph of Q against time, representing the subsiding of the stream's spate after the storm is well past. The peak thus "excised" represents quickflow (Dietrich et al. 1982).

How does forest cover influence the total amount and the seasonal distribution of runoff? Deforestation increases total runoff, at least at first (Bruijnzeel 1988, 1990) because evapotranspiration from a forest transfers more water from the soil to the atmosphere than does evaporation from bare soil. This increase in runoff is permanent if tall forest is converted to grassland or fields of shallow-rooted agricultural crops (Bruijnzeel 1988). Runoff is higher from grassland catchments than from forested ones (Table 4.3).

Deforestation has often been blamed for increased flooding downstream. Many areas do suffer ever higher economic losses from floods, but, as Bruijnzeel (1990) remarks, one must ask, are the floods themselves growing worse, or are there simply more people, and more and costlier struc-

Contour Map of Lutz Creek Catchment

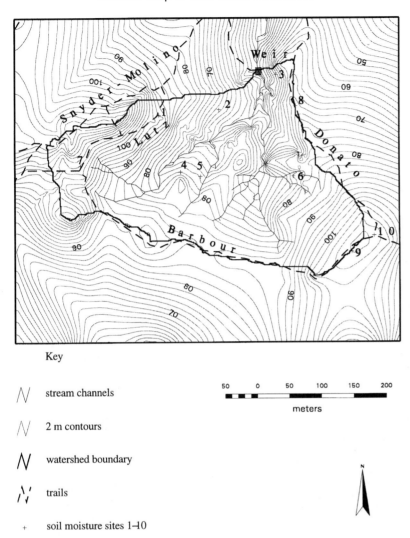

Key

/\/ stream channels

/\/ 2 m contours

/\/ watershed boundary

/\/ trails

+ soil moisture sites 1–10

watershed area: 9.6 ha

Figure 4.2. A contour map of Lutz catchment. The trails were surveyed by Wibke Thies. Catchment topography was surveyed by David Kinner; remaining topography was prepared from the 957 1:25,000 maps of the Panama Canal by David Kinner. Global Positioning System fixes were supplied by David Clark. All these data were reconciled by David Kinner and Robert Stallard.

tures, in the paths of these floods? To be sure, deforestation does decrease evapotranspiration, so that soil saturates more quickly as the rainy season progresses. Thus deforesting a catchment should increase runoff during storms, especially storms occuring rather late in the rainy season: "there is little doubt that, in general, an undisturbed forest moderates quickflow rates and stormflow volumes, although the influence on the magnitude of extreme events is marginal at best" (Bruijnzeel 1988: p. 229). Deforestation is more likely to increase stormflow and flooding if bulldozers and crawler tractors, which compact the soil and reduce infiltration rates, are used to clear the forest. Even in the absence of such machinery, stormflow volumes were markedly higher in catchments near Perinet, Madagascar, which were cleared, cultivated, and now covered by shrubby regrowth (savoka) than in catchments less than 4 km distant covered by native forest or *Eucalyptus* plantations (Bailly et al. 1974: p. 39). On the other hand, due care for the land in cutting the timber and appropriate management thereafter can prevent deforestation from increasing the frequency and severity of flooding (Bruijnzeel 1990).

Many think that clearing the forest diminishes, or even eliminates, streamflow during the dry season (Bruijnzeel 1988: p. 229). Evidence for this generalization is far from decisive (Bruijnzeel 1988, 1990). This generalization is best tested by comparing a "treated" catchment, whose forest is cleared in the course of the study, and a "control" catchment whose forest cover is left intact. In some such cases, deforestation has *increased* dry season flow; in others, deforestation has increased flow during the rainy season and decreased flow during the dry season; in yet others, the effect of deforestation on dry season flow is not clearly marked (Bruijnzeel 1988, 1990). Dry season streamflow is most likely to decline where deforestation or subsequent land use compacts (or paves over) the soil, reducing the proportion of rainfall soaking into the soil (Bruijnzeel

Table 4.1 Monthly rainfall, quickflow, delayed flow, and baseflow (mm) from a 10–ha catchment on Barro Colorado, 1994, as inferred from bubble gauge and chart records

	Jan.	Feb.	Mar.	Apr.	May	Jun.	Jul.	Aug.	Sep.	Oct.	Nov.	Dec.	Total
Rainfall	86	3	61	28	270	287	355	302	241	236	376	41	2285
Bubble Gauge													
Quickflow	5	0	0	0	2	14	60	44	28	18	81	0	253
Delayed Flow	2	0	0	0	3	7	11	16	13	9	20	0	82
Baseflow	25	1	2	1	3	21	52	81	81	63	132	34	495
Total Runoff	31	1	2	1	9	42	123	140	123	90	233	34	829
Chart													
Quickflow	4	0	0	0	2	14	58	43	27	19	80	0	248
Delayed Flow	1	0	0	0	2	7	9	13	11	9	18	1	71
Baseflow	20	4	2	1	4	21	53	87	82	64	139	31	507
Total Runoff	25	4	2	1	8	42	120	143	120	92	237	32	826

Data from Smithsonian Institution Environmental Monitoring Program, analyzed by S. Paton.

1990). Perhaps sticking to the best evidence underestimates the damage from deforestation, for the carelessness manifested by abuse of the land is often reflected by indifference to the consequences of such abuse. On the other hand, clearing the forest in a way that avoids compacting the soil, and appropriate husbandry of the land thus cleared, will preserve the flow of water during the dry season (Bruijnzeel 1988, 1990). Unfortunately, the techniques of forest clearing that are best for the land are the most labor intensive (Bruijnzeel 1990: p. 70). Good husbandry presupposes a real understanding and, yes, love, of the land, which is rarely met with in these days of frequent, often involuntary, deracination and translocation.

RUNOFF AND WEATHERING

The more rapidly runoff leaves a catchment, the higher the concentration of sediment suspended in it. Sediment export increases disproportionately with runoff, as anyone knows who has compared the clear water of a small rainforest stream under normal circumstances with the muddy torrent it becomes during a severe storm. Sediment export is greatest during intensely heavy, prolonged rains and in catchments where such rain yields copious overland flow. In Lutz catchment, the year's single worst storm can account for a third or more of the year's export of sediment. In many North American rivers, the most sediment-laden 1% of the year accounts for about half the river's annual output of suspended sediments (Meade et al. 1990). Erosion is a process dominated by rare events.

Erosion and Soil on Barro Colorado

The relationship between the concentration, C, of suspended sediments in a stream, in milligrams per liter or grams per cubic meter, and the rate, dV/dt, at which water is leaving the catchment, in cubic meters per second, can be calculated from the regression of $\log C$ on dV/dt. In Lutz catchment, R. F. Stallard (p. c.) finds that the concentra-

Table 4.2 Monthly rainfall, P; runoff, R; and quickflow, Q, from Lutz Creek catchment (Barro Colorado, Panama) and Barro Branco catchment (Brazil). Runoff is expressed as rainfall equivalent, mm, (assuming a catchment area of 9.6 ha for Lutz on BCI)

		Jan.	Feb.	Mar.	Apr.	May	Jun.	Jul.	Aug.	Sep.	Oct.	Nov.	Dec.	Total
Barro Colorado 1995	P	150	35	8	126	327	330	275	198	231	262	304	271	2516
	R	20	4	2	2	22	109	113	71	57	88	206	136	830
	Q	5	0	0	0	8	69	28	12	12	22	48	40	244
Barro Colorado 1996	P	359	103	92	23	412	390	245	327	298	354	467	159	3228
	R	281	26	10	4	68	200	147	189	159	201	334	127	1746
	Q	135	3	0	0	41	71	33	73	56	88	137	24	661
Barro Branco 1982	P	286	308	285	379	295	95	100	48	150	157	90	180	2372
	R	57	99	112	135	129	82	54	52	54	55	35	45	909
	Q	6	16	11	19	15	1	2	1	4	6	2	3	85
Barro Branco 1983	P	39	67	282	175	243	123	97	83	123	159	75	438	1906
	R	32	30	39	45	53	47	36	30	28	35	26	60	482
	Q	1	1	7	3	4	5	1	1	2	2	1	12	41

Data for Barro Colorado from the Smithsonian Institution Environmental Monitoring Program, analyzed by S. Paton; data for Barro Branco from Franken and Leopoldo (1987).

Table 4.3 Annual rainfall, P; runoff, R; and evapotranspiration, $P - R$: average, standard deviation (SD), and coefficient of variation (CV = Avg/SD), from a forested (Severn) and a grassland (Wye) catchment in Wales [data from Newson (1979), quoted in Marshall and Holmes (1988: Table 12.4)], 1970–77

	Forest Catchment			Grassland Catchment		
	P	R	$P - R$	P	R	$P - R$
Avg	2213	1364	849	2349	1944	405
SD	385	361	84	413	386	65
CV	0.17	0.26	0.10	0.18	0.20	0.16

For Severn, $P - R = 666 + .0826P$ ($r^2 = 0.145$) and $R = 0.9174P - 666$ ($r^2 = 0.955$). For Wye, $P - R = 0.07414P + 231$ ($r^2 = 0.222$); $R = 0.926P - 232$ ($r^2 = 0.979$)

tion, C, of solids suspended in the outflow over the weir is $3030(dV/dt)^{0.933}$, a relation rather similar to that for the Icacos basin in the Luquillo Mountains of Puerto Rico. According to Stallard's relationship, 42 tons of sediment leaves the catchment in an "average" year with a runoff of 1106 mm: 453 g/m² catchment. By contrast, Dietrich et al. (1982: p. 43, caption to fig. 17) found $C = 3439(dV/dt)^{0.768}$ in the stream above the weir and concluded that annual sediment export from Lutz was about 56 tons. These calculations are a rough and ready business.

At Conrad Stream, concentrations of suspended (and dissolved) solids have been measured every 2 weeks for 4 years. Weighting each measurement by the flow rate when it was made, the average concentration of suspended sediments in Conrad Stream was 7.0 g/m³ water. Assuming that, as at Lutz, annual runoff averages 1106 mm, each square meter of catchment contributed 1.1 m³ of runoff, and 7.7 g of suspended sediment, per year. In Conrad Stream, concentration of suspended solids only begins to increase at the highest flows for which measurements were taken, and no really high flows were sampled. The true erosion rate may well be five times higher, about 40 g suspended sediment/m² of catchment per year (R. F. Stallard, p. c.). Nevertheless, this is less than one-tenth of the 450 g suspended sediment exported per square meter of catchment per year at Lutz, a denudation rate of 0.015 mm/year by "mechanical" weathering, compared to 0.17 mm/year at Lutz (R. F. Stallard, p. c.).

Erosion involves not only mechanical weathering, leading to the export of suspended sediments, but also chemical weathering, the dissolution of soil and bedrock by percolating groundwater, which enhances the removal of suspended sediment but also contributes dissolved matter, solutes, to the runoff. The concentration of solutes behaves differently from the concentration of suspended sediments: the concentration of dissolved substances *decreases* when the stream is in spate. Our concern here is solutes derived from the bedrock, not brought in from the sea with the rain. All the calculations of solute content and chemical weathering assume that the chloride in the runoff is entirely derived from the sea and that, for every mole of chloride in the runoff, 0.84 mole of sodium, 0.021 mole of potassium, 0.1086 mole of magnesium, 0.0185 mole of calcium, and 0.127 mole of sulfur in the runoff are derived from the atmosphere (Stallard and Edmond 1981: Table 6), and must be subtracted from the contents of the runoff. Making this correction for "cyclic sea-salts," the concentration CS, of dissolved solids in the outflow from Lutz weir is 398 g/m³, or $92.5 + 2.299/(dV/dt)$ g/m³, whichever is lower. The second alternative is lower when $dV/dt > 0.00748$ m³/sec. In a year with 1106 mm of runoff, 22 tons of dissolved solids leave the catchment, 233 g/m² · year, including 58 g Ca and 16 g Si. This amounts to 0.09 mm/year of chemical erosion. The total loss of matter from the catchment in such a year would be 687 tons/km² · year, a total rate of erosion of 0.25 mm/year.

Chemical erosion from Barro Colorado's central plateau is much lower. Using the same averaging technique as for suspended sediments, R. F. Stallard (p. c.) finds that 13.3 g/m² of dissolved matter leaves Conrad catchment in the average year, including 3.3 g calcium and 4.2 g silicon, a chemical erosion rate of 0.0055 mm/year. These numbers are upper limits, as solute concentration is lower when flow rate is high.

Why should there be such an enormous difference between erosion rates in Lutz catchment and on the central plateau? First, in Lutz catchment, calcium carbonate serves as the glue that holds the bedrock together (Johnson and Stallard 1989: p. 769). During the rainy season, partial pressure of CO_2 in soil water is over a hundredfold that in the atmosphere (Johnson and Stallard 1989: p. 778), thanks to the respiration of roots, mycorrhizae, and soil micro-organisms (Berner 1992); moreover, plant roots also release organic acids (Schlesinger 1991). The calcium carbonate in the bedrock of Lutz catchment dissolves readily under the onslaught of soil water acidified by such abundant CO_2. The combination of chemical weathering, which dissolves the bedrock's limestone glue, and mechanical weathering, by which the disaggregated rock slides or washes away, leads to rapid erosion. The integrity of the plateau's andesite does not depend on such readily dissolved glue.

Moreover, at Lutz erosion is "weathering limited" (Johnson and Stallard 1989); that is to say, matter is removed as soon as it "weathers out," while, at Conrad, erosion is "transport limited," and weathered material accumulates faster than it can be removed. The distinction between weathering-limited and transport-limited erosion is fundamental (Stallard 1985, 1988, 1995). At Lutz, landslides are frequent (Dietrich et al. 1982), and its steep slopes allow the rapid removal of unconsolidated sediments. Soils are therefore thin, often no more than 50 cm thick (Dietrich et al. 1982). In Conrad catchment, however, as in the other parts of Barro Colorado's central plateau, the products of weathering remain in place much longer, so the soil is deeper, > 2 m thick.

The consequences for soil characteristics and stream chemistry of the contrast between Lutz's weathering-limited erosion and Conrad's transport-limited erosion are remarkable and sometimes surprising. At Lutz, residence time of any given mineral fragment in the thin, rapidly replenished soil is short, and weathering yields incompletely leached, cation-rich clays such as smectites (Johnson and Stallard 1989: p. 779). In Lutz catchment, moreover, weathering of the carbonate "glue" that holds the bedrock together produces an abundance of calcium and magnesium ions that enhance the formation of smectites. As the dry season progresses, the concentration of these cations in water occupying minute pores in the soil increases, and smectite clays form more readily. Thus, in Lutz catchment, the dry season restores soil fertility (Johnson and Stallard 1989).

On Barro Colorado's central plateau, where erosion is transport-limited, the residence time of any given mineral fragment in the soil is much longer, leaching is extensive, and the soil is relatively poor in cations. Iron and aluminum are retained in gravelly concretions called pisolites. Weathered sand grains are rich in quartz, even where the plateau's bedrock is poor in silica (Johnson and Stallard 1989).

More generally, soils on the andesitic cap of the island's central plateau, the andesitic outcrops north of this cap, and the volcaniclastic facies of the Caimito formation covering the southeastern third of the island are kaolinitic (Fig. 4.3), with a ratio of silicon to aluminum atoms close to 1. These are well-weathered oxisols with low cation exchange capacity. Except possibly for flat hilltops whose soils may be more completely weathered, soils on the cation-rich, calcareous Caimito marine facies along the island's western and southern margins (Fig. 4.3) are montmorillonitic, with a ratio of roughly two silicon atoms per aluminum atom. These montmorillonitic soils are alfisols with high cation exchange capacity. The Bohio conglomerate, which covers Barro Colorado's northern and western margin, and separates the andesitic cap from the volcaniclastic facies of the Caimito (Fig. 4.3), weathers rapidly, creating a steep topography. On slopes steeper than 15°, erosion is weathering-limited, and soils tend to be montmorillonitic, with high cation exchange capacity. On gentler slopes, erosion tends to be transport-limited, producing kaolinitic soils.

Comparing the chemistry of Lutz with Conrad Stream (R. F. Stallard, quoted in Leigh and Wright 1990) suggests that Barro Colorado's central plateau is much less fertile than Lutz catchment. Conrad Stream, which drains the plateau, is hardly richer in nitrogen and phosphorus than streams of the nutrient-poor regions of central and eastern Amazonia, although it is slightly richer in cations than streams draining the more fertile plains of western Amazonia (Table 4.4). On the other hand, Lutz Stream is richer in cations than the Amazon at Iquitos, which drains the rapidly eroding eastern slopes of the Andes. Lutz Stream is also richer in inorganic phosphorus, although no richer in inorganic nitrogen, than Conrad Stream (Table 4.4). Nonetheless, it is dangerous to infer that soils of the plateau are infertile simply because nutrient levels are low in the streams that drain them (Parker 1994: p. 59). Nortcliff and Thornes (1989) draw attention to the prevalence of "biphasic flow" in many tropical soils, especially oxisols. Streams draining such soils are supplied mostly by water that moves rapidly through "macropores" opened in the soil by animal activity and the turnover of woody roots (Sollins 1989). This water moves too quickly

Table 4.4 Nutrient levels of streams and rivers (g/m^3) at selected tropical sites

Site	Ca^{2+}	Mg^{2+}	K$^+$	Inorganic N	Inorganic P
Amazonia					
Lower Amazon River (Obidos)	5.0	0.9	0.9	0.10	0.009
Eastern Amazon tributaries (Xingu, Tapajos, Trombetas)*	1.6	0.5	0.8	0.04	0.0012
Central Amazonia (Rio Negro at Manaus)	0.35	0.14	0.4	0.06	0.0022
Western Amazon tributaries (Purus, Jurua, Javari)*	4.0	0.8	0.9	0.08	0.007
Andean region (Iquitos)	20.0	2.2	1.2	0.16	0.022
Amazon rain: showers over river†	0.03	0.012	0.03	0.04	—
Barro Colorado Island, Panama					
Conrad Stream	2.6	1.0	0.5	0.07	0.002
Lutz Stream (at weir)	50.0	4.0	2.2	0.08	0.010
Rainfall over Barro Colorado†	0.10	0.08	0.09	0.20	0.0026
La Selva, Costa Rica					
Watershed 3, Sarapiqui Annex	3.9	1.9	1.2	0.09	0.063
Watershed 4, Sarapiqui Annex‡	2.1	0.9	0.9	0.25	0.0006

Data for Amazonia and Barro Colorado from R. F. Stallard (p. c.), as given in Leigh and Wright (1990). Data for La Selva from Parker (1994: p. 59).

*Lakes at the mouths of these tributaries may explain low values of N and P.
†These figures are based on a limited number of samples.
‡Chemistry of Watershed 4 more typical of streams in this neighborhood than that of Watershed 3.

Preliminary Soils Map of Barro Colorado Island

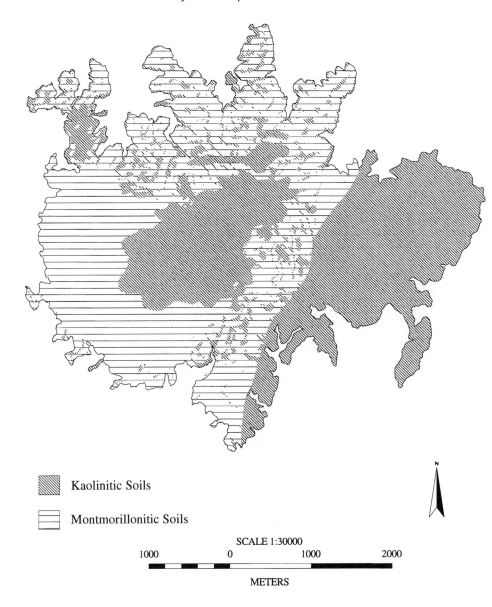

Figure 4.3. A preliminary soils map of Barro Colorado Island. This map was prepared by David Kinner from the geologic map of Barro Colorado Island of Woodring (1958), the 1957 1:25,000 map of this part of the Panama Canal, soil analyses by Robert Stallard and Michael Keller, and his own observations. Soils with a ratio of silicon to aluminum > 1.5 were classified as montmorillonitic, and those with lower silicon/aluminum ratio were classified as kaolinitic.

Based on correlations of soil chemistry with parent rock type and topography, soils of the andesitic cap and outcrops, and those of the Caimito volcaniclastic facies, were assumed to be kaolinitic; those of the Caimito marine facies were assumed to be montmorillonitic, except on flat well-weathered hilltops; those of the Bohio conglomerate were assumed to be montmorillonitic when on slopes > 15° and kaolinitic otherwise.

through the soil to leach nutrients from it, so nutrient levels are low in the streams supplied by such waters. There is relatively little exchange between the water moving through these macropores and the water in micropores, from which plants draw their nutrients. Consequently, especially when the soil is saturated, there is not necessarily a relationship between the nutrient content of water moving through macropores and that caught in micropores (Nortcliff and Thornes 1989).

In fact, the soils of Barro Colorado's central plateau are relatively fertile. The total phosphorus content in the top 15 cm of soil is higher on this plateau than on the surrounding sedimentary slopes (Yavitt and Wieder 1988): the total phosphorus content of the soil on this plateau is, moreover, five times higher than that in well-weathered oxisols at San Carlos de Rio Negro, or near Manaus (Table 4.5). There seems to be no clear-cut contrast between the oxisols of the plateau and the alfisols of the surrounding slopes with respect to their concentration of forms of nitrogens and phosphorus considered readily available to plants, or in the rate at which these soils release phosphorus and nitrogen under incubation in the laboratory (Yavitt and Wieder 1988). Phosphorus appears to be more readily available from both oxisols and alfisols of Barro Colorado than from most tropical soils (Yavitt and Wieder 1988). The fertility of the plateau's soils is reflected in its vegetation. The litter that falls on Barro Colorado's plateau is unusually rich in nitrogen, phosphorus, and calcium (B. Haines and R. B. Foster, quoted in Leigh and Windsor 1982), while the vegetation of that plateau, like that of the rest of the island, is considered typical of neotropical forest on fertile soil (Foster and Brokaw 1982).

Table 4.5 Mineral content of selected tropical soils (percent)

	Barro Colorado Plateau	Central Amazonia	Paleosol
P_2O_5	0.25	0.05	0.20
K_2O	0.14		0.36
CaO	0.23		0.08

Data for Barro Colorado and Amazonia from R. F. Stallard (p. c.), data for paleosol (Pennsylvanian of Missouri) from Retallack and Germán-Heins (1994).

Implications for Other Sites

Topography and bedrock both influence the natural erosion rate. In Amazonia, the median erosion rate from lowland soils on sedimentary rocks is about 0.015 mm/year, while median erosion rates from Andean soils derived from purely silicic rocks is about 0.06 mm/year, and median erosion rates from Andean soils derived from shales or carbonates is about 0.2 mm/year (Stallard 1985: fig. 3). Erosion from Barro Colorado's Conrad catchment is typical of lowland Amazonia: erosion from Lutz is characteristic of Andean soils derived from carbonates. Thus Barro Colorado spans the extremes of natural erosion rates.

It should be noted that the observed ratio of chemical to mechanical weathering can depend on the time interval over which the measurement is made. Bedrock of the 38–ha Pond Branch catchment 15 km north of Baltimore, Maryland, is mostly silicic. In this catchment the weathering of plagioclase to kaolinite and gibbsite and the weathering of biotite to vermiculite and kaolinite led to the removal of 5.9 g/m² of rock per year from this catchment in solution, while only 1.1 g/m² of particulate matter left the catchment as suspended sediment (Cleaves et al. 1970). However, the chemical composition of soil and bedrock in the Pond Branch catchment can only be explained if, over the long run, as much material is weathered mechanically as chemically: here, mechanical weathering appears to act primarily in occasional sudden landslides (Cleaves et al. 1970).

So far, we have measured erosion by the outflow of dissolved and suspended sediment in runoff. This is acceptable for catchments that have reached an equilibrium where accumulation of sediments in valleys is balanced by the re-erosion and export of such sediments from the catchment. This approach, however, grossly underestimates erosion in catchments subject to progressive human disturbance (Meade et al. 1990). In Puerto Rico, 500 ± 250 g/m² · year of sediment leaves the forested Icacos catchment (Brown et al. 1995)—much the same rate of export as Lutz—but 2000 ± 1000 g/m² · year leave the agriculturally "developed" Cayaguas catchment, which has the same bedrock (quartz diorite) as the Icacos basin (Larsen 1996). These figures understate the real contrast. In Cayaguas catchment, landslides moved 1.2 million tons of sediment in 68 years, 17.6 kg/m² · year (Larsen 1996). This is higher than the erosion rate of bare soil at construction sites. Cayaguas catchment has 100 landslide scars/km², with an average volume per scar of 8600 m³; Icacos catchment has 6 landslide scars/km², with an average volume per scar of 300 m³ (Larsen 1996), 0.2% the total volume of scar/km² as Cayaguas. Most of the Cayaguas landslide debris accumulates as colluvium at the base of hillslopes or as alluvium in stream channels. This material continues to be a source of suspended sediments during rainstorms. The Cayaguas story reminds us that eroded sediments do not flow straight to the ocean. They accumulate, and erode again, perhaps repeatedly, before reaching the sea (Meade et al. 1990).

Indeed, landslides are the quickest, most drastic form of erosion. The total rainfall, R, in millimeters, needed to trigger landslides in the central, mountainous area of Puerto Rico, where mean annual rainfall is about 2 m/year, is $91.46 D^{0.18}$ if rainfall duration, D, is expressed in hours, or $162.06 d^{0.18}$ if rainfall duration, d, is expressed in 24-hr days (Larsen and Simon 1993, Larsen and Torres-Sánchez 1996; also see Table 4.6). Hurricane Hugo, which in 1989 dropped 225 mm of rain in 6 hr, like other short, intense storms, triggered shallow landslides whose average thickness was 1.5 m and whose median area was 161 m². Longer-lasting storms, such as 1979's Hurricane David, which dropped 502 mm of rain over 72 hr, or a tropical depression which dropped 976 mm of rain in 144 hr in 1970, cause deep-seated debris avalanches with an average thickness of 10 m and an average surface area of over a hectare. A landslide-causing storm strikes Puerto Rico every 1.2 years (Larsen and Simon 1993).

In the temperate zone, much less intense thunderstorms can trigger landslides. There, the dependence of threshold rainfall, R, on duration, D, in hours, is $14.82 D^{0.61}$. The threshold rainfall required to trigger landslides elsewhere in the tropics, however, appears similar to Puerto Rico's (Larsen

Table 4.6 Minimum rainfall, R (mm), for various durations, D (hr), that trigger landslides in Puerto Rico (data from Larsen and Simon 1993)

D	Minimum R Predicted	Minimum R Observed	No. of slide-causing storms observed of duration D
2	104	142	2
3	111	127	2
4	117	102	1
6	126	127	4
9	136	225	2
12	143	168	1
24	162	144	7
48	184	232	6
72	197	254	4
96	208	268	5
144	224	254	3
312	257	303	1

and Simon 1993). Forest cover decreases a catchment's susceptibility to shallow landslips but does not affect the likelihood of deep-seated slides (Bruijnzeel 1990).

Forest plays a fundamental role in protecting its catchment from erosion (Bruijnzeel 1988, 1990, Meade et al. 1990). Land use of a type where the soil is protected by an adequate layer of leaf litter and ground herbs does not increase the rate of soil erosion relative to native forest. Erosion increases by 10 to a 100-fold, however, when the litter layer is destroyed or removed, as in farms, completely weeded tree plantations, or forest plantations where litter is removed or burned (Bruijnzeel 1990: p. 117). In tropical sites where manual clearing of the forest leads to the loss by erosion of 40 g/m^2 of soil during the following year, clearing the forest by a shear blade mounted on a crawler tractor leads to the loss of 400 g/m^2 of soil during the following year, while using a crawler tractor with a tree pusher/root rake attachment to clear the forest leads to a loss of 1500 g/m^2 of soil during the following year (Lal 1981, quoted by Bruijnzeel 1990: p. 70). Forest in the temperate zone plays the same role. In a 38-ha forested catchment ranging in altitude from 120 to 190 m about 15 km northwest of Baltimore, Maryland, where natural weathering removes 7 g/m^2 of matter per year, building a pipeline across the catchment led to the deposit in a pond at the base of the catchment of 14 tons of sediment from the 1.2-ha construction area during an 8-month period. This sedimentation rate of 1150 g/m^2 in 8 months represents a sharp increase over the background erosion rate (Cleaves et al. 1970).

SOIL AND VEGETATION

The distinction between settings where soils are so deeply weathered that bedrock contributes no nutrients to the vegetation, and settings such as Barro Colorado, where bedrock does contribute to the vegetation's nutrient budget, is currently receiving the emphasis it deserves (Baillie 1989, Burnham 1989). Forests whose nutrient supply is replenished only through rainfall and dust from the atmosphere show especially tight nutrient cycling (Baillie 1989).

Especially where soils are poor, the forest protects its nutrients. The lower the phosphorus content of the soil, the lower the phosphorus content per unit mass of litterfall (Silver 1994), as if, on poor soils, plants are more likely to withdraw phosphorus from their leaves before dropping them. Compared to the nutrients arriving with rainfall, or reaching the forest floor in leaf drip, not to speak of the nutrients contained in falling leaves, fruit, twigs, etc., the quantity of nutrients exported per hectare from small catchments near Manaus is remarkably low (Table 4.7), as are exports of nitrogen and phosphorus from temperate-zone catchments covered by mature forest (Table 4.8). Nutrient export in the Rio Caura of Venezuela is thought to be so high because the river drains subsoil as well as soil and carries nutrients weathered from bedrock far below where plant roots can reach (Bruijnzeel 1991), as is often the case in rivers draining large catchments. The conservation of nutrients is a biotic adaptation, which occurs only where plants can reach. Moreover, this adaptation is costly and is perfected most highly on forests with poor soils: where

Table 4.7 Nutrient gains, transfers, and exports (kg/ha · year) in three South American rainforest catchments on poor, sandy soil: two small catchments near Manaus and the Rio Caura catchment in Venezuela

	N	P	K	Ca	Mg
Nutrient gains					
Rainfall, Barro Branco, Brazil, 1977	6.0	0.1	2.1	nd	nd
Rainfall, Bacia Modelo, Brazil, 1981	nd	0.23	4.8	nd	nd
Rainfall, Rio Caura, Venezuela	2.3	0.14	1.0	1.3	0.3
Leaf drip, Barro Branco, 1977	7.4	0.27	22	1.0	7.8
Leaf drip, Bacia Modelo, 1980–81	5.6	0.42	18	7.2	3.1
Nutrient transfers					
Litterfall, Egler Reserve, 1963	114	2.3	13	18	14
Litterfall, Egler Reserve, 1964	97	2.0	12	19	11
Nutrient exports					
Runoff, Barro Branco, 1977	0.2	0.01	0.4	nd	nd
Runoff, Rio Caura, Venezuela	6.3	0.24	14.6	15.5	6.0

Litterfall data from Klinge and Rodrigues (1968a,b); other Amazon data from Franken and Leopoldo (1984); Rio Caura data from Bruijnzeel (1991).

Barro Branco is a 1.3-km^2 catchment in the Ducke Reserve, Bacia Modelo is a 23.5-km^2 catchment, also near Manaus, and R. Caura drains a 47,500-km^2 catchment in Venezuela.

The nitrogen and phosphorus measured in rainfall, runoff, and leaf drip in the Manaus catchments is NH_4^+ and PO_4^{3-}. Bacia Modelo exported 0.035 kg/ha of P as phosphate ion in 1982.

Table 4.8 Nutrient gains, transfers, and losses (kg/ha · year) in Watershed 18 of Coweeta, North Carolina (mixed hardwoods, 860 m)

	NO_3	NH_4	PO_4	K	Ca	Mg
Nutrient gains						
Rainfall, 1972–73	3.765	2.359	0.151	2.359	4.317	1.004
Rainfall, 1974–75	2.624	1.476	0.062	1.558	4.367	0.820
Nutrient transfers						
Litterfall, 1970–71	33.87 (total N)		5.03 (total P)	18.07	44.49	6.55
Nutrient exports						
Runoff, 1972–73	0.030	0.045	0.030	6.675	9.435	4.455
Runoff, 1974–75	0.078	0.065	0.013	5.642	7.943	4.108

Rainfall and runoff data are for June–May water years and are from Swank and Douglass (1977); litter data are from Cromack and Monk (1975).

soils are fertile, nutrients are far more abundant in the runoff (Bruijnzeel 1991).

In weathering-limited settings, where soils are shallow—less than 2 m deep—the forest's species composition and cation content of the soil, depends markedly on the constitution of its bedrock (Baillie and Ashton 1983, Burnham 1989). In some transport-limited settings, where soils are deep, weathering is intense, and the vegetation receives few nutrients from weathering bedrock, so the species composition of the forest and the cation content of its soil are remarkably independent of the constitution of its bedrock (Burnham 1989). Weathering in tropical soils tends to produce an abundance of aluminum and iron sesquioxides, which form clays capable of binding phosphorus nearly irreversibly. Thus tropical forests on "typical" well-weathered oxisols and ultisols is limited by lack of phosphorus rather than by lack of nitrogen (Vitousek 1984: p. 294). Such forests cycle phosphorus (and calcium) far more efficiently than they do nitrogen (Vitousek and Sanford 1986).

Such soils have been around a long time. Retallack and Germán-Heins (1994) described a rain forest paleosol from the middle Pennsylvanian of Missouri. This was a deeply weathered, oxidized, clayey soil, poor in weatherable bases and rich in kaolinite. This paleosol is capped by a dense root mat 3 cm thick, and it is penetrated to a depth of 2 m by apparently woody roots up to 6 mm thick. It appears that deeply weathered tropical lowland soils often converge on a common type.

Not all tropical soils, however, converge on this common type, even where there is ample time to do so. In Venezuela, the Guayana Shield lies exposed north and east of the Orinoco. On the northern flank of this shield, bordering the Orinoco near the Rio Aro, the bedrock is pre-Cambrian crust, formed more than a billion years ago (Edmond et al. 1995). Where this bedrock is basic (basalt, basaltic andesite, or greenstone belts), the forest is tall and farming is practicable. Nearby granites, however, weather into a soil so wretched that they support neither crops nor pastures. On these granites, the natural vegetation is an open grassland, with occasional small trees. A decisive difference is that the soil derived from basic rocks holds more water (R. F. Stallard, p. c.).

The impact of rainforest on its soil is a complex story. This impact is usually beneficial. In deeply weathered soils such as those near Manaus, Brazil, organic acids released by roots and the decomposition of litter promote the deposition of kaolinite in the top 3 m or so of soil, in part by facilitating the downward leaching of iron and aluminum (Stallard 1988). By preventing the transformation of kaolinite into gibbsite (pure aluminum hydroxide) in the upper soil, these organic acids protect the forest from aluminum poisoning and preserve at least a remnant of the soil's cation exchange capacity. Cation exchange capacity, the readiness with which the soil exchanges cations with the surrounding water, is an aspect of soil fertility. In central Amazonia, plant products enhance cation exchange capacity more directly. On Barro Colorado, most of the soil's cation exchange capacity is contributed by clays such as montmorillonite, but in well-weathered Amazonian soils, most of the cation exchange capacity is provided by organic matter bound to particles of soil (R. F. Stallard, p. c.).

On the other hand, in cloud forest settings with low evapotranspiration, the high content of phenolics and tannins, toxins with which the trees defend their leaves from herbivores, in falling leaves, poison the soil, for these long-lived toxins block decomposition and recycling of nutrients (Bruijnzeel and Proctor 1995, Kitayama 1995). In Sarawak, *Casuarina nobilis* planted on a sandy clay-loam suitable for mixed dipterocarp forest soon produced a layer of raw humus several centimeters thick that led to the bleaching of the top 5 or 10 cm of soil (Bruening 1996: p. 23). Persistent pesticides evolved by natural selection can be as bad for the soil as their counterparts of human invention. When it pays plants to benefit their soil and when it does not is a question that will be taken up in chapter 9.

CONCLUDING REMARKS

Runoff can provide a reliable water supply, or its variation can lead to an alternation of severe water shortage with violent floods. Deforestation need not worsen seasonal water shortage or enhance the violence of floods. It is more likely to cause such inconvenience, however, the less care that is taken of the land during its clearing and subsequent use.

Runoff also causes erosion. Its content of dissolved and suspended sediments reflects the process of soil formation. On Barro Colorado, one can see the contrast between the rapid weathering-limited erosion resulting from the rugged topography of Lutz catchment and the much slower, transport-limited erosion on the level surfaces of the central plateau, which illustrates the contrast between erosion of mountain landscapes and well-weathered plains.

Forest cover usually slows erosion. Again, the influence of deforestation on erosion reflects how much care was taken to protect the land when clearing it, which in turn reflects the degree to which the people involved understand the land, and how much, and how obviously, they benefit from its future health. The linkage between greed for immediate profit, regardless of the land's future prospects, social injustice toward the original inhabitants of the forest, and natural disaster, outlined by the Hebrew prophets of the Old Testament (Northcott 1996), finds clear illustration in the relation between land use, erosion, and runoff.

Do tropical forest soils converge on a common type? Not always. The proportion of tropical soils likely to converge in relevant properties and the extent to which this convergence explains the relatively uniform appearance of tropical rainforest commented upon by Schimper (1903: p. 378) is an open question.

Acknowledgments This chapter owes an incalculable debt to the patient guidance of R. F. Stallard, who has answered a host of questions and generously supplied me with a wealth of unpublished data. I am equally indebted to the writings of L. A. Bruijnzeel. Although I could not have done this job without them, any mistakes here are my own.

References

Baillie, I. C. 1989. Soil characteristics and classification in relation to the mineral nutrition of tropical wooded ecosystems, pp. 15–26. In J. Proctor, ed., *Mineral Nutrients in Tropical Forest and Savanna Ecosystems*. Blackwell Scientific Publications, Oxford.

Baillie, I. C., and P. S. Ashton. 1983. Some soil aspects of the nutrient cycle in mixed Dipterocarp forests in Sarawak, pp. 347–356. In S. L. Sutton, T. C. Whitmore, and A. C. Chadwick, eds., *Tropical Rain Forest: Ecology and Management*. Blackwell Scientific Publications, Oxford.

Bailly, C., G. Benoit de Coignac, C. Malvos, J. M. Hingre, and J. M. Sarrailh. 1974. Étude de l'influence du couvert naturel et de ses modifications á Madagascar. *Bois et Forêts des Tropiques* (supplement no. 4: 1–115).

Berner, R. A. 1992. Weathering, plants, and the long-term carbon cycle. *Geochimica et Cosmochimica Acta* 56: 3225–3231.

Brown, E. T., R. F. Stallard, M. C. Larsen, G. M. Raisbeck, and F. Yiou. 1995. Denudation rates determined from the accumulation of in situ-produced ^{10}Be in the Luquillo Experimental Forest, Puerto Rico. *Earth and Planetary Science Letters* 129: 193–202.

Bruenig, E. F. 1996. *Conservation and Management of Tropical Rainforests*. CAB International, Wallingford, UK.

Bruijnzeel, L. A. 1988. Deforestation and dry season flow in the tropics: a closer look. *Journal of Tropical Forest Science* 1: 229–243.

Bruijnzeel, L. A. 1990. *Hydrology of Moist Tropical Forests and Effects of Conversion: A State of Knowledge Review*. Free University, Amsterdam.

Bruijnzeel, L. A. 1991. Nutrient input-output budgets of tropical forest ecosystems: a review. *Journal of Tropical Ecology* 7: 1–24.

Bruijnzeel, L. A., and J. Proctor. 1995. Hydrology and biogeochemistry of tropical montane cloud forests: what do we really know? pp. 38–78. In L. Hamilton, J. O. Juvik, and F. N. Scatena, eds., *Tropical Montane Cloud Forests*. Springer-Verlag, New York.

Burnham, C. P. 1989. Pedological processes and nutrient supply from parent material in tropical soils, pp. 27–41. In J. Proctor, ed., *Mineral Nutrients in Tropical Forest and Savanna Ecosystems*. Blackwell Scientific Publications, Oxford.

Cleaves, E. T., A. E. Godfrey, and O. P. Bricker. 1970. Geochemical balance of a small watershed and its geomorphic implications. *Geological Society of America Bulletin* 81: 3015–3032.

Cromack, K. Jr., and C. D. Monk. 1975. Litter production, decomposition, and nutrient cycling in a mixed hardwood watershed and a white pine watershed, pp. 609–624. In F. G. Howell, J. B. Gentry, and M. H. Smith, eds., *Mineral Cycling in Southeastern Ecosystems*. ERDA Symposium Series. National Technical Information Service, Springfield, VA.

Dietrich, W. E., D. M. Windsor, and T. Dunne. 1982. Geology, climate and hydrology of Barro Colorado Island, pp. 21–46. In E. G. Leigh, Jr., A. S. Rand, and D. M. Windsor, eds., *Ecology of a Tropical Forest*. Smithsonian Institution Press, Washington, DC.

Edmond, J. M., M. R. Palmer, C. I. Measures, B. Grant, and R. F. Stallard. 1995. The fluvial geochemistry and denudation rate of the Guayana Shield in Venezuela, Colombia, and Brazil. *Geochimica et Cosmimica Acta* 59: 3301–3325.

Foster, R. B., and N. V. L. Brokaw. 1982. Structure and history of the vegetation of Barro Colorado Island, pp. 67–81. In E. G. Leigh, Jr., A. S. Rand, and D. M. Windsor, eds., *The Ecology of a Tropical Forest*. Smithsonian Institution Press, Washington, DC.

Franken, W., and P. R. Leopoldo. 1984. Hydrology of catchment areas of Central-Amazonian forest streams, pp. 501–519. In H. Sioli, ed., *The Amazon: Limnology and Landscape Ecology of a Mighty Tropical River and Its Basin*. Dordrecht: W. Junk.

Franken, W., and P. R. Leopoldo. 1987. Relações entre fluxos de água subterrânea e superficial em bacia hidrográfica caracterizada por cobertura florestal amazônica. *Acta Amazônica* 16/17: 253–262.

Hertzler, R. A. 1938. Determination of a formula for the 120-deg V Notch Weirs. *Civil Engineering* 8: 756.

Johnsson, M. J., and R. F. Stallard. 1989. Physiographic controls on the composition of sediments derived from volcanic and sedimentary terrains on Barro Colorado Island, Panama. *Journal of Sedimentary Petrology* 59: 768–781.

Kitayama, K. 1995. Biophysical conditions of the montane cloud forests of Mount Kinabalu, Sabah, Malaysia, pp. 183–197. In L. Hamilton, J. O. Juvik, and F. N. Scatena, eds., *Tropical Mountain Cloud Forests*. Springer-Verlag, New York.

Klinge, H., and W. A. Rodrigues. 1968a. Litter production in an area of Amazonian terra firme forest. Part I. Litter-fall, organic carbon and total nitrogen content of litter. *Amazoniana* 1: 287–302.

Klinge, H., and W. A. Rodrigues. 1968b. Litter production in an area of Amazonian terra firme forest. Part II. Mineral nutrient content of the litter. *Amazoniana* 1: 303–310.

Larsen, M. C. 1996. Mass wasting in the humid tropics: the agriculturally developed Cayaguás watershed, Puerto Rico—an extreme case. *Geological Society of America, Abstracts and Program* 28(7): A–80.

Larsen, M. C., and A. Simon. 1993. A rainfall intensity-duration threshold for landslides in a humid-tropical environment, Puerto Rico. *Geografisker Annaler* 75A: 13–23.

Larsen, M. C., and A. J. Torres-Sánchez. 1996. Geographic relations of landslide distribution and assessment of landslide hazards in the Blanco, Cibuco, and Coamo Basins, Puerto Rico. *U. S. Geological Survey Water-Resources Investigations Report* 95–4029: 1–56.

Leigh, E. G. Jr., and D. M. Windsor. 1982. Forest production and regulation of primary consumers on Barro Colorado Island, pp. 111–122. In E. G. Leigh, Jr., A. S. Rand, and D. M. Windsor, eds., *The Ecology of a Tropical Forest*. Smithsonian Institution Press, Washington, DC.

Leigh, E. G. Jr., and S. J. Wright. 1990. Barro Colorado Island and tropical biology, pp. 28–47. In A. H. Gentry, ed., *Four Neotropical Rainforests*. Yale University Press, New Haven, CT.

Leopoldo, P. R., W. Franken, and E. Matsui. 1984. Hydrological aspects of the tropical forest in central Amazon. *Interciencia* 9: 125–131.

Marshall, T. J., and J. W. Holmes. 1988. *Soil Physics* (2nd ed). Cambridge University Press, Cambridge.

Meade, R. H., T. R. Yuzyk, and T. J. Day. 1990. Movement and storage of sediments in rivers of the United States and Canada, pp. 255–280. In M. G. Wolman and H. C. Riggs, eds., *Surface Water Hydrology*, vol. 0–1, *The Geology of North America*. Geological Society of America, Boulder, CO.

Newson, M. D. 1979. The results of ten years' experimental study on Plynlimon, mid-Wales, and their importance for the water industry. *Journal of the Institute of Water Engineers and Scientists* 33: 321–333.

Nortcliff, S., and J. B. Thornes. 1989. Variations in soil nutrients in relation to soil moisture status in a tropical forested ecosystem, pp. 43–54. In J. Proctor, ed., *Mineral Nutrients in Tropical Forest and Savanna Ecosystems*. Blackwell Scientific Publications, Oxford.

Northcott, M. S. 1996. *The Environment and Christian Ethics*. Cambridge University Press, Cambridge.

Parker, G. G. 1994. Soil fertility, nutrient acquisition, and nutrient cycling, pp. 54–63. In L. A. McDade, K. S. Bawa, H. A. Hespenheide, and G. S. Hartshorn, eds. *La Selva*. University of Chicago Press.

Retallack, G. J., and J. Germán-Heins. 1994. Evidence from paleosols for the geological antiquity of rain forest. *Science* 265: 499–502.

Schimper, A. F. W. 1903. *Plant-Geography upon a Physiological Basis*. Clarendon Press, Oxford.

Schlesinger, W. H. 1991. *Biogeochemistry*. Academic Press, San Diego, CA.

Silver, W. L. 1994. Is nutrient availability related to plant nutrient use in humid tropical forests? *Oecologia* 98: 336–343.

Sollins, P. 1989. Factors affecting nutrient cycling in tropical soils, pp. 85–95. In J. Proctor, ed., *Mineral Nutrients in Tropical Forest and Savanna Ecosystems*. Blackwell Scientific Publications, Oxford.

Stallard, R. F. 1985. River chemistry, geology, geomorphology, and soils in the Amazon and Orinoco basins, pp. 293–316. In J. I. Drever, ed., *The Chemistry of Weathering*. Reidel, Dordrecht.

Stallard, R. F. 1988. Weathering and erosion in the humid tropics, pp. 225–246. In A. Lerman and M. Maybeck, eds., *Physical and Chemical Weathering in Geochemical Cycles*. Kluwer Academic Publishers, Dordrecht.

Stallard, R. F. 1995. Tectonic, environmental and human aspects of weathering and erosion: a global review using a steady-state perspective. *Annual Review of Earth and Planetary Sciences* 23: 11–39.

Stallard, R. F., and J. M. Edmond. 1981. Geochemistry of the Amazon 1. Precipitation chemistry and the marine contribution to the dissolved load at the time of peak discharge. *Journal of Geophysical Research* 86: 9844–9858.

Swank, W. T., and J. E. Douglass. 1977. Nutrient budgets for undisturbed and manipulated hardwood forest ecosystems in the mountains of North Carolina, pp. 343–364. In D. L. Correll, ed., *Watershed Research in Eastern North America*, vol. 1. Chesapeake Bay Center for Environmental Studies, Smithsonian Institution, Edgewater, MD.

Vitousek, P. M. 1984. Litterfall, nutrient cycling, and nutrient limitation in tropical forests. *Ecology* 65: 285–298.

Vitousek, P. M., and R. L. Sanford Jr. 1986. Nutrient cycling in moist tropical forest. *Annual Review of Ecology and Systematics* 17: 137–167.

Woodring, W. P. 1958. Geology of Barro Colorado Island, Canal Zone. *Smithsonian Miscellaneous Collections* 135(3): 1–39.

Yavitt, J. B., and R. K. Wieder. 1988. Nitrogen, phosphorus and sulfur properties of some forest soils on Barro Colorado Island, Panama. *Biotropica* 20: 2–10.

FIVE

Telling the Trees from the Forest

Tree Shapes and Leaf Arrangement

What are tropical forests like? A proper lowland rainforest is tall, but not as tall as a redwood forest; its understory is dark, but not much darker than that of a mature deciduous forest of Maryland in summertime. In many forests, the visitor will remark how many trees have relatively slender, straight trunks, and smooth, pale bark (Richards 1952, 1996). Some of the trees are encumbered by huge lianes, woody vines sometimes more than 10 cm thick; some of the larger trees may have branches festooned with epiphytes. Even the leaves may become covered by "epiphylls," liverworts or lichens forming a thin coating that may preempt up to 30% of a leaf's light (Coley et al. 1993, Coley and Kursar 1996: p. 350). The appearance of tropical forest seems dominated by the intense struggle for light (Schimper 1903: p. 291), and the abundance of plants which take advantage of other trees' wood to place themselves in the sun. Indeed, smooth bark seems to deter epiphytes (Campbell and Newbery 1993). More generally, some trees are designed to avoid or shed lianes (Putz 1984, Campbell and Newbery 1993).

Especially on soggy ground, many of the canopy trees have flanges jutting out from the base. These "buttresses" are best developed on the side of the tree facing the wind: they serve as "tension members" (Richter 1984, Richards 1996), like the support cables that keep radio towers from blowing over, and in contrast to the flying buttresses of medieval cathedrals, which are meant to resist compression. Many understory leaves have "drip tips," long slender, pendant tips, which allow water to drain more rapidly from the leaf (Dean and Smith 1978): drip tips are among the many ways tropical plants avoid or shed rainwater (Dean and Smith 1978).

The overwhelming impression of many first-time visitors to tropical forest, however, is the difficulty of distinguishing individual kinds of tree from the overwhelming mass of greenery before them. They cannot see the trees for the forest. The difficulty is compounded because the visitor from Europe or North America is accustomed to distinguish such trees as pines, maples, oaks, tulip poplars, elms and holly by a glance at their leaves. In a moist or wet tropical forest, however, most trees have leaves or leaflets with entire (smooth) margins (Bailey and Sinnott 1915, 1916; Wolfe 1971, 1978; Givnish 1979), and the silhouettes of leaves of different species can be quite confusingly similar (see Richards 1952: p. 81). The relation between a forest's climate and the proportion of its species with entire leaves is a useful tool for inferring the climates of fossil floras (Wolfe 1971), but it does not help us tell apart the tree species in tropical forest.

On the other hand, tropical trees exhibit an extraordinary variety of branching patterns. Unlike many desert plants and some epiphytes, where most or all of the photosynthesis occurs in stems, tropical forests have a fairly sharp division of labor between leaves, which photosynthesize, and twigs, branches and stems, which support these leaves. Some rainforest plants have green twigs, or green twig-ends, whose contribution to photosynthesis have not been studied, but the division of labor between photosynthesis and support is much sharper in forest than in deserts or grassland.

Looking at the branching patterns ("architectures") of different tropical trees and the arrangements of their leaves helps bring these trees into sharp enough focus to allow different kinds to be distinguished. Attending to the architecture of trees and the sizes, shapes, textures, and arrangements of their leaves also brings other advantages. Comparing the branching patterns and leaf arrangements of canopy with understory plants or pioneer with mature forest species reveals something about the problems faced by plants in these different stations. This focus allows us to begin to make sense of a forest, or to compare different forests, before we have learned to identify any of the trees.

Comparing the spectrum of sizes, shapes, and arrangements of leaves in forests of different climates provides a tool for inferring the climates of fossil floras (Bews 1927). Comparing how the spectrum of tree architectures and leaf characteristics of tropical forest vary with climate and soil in different biogeographic realms tells us something about the predictability of evolution—if we can identify the relevant features of climate and soil.

In this chapter, I first distinguish different tree architectures. To learn whether different architectures have different advantages, I first review some principles of tree engineering and design. Then I resort to comparisons: the growth forms of herbs on the floors of different tropical forests, the contrast between the architectures of understory versus canopy plants, the relation between climate and leaf size, leaf arrangement, and forest structure in different biogeographic realms. Finally, I employ a variety of comparisons to learn what factors stunt the "elfin woodlands" of foggy tropical mountaintops. Exploring this problem will raise the issue of what governs the structure and limits the productivity of tropical rainforest.

DIFFERENT ARCHITECTURES OF TROPICAL TREES

The Art of Leaf Arrangement

There is an obvious contrast between plants that arrange their leaves along the sides of horizontal twigs to form horizontal sprays of foliage—plants with "distichous" leaves, like beeches and elms—and plants which arrange their leaves around erect or ascending twigs, as do poplars, rhododendrons, and red mangroves. Some plants with horizontal twigs and distichous leaves have opposite leaves, each leaf-node carrying a leaf on each side of the twig. Others have alternate leaves, one leaf per node, with successive leaves on alternating sides of the twig (Fig. 5.1). Some plants with erect or ascending twigs, like ashes or maples, also have opposite leaves, with each pair rotated 90° around the stem from its predecessor: this leaf-arrangement is called decussate. Others, like rhododendrons and palms, set each leaf a fixed angle around the stem from its predecessor: this leaf arrangement is called spiral because a line drawn through successive leaf bases forms a spiral around the stem.

Plants with distichous leaves are more common in the understory than in the canopy, as we shall see later on. In shady conditions, it is desirable to minimize leaf overlap, so that each leaf may capture as much light as possible. And, indeed, distichous leaf arrangements facilitate the avoidance of leaf overlap (Fig. 5.1). Yet many understory plants carry a tuft of spirally arranged leaves atop an unbranched stem (Fig. 5.1, 5.3). How do these plants avoid leaf overlap?

Consider a plant that sets horizontal leaves around a vertical stem, each the same fraction, α, of a revolution around the stem from its predecessor. What is the "divergence angle" α that minimizes leaf overlap? If, for some whole number, n, $n\alpha$ is equal to some whole number, m, then leaf n stands precisely over leaf 0: their overlap is complete. The more $n\alpha$ differs from a whole number, the less leaf n overlaps leaf 0.

In mathematical terms, we wish to calculate that angle α such that finding a rational number m/n satisfying the condition $|n\alpha - m| < \varepsilon$ requires as large a denominator n as possible, whatever positive value of ε less than 0.5 one chooses. The solution to this problem links the golden section of the Greeks, Fibonacci series, the numbers of intersecting spirals which we see in the scales of pine cones or the leaf scars on the trunks of palm or papaya, and the best way to approximate irrational numbers. To begin, let us see how the method of continued fractions (Khinchin 1964, Hardy and Wright 1979: pp. 129–177) enables the approximation of the irrational number $1/\pi = 0.318309886 \ldots$, by rational fractions. First, set

$$\frac{1}{\pi} = \frac{1}{3 + 0.141592654\ldots}. \quad (5.1)$$

Continuing the process, set

$$0.141592625\ldots = \frac{1}{7 + 0.06251330\ldots}, \quad (5.2)$$

$$0.06251330\ldots = \frac{1}{15 + 0.9965944}, \quad (5.3)$$

$$0.9965944\ldots = \frac{1}{1 + 0.0034172}, \quad (5.4)$$

and so forth, as far as one chooses. This process yields successive "convergents" or approximations to $1/\pi$:

$$\frac{1}{3} = \frac{p_1}{q_1}, \frac{1}{3 + 1/7} = \frac{7}{22} = \frac{p_2}{q_2} \quad (5.5)$$

$$\frac{1}{3 + \cfrac{1}{7 + \cfrac{1}{15}}} = \frac{106}{333} = \frac{p_3}{q_3}, \frac{1}{3 + \cfrac{1}{7 + \cfrac{1}{15 + \cfrac{1}{1+\ldots}}}} = \frac{113}{355} = \frac{p_4}{q_4}. \quad (5.6)$$

More generally, the continued fraction of a number x between 0 and 1 is

$$x = \cfrac{1}{a_1 + \cfrac{1}{a_2 + \ldots}}, \quad (5.7)$$

where

$$a_1 = \left[\frac{1}{x}\right], a_2 = \left[\cfrac{1}{\cfrac{1}{x} - \left[\cfrac{1}{x}\right]}\right]. \quad (5.8)$$

Here, $[x]$ denotes the largest integer less than x.

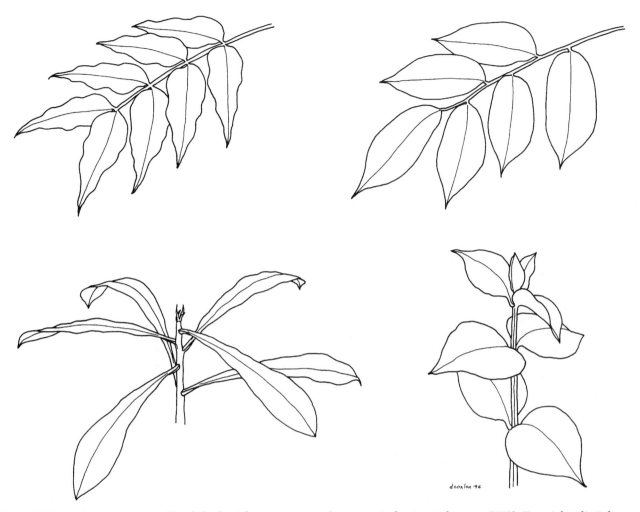

Figure 5.1. Leaf arrangements. Top left: distichous, opposite (opposite in horizontal spray, OHS). Top right: distichous, alternate (alternate in horizontal spray, AHS). Bottom left: spiral. Bottom right: decussate. Drawing by Donna Conlon.

Successive q_i/p_i approximate π quite rapidly (Table 5.1). These successive "convergents" have the property that $q_i(1/\pi) - p_i$ is closer to 0 than $n(1/\pi) - m$ where n and m are any pair of whole numbers less than q_i (Khinchin 1964: pp. 26–28; Hardy and Wright 1979: pp. 136–152). Notice that for each convergent p_i/q_i, the larger the remainder and the smaller its inverse, the less accurately $q_i(1/\pi) - p_i$ approximates 0 (Table 5.2). Thus, to minimize leaf overlap, the divergence angle α between successive leaves should have a continued fraction whose denominators a_1, a_2, etc., are all 1's (Linsbauer et al. 1903, Leigh 1972, Hardy and Wright 1979: pp. 163–166):

$$\alpha = \cfrac{1}{1+\cfrac{1}{1+\cfrac{1}{1+\cfrac{1}{1+..}}}} = \frac{1}{1+\alpha} = \frac{1}{2}(\sqrt{5}-1). \quad (5.9)$$

This fraction is the "golden section." The Greeks felt that rectangles with sides in the ratio $1:\alpha$ were especially well proportioned. The successive "convergents" for α are $p_1/q_1 = 1$, $p_2/q_2 = 1/2$, $p_3/q_3 = 2/3$, $p_4/q_4 = 3/5$, $p_5/q_5 = 5/8$, and so forth. Each convergent is the ratio of successive members of the Fibonacci series: $p_1 = p_2 = 1$, $p_n = p_{n-1} + p_{n-2}$ for $n > 2$, and $q_n = p_{n+1}$. If the vertical displacement along the stem of leaf n from leaf $n + q_k$ is comparable to the horizontal distance between them, the eye will readily pick out the spiral through leaves n, $n + q_k$, $n + 2q_k$, $n + 3q_k$ and its $q_k - 1$ parallels; passing in the other direction, these same leaf scars will sort out into p_k or q_{k+1} spirals. This is the feature connecting the divergence angle yielding minimum overlap to Fibonacci numbers of intersecting spirals.

A divergence angle near 137.5°, which represents the fraction $1 - \alpha$ of a revolution, is characteristic of most plants with spirally arranged leaves. This angle is most precisely matched by plants with little vertical displacement between successive leaves. Where there are many leaves in one spiral, petioles of older leaves are long, or leaf bases are narrow, to further minimize leaf overlap close to the stem (Leigh 1972, Blanc 1992).

Table 5.1 Approximating π by successive convergents

$(q/p)1$	$(q/p)2$	$(q/p)3$	$(q/p)4$	$(q/p)5$
3/1	22/7	333/106	355/113	103993/33102
3.0000000	3.142857...	3.141509...	3.1415929...	3.14159265...

How Many Ways Are There to Build a Tree?

Especially in tropical forest, trees come in many shapes (Corner 1940, 1964). An unbranched trunk may support a huge rosette of leaves at its top, as does a palm or a tree fern. Other trees may bear a few branches, each ending in a rosette of large leaves, like a papaya with side branches. Other trees, with distichous leaves, bear many branches that ramify into horizontal sprays of foliage. In some of these species, like elms, the leader shoot leans over to cast a horizontal spray of foliage, and a relay sprouts from the bend in the leader to form another, and so forth; others, like holly, have straight, erect trunks with horizontal branches spaced evenly along it. Other trees, with spirally arranged leaves, produce branches forming horizontal sprays of leaf-rosettes, *Terminalia* fashion. Yet others arrange their leaves around ascending twigs set on ascending branches; some, like poplars, have straight trunks, while the trunks of others, such as *Schefflera*, may bifurcate repeatedly.

Spiral and alternate (distichous) leaf arrangements are intimately related. As a seedling or young sapling, a plant may first set leaves spirally around an erect stem and then produce horizontal branches in the axils of these leaves (that is to say, a branch leaves the stem just above the base of each stem leaf). These horizontal branches bear distichous leaves to form horizontal sprays of foliage. Here, the divergence angle that minimizes self-shading among the stem leaves also minimizes overlap among the branches. This is especially true in plants like *Virola* (Fig. 5.2), which, every so often, produce a close-set spiral of leaves (that is to say, a spiral with little vertical displacement between successive leaves). Branches forming a horizontal layer of foliage then sprout from the axils of these leaves.

Terminalia spp. use yet another approach to forming horizontal sprays of foliage. A horizontal branch turns upward to support a spiral of leaves at its top. One or more horizontal relay twigs sprout from the bend in this branch to repeat the process (Fig. 5.2). The branches of the pagoda tree, *Terminalia catappa*, characteristic of Southeast Asian beaches and Neotropical golf courses, are designed to intercept nearly all the light coming their way with a minimum of leaf overlap. Computer simulations have been used to calculate the average ratio of twig length to leaf rosette diameter and the average angles between relay twigs and their predecessors, which yield the most efficient coverage of space by leaf rosettes (Fisher and Honda 1979a, 1979b, Honda and Fisher 1979). Overlap among the (usually five) branches within a layer, like the overlap of leaves within a rosette, is minimized by setting each branch a golden section of a revolution around the stem from its predecessor. Like honeybees adding cells to their comb, *Terminalia* trees are good geometers.

Hallé and Oldeman (1970) provided an orderly framework by which to classify tree architecture. Their classification was based on both tree shape and its mode of genesis, the mode of growth and branching which produced it: especially the relation of flowering to branch origin (Table 5.3). Theirs is much the best approach if one can watch trees grow from seed to adult. A field worker, however, would be hard put to distinguish the models of Holttum, Corner, and Chamberlain (IA, IB, and IIIC1 in Table 5.3) during a brief forest survey: trees of all three models carry a tuft of leaves atop an unbranched stem. In addition, users of the Hallé and Oldeman scheme face two other problems.

First, as the tree matures, it may produce supplementary shoots on its branches or trunk which "reiterate" the architecture of the young tree (Oldeman 1974, Hallé et al. 1978: pp. 269–284). Reiteration plays an essential role in the growth of many canopy trees, providing them with the flexibility to exploit the distribution of available light (Hallé 1986b, Edelin 1990). Sometimes reiteration is so regular that it seems to represent a higher-order architec-

Table 5.2 Values of $q_k X - p_k$ for successive convergents to different irrational fractions X

$\sqrt{2} - 1$*		$(\sqrt{5} - 1)/2$		$1/\pi$	
$2X - 1$	$= -0.17157$	$2X - 1$	$= 0.23607$	$3X - 1$	$= -0.045070$
$5X - 2$	$= 0.07107$	$5X - 3$	$= 0.09017$		
$12X - 5$	$= -0.02944$	$13X - 8$	$= 0.03444$		
$29X - 12$	$= 0.01219$	$34X - 21$	$= 0.01316$	$22X - 7$	$= 0.002817$
$70X - 29$	$= -0.00505$	$89X - 55$	$= 0.00503$		
$169X - 70$	$= 0.00209$	$144X - 89$	$= -0.00311$		
$408X - 169$	$= -0.00087$	$377X - 233$	$= -0.00119$	$333X - 106$	$= -0.002808$
$985X - 408$	$= 0.00036$	$987X - 610$	$= 0.00045$	$355X - 113$	$= 0.000010$

*$\sqrt{2} - 1$ has a continued fraction whose denominators a_i are all 2s.

Figure 5.2. Top. A layer of foliage of *Virola sebifera* (Myristicaceae) 60 cm wide, Lutz ravine, BCI. Drawing by Marshall Hasbrouck. Middle. A horizontal spray of foliage of *Terminalia amazonica* (Combretaceae) 60 cm long, "Pipeline Road," Parque Soberania, Panama. Drawing by Alex Murawski. Bottom. A horizontal spray of foliage of *Faucheria parvifolia* (Sapotaceae) 40 cm long, Analamazaotra, Madagascar. Drawing by George Angehr.

Table 5.3 Architectural models of Hallé and Oldeman (1970)

I. Unbranched
   A. Holttum. Flowers once at apex, fruits and dies (monocarpic)
   B. Corner. Repeatedly produces flowers and fruits under its leaves (lateral, not terminal, flowering)
II. Branches underground to form clonal clumps
   A. Tomlinson. Shoots do not branch above ground (cf. clonal palms)
   B. McClure. Shoots produce lateral, horizontal branches with distichous leaves, as in clonal bamboos (distichous leaves are arranged along the sides of twigs, as in elm or nutmeg)
III. Branches only above ground
   A. Wholly modular construction: each module (trunk or branch segment) flowers at its tip and forks
      1. Leeuwenberg: each module forks symmetrically: no one of its branches is dominant
      2. Kwan Koriba: modules fork asymmetrically: one branch larger than others
   B. Schoute: Trunk forks without flowering by division of the apical meristem
   C. Modular (sympodial) trunk, composed of a succession of "relays," with each relay (segment of the trunk) originating just below tip of predecessor
      1. Chamberlain: each module flowers at top and produces a single, unbranched relay
      2. Module ends in a whorl of branches
         a. Prévost: Branches upturned, at least at tip; leaves spiralled or decussate, disposed around their twig. Each branch segment flowers and forks at tip as in Kwan Koriba, or flowers at tip and propagates relays from just below the tip to repeat the process, as in Petit or Fagerlind (see below)
         b. Nozeran: Branches horizontal, leaves distichous
   D. Trunk built of leaning relays, each sprouting from where its predecessor bends
      1. Champagnat: Relays ascending or upturned at tip, supporting spiral or decussate leaves
      2. Leafy twigs horizontal, at least at tip, leaves distichous
         a. Troll: Branch leans smoothly over to cast a horizontal spray of foliage; successor sprouts from the bend
         b. Mangenot: Each erect trunk relay turns sharply to form a horizontal branch with distichous leaves: succeeding relay sprouts from the corner to repeat the process
   E. Nonmodular, monopodial trunk, relay branching
      1. Each lateral branch flowers and forks as in Kwan Koriba or Leeuwenberg
         a. Scarrone: Primary branches bunched along straight trunk
         b. Stone: Primary branches distributed continuously along trunk
      2. Branches propagate horizontal sprays of leaf rosettes in relays, *Terminalia* fashion: each twig tip turns upward to support a spiral of leaves and sprouts horizontal relays from the bend to repeat the process
         a. Each branch tip eventually ends in an inflorescence
            i. Fagerlind: Primary branches in bunches or layers along the straight trunk
            ii. Petit: Primary branches distributed continuously along stem
         b. Aubreville: Primary branches bunched along straight stem; flowers appear below or among leaves, and do not terminate growth of leafing in twig tips.
   F. Straight "monopodial" trunk, branches not formed of relays
      1. Leaves spiral or decussate around erect or upturned twigs
         a. Attims: Branches distributed continuously along branches of next higher order in spiral or decussate arrangement
         b. Rauh: Branches in bunches or whorls along branches of next higher order
      2. Leaves distichous
         a. Roux: Primary branches horizontal, distributed continuously along trunk in spiral or decussate arrangement. These branches are not shed as units
         b. Cook: Primary branches horizontal, distributed continuously along trunk, but shed as units, as if they were palm fronds
         c. Massart: Primary branches horizontal, in bunches or whorls along the trunk, forming discrete layers of foliage

tural model. The papaya, *Carica papaya*, is assigned to Corner's model, because, like a palm, it can mature sexually and flower repeatedly without branching. Yet many papayas carry ascending branches, some with branches of their own, evenly distributed over their trunks (cf. Fig. 3 in Oldeman 1974); a neophyte might assign this papaya to Attims's model (III F1a in Table 5.3). The rubber tree, *Hevea brasiliensis*, is classified as Rauh because its straight trunk bears quasi-whorls of ascending branches, which in turn may bear quasi-whorls of second-order branches. Mature rubber trees, however, often have Rauh branches distributed evenly along their straight trunks, as if they could be labeled "Attims with Rauh primary branches" or an "Attims with Rauh axes" (see Hallé 1986a: Fig. IX). The *Thevetia* used to illustrate Champagnat's model (see Fig. 5.7) is Leeuwenberg in better-lit situations, as if, in deep shade, it grows by reiteration from shoots that were never able to flower, and leaned over (D. King, p. c.). The point is that a plant's architecture supposes a preprogrammed relationship between branching and leafing, which can be quite rigid, as in the talipot palm, which produces leaves to a metronomic rhythm, each one a fixed angle around the stem from its predecessor, until it has stored enough starch to reproduce, at which point it flowers, fruits, and dies (Corner 1966). Reiteration, on the other hand, involves the facultative awakening of a small proportion of a plant's many dormant buds, to form shoots which may recapitulate the architecture of that plant. For this reason, Hallé et al. (1978) and their followers insist on distinguishing architectural models from the products of reiteration.

Second, in species that normally reiterate when mature, the ultimate reiterations in the upper crown are often so reduced that the architecture of the tree species is no longer recognizable in them (Hallé et al. 1978: p. 277; Hallé 1986b). Assigning a specific architectural model to mature canopy trees of such species can be a meaningless exercise even if saplings of these same species are readily classifiable.

Some species, moreover, change architecture as they grow (Hallé and Ng 1981). On Barro Colorado, shaded understory individuals of *Garcinia* (subgenus *Rheedia*) and *Faramea occidentalis* are Roux, with leaves opposite in horizontal sprays on branches which are decussate about the main stem. The petioles of successive leaf pairs, however, seem to have been twisted to make the leaves horizontal. Tall sunlit individuals of these species are Attims, with leaves decussate on leafy twigs which are in turn decussate on the next higher order of branches, and so forth.

Reiteration can occur in association with change of architecture. Saplings of *Tabebuia rosea* (Bignoniaceae), a common roadside tree in the uplands of Panama and Costa Rica, begin by forking dichotomously and symmetrically, as in Leeuwenberg or Schoute (IIIA1 or IIIB in Table 5.3). Fifth-order branches (branches separated by five forks from the trunk) shift toward Kwan Koriba (IIIA2 in Table 5.3): branches pointing most nearly away from the trunk are favored, and continue to fork, while branches pointing upward or inward cease forking after, at most, one or two bifurcations. The end result is a crown shaped like the curved surface of an inverted cone. Finally, a vertical shoot sprouts from the basal fork of this cone and repeats the whole process, complete with both Leeuwenberg and Kwan Koriba phases (Borchert and Tomlinson 1984). The end result is often a crown as neatly layered as that of a young *Terminalia*.

Following Leigh (1990), I have modified Hallé and Oldeman's classification (Table 5.4) to cope with these ambiguities, merging some models that cannot be distinguished without knowing how branching is related to flowering, and adding some "wastebasket categories" for plants which the vagaries of branch shedding, or reiteration and reduction, render otherwise unclassifiable.

DESIGNING TREES: GOALS AND CONSTRAINTS

How can we assess the advantages and disadvantages of different plant architectures? A proper answer demands an understanding of "tree engineering." What problems must tree design solve? How can a tree be designed so as to best solve these problems?

A Fundamental Constraint

A tree must apportion its resources among roots, which secure water and nutrients from the soil; leaves, which when adequately lit and supplied transform CO_2 into carbohydrates; and stems, which give these leaves a suitable place in the sun. The more energy a plant devotes to competing for light, the less energy is available for securing water and nutrients or for defense against herbivores (Tilman 1988: p. 16). The less wood a tree employs to support its leaves at a given height, consistent with its ability to resist wind, rain, and woodborers, the more effective it can be in other aspects of competition (King and Loucks 1978).

How tall can a plant of stem diameter D be before it buckles under its own weight? To see what is involved, imagine a vertical wooden cylinder, fixed at its base, of diameter D and height H (Appendix 5.1). Let its tip be deflected. Then one side of the cylinder is stretched, and the other compressed: the pull of the stretched fibers and the push of the compressed ones create a torque (Fig. 5.13) tending to restore the cylinder to the vertical. The restoring torque is stronger the fatter the cylinder (the larger D) and the stiffer its wood. Stiffness is measured by Young's modulus of elasticity, E. If a force pushes downward on the cylinder's top, the resulting decrease, ΔH, in its height is related to this force, F, by the equation $F/(\pi D^2/4) = E\Delta H/H$, provided that the cylinder springs back to its original height when the force ceases. The same constant, E, measures the proportionate stretch when the top of the cylinder is pulled upward. Notice that the larger E, the smaller $\Delta H/H$: E really does measure stiffness. The units of E are pressure (Pa, kg/m · sec^2), for E relates a force per unit cross-sectional area to the proportion by which it shrinks or stretches the column. On the other hand, deflecting the tip provides a lever arm by which gravity can pull the column down. The greater the deflection and the heavier the column, the stronger gravity's pull. When can gravity's destabilizing torque overcome the column's elastic restoring torque?

Deflecting the column's tip bends the column so that its centerline forms a segment of a circle of radius R (Fig. 5.13). If $z(x)$ is the distance the column is displaced from the vertical at height x, then $1/R = d^2z/dx^2$ and $z(H) = H^2/2R$: the tip's deflection is proportional to the square of the column's height. Gravity's destabilizing torque is proportional to its lever arm (the tip's deflection, itself proportional to H^2, times the column's weight, proportional to D^2H: the destabilizing torque is thus proportional to D^2H^3. The restoring torque's lever arm is proportional to the column's diameter. The pull on the column's stretched side is proportional to the degree of stretching, itself proportional to D, times the cross-sectional area to which this stretch applies, which is proportional to D^2. The same applies to the push from the compressed side. Thus a restoring torque proportional to D^4 opposes a destabilizing

Table 5.4 Tree architecture classification of Leigh (1990)

I. Trunk lacks above-ground branches: leaves disposed around erect stem, spiral or decussate: Palm model (Fig. 5.3)
II. Above-ground branches present. Trunk sympodial, consisting of successive "modules" or "relays"
   A. Each module forks into two or more successors
      1. Branching, symmetric leaves disposed around erect or upturned twigs to form umbrella crown: Frangipani model (Fig. 5.4) (cf. *Schefflera morototoni, Psychotria acuminata, Plumeria* spp.)
      2. One branch dominates others
         a. All branches ascending, leaves spiral or decussate around their twigs. Kwan Koriba (Fig. 5.5) (cf. *Hura crepitans*)
         b. Each "trunk" relay forks into one nearly errect successor relay and one nearly horizontal branch with distichous leaves. Deflexa model (Fig. 5.6) (cf. *Psychotria deflexa, Psychotria furcata*)
   B. "Trunk" module leans over, sprouts relay from bend which repeats the process
      1. Trunk relays curve evenly, twig tips ascending, leaves decussate or spiral: Champagnat (Fig. 5.7, left) (cf. *Thevetia ahouvai*)
      2. Trunk relays curve evenly, twig tips horizontal, leaves distichous: Troll (Fig. 5.7, right) (cf *Olmedia aspera, Hybanthus prunifolius, Eugenia coloradensis*)
      3. Erect trunk relay bends sharply to become a horizontal branch with distichous leaves: vertical relay sprouts from corner to repeat the process. Mangenot (Fig. 5.8, left) (cf. *Unonopsis pittieri*)
   C. "Trunk" relay ends in a whorl of branches, vertical successor relay sprouts from just below this whorl to repeat the process
      1. Twig tips ascending, leaves spiral or decussate: Prévost
      2. Branches horizontal, leaves distichous: Nozeran (Fig 5.9) (cf. *Cordia bicolor, Mabea occidentalis*)
III. Straight, "monopodial" trunk
   A. Modular or relay branching
      1. Branches in groups or layers along the trunk, each branch forks as in Kwan Koriba, leaves spiral or decussate around twig tips: Scarrone (Fig 5.10, left) (cf. *Anacardium excelsum, Hasseltia floribunda*)
      2. Relay branching to form horizontal sprays of leaf rosettes: each twig turns up to support a rosette of leaves, spiral or decussate, around its tip: horizontal relay twigs sprout from the bend to repeat the process: Terminalia model (Fig. 5.2) (cf. *Terminalia amazonica, Macrocnemum glabrescens*)
   B. Branches not constructed of relays
      1. Leaves spiral or decussate around erect or upturned twigs
         a. Branches distributed evenly about branches of next higher order, spiral or decussate. Attims (Fig. 5.10, right) (cf. *Calophyllum longifolium, Rhizophora* spp.)
         b. Branches bunched or whorled about branches of next higher order: Rauh (cf. *Cavanillesia platanifolia*)
      2. Leaves distichous, disposed along horizontal axes
         a. Primary branches horizontal, decussate or spiral around main stem, not layered: Casearia model (Fig 5.8, right) (cf. *Zuelania guidonia*)
         b. Primary branches horizontal, grouped in bunches or whorls: Virola model (Fig. 5.11) (cf. *Quararibea asterolepis, Virola* spp.)
IV. Leaves spiral or decussate about ascending or upturned twigs, branching pattern not readily classifiable
   A. Umbrella crown: Almendro model (Fig. 5.12) (cf. *Dipteryx panamensis*)
   B. Rambling crown, often tall and narrow, few vertical axes: Psychotria model (cf. *Psychotria horizontalis*)
   C. Rather rambling crown, numerous vertical or inclined axes: Coussarea model (cf. *Coussarea curvigemmia*)

torque proportional to D^2H^3. Thus the maximum stable height, H_{cr}, should vary with diameter D as H_{cr}^3 to D^2 (Greenhill 1881, McMahon 1973).

Now model a free-standing plant of height H and stem diameter D by a trunk of the same basal diameter, D, made of wood of the same density ρ (kg/m³) and stiffness E (kg/m · sec²), capped by a point mass of the same weight W (kg) as the crown. Let the square of the trunk's diameter $D(x)$ at height x decrease linearly as x increases. This diameter becomes 0 at the "ideal trunk's" top: replace the top tenth of this trunk's length by the point mass representing the crown. The buckling height H_{cr} for this model tree is

$$H_{cr} = 2^{-\frac{2}{3}} \left(\frac{CE}{\rho g}\right)^{\frac{1}{3}} D^{\frac{2}{3}}, \quad C = \frac{5.33 + 60.6K + 234K^2}{1 + 20.4K + 119K^2 + 429K^3} \quad (5.10)$$

(King and Loucks 1978, King 1987; see Appendix 5.1). Here, g is gravitational acceleration, 9.8 m/sec², and K is the ratio of crown mass to trunk mass, where the fictional top tenth of the trunk is assigned to the crown's point mass. To calculate H_{cr}, crown weight, W, must be kept fixed (cf. Holbrook and Putz 1989: p. 1747).

On the average, the actual height is one-fourth the height at which its model would buckle (Niklas 1994), although many understory plants have much narrower safety margins (King 1996). In equation 5.10, C decreases as R increases, while, for larger trees at least, R increases as the tree grows. Therefore H_{cr} and H increase more slowly than $D^{2/3}$ (Niklas 1994). Equation 5.10 works surprisingly well, considering its naive assumption that wood is a homogeneous and isotropic substance: a gross oversimplification, to say the least (Wilson and Archer 1979). In sum, for herbs,

Figure 5.3. An unbranched sapling of *Gustavia superba* (Lecythidaceae), with crown 50 cm wide, Barro Colorado Island. Drawing by Daniel Glanz.

Figure 5.5. *Hura crepitans* (Euphorbiaceae), 3 m tall sapling, laboratory clearing, Barro Colorado. Drawing by George Angehr.

Figure 5.4. *Psychotria acuminata* (Rubiaceae), 1.5 m tall, Barro Colorado. Drawing by Marshall Hasbrouck.

Figure 5.6. *Psychotria deflexa* (Rubiaceae), 2 m tall, Barro Colorado. Drawing by Karen Kraeger.

Figure 5.7. Left. *Thevetia ahouvai* (Apocynaceae), 3 m tall. Right. *Olmedia aspera* (Moraceae), 4 m tall. Drawings by George Angehr.

Figure 5.8. Left. *Unonopsis pittieri* (Annonaceae), 1 m tall sapling, Lutz ravine, Barro Colorado. Right. *Zuelania guidonia* (Flacourtiaceae), 1.3 m tall sapling, edge of laboratory clearing, Barro Colorado. Drawings by George Angehr.

Figure 5.9. L. *Cordia bicolor* (Boraginaceae), 3 m tall sapling and sprig of leaves, young forest of Barro Colorado (WMW 5). Drawing by Marshall Hasbrouck. R. *Mabea occidentalis* (Euphorbiaceae), 3 m tall, Barro Colorado. Drawing by George Angehr.

Figure 5.10. L. Sprig of canopy foliage of *Anacardium excelsum* (Anacardiaceae), 1 m wide, as seen from tower in Lutz ravine. Drawing by George Angehr. R. *Calophyllum longifolium* (Guttiferae), sapling 5 m tall, Barro Colorado. Drawing by Scott Gross.

90 TROPICAL FOREST ECOLOGY

Figure 5.11. *Quararibea asterolepis* (Bombacaceae), 15 m tall, Lutz ravine. Drawing by Daniel Glanz.

like the aspens, they taper to points 1–3 m above the tops of their actual crowns.

Long et al. (1981) suggested that tree trunks are designed to keep the stress on their most strained fibers below a fixed limit when wind blows on their crowns. Wind stress matters. The diameter of a tropical tree increases rapidly when its crown enters the canopy, where it becomes exposed to the wind (King 1996). Many canopy trees of tropical rainforest have buttresses to prevent windthrow: these buttresses grow allometrically, so that large trees with exposed crowns have disproportionately large buttresses (Richter 1984). How should a tree trunk taper so as to most effectively resist the force of the wind on its crown? A wind applying force F on a crown's center of mass causes a torque yF about a point on the trunk y meters below. The strain on the most stretched fibers at level y, the proportion σ by which fibers on the outside of the bent trunk are stretched, is proportional to the torque yF imposed by the wind, times the trunk's diameter $D(y)$ at level y, divided by the restoring torque $D^4(y)$ about level y. For this strain to be independent of y, $yFD(y)/D^4(y)$ must be constant, in which case $y/D^3(y)$ is constant, and $D(y) = \sqrt[3]{y}$ (Dean and Long

saplings, and large trees, height varies with diameter in such a way that providing an adequate margin of safety against buckling appears to be a fundamental principle of tree design (Niklas 1994).

Trunk and Branch Design

Two governing principles have been suggested for trunk design. King and Loucks (1978: p. 149) suggested that a trunk should taper in such a way that it uses the least wood to support a crown of given weight at the height required without buckling under the load. A trunk that tapers according to the rule $D(x) = \sqrt{x}$, where $D(x)$ is trunk diameter a distance x below its (theoretical) tip, needs 20% less wood to support a crown of given weight (about one-fifth that of the trunk) than a trunk with $D(x) = x^{1/4}$ or one with $D(x) = x$. Aspen trunks taper according to the rule $D(x) = \sqrt{x}$ (King and Loucks 1978: pp. 155–156). Similarly, the trunk cross-sectional areas of two *Schefflera morototoni* in the Barro Colorado Nature Monument decline similarly with height over most of their length (Table 5.5), but, un-

Figure 5.12. *Dipteryx panamensis* (Leguminosae: also known as *D. oleifera*), canopy tree, view from tower in Lutz ravine, Barro Colorado. Drawing by Marshall Hasbrouck.

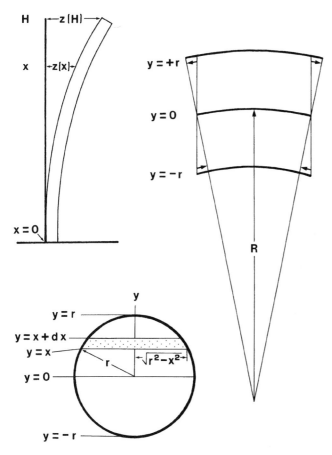

Figure 5.13. Stresses in a bent column. Top left: column of height H and diameter $2r = D$. Its tip is deflected a distance $z(H)$ from the vertical, while at height x above the ground, the column is deflected a distance $z(x)$ from the vertical. Right: a portion of the bent column, viewed horizontally, showing the stretched outer edge of the column, at $y = +r$, the compressed inner edge, at $y = -r$, and the "neutral surface" which is neither compressed nor stretched at $y = 0$. This curved bit of beam can be viewed as forming part of a circle: the neutral surface, which bisects the beam longitudinally, perpendicular to the radius of the circle, lies a distance R from this circle's center. Bottom left: cross-section of bent beam. The pull of the stretched portions above the column's neutral surface and the push of the compressed portions below exert a torque about the line representing the intersection of the neutral surface with this cross-section. The bending moment contributed by the portion of the beam between distances x and $x + dx$ above the neutral surface is the area (stippled in the figure) of the beam's cross-section between these limits, $2dx(r^2 - x^2)^{1/2}$, times the distance, x, of this strip above the neutral surface, times the pull per unit cross-sectional area resulting from the stretching of this portion of the beam. Drawing by Donna Conlon.

1986). Trunks of dominant and codominants in a 45-year-old stand of Douglas fir, *Pseudotsuga menziesii*, in western Washington, and of mature lodgepole pine, *Pinus contorta*, in northern Utah, taper according to this rule (Long et al. 1981, Dean and Long 1986).

Branch design is less well understood. It is often asserted that "all the branches of a tree at every stage of its height" are equal in cross-sectional area to the trunk below them (Zimmermann 1983: pp. 66–67). In the upper crowns of Douglas fir, a branch's cross-sectional area is proportional to the dry weight of foliage it supports (Long et al. 1981), as this rule would imply. In a sense, this rule is obeyed by other trees. Tyree et al. (1991) sectioned a *Schefflera morototoni* of 22.5 cm dbh (diameter at breast height) and height 19.4 m, near Barro Colorado. This species branches dichotomously according to Leeuwenberg's model (Hladik 1970). In the sectioned tree, the total cross-sectional areas of the primary branches, the secondary branches, and the tertiary branches, measured just below their forks, was roughly equal to the cross-sectional area of the bole just below its first fork (Table 5.6), perhaps because, at these points, more than 80% of the cross-sectional area was sapwood. The total cross-sectional area of trunk, primary branches and secondary branches, measured just below their forks, was likewise nearly equal for a *Schefflera pittieri* of dbh 14.6 cm and height 4.5 m (Fig. 5.14) on the windswept ridge above Monteverde in Costa Rica (Table 5.6). Similarly, for an understory shrub on Barro Colorado, also Leeuwenberg, a *Psychotria acuminata* of diameter 1.32 cm at one-tenth its height and height 1.3 m, the sum of the cross-sections of the main stem, primary, secondary and tertiary branches just below their forks were 1.37, 1.34, 1.43, and 1.39 cm², respectively (Leigh, unpublished data).

On the other hand, even in these plants no such regularity related the diameter of a branch at its tip to the basal diameters of its resulting forks. Moreover, in the *Psychotria*, the sum of the cross-sections of branches of order higher than three tended to increase with order number. As yet, our understanding of branch design is far too rudimentary to permit deduction of the distinctive advantages of different architectures.

Empirical Measures of Tree-Performance

Another approach to tree-engineering is to devise empirical measures for assessing the aptness of different architectural models (King 1994). How effectively a plant deploys its leaves is measured by its rate, G, of above-ground dry matter production per unit leaf area, the plant's annual production of dry matter, dM/dt, divided by its total leaf area, LA. A plant's height growth efficiency, HE, is its height gain per year, dh/dt, divided by its above-ground dry matter production per square meter of leaf, $G = (dM/dt)/LA$. Thus $HE = (dh/dt)/G$, $dh/dt = G(HE) = (dM/dt)(HE/LA)$ and $dh/dM = HE/LA$. Suppose now that the total dry weight, M_s, of a plant's stems and branches is related to its height, h, by the law $M_s = ah^\alpha$, where α is an allometric parameter and a is a constant, then $dM_s = a\alpha h^{\alpha-1} dh = \alpha M_s dh/h$, and we obtain

$$\frac{h}{\alpha M_s} = \frac{dh}{dM_s} = \frac{1}{1-F}\left(\frac{dh}{dM}\right) = \frac{1}{(1-F)}\left(\frac{HE}{LA}\right). \quad (5.11)$$

Table 5.5 Trunk taper in *Schefflera morototoni*

Tree felled by Tyree et al. (1991)							
Height above ground, x (m)	0.26	1.38	2.93	5.52	8.52	11.73	14.46
Trunk c.s. at x (cm²)							
Observed	495	398	338	292	240	183	135
Predicted	385	365	<u>338</u>	291	239	183	<u>135</u>
Fallen tree near laboratory clearing							
Height above ground, x (m)	0.38	1.22	3.66	6.71	8.23	11.28	13.41
Trunk c.s. at x (cm²)							
Observed	1483	1098	858	741	689	542	462
Predicted	991	957	<u>858</u>	734	672	549	<u>462</u>

Trunk cross-section (c.s.) at various heights x above the ground; observed and predicted extrapolating linear decline with height from the two underlined values. Data from M. T. Tyree (unpublished) and Leigh (unpublished).

A plant's leaf support efficiency, LSE, is its leaf area divided by the weight of its stems and branches per unit height, M_s/h: $LSE = LA/(M_s/h)$. $M_s/h = \rho A_w$, the density, ρ, of the plant's wood multiplied by the average cross-sectional area, A_w, of wood required to support its leaves. We may therefore set

$$HE = LA\frac{dh}{dM} = LA(1-F)\frac{dh}{dM_s} = (1-F)\, LA\frac{h}{\alpha M_s}$$
$$= \frac{(1-F)\, LSE}{\alpha}$$ (5.12)

$$\frac{dh}{dt} = G(HE) = \frac{G(1-F)\, LSE}{\alpha}. \qquad (5.13)$$

This last equation shows how above-ground dry matter production per unit leaf area, G, the fraction $1-F$ of this production devoted to stems and branches, the leaf support efficiency LSE, and the allometric parameter α affect a plant's height growth.

To illustrate the use of these parameters, let us compare their values for representative saplings with those for a whole forest (Table 5.7). As one might expect, understory saplings produce much less dry matter per unit leaf area, even when relatively well lit, than does the forest at large, and only the most light-demanding species of those measured grows much faster than the forestwide average. The height growth efficiency and the leaf support efficiency of these saplings is higher than the forest-wide average: canopy trees protect the saplings from wind, so saplings expend less on support, but these canopy trees also shade them, imposing a premium on height growth. Understory saplings devote more of their energy to making leaves. Measured values of sapling α are disturbingly low, ranging from 1.4 to 2.8 (King 1994): if normally, a tree's diameter, D, increases at least as rapidly as $h^{3/2}$, while its timber volume and weight varies as D^2h, the production-weighted arithmetic average of α for the forest as a whole should be 4 or higher.

Finally, we measure a plant's shade tolerance by the dry matter production required for it to replace senescing leaves. If M_L is the total dry weight of a plant's leaves, and L is their average lifetime, then the dry matter a plant loses to senescing leaves is M_L/L. The dry weight of leaves a plant produces per unit time is its total above-ground dry matter production, $G(LA)$, times the proportion, F, devoted to leaf making. Leaf making balances leaf loss when

$$\frac{M_L}{L} = FG(LA); \quad G = \frac{1}{FL}\left(\frac{M_L}{(LA)}\right). \qquad (5.14)$$

A plant can survive better in deep shade if it can devote most of its above-ground production to foliage when light is in short supply. Light-demanding plants such as *Cecropia* cannot increase F when light is short. Moreover, their poorly defended leaves are necessarily short lived (Coley 1983, King 1994), so that their ratio of $(M_L/LA)/L$ is three times higher than that of any of King's shade-tolerant species (Table 5.7). Thus, even though at 3% of full skylight *Cecropia* produces twice as much dry matter per unit leaf area as nearly any shade-tolerant species, it cannot "make ends meet" under such shady conditions (King 1994).

King's (1994) approach provides an excellent framework for comparing the performance of different plants; however, King only applied his method to plants < 3 m tall. As most saplings do not develop their characteristic architecture until they are taller, we do not yet know how a

Figure 5.14. Windplaned *Didymopanax pittieri* (Araliaceae), 5 m tall, at Monteverde Cloud Forest Reserve, Costa Rica. Drawing by Alex Murawski.

Table 5.6 Trunk and branch diameters, basal and distal, and area of foliage (or number of leaves) supported by trunk and branches in two species of *Schefflera*

	Height, basal end (m)	cs(cm²)	Height, distal end (m)	cs(cm²)	Total leaf area (m²) or total no. of leaves* supported
Schefflera morototoni, Barro Colorado Nature Monument, Panama					
Third-order branches					
0000	17.2	28.3	18.1	20.4	5.94
0001	17.2	15.9	17.9	15.2	2.75
0010	16.9	16.6	18.4	8.8	2.79
0011	16.9	15.9	17.2	12.9	3.50
0100	18.5	8.3	19.4	3.8	1.26
0101	18.5	9.9	19.2	7.6	1.84
0110	17.4	34.2	18.6	27.8	4.41
0111	17.4	19.6	17.7	20.4	3.71
Second-order branches					
000	16.1	66.5	16.9	43.0	8.69
001	16.1	33.2	16.7	25.5	6.29
010	16.2	25.5	18.2	14.9	3.10
011	16.2	56.7	17.1	47.8	8.12
First-order branches					
00	15.1	95.0	15.8	66.5 (52.5)	14.98
01	15.1	86.6	15.9	69.4	11.22
Trunk 0	1.4	397.6	14.5	134.8 (113.0)	26.20
Schefflera pittieri, Monteverde Cloud Forest Reserve, Costa Rica					
Second-order branches					
000	3.0	9.6	3.3/3.3†	10.5	31
001	3.0	9.6	3.5/3.5†	9.6	14
002	3.0	17.9	3.5/3.7†	15.6	74
003	3.0	13.4	3.3/3.8	12.4	23
004	3.0	28.7	3.5/3.6	24.4	23
012	2.9	15.6	3.4/3.5	15.6	59
013	2.9	15.6	—/3.5	—	35
014	2.9	13.4	—/3.4	—	66
022	2.9	11.5	—/3.6	—	26
023	2.9	8.0	—/3.3	—	9
031	2.9	8.0	3.3/3.4	11.5	19
032	2.9	28.7	3.3/3.5	35.1	177
First-order branches					
00	1.9	75.0	2.8/2.9	59.0	165
01	1.9	54.0	2.7/2.8	42.0	160
02	1.9	24.0	2.7/2.8	18.0	35
03	1.9	47.0	2.6/2.8	42.0	196
Trunk 0	1.0	212.0	1.4/1.7	168.0	556

*For *S. morototoni*, total leaf area supported is given; for *S. pittieri*, total number of leaves.
†number preceding slash is height at which cross section was measured; number following is height of distal tip of branch.

The trunk is designated as 0; primary branches originating from the trunk are designated 00, 01, etc.; secondary branches originating from primary branch 00 are designated 000, 001, etc.; secondary branches originating from primary branch 01 are designated 010, 011, etc. Dead branches were given the earliest numbers, but not recorded here.

Cross-sectional area (cs) is given for distal and basal portions of trunk and branches of the first few orders. "Height" is distance of measured point from ground, as measured along trunk (and branches). Figures in parentheses after cs are sapwood cross-sectional areas.

In *Schefflera morototoni*, total of distal cross-sectional areas of third-order branches, second-order branches, and primary branches are 117, 131, and 131 cm² (the last is the cross-section at 14.7 m, not the cross-section at 14.5 m given in the body of the table), while in *S. pittieri*, these totals are 173 and 161 cm² for secondary and primary branches respectively.

plant's architectural model affects these aspects of plant performance.

Shaping Plants for Different Niches

Another way to interpret tree architecture is to deduce growth forms appropriate for different stations in the forest. Horn (1971) distinguished two extreme types of crown: the monolayer, with a single layer of foliage taking up all the light which reaches its crown, and the multilayer, which cascades incoming light through several layers of foliage.

Horn (1975) argued that, where light is abundant, multilayers use it more effectively. Consider a four-layer tree with each layer spanning the same area as the crown of a competing monolayer. If the leaves of these trees attain

Table 5.7 Empirical assessment of different aspects of plant performance

	G (kg/m²·year)	dh/dt (m/year)	HE (m³/kg)	LSE (m³/kg)	1−F	α	LA/A_w	SLM/L (g/m²·year)
Pasoh								
1 hectare of forest	0.302	0.33	1.09	5.0	0.65	3.0	2333	100
Barro Colorado Island: shade tolerant species								
Tachigali	0.090	0.15	1.9	8.0	0.47	1.9	4640	27
Alseis	0.084	0.12	1.5	9.0	0.43	2.7	4320	33
Trichilia	0.109	0.32	3.3	10.0	0.62	2.1	6400	21
Calophyllum	0.127	0.36	2.7	11.0	0.57	2.2	5610	26
Barro Colorado Island: gap specialist								
Cecropia	0.224	0.71	3.3	19.0	0.47	2.3	2850	109

G, dry matter production per unit leaf area; dh/dt, height growth rate; HE, height growth efficiency; LSE, leaf support efficiency; 1−F, fraction of aboveground dry matter production devoted to stems and branches; α, allometric parameter scaling the dry weight M_s of a sapling's stem and branches against its height (so that $dM_s/M_s dt = \alpha\, dh/hdt$); LA/$A_w$, leaf area per unit cross-sectional area of trunk, calculated as ρ(LSE), where ρ is the density of the sapling's woody parts, expressed as kg (dry weight)/m³; and SLM/L, grams dry weight of leaf replaced per unit area of leaf per year, for 1 ha of forest at Pasoh, Malaysia (data from Kira 1978) and selected species of understory saplings 1–2 m high growing at 6% of full skylight on Barro Colorado (data from King 1994). SLM, specific leaf mass, g dry weight/m² leaf; L, average leaf lifetime, years. The numbers for Pasoh are calculated assuming that 1 ha of forest has 70,000 m² of leaves, 420,000 kg dry weight of stems and branches, and produces 21,140 kg above-ground dry matter per year, of which 7400 kg is foliage (Kira 1978). The "median" leaf in this forest is assumed to be on a tree 30 m tall. Tree mortality is assumed to be 1.1% per year (Manokaran et al. 1992), regardless of size class, so that average growth rate is 1.1% of 30 m.

maximum photosynthesis at 20% of full sunlight, if each layer of leaves is over 100 leaf-widths below the next (so that leaves cast no shadows on the layer below), and if each leaf layer has half the monolayer's area of leaves and takes up half the light reaching it, the top three layers will photosynthesize at full capacity, and the fourth at half capacity, for a total photosynthesis 1.75 times the monolayer's. On the other hand, where, as at the forest floor, there is little light, only monolayers can grow. Horn (1971) predicted that early successional plants, which receive abundant light, would be multilayers, striving for maximum growth even if this allows more light to reach their competitors below, while plants of mature forest, which must grow up in their predecessors' shade, would be monolayers. This picture suggests that a forest of multilayers maximizes community productivity, while a monolayer crown enhances its bearer's competitive influence on plants below it at the expense of both the bearer's growth rate and the productivity of the forest as a whole. And indeed, in deciduous forests of the temperate zone, a point on the forest floor shaded by an old-growth canopy tree is covered by relatively few leafy branches of that tree, while a point shaded by a successional tree will be covered by more of its leafy branches (Horn 1971, 1975). Here, monolayers appear to replace multilayers in the course of succession.

Since the same architectural model can serve both monolayer and multilayer tree species, while both monolayer and multilayer crowns are consistent with many tree architectures (Horn 1975), Horn's schema cannot help us grasp the distinctive advantages of different architectural models. Moreover, this schema does not seem to fit tropical forest. True, most canopy trees of mature tropical forest are monolayers. But many prominent pioneer trees, such as *Cecropia* and *Schefflera* (*Didymopanax*) are also monolayers (Givnish 1984: p. 73). Although the fast-growing pioneer *Trema* is multilayer, like *Muntingia* and *Xylopia frutescens*, most tropical multilayers are side-lit, understory trees. In the rainforest understory of western Amazonia, monolayer treelets, *Neea chlorantha* (Nyctaginaceae) are lit from above, while multilayer treelets, *Rinorea viridifolia* (Violaceae) are lit from the side (Terborgh and Mathews 1993).

What of Horn's more general suggestion that the crown shapes of mature forest trees enhance competitiveness at the expense of forest productivity? This suggestion seems prescient in view of the interest now attracted by the declining productivity of stands of older forest (Gower et al. 1996). Where this decline occurs, however, it represents some combination of declining availability of soil nutrients and the greater resistance taller trees must overcome to bring soil water to their leaves, so that their stomates are closed more of the time (Gower et al. 1996). This decline has nothing to do with crown architecture. The leaf area index (total leaf area per unit ground area) of *tropical* forest appears not to decrease during the later stages of succession (chapter 6), as if succession does not compromise the forest's ability to maintain high leaf area.

COMPARING LEAF ARRANGEMENTS, TREE ARCHITECTURES, AND FOREST STRUCTURE

As yet, theory appears to be too crude to help us understand much about tree architecture or forest structure. Let us now see what can be learned from comparison, that most basic, distinctive, and effective of the tools in the biologist's intellectual armory.

Herbs of the Forest Floor

Blanc (1992) compared the growth forms of herbs and woody perennials of tropical forest floors around the world. These plants live where light is in short supply, and they must economize on support tissue and minimize leaf overlap. Blanc distinguished 20 forms among these plants of the forest floor (Fig. 5.15), including stemless disks (disks of foliage resting on the forest floor), monolayer disks atop stems (such as a *Virola* seedling with one functional layer of foliage), stems supporting several superposed disks (such as a two-layer *Virola*), and domes of disks (as in trunkless palmate-leafed palms); strips or layers ("lames,") lying on the forest floor, strips inclined to the forest floor, strips atop stems, monolayer and multilayer; stemless ovals, ovals atop stems; stemless tori (rosettes of reflexed leaves which together form a semicircle revolved around one extremity to form the top half of a doughnut), similar tori atop stems, funnels (rosettes of inclined leaves), and flat rings (rosettes of horizontal leaves with long inclined petioles); cylinders (erect stems with leaves along enough of their length to form cylinders of foliage), ascending spiral stems lined (on one side) by leaves; and sheets, ground cover forming a horizontal expanse of foliage, with irregular boundary. All these plant forms occur in all the forests Blanc studied. A single one of these plant forms contains plants of several architectural models, whereas different species classified according to the same architectural model may have different forms. There is remarkable convergence in the forms of plants of the floors of different forests (Blanc 1992). Notice, however, that even though a plant's architecture limits its range of possible shapes, a plant's architectural model does not determine its shape.

Tree Architecture and Leaf Arrangement

On Barro Colorado, as in other tall, diverse forests, the majority of understory plants with dbh > 6 mm either carry distichous leaves on horizontal twigs or spiral their leaves atop unbranched stems, while nearly all the canopy trees have ascending ("orthotropic") branches and arrange their leaves around ascending twigs (Table 5.8, Fig. 5.16). Such "orthotropic" architectures are far more prominent in the understory of montane forests of low stature than in the understory of tall lowland forest (Table 5.9).

More generally, a survey of roughly 30 tropical rainforests (Appendix 5.2) shows that, in forests of lower stature, spiral and decussate leaf arrangements are more prevalent among understory saplings between 6 and 25 mm dbh (Table 5.10). The prevalence of orthotropic leaf arrangements among these saplings can be measured by AI, the leaf arrangement index, which is 180 when all sampled saplings and treelets arrange their leaves around ascending twigs, and 0 when all have perfectly distichous leaves

Table 5.8 Relative abundance of trees with different architecture at various levels in three tall forests

	Barro Colorado, Panama					Monteverde, Costa Rica					Montagne d'Ambre, Madagascar				
	Understory					Understory					Understory				
	0–3 m	3–10 m	10+ m	Canopy	S	0–3 m	3–10 m	10+ m	Canopy	S	0–3 m	3–10 m	10+ m	Canopy	S
Distichous leaves															
Troll	3	6	0	1	4	2	7	1	0	5	3	3	1	1	7
Mangenot	2	4	2	0	5	1	1	0	0	1	0	0	0	0	0
casearia	1	0	2	0	2	5	3	3	0	8	4	4	1	0	5
virola	0	2	1	0	3	0	0	0	0	0	0	0	1	0	1
irregular	1	1	5	0	6	2	2	1	1	5	0	1	0	0	1
faramea	1	1	0	0	2	5	1	1	0	3	4	7	5	1	10
deflexa	0	0	0	0	0	0	1	0	0	1	0	0	0	0	0
Spiral/decussate leaves															
terminalia	0	0	0	0	0	0	0	0	0	0	1	0	0	0	1
palm	6	3	3	0	9	1	0	0	0	1	6	4	0	0	5
Champagnat	3	0	0	0	3	0	0	1	0	1	4	0	0	1	3
Attims	0	0	0	0	0	1	1	2	1	5	0	0	2	5	3
Rauh	0	0	0	1	1	0	0	1	0	1	2	1	0	0	2
Scarrone	0	0	0	2	1	1	0	0	0	1	0	0	0	0	0
Kwan Koriba	0	0	0	0	1	1	0	0	0	2	0	0	0	0	0
psychotria	3	3	3	0	9	4	4	2	1	8	2	2	1	1	4
coussarea	0	1	0	0	1	1	2	0	0	3	0	2	1	7	6
almendro	0	0	2	1	3	0	0	1	4	5	0	0	0	0	0

Each row gives, for each forest transect, the number of plants of the model concerned 0–3 m high, 3–10 m high, > 10 m high but still shaded, and in the canopy with crown exposed to the sun, and the total number of species, S, represented by these plants.
Forest altitudes and canopy heights are as follows: Barro Colorado, 150 m and 35 m; Monteverde, 1350 m and 31 m; Montagne d'Ambre, 900 m and 20 m respectively.
Data from Leigh (1990) and V. Jeannoda and E. G. Leigh, Jr. (unpublished).

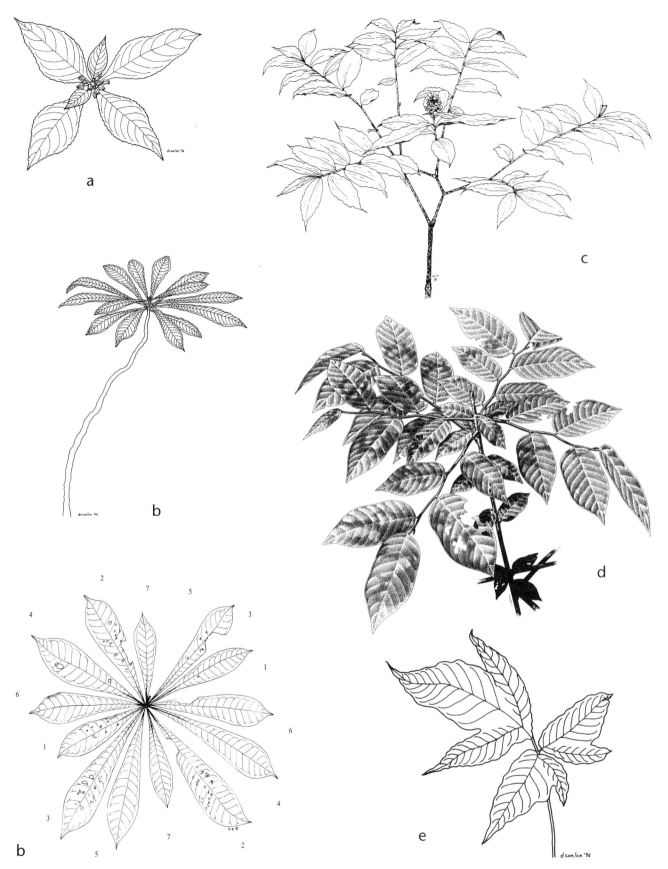

Figure 5.15. Growth forms of understory herbs and saplings. (1) Disk: a. *Chrysothemis friedrichsthaliana* (Gesneriaceae), edge of laboratory clearing, Barro Colorado. Drawing by Donna Conlon. (2) Elevated disk: b. *Alseis blackiana* (Rubiaceae). Crown (leaves of the same pair have the same number), 40 cm wide (drawing by F. Gattesco) and the sapling, 1.3 m tall, Lutz ravine, Barro Colorado (drawing by Donna Conlon). c. *Dichorisandra hexandra* (Commelinaceae), 1 m tall, Lutz ravine, Barro Colorado. Drawing by F. Gattesco. d. *Virola sebifera* (Myristicaceae), 60 cm wide, Lutz ravine, Barro

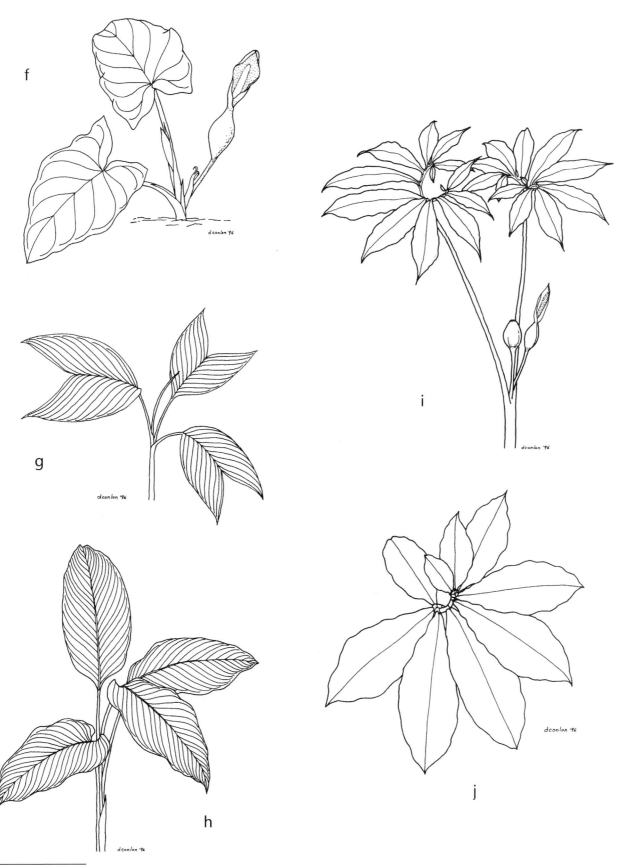

Colorado. Drawing by Alex Murawski. (3) Oval: e. *Tectaria incisa* (Polypodiaceae), frond 20 cm long, Lutz ravine, Barro Colorado. Drawing by Donna Conlon. (4) Paired ovals: f. *Xanthosoma pilosum* (Araceae), leaves 15 cm long, laboratory clearing, Barro Colorado. Drawing by Donna Conlon. (5) Torus: g. juvenile palm, crown 20 cm wide, Lutz ravine, Barro Colorado. Drawing by Donna Conlon. h. *Calathea lutea* (Marantaceae), juvenile, leaves 25 cm long, laboratory clearing, Barro Colorado. Drawing by Donna Conlon. (6) Monolayer helix: i. *Xanthosoma helleboricum* (Araceae), crown 30 cm

wide, laboratory clearing, Barro Colorado. j. *Costus pulverulentus* (Zingiberaceae), crown 20 cm wide, young forest, Barro Colorado. View from above and view from side. Drawings by Donna Conlon. (7) Two-layer helix. k. *Costus guanaiensis* (Zingiberaceae), 1.2 m tall, near edge of remnant forest, laboratory clearing, Barro Colorado. Drawing by Donna Conlon. (8) Layer of ground cover (branched runners): l. *Geophila repens* (Rubiaceae), leaves up to 3 cm long, young forest, Barro Colorado. Side view of uprooted runner, and view from above. Drawing by Donna Conlon.

Figure 5.16. Drawing of old forest from treefall gap, Barro Colorado Island (near AVA 14). Tall tree at left is *Prioria copaifera* (Leguminosae), large-crowned tree at center is *Anacardium excelsum* (Anacardiaceae), and tall tree at right is *Quararibea asterolepis* (Bombacaceae). Drawing by Daniel Glanz.

Table 5.9 Relative abundance of trees with different architecture at various levels in three stunted montane forests

| | Avalanchi, Nilgiris, India | | | | Ambohitantely, Madagascar | | | | El Yunque, Puerto Rico | | |
| | Understory | | | | Understory | | | | | | |
	0–3 m	3+ m	Canopy	S	0–3 m	3+ m	Canopy	S	Understory	Canopy	S
Distichous leaves											
Troll	0	0	0	0	0	3	0	2	0	0	0
casearia	0	0	0	0	3	3	0	2	0	0	0
virola	0	0	0	0	4	4	1	2	0	0	0
irregular	1	0	0	1	0	0	0	0	0	0	0
faramea	3	2	0	2	6	1	0	6	0	0	0
Spiral/decussate leaves											
terminalia	0	0	0	0	0	0	0	0	7	0	1
palm	0	0	0	0	0	1	0	1	3	0	2
Attims	24	25	7	8	0	6	10	7	1	0	1
Rauh	0	8	1	2	1	3	8	4	0	0	0
Leeuwenberg	0	0	0	0	1	2	2	2	0	13	5
Kwan Koriba	0	4	0	1	0	0	0	0	0	0	0
psychotria	0	4	0	2	3	2	0	3	4	1	4
coussarea	0	1	5	3	1	2	2	3	3	0	1
Almendro	0	0	0	2	0	0	5	2	8	5	3

Each row gives, for each forest transect, the number of plants of the model concerned < 3 m high, > 3 m high but shaded, and in the canopy with crown exposed to the sun, and the total number of species S included among these plants.
Forest altitudes are as follows: Avalanchi, 2000 m; Ambohitantely, 1550 m; and El Yunque, 1000 m.
Data from Leigh (1990) and V. Jeannoda, R. Benja and E. G. Leigh Jr., (unpublished).

on horizontal twigs. Omitting the "outlier" of Costa Rica's Monteverde cloud forest 1, as we shall do in all this section, the regression of AI on the natural logarithm of forest height (FH, m) is

$$AI = 217 - 41.837 \ln FH, \quad r = -0.88456. \quad (5.15)$$

The regression of AI on the average temperature of the forest's coolest month is

$$AI = 211.5 - 6.020545t, \quad r = -0.744185. \quad (5.16)$$

Does the average temperature, t, of the coolest month have a significant effect on the leaf arrangement index AI aside from its effect on ln FH? The "partial correlation coefficient" (Fisher 1958: pp. 187–189) can be used to answer this. The partial correlation between AI and t, eliminating the effect of FH, is (Fisher 1958: p. 188)

$$\frac{r(AI, t) - r(AI, \ln FH)\, r(t, \ln FH)}{\sqrt{[1 - r^2(AI, \ln FH)]\,[1 - r^2(t, \ln FH)]}}. \quad (5.17)$$

The correlation $r(t, \ln FH)$ is 0.691913, so the partial correlation between AI and t, eliminating the effect of ln FH, is 0.413415. This correlation's "significance" ($0.02 < p < 0.05$) derives from the high AI in the tall but very cold forest atop Cerro de la Muerte.

If we treat each geographic region separately [Malaysia, Peru, the West Indies (Puerto Rico and Dominica) and Madagascar], we find that in each region, shorter forests have higher AI (Table 5.10). The same is true in Central America if one sets aside the high AI in the forest on Cerro de la Muerte.

Why do most canopy species and most understory plants in forests of low stature arrange their leaves around ascending twigs, while far more of the understory species in tall forests carry distichous leaves on horizontal twigs? Orthotropic branches bearing spiral or decussate leaves are more appropriate to canopy conditions and they also help saplings attain the canopy more quickly. In bright light, the greater self-shading associated with spiral or decussate relative to distichous leaves presumably distributes heat load over a greater area of leaf without materially reducing photosynthesis per unit leaf area, while in the shaded understory, reducing leaf overlap increases photosynthesis without significantly increasing heat load, making distichous leaf arrangements (which do reduce leaf overlap) a viable alternative (Givnish 1984: p. 66). Moreover, it pays a sapling of a canopy species to keep its crown narrow and its stem slender in order to concentrate its resources as far as possible on fast growth toward the canopy (King 1990, 1996). Growth toward the canopy is managed with a minimum of branch shedding by setting leaves around ascending twigs or branches (King 1991b: pp. 350–351, Kohyama 1991). On the other hand, an understory shrub should make the best of its current, shaded station, spreading its crown, which can be done by extending horizontal branches bearing distichous leaves, and thickening its trunk for safety's sake (King 1990). Among six species of woody plant studied on the forest floor of Barro Colorado, 2.5 m tall individuals of the two species that are smallest when adult, *Hybanthus prunifolius* and *Desmopsis panamensis*, had

Table 5.10 Relation between forest height, temperature, and arrangement of understory leaves

Site	t (°C)	FH	No. of Small and Medium Saplings with leaf arrangement type				AI	
			0	1	2.5	4	Observed	Predicted
Costa Rica, Monteverde Elfin 79†	15	4	1	0	1.5	15.5	146	159
Puerto Rico, El Yunque Elfin 71†	17	5	2	0.5	0	14.5	137	150
Puerto Rico, El Yunque Elfin 79†	17	5	1	0	0	14	150	150
Costa Rica, Monteverde Elfin 75†	15	6	0	0	0.5	13.5	167	142
Malaya, Gunong Brinchang, Elfin†	13	6	5	1	4.5	25.5	126	142
India, Nilgiris, Avalanchi 4&5	12	6	0	1.5	1.5	29	153	142
Malaya, Gunong Ulu Kali B	15	7	2.5	1.5	2	49	147	136
Costa Rica Monteverde cloud for 1†*	15	9	3	0	1.5	28.5	142	125
India, Nilgiris, Avalanchi 2&3	12	9	3	4.5	9.5	36	129	125
Madagascar, A'hitantely plateau for	13	11	9	13.5	4	35.5	110	117
Costa Rica, Monteverde cloud for 2†	15	13	5	0	8	20	121	110
Peru, Vilcabamba 2.5, cloud forest†	17	14	9	1.5	4	34.5	122	107
Madagascar, Mtne d'Ambre, young forest	18	17	23	2	0	5	51	98
Madagascar, Analamazaotra/Perinet	15	18	8	7.5	4	9.5	88	96
Madagascar, Mtne d'Ambre, older forest	18	21	19.5	11.5	5	24	90	90
Malaya, Fraser's Hill, montane	18	21	26	7.5	1	29.5	90	90
Puerto Rico, lowland tabonuco forest	23	22	14.5	2.5	9.5	28.5	106	88
Peru, Panguana, R. Llulapichis	24	23	43	6	6	24	75	86
Madagascar, Mananara Nord, lowland	20	24	25	7	1	31	92	84
Papua New Guinea, Gabensis R.	25	24	67.5	0	1	23.5	62	84
Madagascar, Masoala, 600 m	18	25	33	9.5	2	9.5	59	82
Costa Rica, C. de la Muerte†	12	25	2	1	3.5	29.5	141	82
Dominica, Palmist Ridge	22	29	29.5	1	4.5	13	71	76
Peru, Sira, along ridge	23	29	39	6.5	4.5	20	72	76
Peru, Vilcabamba 0, lowland forest	23	30	35	6.5	9.5	31	87	75
Costa Rica, Monteverde ravine, tall	17	31	10	7.5	3	11.5	87	73
Panama, BCI, Brokaw Ridge (B10)	26	35	20	1	1	6	59	68
Malaya, Pasoh, mixed dipterocarps	25	41	45	2	0	9	49	62
Malaya, Ulu Gombak, Shorea curtisii	23	45	23	0.5	0.5	7	58	58

Average temperature of coolest month, t; forest height, FH; the number of sampled small and medium saplings, 6 mm < dbh < 26 mm, with distichous leaves (leaf arrangement score 0); with spiral or decussate leaves flattened into distichy (leaf arrangement score 1); the leaves arranged around erect twigs, spiral, decussate, or whorled (leaf arrangement score 4); and with leaf arrangements intermediate between score 1 and 4 (score 2.5); AI, the leaf arrangement index for this set of saplings [arccos $(1 - z/2)$, where z is the average score for these saplings], observed, and as predicted from regression of AI on ln FH. Data are from Leigh (1990 and unpublished).
  Regressions among these variables are:
  AI = 216.93 − 41.8369 ln FH, $r = -0.8446$; AI = 211.48 − 6.0205t, $r = -0.7442$
*Outlier omitted from all regressions.
†Cloud forest.

thicker trunks, wider and heavier crowns, and greater total leaf area than equal-sized saplings of mid- and overstory species (King 1990: Table 4).

Finally, why are distichous leaf arrangements so much rarer in stunted montane forests, especially fog-bound ones, than in taller forest? A forest may be stunted either because wind prevents trees from growing tall, as on the ridge crest above Monteverde, Costa Rica (Lawton 1982, Lawton and Dryer 1980), or because the soil is poor enough to slow tree growth or limit tree height (van Steenis 1972). In the first case, plants must build stoutly to resist the wind (Lawton 1982). Here, architectures that minimize the amount of twig and branch shedding and rebuilding required for further growth are at a premium (Leigh 1990): why waste such stout work? Trees of montane forest tend to have monolayer crowns; in such a crown, erect or inclined twigs can be extended upward as the trees grow, while horizontal twigs and branches become useless when newer branches of the growing tree shade them. When the soil is poor, wasting previous construction is equally undesirable. Moreover, when there are few resources available for growth, species with spiral or decussate leaves need to devote the least energy to tasks other than renewing their leaves. In the shady understory of La Selva, Costa Rica, slow-growing individuals of Pourouma aspera and Pentaclethra macroloba, which carry spiral leaves on ascending twigs, devote more than 80% of their aboveground production to replacing leaves, while equally slow-growing exemplars of their distichous-leaved counterparts devote an average of only 59% of their above-ground production to leaf replacement (King 1991a: Table 4). Architectures that permit plants to tread water so economically when the need arises must be advantageous in settings of limited productivity (cf. equation 5.14).

Leaf Size, Leaf Physiognomy, and Their Correlates

Paleontologists normally use leaf size to infer the annual rainfall of fossil settings (Wolfe and Upchurch 1987). The higher the annual rainfall and the shorter and milder the dry season of a tropical forest, the larger that forest's average leaf size (Givnish 1984). In general, large leaves occur where adequate water and nutrients are available to them (Wolfe 1993: p. 62). Among the rainforests listed in Appendix 5.2, however, rainfall should have no influence because they are all thought to receive more water than they need. Among rainforests, temperature is the dominant influence on leaf size, probably because of the effect of temperature on rainforest productivity. Nutrient supply exerts a distinctly subordinate influence on the rainforests of Appendix 5.2, since, elfin forests excepted, edaphic extremes such as heath or swamp forests were avoided.

In the 31 rainforests in Appendix 5.2, the average length, LL, of litter leaves, fallen leaves collected from the forest floor, correlates most closely with the average temperature, t, of the forest's coolest month (Table 5.11). The regression of LL on t is

$$LL = 5.1918t - 10.2708, \quad r = .8336. \quad (5.18)$$

The regression of LL on forest altitude, A, is

$$LL = 112.35 - 0.02698A, \quad r = -0.7083. \quad (5.19)$$

Apart from its effect on t, altitude exerts no significant influence on litter leaf length (Leigh 1975, 1990).

The average length, $L(S + M)$, of leaves from sampled small and medium saplings 6-25 mm dbh also declines with t (Table 5.12):

$$L(S + M) = 8.57907t - 2.28237, \quad r = 0.7247. \quad (5.20)$$

This correlation is weaker for sapling leaves than for leaves sampled from the forest floor. Aside from its correlation with t, forest height has no significant influence on the average length of sapling leaves.

Litter leaf size declines with the average temperature of the coolest month in Central America (Costa Rica and Panama), in the West Indies (Puerto Rico and Dominica), in Peru, in Papua New Guinea, in Malaysia, and in Madagascar (Table 5.11); this decline in litter leaf size is a general rule of altitudinal gradients in tropical rainforest. In most of these regions, sapling leaf size also declines with t, but the relationship is questionable for Central America, and quite opposite for the forests shown from Peru (Table 5.12).

In Table 5.11, the forest whose average litter leaf size falls the shortest of the prediction from the average temperature of the coolest month is Gunong Ulu Kali in Malaysia, where the soil is remarkably poor (Whitmore and Burnham 1969). Unfortunately, information on soil is lacking for most of the forests in Tables 5.10 and 5.11. How can we judge whether, in general, leaf size is smaller on poorer soils?

When soil is poor, it is harder for a tree to mobilize the resources required to overtop neighbors. Thus, other things being equal, tree density is higher on poorer soils (Stocker and Unwin 1989: p. 256). Density of trees ≥ 10 cm dbh is also higher where forest height is lower: among the forests in Table 5.12, the number, D, of square meters per tree varies with forest height, FH, according to the regression

$$\ln D = 0.61303 \ln FH - 0.31426, \quad r = 0.84128. \quad (5.21)$$

I consider a forest short of nutrients if, in that forest, D is lower than this regression predicts for that forest's height, and vice versa. Among the 25 forests with information available on both litter leaf length and forest height, the ratio of observed litter leaf length, LL, to that predicted from its regression on the average temperature t of the coolest month is significantly correlated with the ratio of D to that predicted from forest height: $r = 0.47173$ ($p < 0.05$). There is, however, no correlation between the ratio of the average length, $L(S + M)$, of sapling leaves, to that predicted from t, and the ratio of observed D to that predicted from forest height. Litter leaves seem to be a better index than sapling leaves of a forest's nutritional state.

Why should leaves be larger in rainforests with warmer climates or more fertile soils? Givnish (1984) explained why sun-leaves should be larger in more favorable habitats. Increasing leaf size increases the leaf's temperature because it slows the convective cooling of the leaf (Givnish and Vermeij 1976). Transpiration rate is higher in warmer leaves, and photosynthesis is enhanced when the leaf is warmer and higher transpiration admits more CO_2, provided that the water supply can support high transpiration and the nutrient supply can support high photosynthesis. On the other hand, a leaf overheats and is damaged if there is too little water to permit high transpiration, while the costly structures required to ensure an abundant water supply are pointless if there are not enough nutrients or warmth to support high photosynthesis. Where climates are dry or soils poor, plants cannot supply their leaves so abundantly with water or nutrients, even though they devote a higher proportion of their resources to making roots (Table 6.5). Thus, optimum leaf size is smaller in drier or cooler climates or where soils are poorer. Under these circumstances, leaves are also more stoutly built (have more dry matter per unit area of leaf), presumably because, being harder to replace, it is desirable that they last longer. Finally, plants of dry climates or poor soil often incline their leaves, or store water in them, to prevent overheating (Medina et al. 1990).

The most extraordinary aspect of Givnish's (1984) discussion is the similarity he predicts between leaves on poor

Table 5.11 Altitudinal gradient in size of canopy leaves in tropical rainforests

Site	A	t (°C)	LL O	LL P	√LW	n/s
Costa Rica, Cerro de la Muerte†	2700	12	59	52	38	70/7
Peru, Vilcabamba Camp 4†	2600	12	69	52	42	41/5
Papua New Guinea, Mt. Kaindi†	2340	12	41	52	27	64/7
Avalanchi, Nilgiri Hills, India	2100	12	61	52	40	114/12
G. Brinchang, Malaya†	2000	13	42	57	30	52/5
G. Ulu Kali, Malaya†	1750	15	32	68	21	40/4
Peru, Vilcabamba Camp 2.5†	1750	17	103	80	64	65/7
Madagascar, Andringitra†	1600	11	53	47	31	34/4
Madagascar, Ambohitantely	1550	13	53	57	34	150/9
Costa Rica, Monteverde Elfin 78†	1500	15	74	68	52	45/4
Costa Rica, Monteverde†	1500	15	81	68	59	99/10
Costa Rica, Monteverde ravine, tall forest	1350	17	74	78	49	40/4
Malaya, Fraser's Hill	1200	18	69	83	46	80/8
Madagascar, Montagne d'Ambre†	1200	16	52	73	32	41/4
Madagascar, Montagne d'Ambre†	1100	17	50	78	32	40/4
Venezuela, Rancho Grande†	1100	18	117	83	84	122/12
Puerto Rico, El Yunque elfin forest†	1050	17	90	78	68	49/5
Madagascar, Analamazaotra-Perinet	1000	15	62	68	41	41/6
Madagascar, Masoala†	1000	15	70	68	52	67/8
Madagascar, Mtne d'Ambre mature	1000	18	64	83	39	84/8
Madagascar, Mtne d'Ambre younger	1000	18	87	83	57	36/4
Peru, Sira, 680 m	680	23	112	109	78	100/10
Peru, Vilcabamba Camp 0	660	23	124	109	80	47/5
Madagascar, Masoala (Hiarakh)	600	18	99	83	62	64/7
Malaya, Ulu Gombak *Shorea curtisii*	515	23	115	109	73	51/8
Madagascar, Masoala (Hiarakh)	500	18	92	83	60	81/9
Papua New Guinea, Gabensis	300	25	129	120	87	85/9
Dominica, Palmist Ridge	240	22	94	104	64	61/7
Puerto Rico, tabonuco forest	220	23	104	109	71	109/11
Madagascar, Mananara Nord	200	20	89	94	58	101/10
Papua New Guinea, Lae (Busu R)	200	25	129	120	87	86/16
Peru, R. Llulupichis (Panguana)	200	24	122	114	80	117/12
Panama, Barro Colorado, "plateau"	140	26	119	125	75	79/8
Malaya, Pasoh Reserve	100	25	120	92	63	80/8

†Cloud forests.
Altitude, A; average temperature of year's coolest month, t; average length of litter leaves sampled from the forest floor, LL (O, observed, and P, predicted from the regression on t); observed average size (square root of length times width) of these same leaves, $\sqrt{LW}$, and number of leaves sampled/number of points from which they were sampled along a transect line, n/s (in tall forest, one point was sampled per 8 m along one or more 30-m transect lines). Data are from Leigh (1990) and unpublished.
  Regressions among these variates are
  $LL = 112.3477 - 0.02698A$, $r = -0.7083$; $LL = 5.1918t - 10.2708$, $r = 0.8336$
  $\sqrt{LW} = 75.0855 - 0.01832A$, $r = -.7025$; $\sqrt{LW} = 3.4770t - 7.3080$, $r = 0.8154$
  $t = 23.8537 - 0.005414A$, $r = -0.8852$

soils and sclerophyllous leaves in dry climates. And, indeed, structural features of leaves from the nutrient-starved (but very wet) "low bana" of Venezuela's San Carlos de Rio Negro, such as dry matter content per unit leaf area, leaf thickness, and sunken stomata, are closely parallelled by those of sclerophyllous leaves from dry Mediterranean climates (Medina et al. 1990). Indeed, this parallel between "peinomorphic" and "xeromorphic" leaves led many to infer that plants of heath forests, peat bogs, and even the seemingly perpetually soggy elfin forests of tropical mountaintops, are subject to occasional drought (Schimper 1903, Brunig 1970). In fact, however, drought reduces transpiration just as much in mixed dipterocarp forest on sandy loam as in nearby heath forest on bleached sand (Becker 1996). Leaves of bog, heath forest, and elfin forest are not particularly subject to water shortage, nor do they have any special capacity for drought resistance (Schlesinger and Chabot 1977, Peace and MacDonald 1981, Buckley et al. 1980).

The parallel between leaves of dry climates and those of problematic habitats is illustrated by mangroves (Ball et al. 1988, Ball 1996). Mangroves live in a difficult habitat: like elfin forest plants (Table 5.9), mangroves tend to have ascending branches and to arrange their leaves around erect twigs (Fig. 5.17). Mangroves exclude nearly all the

Table 5.12 Relation between forest height, tree density, and size and arrangement of understory leaves

Site	t	FH	D O	D P	L(S&M) O	L(S&M) P	AI O	AI P
Costa Rica, Monteverde Elfin 79†	15	4	7	5	137 (19)	126	146	159
Puerto Rico, El Yunque Elfin 71†	17	5	5	6	100 (18)	144	137	150
Puerto Rico, El Yunque Elfin 79†	17	5	7	6	71 (14)	144	150	150
Costa Rica, Monteverde Elfin 75†	15	6	5	6	132 (14)	126	167	142
Malaya, Gunong Brinchang, Elfin†	13	6	6	6	83 (38)	109	126	142
India, Nilgiris, Avalanchi 4&5	12	6	5	6	101 (39)	101	153	142
Malaya, Gunong Ulu Kali B†	15	7	4	7	98 (58)	126	147	136
Costa Rica Monteverde 1†*	15	9	21	(8)	273 (32)		142	125
India, Nilgiris, Avalanchi 2&3	12	9	10	8	95 (56)	101	129	125
Madagascar, A'hitañtely plateau forest	13	11	7	9	110 (63)	109	110	117
Costa Rica, Monteverde 2†	15	13	13	10	175 (32)	126	121	110
Peru, Vilcabamba 2.5†	17	14	13	10	233 (40)	144	122	107
Madagascar, Mtne d'Ambre, young forest	18	17	24	12	101 (30)	152	51	98
Madagascar, Analamazaotra/Perinet	15	18	9	12	110 (30)	126	88	96
Madagascar, Mtne d'Ambre, older forest	18	21	13	13	123 (60)	152	90	90
Malaya, Fraser's Hill, montane	18	21	8	13	209 (64)	152	90	90
Puerto Rico, lowland tabonuco forest	23	22	16	14	241 (62)	195	106	88
Peru, Panguana, R. Llulapichis	24	23	16	14	209 (80)	204	75	86
Madagascar, Mananara Nord, lowland	20	24	12	15	175 (64)	169	92	84
Papua New Guinea, Gabensis R.	25	24	17	15	231 (92)	212	62	84
Madagascar, Masoala, 600 m	18	25	14	15	153 (54)	152	59	82
Costa Rica, C. de la Muerte†	12	25	20	15	115 (36)	101	141	82
Dominica, Palmist Ridge	22	29	16	16	198 (50)	186	71	76
Peru, Sira, along ridge	23	29	13	16	216 (72)	195	72	76
Peru, Vilcabamba 0, lowland forest	23	30	16	17	227 (82)	195	87	75
Costa Rica, Monteverde ravine, tall	17	31	16	17	153 (32)	144	87	73
Panama, BCI, Brokaw Ridge (B10)	26	35	28	18	168 (30)	221	59	68
Malaya, Pasoh, mixed dipterocarps	25	41	16	20	178 (64)	212	49	62
Malaya, Ulu Gombak, *Shorea curtisii*	23	45	20	21	178 (32)	195	58	58

Average temperature of coolest month, t; forest height, FH; tree density D (average m^2/tree $\geq$ 10 cm dbh), observed (O) and predicted (P) from ln FH; average length of leaves of sampled small and medium saplings 6mm < dbh < 26 mm, $L(S\&M)$, with number of plants sampled in parentheses (one large representative leaf was sampled from each sapling); and the leaf arrangement index AI for this set of saplings (arccos $(1 - z/2)$, where z is the average score for this set of saplings). Data are from Leigh (1990 and unpublished).

Regressions among these variates are

$D = 0.7303(FH)^{0.61303}$, $r(\ln FH, \ln D) = 0.8413$; $L(S\&M) = 46.0277 \ln FH + 27.3554$, $r = 0.6350$; $L(S\&M) = 8.5791t - 2.2824$, $r = 0.7247$

*Outlier omitted from all averages and regressions.
†Cloud forest.

salt from the water entering their roots, so their transpiration rate must be restricted to prevent the excessive buildup of salt around the roots (Passioura et al. 1992); even at normal salinities, mangrove roots absorb water against a fierce osmotic gradient (Rada et al. 1989). Thus leaves must be kept small, succulent, and inclined relative to the horizontal to avoid overheating without resort to excessive transpiration. In a given area, the most salt-tolerant mangroves have the smallest, most steeply inclined, most succulent, and most stoutly built (most dry matter per unit area) leaves, and they use the least water per unit CO_2 fixed (Table 5.13). Mangroves face a trade-off between surviving high salinities and growing faster in less salty environments, which can cause different mangrove species to sort out in distinctive "zones" along a salinity gradient (Ball et al. 1988).

Canopy Roughness

Where climates are warm and wet and soils are fertile, the canopy of the entire forest seems designed to enhance transpiration and gas exchange (Fig 5.18; Brunig 1970, 1983). The canopy of such forests is uneven and "aerodynamically rough," as if to enhance the turbulent eddying of air through all levels of the forest (Brunig 1970); indeed, Schimper (1903: p. 286) commented that tropical forest canopy was far more irregular and uneven than its temperate-zone counterpart. This uneven canopy also enables light to penetrate at many angles and thereby support many layers of vegetation (Terborgh 1992). This irregular canopy results from a combination of frequent tree falls, crown shyness (avoidance of contact between adjacent crowns) and the crown shapes of individual trees.

Figure 5.17. Sprig of canopy foliage of red mangrove, *Rhizophora mangle* (Rhizophoraceae) from sapling 3 m tall, from STRI's Galeta Marine Laboratory, on Panama's Caribbean coast. Drawing by Roxane Trappe.

Bruenig (1996: pp. 35–36) remarked that

An open, bunched architecture of the crown, such as in rainforest trees and temperate oaks and pines, improves ventilation and facilitates air exchange even during almost calm, bright weather by active free and forced convective air exchange. Rapid removal of the atmospheric boundary layer is facilitated by small, longish and more-or-less upright leaves.

What evidence is there that a forest's canopy is designed to facilitate that forest's productivity? In boreal forest, root function is limited by the coldness of the soil (Steele et al. 1997: p. 585). In Canada, boreal jack pine forest has a sparse, open canopy with less than 2.3 ha of leaves/ha of ground (most tropical forests have 7 or more: see chapter 6) and high stomatal resistance (low rate of release of water vapor from a leaf per unit vapor pressure differential between the leaf's interior and the air just outside). As a result, transpiration of this forest accounts for a third of its incoming solar radiation (Saugier et al. 1997: p. 517) compared to over half in tropical forest (see above, chapter 3). Is the open canopy of boreal forest designed to promote warming of the soil? Devastating fires may be the proximate cause of the open canopy and ample spacing between trees of boreal forest. The question still remains: Does the forest "make itself flammable" to promote a canopy structure that permits rapid warming of the soil?

Boreal forest, however, is a bit removed from the object of this enquiry. What evidence is there for the adaptive organization of tropical forest canopy? in Malaysia, especially Sarawak, one often sees gradients from mixed dipterocarp forest on loam soils to heath forest on progressively poorer soil, more and more limited in its capacity to store water and support copious photosynthesis. Emergent trees of mixed dipterocarp forests have "rough" crowns permitting ample ventilation and light penetration (Brunig 1970). On poorer soils, the forest's canopy is smoother, rough-crowned emergents are rarer, and canopy leaves tend to be smaller, more steeply inclined, and more sclerophyllous or coriaceous (Brunig 1970, Bruenig 1996:

Table 5.13 Characteristics of sun-leaves and water use efficiency of mangroves with different degrees of salt tolerance (data from Ball 1996)

Species	SLM (g/m²)	% water	Average area of leaf rosette (cm²)	Area of avg leaf (cm²)	LIA	WUE
Bruguiera gymnorhiza	133	66	635	58	56	0.072
Rhizophora apiculata	149	70	553	69	68	0.096
Rhizophora stylosa	169	70	419	44	73	0.101
Ceriops tagal	189	71	102	8	69	0.113

SLM, specific leaf area, grams dry matter per unit leaf area; LIA: leaf inclination angle, arccos (projection of leaf area on horizontal surface, divided by actual leaf area); WUE: water use efficiency, ratio of photosynthesis (μmol CO_2 fixed/m² leaf·sec) to stomatal conductance (mmol H_2O/m² leaf·sec).

Area of leaf rosette may be the appropriate unit for assessing the susceptibility of leaves to heating.

The first species in the table is the least salt tolerant, and salt tolerance increases toward the bottom of the table.

p. 32). Brunig argued that, on poorer soils, the forest's canopy is smoother, and the leaves smaller and more steeply inclined, in order to reduce transpiration without causing the leaves to become overheated.

Brunig (1983) proposed to measure the aerodynamic roughness of a forest's canopy by the parameter z_0, where

$$\log(z_o) = \log[(h_e - h)DS] - 2.94. \quad (5.22)$$

Here, $h_e - h$ is the height of emergent tree crowns above the main canopy, D is the diameter of an emergent's crown, or of a group of contiguous emergent crowns, and S is the (edge-to-edge) distance between successive groups of emergents. Brunig (1983) found that z_0 was lower in forest on poorer soil.

Is the canopy structure of tropical forest organized to regulate transpiration? Low crown resistance—high transpiration per unit difference of vapor pressure between the interior of leaves and the air above the canopy—promotes a tree's transpiration.

Resistance of a tree's crown to transpiration, R_e, has two components: stomatal resistance to movement of water

Figure 5.18. Above. View of forest canopy from the top of the tower in Lutz ravine, then 27 m tall, Barro Colorado, 1976. The emergent straight-trunked tree in the center is *Virola nobilis* (Myristicaceae), otherwise known as *V. surinamensis*, the "lawn" of upright leaves in front of it, and masking its lower parts, belong to *Ficus insipida* (Moraceae), and the distichous leaves in the foreground to the right belong to the liana *Uncaria tomentosa* (Rubiaceae). Drawing by Lynn Siri Kimsey. Right. View (in a different direction) from top of this same tower, now 40 m tall. Tree with drooping branches and leaves, center left, is *Quararibea asterolepis* (Bombacaceae), foreground foliage and tree above *Quararibea* to the right is *Anacardium excelsum* (Anacardiaceae), tree at top left with upright leaves if *Ficus yoponensis* (Moraceae). Drawing by Donna Conlon.

vapor from inside to outside the leaf, R_s, and "boundary layer resistance" to the movement of water vapor from near the leaf to the air above the canopy, R_b. Let the vapor pressure differential between leaf interiors and the air above the canopy be ΔP. Then transpiration rate, E, is $\Delta P/R_c = \Delta P/(R_b + R_s)$. The higher R_b/R_s, the less a leaf's stomata influence transpiration rate. To see this, let R_s be increased by an amount δR_s. Then the transpiration rate will change by an amount δE, where

$$E + \delta E = \frac{\Delta P}{R_b + R_s + \delta R_s} = E(1 - \frac{\delta R_s}{R_b + R_s}) \quad (5.23a)$$

$$\frac{\delta E}{E} = -\frac{\delta R_s}{R_b + R_s} = -(1 - \Omega)\frac{\delta R_s}{R_s} \quad (5.23b)$$

where Ω, the "decoupling" between stomatal resistance and transpiration rate (Meinzer 1993) is $R_s/(R_s + R_b)$. Especially among understory plants, and canopy trees in still air, Ω usually exceeds 0.5 (Meinzer et al. 1995, 1997).

Does the resistance, R_c, of a tree's crown to transpiration reflect other aspects of its physiology? Meinzer et al. (1997) compared the dependence of R_c on ΔP in four species of canopy tree in Panama's metropolitan park, using a construction crane to reach their crowns. These trees were strikingly different: *Cecropia longipes* had an open, shallow crown of large, lobed leaves; *Ficus insipida* had a split-level monolayer crown, each half a smooth, dense lawn of upright leaves spiralled about a multitude of twigs; *Luehea seemanii* had a multilayer crown with distichous leaves on leaning twigs (Troll's model); *Spondias mombin* spiralled pinnately compound leaves around inclined branches. In all four species, R_c was higher the higher ΔP, but the dependence of R_c on ΔP differed from species to species. Nonetheless, all four species had the same value of $R_c(SA/LA)$ for the same value of ΔP (Meinzer et al. 1997), where SA is sapwood area, the cross-sectional area of functional xylem, of a canopy branch 4 to 6 cm in diameter, and LA is the total area of leaves on that branch. If $(SA/LA)R_c(\Delta P) = $ const., then LA/SA is proportional to R_c. If LA/SA reflects a tree's "hydraulic resistance," resistance to the movement of water from the soil to its leaves (see below, chapter 7), then crown resistance and hydraulic resistance vary in parallel. This parallel is also reflected in plant behavior: Meinzer et al. (1995) found that stomatal resistance of a sapling in a tree fall gap was governed more by the resistance per unit leaf area to movement of water from its soil to its leaves than by water tension in these leaves. Crown resistance is greater and photosynthetic capacity per unit leaf is lower in mangroves adapted to more saline conditions (Ball 1996). How poor soil must be to favor high crown and whole-canopy resistance, and how leaf arrangement and branching pattern affect this resistance, we do not know.

Why Are Elfin Forests Stunted?

The density of trees ≥ 10 cm dbh and the spectrum of architectures and leaf arrangements of understory saplings seem to be governed largely by the height of the forest canopy. What, however, governs forest height?

One way to approach such a question is to look at an extreme case. On the windy, foggy summits of Puerto Rico's Luquillo Mountains, the forest is dwarfed, only 3–5 m tall (Howard 1968, Weaver and Murphy 1990). Similarly dwarfed forest occurs on wind-planed ridge crests at 1500 m in the Monteverde Cloud Forest Reserve (Lawton and Dryer 1980), from 2500 m upward in Peru's Cordillera Vilcabamba (Terborgh 1971), from 800 m upward in Venezuela's Isla Margarita (Sugden 1986), and on many other tropical mountaintops. An extraordinary pygmy forest of *Leptospermum flavescens* (Fig. 5.19) and the conifer *Dacrydium comosum* grows atop the summit of Malaysia's Gunong Ulu Kali (Leigh and Lawton 1981). What stunts these elfin forests?

The answer is by no means obvious. There is a multitude of possible explanations. The soils of elfin forest are usually waterlogged. But who can forget such sights as a

Figure 5.19. *Leptospermum flavescens* (Myrtaceae), 2 m tall, summit heath of Gunung Ulu Kali, west Malaysia. Drawing by Marshall Hasbrouck.

virgin swamp of tall bald cypress (*Taxodium distichum*), or the swamps of great water tupelos (*Nyssa aquatica*) through which one passes when crossing southern Louisiana by railway? Their soils must also be waterlogged. Elfin forest usually grows in cold, foggy climates, yet the cloud forests at 2450 m in New Guinea (Edwards and Grubb 1977) and the oak forest at 3000 m on the deathly cold and foggy heights of Costa Rica's Cerro de la Muerte (Holdridge et al. 1971) are about 30 m tall. The redwood forests of northern California, which grow best where coastal fog is most prevalent, are among the tallest in the world. The elfin forest at Monteverde is stunted even though its soil is fertile, and saplings in its tree fall gaps grow at least as fast as saplings in clearings in the much taller cloud forest of Cerro de la Muerte (Lawton and Putz 1988). The elfin forests of the Luquillos are wreathed in fog all night and most of the day, and their temperature varies little over the day (Baynton 1968), but in the Uluguru Mountains of Tanzania, forests wholly within the cloud belt are taller, while at night the cloud belt drops below the elfin forests, which therefore cool rapidly: they suffer a diurnal temperature range of 15°C (Pócs 1976). One could compose a "*Sic et Non*" about the causes of stunting in elfin forest every bit as full of contradictions as the *Sic et Non* with which Peter Abelard infuriated the Christian theologians of his day. How can we resolve this dilemma?

Let us first try to understand the common features and the common problems of cloud forests, regardless of their height. Cloud forests spend a substantial proportion of their time wreathed in cloud and fog: they appear at the "cloud line" which forms at a characteristic altitude on a given tropical mountainside—lower down on slopes, especially of isolated hilltops, close to the sea, and higher up on continental massifs. The cloud line is at 525 m on the windward slope of the 865 m Cerro Macuira in Colombia's Guajira Peninsula (Sugden 1982), 1380 m in Peru's Cordillera Vilcabamba (Terborgh 1971), and slightly over 2000 m on Sabah's 4100 m Mt. Kinabalu (Kitayama 1995). Where the cloud line is clearly defined, so is the lower boundary of the cloud forest. At Macuira, a 150 m walk involving a 50 m gain of altitude brings one from deciduous dry forest to stunted, gnarled evergreen forest with an abundance of vascular epiphytes (Sugden 1982, Cavelier and Mejia 1990). At 1380 m in the Cordillera Vilcabamba, the lower limit of cloud forest is marked by a sudden appearance of an abundance of vascular epiphytes on the branches of canopy trees, thick jackets of moss on their trunks, and an acidic mat of litter and humus on the forest floor (Terborgh 1971). Fog, and the moisture it brings, is the crucial agent. On the Serrania de Macuira, solar radiation is lower at 300 m than in the cloud forest at 700 m (Cavelier and Mejia 1990).

Cloud cover in these forests is often thick and persistent. The elfin forest atop Puerto Rico's 1051-m Pico del Oeste was wreathed in cloud for the full 24 hr on 138 days of the year which began 1 March 1966. It was in fog all of nearly every night and for an average of 60% of the daylight hours (Baynton 1968). Unlike rain gauges, trees extract water from clouds. One year, the cloud forest at Macuira received 853 mm of rain and 796 mm of fog drip (Cavelier and Goldstein 1989b). This soggy atmosphere can limit temperature variation. On the foggy summit of Pico del Oeste, the average diurnal temperature range is only 2.5°C (Baynton 1968); in the cloud forests of Tanzania's Uluguru Mountains, diurnal temperature range was 4°C, compared to 9°C in the lowland forest of Barro Colorado. Cloud forests also receive little solar radiation: Macuira (700 m) receives 92 W/m^2 (Cavelier and Mejia 1990), Pico del Oeste 127 W/m^2 (Baynton 1968), the nearby El Yunque (1050 m) also receives 127 W/m^2 (Briscoe 1966), and Monteverde 107 W/m^2 (Lawton 1990), compared to Barro Colorado's 187 W/m^2 (calculated from Windsor 1990).

This soggy, dripping climate sometimes supports an enormous buildup of bryophytes. In a cloud forest of Tanzania's Uluguru Mountains, there were 11 tons dry weight of bryophytes per hectare of forest (Gradstein and Pócs 1989)—more biomass than tropical forests devote to tree leaves (Table 5.14). The moss was so thick that this forest intercepted 50% of the incoming rainfall (Gradstein and Pócs 1989), compared to 9% intercepted by a lowland Amazonian rain forest, and the 20% measured for most lowland rainforests by less accurate, but more frequently employed, techniques (Lloyd and Marques 1988).

What do the structure and physiognomy of cloud forests suggest about the problems they face? These forests have the physiognomy Givnish (1984) predicted for forests where rapid photosynthesis and/or transpiration is not feasible, so that leaves must be kept cool and their transpiration and gas exchange strictly limited. Many cloud forests—tall forests like Cerro de la Muerte's as well as elfin forests like Puerto Rico's—have a remarkably even, aerodynamically smooth, canopy of sclerophyllous or pachyphyllous leaves, held stiffly inclined at the tips of erect twigs. Canopy crowns may be intermingled in the most extraordinary manner, but they fit together like patches in a quilt to form a smooth canopy (Fig. 5.20). In Puerto Rico, Gleason and Cook (1927: pp. 132–133) observed that, although at lower altitudes, tree "crowns are more or less rounded, but with greater exposure to the wind the general crown level is flattened, until at the summit it presents a smooth expanse of dense foliage over which the only contrast is formed by the occasional plants of epiphytic bromeliads, perching on the tips of the branches." Even in cloud forests with more ragged canopy, an individual tree crown usually consists of a single dense layer of upright leaves, spread in a shallow dome over the often dense array of supporting branches (Fig. 5.21). The architecture of these trees seems designed not only to assure aerodynamic smoothness but to minimize the amount of shedding and rebuilding of branches required by the growth of these trees, as if it were important to husband resources. Cloud forests lack canopy trees with distichous leaves on hori-

Table 5.14 *Weight and surface area of foliage and total annual weight of falling leaves per hectare in different forests*

Site	Altitude (m)	Foliage (tons/ha)	LAI	SLW (tons/ha)	Leaf Fall (tons/ha·year)	Leaf lifetime (years)	Reference
Pasoh, Malaysia	90	7.3	7.4	0.99	7.00	1.04	Kira (1978)
Elfin forest, Puerto Rico	1050	2.9	2.0	1.45	2.45	1.18	Weaver et al. (1986)
Mull ridge, Jamaica	1600	6.8*	5.7*	1.24	5.32	1.28	Tanner (1980a,b)
Mor ridge, Jamaica	1600	8.1*	4.3*	1.88	4.99	1.62	Tanner (1980a,b)
Papua New Guinea	2450	8.9	5.5	1.62	6.20	1.44	Edwards and Grubb (1977), Edwards (1977)
Rio Negro, Oxisol	Lowland	9.8	7.5	1.31	6.50	1.51	Medina and Cuevas (1989)
Rio Negro, Tall caatinga	Lowland	6.9	4.5	1.53	5.20	1.33	Medina and Cuevas (1989)

*Includes leaves of trees, saplings, seedlings and climbers.
Foliage is given as dry weight per hectare of forest; LAI, leaf area index, is hectares of leaves per hectare of ground; SLW, specific leaf weight, is average leaf weight per hectare of leaf surface; leaf fall is the dry weight of leaves falling per hectare of forest per year; and average leaf lifetime is the forest's total weight of foliage divided by leaf fall.

Figure 5.20. A patch of elfin shrubbery 3 m tall near the roadside in paramo at 3000 m near Totoró, Colombia. Drawing by Gerardo Ravassa.

Figure 5.21. Sprig of canopy foliage of *Eugenia wrayi* (Myrtaceae), with 3 cm leaves, Gunong Batu Brinchang. Drawing by George Angehr from slide of Elizabeth Leigh.

zontal branches (Table 5.9) because, in the monolayer trees that elaborate the canopy of cloud forest, such horizontal branches would inevitably be shaded and rendered useless as the tree grows (Leigh 1990).

As one expects in settings where resources are precious, elfin forests are built to last. A hectare of Puerto Rico's elfin forest, with 4000 trees ≥ 10 cm dbh of total basal area (total cross-sectional area of trunks at breast height) of 44 m^2, uses 90 tons dry weight of wood to support 2 ha of leaves 4 m above the ground (Weaver et al. 1986, Weaver and Murphy 1990); this forest uses 1.1 kg dry weight of wood for each square meter of leaves lifted an additional meter above the ground (a leaf support efficiency, sensu King 1994, of 0.9 m^3/kg). For comparison, the 500+ trees ≥ 10 cm dbh on 1 ha of Malaysia's Pasoh Reserve, with basal area about 30 m^2 and average height (weighted by basal area) about 30 m, use 420 tons dry weight of wood to support their leaves—200 g of wood to lift each square meter of leaves an extra meter above the ground, for a leaf support efficiency of 5 m^3/kg (Table 5.7). The contrast is particularly extraordinary if one recalls that taller trees require more wood to lift a square meter of leaves an additional meter above the ground to avoid buckling under the weight of their crown (Niklas 1994). The effect of height is suggested by the contrast between Tyree et al.'s (1991) *Schefflera morototoni*, which used 462 l of wood to carry 26.2 m^2 of leaves at an average height of 19.4 m—0.9 l of wood to lift each square meter of leaves another meter above the ground, and my *Psychotria acuminata*, of similar dichotomous architecture, which used 0.18 l of wood to support 1.2 m^2 of leaves at an average height of 1.3 m— 0.117 l of wood to lift each square meter of leaves another meter above the ground. The effect of wind is reflected by the *Schefflera pittieri* I measured on the windswept ridge above Monteverde, which used 81 l of wood to support 13 m^2 of leaves at an average height of 4.5 m—1.4 l of wood to lift each square meter of leaves another meter above the ground.

Leaves of cloud forest, like those of trees on poor soil, are stoutly built, as if meant to last, and somewhat longer-lived than leaves of normal lowland forest (Table 5.14). Moreover, the leaves of cloud forest trees are usually simple, each leaf consisting of a single blade. As one passes from Macuira's deciduous dry forest into its cloud forest, trees with compound leaves (many leaflets per leaf) drop out. Some cloud forests are well stocked with palms and/ or tree ferns. Otherwise, except on the most fertile soils, the leaves of those cloud forest trees possessing compound leaves, such as *Weinmannia* (Fig. 5.22), are no larger than the simple leaves of their neighbors. In contrast, the leaflets of lowland forest trees are often as big as the simple leaves of their neighbors (Givnish 1984). The rachis of a compound leaf is a "throwaway twig," quite appropriate for the flimsy, ephemeral leaves of deciduous trees, but apparently too wasteful for the restrictive conditions of cloud forest (Givnish 1978).

Not only are cloud forest plants designed to avoid needless shedding and rebuilding; their stoutly built, succulent leaves (Fig. 5.23), which often have downrolled margins and which are nearly always held stiffly inclined in clusters at the tips of their twigs, remind the biologist of the mangrove leaves described by Ball et al. (1988), which are designed and arranged to avoid overheating in a habitat that severely restricts the rate of transpiration. Can we conclude that cloud forests, and particularly elfin forests, live in unproductive habitats?

Some elfin forests support this conclusion. In Malaysia, the summit heath of Gunong Ulu Kali and the elfin forest atop Gunong Brinchang abound in *Nepenthes*, vines that form insect-trapping pitchers at the ends of their leaves, as if it were easier to obtain nutrients from trapped insects than from a soil poisoned by the long-lived chemical defenses of this forest's leaves. In the coniferous forests of the north, the phenols and lignins used by the conifers to defend their needles create an acid mor humus, which is hostile to earthworms and slows or blocks decomposition when these needles fall to the ground (Waring and Schlesinger 1985); indeed, "evergreen conifers are found on, adapted to, and 'drive' sites toward resource-poor conditions" (Reich et al. 1995: p. 29). Similarly, in the forests of Gunung Ulu Kali and Gunung Brinchang in Malaysia, litter decomposition and nutrient recycling appear to be poisoned by the chemicals plants use to defend their precious, stoutly built leaves, and substantial layers of peat have formed there (Whitmore and Burnham 1969). Even "natural" pest repellants harm the soil if they are too long-lived; here they have created a vicious circle which keeps the soil poor.

In the elfin forest of Puerto Rico, above-ground productivity is very low, about 4 tons of dry matter per hectare per year, of which about 400 kg is wood production (Weaver et al. 1986); this is one-tenth the wood production on Barro Colorado. (Wood production on a mor ridge cloud forest in Jamaica is equally low; Tanner 1980a). In the elfin forest of the Luquillos, recovery after disturbance is glacially slow. Eighteen years after a small plane crashed, opening a clearing in this elfin forest, the clearing contained only 800 g dry weight of regrowth/m^2, of which only 240 g consisted of woody dicots (Weaver 1990). The rate of forest recovery amounted to only 150 kg/ha · year, less than 1% of the productivity of a normal lowland rainforest (Kira 1978). In this clearing, most of the woody vegetation had regenerated from stump sprouts (Byer and Weaver 1977). In the nutrient-starved heath forests of Borneo, regeneration in a new clearing consists mostly of stump sprouts because seedlings cannot gain quick enough access in such poor soil to the nutrients they need. In mixed dipterocarp forests, which grow on better soil, regrowth in a new clearing is driven by seed fall and seedling growth (Riswan and Kartawinata 1991). The elfin forest of the Luquillos appears as starved of nutrients as the heath forests of Borneo.

Figure 5.22. *Weinmannia* sp. (Cunoniaceae). Canopy tree 15 m tall, leafless branch 1 m tall, and sprig of foliage 20 cm tall, from Montagne d'Ambre, Madagascar. Drawing by George Angehr.

Not all cloud forests, however, appear so starved of nutrients. In cloud forests such as Vilcabamba Camp 2.5 and Monteverde Cloud Forest 1, leaves are unexpectedly large, as if nutrients were abundant. Even though the trees of Monteverde's cloud forest send roots into the organic matter which collects around the bases of epiphytes (Nadkarni 1981), as if nutrients were at a premium, the elfin forest there is dynamic, tree fall gaps refill rapidly, and their regeneration originates primarily from seedlings rather than stump sprouts (Lawton and Putz 1988). At Monteverde, the forest is stunted by the wind rather than by shortage of nutrients (Lawton 1982, 1984).

Moreover, other explanations besides those of Givnish (1984) might account for the structure of elfin forest canopy. Leigh (1975) argued that in a habitat with perpetual wind and high humidity, the only chance a leaf has to warm

Figure 5.23. *Fagraea* sp. (Loganiaceae), from summit heath of Gunung Ulu Kali, west Malaysia, showing leathery leaves up to 9 cm long with downrolled edges, typical of many cloud forest plants. Drawing by Marshall Hasbrouck.

up above ambient temperature (desirable for photosynthesis and transpiration alike) is to hide behind another leaf from the cooling wind while remaining as exposed as possible to the sun. This combination of features is best achieved by a dense, level canopy of upright leaves. Cavelier and Goldstein (1989b: p. 317) showed that upright leaves are better at extracting water from fog—an essential asset for plants at Macuira, with its low rainfall, but not very useful on Pico del Oeste. Pócs (1976) argued that the dense canopy of elfin forest reduces the escape of heat to the atmosphere during clear nights—an advantage at high elevations in Tanzania's Uluguru Mountains, but almost irrelevant for elfin forest in the Luquillos. Cavelier and Goldstein (1989a) argued that elfin forest trees stored water in their leaves so that, during the brief spells of sunlight so characteristic of the changeable weather of cloud forests, leaves could transpire readily when the sun did come out, avoiding the delay involved in bringing water up from the cold roots: an advantage that would be helpful for plants in most elfin forests.

We still do not know what stunts elfin forests. Our ignorance suggests that we have a great deal yet to learn about the functional morphology of tropical plants and the factors that govern or restrict the productivity of tropical forest.

CONCLUDING REMARKS

Tropical trees come in a stunning variety of shapes. Saplings and shrubs can often be classified by their "architecture," their arrangement of branches and leaves. Adult trees are more difficult to classify.

Three theoretical principles help sort out this variety. First, plant diameter varies with height so as to provide a margin of safety against a stem buckling under its crown's weight. Second, shaded understory plants that are lit from above rather than from the side display their leaves in flat, horizontal monolayers with minimum overlap to catch as much light per unit area of leaf as possible. Canopy trees, on the other hand, spread the abundant light over as much leaf surface as possible: their leaves are often inclined and/or curved about the midrib, and there is much more self-shading. Third, where water and nutrients are readily available, tree crowns are organized to promote maximum photosynthesis and transpiration, while, where nutrients or water are scarce, tree crowns are organized to restrict photosynthesis and/or transpiration to "affordable" levels.

In rainforest, average leaf size declines with average temperature of the coolest month. For given temperature, leaves are smaller on poorer soils.

The most basic distinction in tree architecture is between plants with distichous leaves on horizontal twigs and branches and plants arranging leaves around ascending twigs on ascending branches. Plants with distichous leaves and horizontal branches are most common in the understory of tall, lowland forest, presumably because they can easily and economically widen their crowns to catch more light. Such plants are rarer in shorter forests. They are entirely absent from the elfin forests of fog-bound tropical mountaintops.

What governs a rainforest's height? One way to find out is to ask why the elfin forests of windblown tropical mountaintops are so stunted. Elfin forest trees must usually build stoutly to resist the wind, which slows their growth. In many (but not all) elfin forests, productivity is low, so resources for tree-building are scarce. In either case, elfin forest trees require architectures that demand a minimum of shedding and rebuilding in the course of tree growth. Thus they have ascending branches that can be extended as the tree grows, rather than horizontal ones that must be replaced when shaded.

In sum, the study of tree-design is still an art and very far from the quantitative, predictive science biologists need to understand the meaning of different tree shapes. Nevertheless, a plant's shape and the size and arrangement of its leaves are obviously adjusted in accord with the availability of light and nutrients in its habitat and the need to resist such factors as windthrow. The next chapter discusses the biomass and productivity of tropical forest and the factors that govern these features.

APPENDIX 5.1

How Tall Can a Tree Be?

When the tip of an upright beam with a fixed base is bent, the stretched fibers on the curve's outside exert a pull, while the compressed fibers on the inner side exert a push, which jointly create a restoring *torque* about each straight line x = constant (x being the distance above the base) across the unstretched "neutral surface" that bisects the beam longitudinally (Feynman et al. 1964: ch. 38). Let y be the perpendicular distance from the neutral surface of a point in the beam. Suppose that the bent beam forms a segment of a circle of radius R (Fig. 5.13). Then the proportion, $\Delta l/l$, by which a fiber of length l passing through this point parallel to the beam's long axis is stretched, is the ratio of its distance, y, outward from the neutral surface to the radius, R, of the curve imposed on the beam. The force exerted by wood fibers of distance between y and $y + dy$ from the neutral surface is Ey/R times the cross-sectional area $s(y)dy$ of these fibers: E is the modulus of elasticity, and Ey/r denotes the restoring force per unit cross-section of fibers (thus the units are of pressure, pascals or newtons per square meter) by wood fibers stretched a proportion, y/R, beyond their normal length. Since y/R is dimensionless, E also has units of pressure. Fibers x meters above ground and a distance y from the neutral surface thus contribute a torque $y(Ey/R)s(y)dy$ about the line $y = 0$, x = constant. For a beam of thickness $2r(x)$ a distance x from its base, the total torque about this line is

$$\frac{E}{R} \int_{-r(x)}^{r(x)} y^2 s(y)\, dy. \quad (5.24)$$

If, like a tree trunk, the beam is circular, then $s(y)dy = 2\sqrt{(r^2 - y^2)}$, and the integral of equation 5.24 is $\pi r^4/4$. Let $z(x)$ be the displacement of a point on the beam x meters above the ground when the beam is bent. Then $1/R = d^2z/dx^2$. The action of gravity on the bent beam creates a torque about the line x = constant, $y = 0$ through the beam's neutral surface. If the maximum displacement, z, is a small fraction of the beam's height, H, this torque's magnitude is

$$\int_x^H [z(q) - z(x)]\, \rho g \pi r^2(q)\, dq. \quad (5.25)$$

Here, g is gravitational acceleration (9.8 m/sec^2), and ρ is the density of wood (kilograms fresh weight per cubic meter). The equation of the curve $z(x)$ is thus

$$\frac{1}{4} E \pi r^4(x) \frac{d^2z}{dx^2} = \int_x^H [z(q) - z(x)]\, \rho g \pi r^2(q)\, dq. \quad (5.26)$$

If $r(x) = r$, and if there is no weight atop the beam,

$$\frac{E}{4} \pi r^4 \frac{d^2z}{dx^2} = \pi r^2 \rho g \int_x^H [z(q) - z(x)]\, dq. \quad (5.27)$$

This equation can be solved by successive approximations. First set $d^2z/dx^2 = 2c$. Then $z(x) = cx^2$. Substitute cx^2 for $z(x)$ and cq^2 for $z(q)$ in the right-hand side of equation 5.26 to obtain

$$\frac{E}{4\rho g} r^2 \frac{d^2z}{dx^2} = \int_x^H [z(q) - z(x)]\, dq \quad (5.28)$$

$$\quad (5.29)$$

$$\frac{E}{4\rho g} r^2 \frac{d^2z}{dx^2} = \int_x^H c(q^2 - x^2)\, dq = \frac{c}{3}(H^3 - x^3) - cx^2(H - x).$$

If we now set $d^2z/dx^2 = 2c$ on the left-hand side of equation 5.29 and $x = 0$ on the right, this equation becomes $Er^2c/2\rho g = cH^3/3$. The beam buckles if the right-hand side (which is proportional to the torque exerted by gravity) exceeds the left-hand side (which is proportional to the restoring torque). The beam buckles if $H^3 > 3Er^2/2\rho g$, that is to say, if $H > 1.14(Er^2/\rho g)^{1/3}$.

For the next approximation, solve equation 5.29 for d^2z/dx^2 to obtain

$$\quad (5.30)$$

$$\frac{d^2z}{dx^2} = k(H^3 - x^3) - 3kx^2(H - x) = kH^3 - 3kx^2H + 2kx^3$$

$$z(x) = \frac{k}{2} H^3 x^2 - \frac{k}{4} H x^4 + \frac{k}{10} x^5. \quad (5.31)$$

Substituting these expressions for d^2z/dx^2 and $z(x)$ into equation 5.28 and setting $x = 0$, we obtain

$$\quad (5.32)$$

$$\frac{Er^2}{4\rho g} KH^3 = \int_0^H \left[\frac{k}{2} H^3 q^2 - \frac{k}{4} H q^4 + \frac{k}{10} q^5\right] dq = \frac{8kH^6}{60}.$$

We find that the beam buckles if $kH^6/15 > EH^3r^2k/8\rho g$, $H > 1.23(Er^2/\rho g)^{1/3}$. According to Greenhill's (1881) exact solution, the beam buckles if $H > 1.26(Er^2/\rho g)^{1/3}$. This formula is the basis for the familiar assertion that H^3 should vary as r^2 (McMahon 1973).

King and Loucks (1978) set $r^2(x) = r^2(0)(1 - x/H)$ and replace the top tenth of their tapered stem by a point mass of $K[\rho \pi r^2(0)H/2]$. Two approximations analogous to mine led to the formula in the text.

APPENDIX 5.2

Collecting the Data for Tables 5.10–5.12

Tables 5.10–5.12 are based on standard point-quarter transects (Cottam and Curtis 1956) in selected tropical rainforests whose climates are summarized in Table 5.15. For most transects four transect points were chosen on

Table 5.15 Monthly temperature (°C) and rainfall (P, mm) in forests of Tables 5.10–5.12

Forest and reference	Altitude (m)		Jan.	Feb.	Mar.	Apr.	May	Jun.	Jul.	Aug.	Sep.	Oct.	Nov.	Dec.	Total/Avg.
Cerro de la Muerte	3000	T_m	9.9	10.6	11.5	11.1	11.8	11.4	11.3	11.0	10.9	10.6	10.4	9.6	10.8
Holdridge et al. (1971)		P	35	22	26	91	372	487	268	289	448	518	229	156	2941
Mt. Kaindi	2360	$T+$	18.8	19.6	19.5	19.3	19.4	18.2	17.0	17.9	18.5	19.3	19.5	19.0	18.8
		$T-$	11.7	12.4	12.8	12.8	12.8	12.1	11.5	11.2	11.5	11.9	11.8	12.8	12.1
Gressitt and Nadkarni (1978)		P	231	449	274	229	213	180	162	134	146	183	263	240	3104
Avalanchi	2100	P	7	4	15	72	165	384	1075	524	261	197	107	39	2850
Blasco (1971)		T_m	12.4	13.1	14.9	16.4	16.3	14.5	13.8	13.9	14.2	14.1	13.2	12.4	14.1
Ambohitantely	1550	$T+$	24.7	24.9	24.2	23.8	21.6	19.8	19.0	20.5	23.0	25.6	26.3	25.1	23.2
		$T-$	14.7	14.9	14.6	13.2	10.2	8.4	7.9	7.9	9.2	10.8	12.7	14.2	11.5
Bailly et al. (1974)		P	391	310	327	53	21	5	11	9	13	44	199	334	1717
Monteverde, elfin ridgetop	1500	$T+$	16.5	16.8	18.3	21.7	21.8	19.4	19.6	18.3				17.2	18.9
Robert O. Lawton (p. c.)		$T-$	13.0	13.1	14.2	15.2	16.4	15.9	15.6					14.0	14.7
Tanah Rata (for G. Brinchang)	1430	T_m	17.0	17.1	17.7	18.2	18.4	17.9	17.4	17.5	17.6	17.7	17.6	17.2	17.6
Malayan Meteorological Service		P	151	119	215	308	257	133	136	176	250	339	320	223	2626
Fraser's Hill	1200	T_m	18.1	18.9	19.7	20.4	20.6	20.5	20.0	20.1	19.9	19.9	19.3	18.5	19.7
Malayan Meteorological Service		T_r	4.7	5.7	6.0	6.4	6.3	6.4	6.5	6.6	6.7	6.6	5.6	5.0	6.0
		P	299	180	248	270	219	122	111	131	185	286	369	306	2721
Rancho Grande	1100	T_m			18.6	19.8	19.4	18.9	18.5	18.3					
Beebe and Crane (1947)		P	20	25	33	100	188	216	301	286	185	188	135	71	1747
Mt d'Ambre, Roussettes	1100	(Avg. temperatures for the coolest and warmest months are 17° and 21 °C, respectively)													
Humbert and Cours Darne (1965)		P	500	724	500	140	85	90	85	75	60	85	144	210	2600
Pico del Oeste, P. Rico	1051	$T+$	17.5	17.6	19.2	20.0	20.0	20.4	20.7	20.7	21.1	21.6	21.0	18.3	19.9
		$T-$	16.6	16.6	16.5	16.7	18.4	18.8	18.9	18.5	18.4	18.6	16.8	16.3	17.4
		P	303	276	348	385	451	364	439	418	399	256	274	620	4533
Baynton (1968)																
El Yunque, Puerto Rico	1050	T_m	17.4	17.8	17.7	18.6	18.8	19.2	18.3	19.3	18.8	18.8	18.7	17.3	18.4
Briscoe (1966)		P	176	174	172	446	447	287	301	312	687	180	286	375	3813
Analamazaotra	1000	$T+$	27.0	27.1	26.1	25.4	23.2	21.3	20.3	20.6	22.2	24.1	26.3	27.1	24.2
		$T-$	16.9	16.6	16.7	15.4	12.9	11.3	10.5	10.3	11.1	12.4	14.7	16.3	13.8
Bailly et al. (1974)		P	306	320	262	92	61	77	78	67	51	44	112	238	1708
Ulu Gombak (Medway 1972)	515	P	116	128	141	184	167	159	123	187	246	330	298	205	2283
Londonderry (for Palmist Ridge, Dominica 240 m)																
Hodge (1954)		P	142	84	160	106	168	334	292	268	586	709	738	81	3668
Panguana, Peru	220	$T+$	27.0	26.0	27.4	26.8	27.4	26.7	26.9	27.9	28.4	26.8	27.3	28.0	27.2
H. W. Koepcke (unpublished data)		$T-$	22.0	21.4	21.9	21.3	21.1	20.7	19.9	20.1	20.6	21.3	21.6	21.6	21.1
		P	255	267	210	190	92	142	113	78	140	239	258	277	2261
Catalina (Puerto Rico lowland)	150	$T+$	26.1	26.9	27.5	28.1	27.9	28.0	27.1	28.2	27.7	26.9	25.9	27.3	
Briscoe (1966)		$T-$	21.0	20.8	21.2	21.7	22.6	23.5	22.6	23.2	23.0	22.5	21.4	22.1	
		P	89	73	94	260	374	222	268	225	260	188	203	227	2483
Bubia Morobe (for Gabensis)																
Brookfield and Hart (1966)		P	192	196	233	246	219	248	363	366	281	240	167	204	2927
Pasoh Reserve	90	$T+$	29.1	30.0	28.6	29.2	28.4	28.2	28.4	28.2	28.4	25.4	28.2	26.2	28.2
Soepadmo (1973, 1974)		$T-$	21.4	20.0	21.8	22.1	21.0	21.1	20.0	20.3	20.2	20.7	21.2	20.0	20.8
		P	88	132	90	291	200	133	87	101	108	160	263	239	1892
Lae (for Busu River)	0														
Brookfield and Hart (1966)	0	P	267	236	330	414	417	394	472	526	488	335	335	318	4532
Maroantsetra (for Hiarakh)		$T+$	29.8	30.1	29.5	28.1	26.2	24.7	23.5	23.7	24.6	26.5	28.0	29.2	27.0
Lebedev (1970)		$T-$	22.8	22.6	22.7	22.1	20.6	19.0	18.2	17.9	18.1	19.3	20.7	21.9	20.5
Paulian (1961)		P	377	382	445	464	344	304	343	253	133	87	113	288	3536

$T+$, $T-$ are monthly averages of daily maximum and minimum temperatures; T_m is monthly average temperature and T_r is monthly average of diurnal temparature range. Information on Cordillera Vilcabamba is summarized in Terborgh (1971). Under Avalanchi, rainfall figures are for Mukurti lower, 12 km north at the same altitude, and temperature is for Ootacamund, 20 km away at similar altitude (Blasco 1971, annexe II).

alternating sides of a straight 30-m transect line laid along the floor of a likely piece of forest, so as to avoid tree fall gaps. A coin was tossed to decide whether the first transect point (TP1) fell to the right or the left of the transect line. If TP1 fell to the right, it was set 3 m rightward of the transect line, in which case TP2 was 3 m left of 11 m, TP3 3 m to the right of 19 m, and TP4 3 m to the left of this transect line (Fig. 5.24). Some transects were longer, including five or six transect points, in which case TP5 was 3 m to one side of 35 m and TP6 3 m to the other side of 43 m on the transect line.

The space around each transect point was divided into four quadrants by lines through the point parallel to and perpendicular to the transect line (Fig. 5.24). These quadrants were numbered in circuit around the transect point, with quadrant 1 including 0 m on the transect line, and quadrant 4 including this line's end (usually 30 m).

In each quadrant, I sampled the small sapling, S (6–12 mm dbh), the medium sapling, M (13–25 mm dbh), the large sapling, L (26–101 mm dbh), and the tree, T_1 (dbh > 101 mm), nearest to this point, as well as the second nearest tree (T_2). For each plant, I measured diameter at breast height and distance to the transect point, estimated height by eye, and recorded its architecture as best I could according to the schema outlined in Table 5.4. I also recorded the leaf arrangements of sampled S and M and measured the length and breadth and recorded the shape of one good, healthy leaf from each sampled sapling < 26 mm dbh. For T_1 and T_2, I measured the distance to the nearest neighboring tree, regardless of what quadrant it was in and regardless of whether it had been sampled (T_1 and T_2 were often each other's nearest neighbors). If these nearest neighbors, NT_1 and NT_2, had not yet been sampled, I took the same records for them as for other sampled trees, except that I recorded the distance from NT_i to T_i rather than the distance from NT_i to the transect point. The standard 4-point transect sampled 16 small saplings, 16 medium saplings, 16 large saplings, and about 40 trees. Where possible, I tried to do three such transects per forest.

The number of square meters per tree can be estimated either as $\pi R_1^2/4$, where R_1^2 is the mean square distance from the transect point to T_1 for the 16 nearest trees; $\pi R_2^2/8$, where R_2^2 is the mean square distance from TP to T_2 for the 16 second nearest trees; or as πR^2, where R^2 is the mean square distance from a tree to its nearest neighbor. In Tables 5.10 and 5.12, these three estimates were averaged, and this average averaged over the transects of the forest in question, to find the number of square meters per tree.

At 3, 11, 19, and 27 m along the transect line, the ten nearest litter leaves were sampled from the forest floor and their outlines traced in a notebook, from which length and width were measured later. In elfin and other montane forests of low stature, leaf samples were taken at shorter intervals.

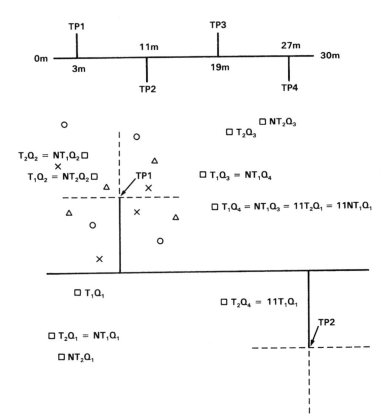

Figure 5.24. Design of standard transect. Top: location of transect points (TP) relative to tape points at 3, 11, 19 and 27 m on the 30-m transect line. Bottom: positions of small sapling (△), medium sapling (○) and large sapling (×) nearest transect point 1 in each of the four quadrants surrounding that point; positions of the nearest and second nearest trees, T_1Q_x and T_2Q_x (□) to that point in each quadrant x, $x = 1, 2, 3, 4$; and positions of the trees nearest to those trees, NT_1Q_x and NT_2Q_x, which might be each other, or trees sampled in the next quadrant, or even at the next transect point.

References

Bailey, I. W., and E. W. Sinnott. 1915. A botanical index of Cretaceous and Tertiary climates. *Science* 41: 831–834.

Bailey, I. W., and E. W. Sinnott. 1916. The climatic distribution of certain types of angiosperm leaves. *American Journal of Botany* 3: 24–39.

Bailly, C., G. Benoit de Coignac, C. Malvos, J. M. Ningre, and J. M. Sarrailh. 1974. Étude de l'influence du couvert naturel et de ses modifications à Madagascar. *Cahiers Scientifiques, Centre Technique Forestier Tropical* 4: 1–115.

Ball, M. C. 1996. Comparative ecophysiology of mangrove forest and tropical lowland moist rainforest, pp. 461–496. In S. S. Mulkey, R. L. Chazdon, and A. P. Smith, eds., *Tropical Forest Plant Ecophysiology*. Chapman and Hall, New York.

Ball, M. C., I. R. Cowan, and G. D. Farquhar. 1988. Maintenance of leaf temperature and the optimisation of carbon gain in relation to water loss in a tropical mangrove forest. *Australian Journal of Plant Physiology* 15: 263–276.

Baynton, H. W. 1968. The ecology of an elfin forest in Puerto Rico, 2. The microclimate of Pico del Oeste. *Journal of the Arnold Arboretum* 49: 419–430.

Becker, P. 1996. Sap flow in Bornean heath and dipterocarp forest trees during wet and dry periods. *Tree Physiology* 16: 295–299.

Beebe, W., and J. Crane. 1947. Ecology of Rancho Grande, a subtropical cloud forest in northern Venezuela. *Zoologica* 32: 43–59.

Bews, J. W. 1927. Studies in the ecological evolution of the angiosperms. *New Phytologist* 26: 1–21.

Blanc, P. 1992. Les formes globales des plantes de sous-bois tropicaux et leur signification écologique. *Revue d'Écologie (La Terre et la Vie)* 47: 3–49.

Blasco, F. 1971. *Montagnes du Sud de l'Inde: Forêts, Savanes, Écologie*. Institut Français de Pondichéry, Pondicherry, India.

Borchert, R., and P. B. Tomlinson. 1984. Architecture and crown geometry in Tabebuia rosea (Bignoniaceae). *American Journal of Botany* 71: 958–969.

Briscoe, C. B. 1966. *Weather in the Luquillo Mountains of Puerto Rico*. Forest Service Research Paper ITF-3. Institute of Tropical Forestry, Rio Piedras, Puerto Rico.

Brookfield, H. C., and D. Hart. 1966. *Rainfall in the tropical southwest Pacific*. Research School of the Pacific Studies, Department of Geography Publication G/3, Australian National University.

Brunig, E. F. 1970. Stand structure, physiognomy and environmental factors in some lowland forests in Sarawak. *Tropical Ecology* 11: 26–43.

Brunig, E. F. 1983. Vegetation structure and growth, pp. 49–75. In F. B. Golley, ed., *Tropical Rain Forest Ecosystems: Structure and Function*. Elsevier, Amsterdam.

Bruenig, E. F. 1996. *Conservation and Management of Tropical Rainforests: An Integrated Approach to Sustainability*. CAB International, Wallingford, UK.

Buckley, R. C., R. T. Corlett, and P. J. Grubb. 1980. Are the xeromorphic trees of tropical upper montane rain forests drought-resistant? *Biotropica* 12: 124–136.

Byer, M. D., and P. L. Weaver. 1977. Early secondary succession in an elfin woodland in the Luquillo Mountains of Puerto Rico. *Biotropica* 9: 35–47.

Campbell, E. J. F., and D. M. Newbery. 1993. Ecological relationships between lianas and trees in lowland rain forest in Sabah, East Malaysia. *Journal of Tropical Ecology* 9: 469–490.

Cavelier, J., and G. Goldstein. 1989a. Leaf anatomy and water relations in tropical elfin cloud forest tree species, pp. 243–253. In K. H. Kreeb, H. Richter, and T. M. Hinckley, eds., *Structural and Functional Responses to Environmental Stresses*. SPB Academic Publishing, The Hague.

Cavelier, J., and G. Goldstein. 1989b. Mist and fog interception in elfin cloud forests in Colombia and Venezuela. *Journal of Tropical Ecology* 5: 309–322.

Cavelier, J., and C. A. Mejia. 1990. Climatic factors and tree stature in the elfin cloud forest of Serrania de Macuira, Colombia. *Agricultural and Forest Meteorology* 53: 105–123.

Coley, P. D. 1983. Herbivory and defensive characteristics of tree species in a lowland tropical forest. *Ecological Monographs* 53: 209–233.

Coley, P. D., and T. A. Kursar. 1996. Causes and consequences of epiphyll colonization, pp. 337–362. In S. S. Mulkey, R. L. Chazdon, and A. P. Smith, eds., *Tropical Forest Plant Ecophysiology*. Chapman and Hall, New York.

Coley, P. D., T. A. Kursar, and J.-L. Machado. 1993. Colonization of tropical rain forest leaves by epiphylls: effects of site and host plant leaf lifetime. *Ecology* 74: 619–623.

Corner, E. J. H. 1940. *Wayside Trees of Malaya*. Government Printer, Singapore.

Corner, E. J. H. 1964. *The Life of Plants*. World Publishing, Cleveland, OH.

Corner, E. J. H. 1966. *The Natural History of Palms*. University of California Press, Berkeley.

Cottam, G., and J. T. Curtis. 1956. The use of distance measures in phytosociological sampling. *Ecology* 37: 451–460.

Dean, J. M., and A. P. Smith. 1978. Behavioral and morphological adaptations of a tropical plant to high rainfall. *Biotropica* 10: 152–154.

Dean, T. J., and J. N. Long. 1986. Validity of constant-stress and elastic-instability principles of stem formation in *Pinus contorta* and *Trifolium pratense*. *Annals of Botany* 58: 833–840.

Edelin, C. 1990. The monopodial architecture: the case of some tree species from tropical Asia. *Research Pamphlet, Forestry Research Institute Malaysia* 105: 1–222.

Edwards, P. J. 1977. Studies of mineral recycling in a montane rain forest in New Guinea II. The production and disappearance of litter. *Journal of Ecology* 65: 971–992.

Edwards, P. J., and P. J. Grubb. 1977. Studies of mineral cycling in a montane rain forest in New Guinea. I. The distribution of organic matter in the vegetation and soil. *Journal of Ecology* 65: 943–969.

Feynman, R. P., R. B. Leighton, and M. Sands. 1964. *The Feynman Lectures on Physics*, vol. 2. Addison-Wesley, Palo Alto, CA.

Fisher, J. B., and H. Honda 1979a. Branch geometry and effective leaf area: a study of Terminalia-branching pattern. 1. Theoretical trees. *American Journal of Botany* 66: 633–644.

Fisher, J. B., and H. Honda. 1979b. Branch geometry and effective leaf area: a study of Terminalia-branching pattern. 2. Survey of real trees. *American Journal of Botany* 66: 645–655.

R. A. Fisher. 1958. *Statistical Methods for Research Workers*. Hafner, New York.

Givnish, T. J. 1978. On the adaptive significance of compound leaves, with particular reference to tropical trees, pp. 351–380. In P. B. Tomlinson and M. H. Zimmermann, eds., *Tropical Trees as Living Systems*. Cambridge University Press, Cambridge.

Givnish, T. 1979. On the adaptive significance of leaf form, pp. 375–407. In O. T. Solbrig, S. Jain, G. B. Johnson, and P. H. Raven, eds., *Topics in Plant Population Biology*. Columbia University Press, New York.

Givnish, T. J. 1984. Leaf and canopy adaptations in tropical forests, pp. 51–84. In E. Medina, H. A. Mooney, and C. Vasquez-Yanez, eds., *Physiological Ecology of Plants of the Wet Tropics*. W. Junk, The Hague.

Givnish, T. J., and G. J. Vermeij. 1976. Sizes and shapes of liane leaves. *American Naturalist* 110: 743–778.

Gleason, H. A., and M. T. Cook. 1927. Plant Ecology of Porto Rico, Parts 1 & 2. *Scientific Survey of Porto Rico and the Virgin Islands* 7: 1–173.

Gower, S. T., R. E. McMurtrie, and D. Murty. 1996. Aboveground net primary production decline with stand age: potential causes. *Trends in Ecology and Evolution* 11: 378–382.

Gradstein, S. R., and T. Pócs. 1989. Bryophytes, pp. 311–325. In H. Leith and M. J. A. Werger, eds., *Tropical Rain Forest Ecosystems: Biogeographical and Ecological Studies*. Elsevier, Amsterdam.

Greenhill, A. G. 1881. Determination of the greatest height consistent with stability that a vertical pole or mast can be made, and of the greatest height to which a tree of given proportions can grow. *Proceedings of the Cambridge Philosophical Society* 4: 65–73.

Gressitt, J. L., and N. Nadkarni. 1978. *Guide to Mt. Kaindi*. Wau Ecology Institute, Wau, Papua New Guinea.

Hallé, F. 1986a. Deux stratégies pour l'arborescence: gigantisme et repetition. *Naturalia monspeliensis—Colloque international sur l'Arbre 1986* pp. 159–170.

Hallé, F. 1986b. Modular growth in seed plants. *Philosophical Transactions of the Royal Society of London* B 313: 77–87.

Hallé, F., and F. S. P. Ng. 1981. Crown construction in mature dipterocarp trees. *Malaysian Forester* 44: 222–233.

Hallé, F., and R. A. A. Oldeman. 1970. *Essai sur l'Architecture et la Dynamique de Croissance des Arbres Tropicaux*. Masson et Cie, Paris.

Hallé, F., R. A. A. Oldeman, and P. B. Tomlinson. 1978. *Tropical Trees and Forests: An Architectural Analysis*. Springer, Berlin.

Hardy, G. H., and E. M. Wright. 1979. *An Introduction to the Theory of Numbers*. Oxford University Press, Oxford.

Hladik, A. 1970. Contribution à l'étude biologique d'une Araliaceae d'Amérique Tropicale: *Didymopanax morototoni*. *Adansonia* (ser. 2) 10: 383–407.

Hodge, W. H. 1954. Flora of Dominica, B. W. I. *Lloydia* 17: 1–238.

Holbrook, N. M., and F. E. Putz. 1989. Influence of neighbors on tree form: effects of lateral shade and prevention of sway on the allometry of *Liquidambar styraciflua* (sweet gum). *American Journal of Botany* 76: 1740–1749.

Holdridge, L. R., W. C. Grenke, W. H. Hatheway, T. Liang, and J. A. Tosi Jr. 1971. *Forest Environments in Tropical Life Zones: A Pilot Study*. Pergamon Press, Oxford.

Honda, H., and J. B. Fisher. 1979. Ratio of tree branch lengths: the equitable distribution of leaf clusters on branches. *Proceedings of the National Academy of Sciences, USA* 76: 3875–3879.

Horn, H. S. 1971. *The Adaptive Geometry of Trees*. Princeton University Press, Princeton, NJ.

Horn, H. S. 1975. Forest succession. *Scientific American* 232(5): 90–98.

Howard, R. A. 1968. The ecology of an elfin forest in Puerto Rico, 1. Introduction and composition studies. *Journal of the Arnold Arboretum* 49: 381–418.

Humbert, H., and G. Cours Darne. 1965. *Notice de la Carte Madagascar*. Institut Français de Pondichéry, Pondicherry, India.

Khinchin, A. Ya. 1964. *Continued Fractions*. University of Chicago Press, Chicago.

King, D. A. 1987. Load bearing capacity of understory treelets of a tropical wet forest. *Bulletin of the Torrey Botanical Club* 114: 419–428.

King, D. A. 1990. Allometry of saplings and understorey trees of a Panamanian forest. *Functional Ecology* 4: 27–32.

King, D. A. 1991a. Correlations between biomass allocation, relative growth rate and light environment in tropical forest saplings. *Functional Ecology* 5: 485–492.

King, D. A. 1991b. Tree size. *National Geographic Research and Exploration* 7: 342–351.

King, D. A. 1994. Influence of light level on the growth and morphology of saplings in a Panamanian forest. *American Journal of Botany* 81: 948–957.

King, D. A. 1996. Allometry and the life history of tropical trees. *Journal of Tropical Ecology* 12: 25–44.

King, D. A., and O. L. Loucks. 1978. The theory of tree bole and branch form. *Radiation and Environmental Biophysics* 15: 141–165.

Kira, T. 1978. Community architecture and organic matter dynamics in tropical lowland rain forests of Southeast Asia, with special reference to Pasoh Forest, West Malaysia, pp. 561–590. In P. B. Tomlinson and M. H. Zimmermann, eds., *Tropical Trees as Living Systems*. Cambridge University Press, New York.

Kitayama, K. 1995. Biophysical conditions of the montane cloud forests of Mt. Kinabalu, Sabah, Malaysia, pp. 183–197. In L. S. Hamilton, J. O. Juvik, and F. N. Scatena, eds., *Tropical Montane Cloud Forests*. Springer-Verlag, New York.

Kohyama, T. 1991. A functional model describing sapling growth under a tropical forest canopy. *Functional Ecology* 5: 83–90.

Lawton, R. O. 1982. Wind stress and elfin stature in a montane rain forest tree: an adaptive explanation. *American Journal of Botany* 69: 1224–1230.

Lawton, R. O. 1984. Ecological constraints on wood density in a tropical montane cloud forest. *American Journal of Botany* 71: 261–267.

Lawton, R. O. 1990. Canopy gaps and light penetration into a wind-exposed tropical lower montane rain forest. *Canadian Journal of Forest Research* 20: 659–667.

Lawton, R. O., and V. Dryer. 1980. The vegetation of the Monteverde Cloud Forest Reserve. *Brenesia* 18: 101–116.

Lawton, R. O., and F. E. Putz. 1988. Natural disturbance and gap-phase regeneration in a wind-exposed tropical cloud forest. *Ecology* 69: 764–777.

Lebedev, A. N. (ed.) 1970. *The Climate of Africa*. Israel Program for Scientific Translations, Jerusalem.

Leigh, E. G. Jr. 1972. The golden section and spiral leaf-arrangement, pp. 163–176. In E. S. Deevey, ed., *Growth by Intussusception*. Archon Books, Hamden, CT.

Leigh, E. G. Jr. 1975. Structure and climate in tropical rain forest. *Annual Review of Ecology and Systematics* 6: 67–86.

Leigh, E. G. Jr. 1990. Tree shape and leaf arrangement: a quantitative comparison of montane forests, with emphasis on Malaysia and South India, pp. 119–174. In J. C. Daniel and J. S. Serrao, eds., *Conservation in Developing Countries: Problems and Prospects*. Bombay Natural History Society/Oxford University Press, Bombay.

Leigh, E. G. Jr., and R. O. Lawton. 1981. Why are elfin forests stunted? pp. 20–31. In B. C. Stone, The summit flora of Gunung Ulu Kali (Pahang, Malaysia). *Federation Museums Journal* 26(1) (n.s.): 1–157.

Linsbauer, K., L. Linsbauer, and L. R. von Portheim. 1903. *Wiesner und seine Schule: ein Beitrag zur Geschichte der Botanik*. Alfred Hölder, Vienna.

Lloyd, C. R., and A. de O. Marques F. 1988. Spatial variability of throughfall and stemflow measurements in Amazonian rainforest. *Agricultural and Forest Meteorology* 42: 63–73.

Long, J. N., F. W. Smith, and D. R. M. Scott. 1981. The role of Douglas-fir stem sapwood and heartwood in the mechanical and physical support of crowns and development of stem form. *Canadian Journal of Forest Research* 11: 459–464.

McMahon, T. A. 1973. Size and shape in biology. *Science* 179: 1201–1204.

Manokaran, N., A. R. Kassim, A. Hassan, E. S. Quah, and P. F. Chong. 1992. Short-term population dynamics of dipterocarp trees in a lowland rain forest in peninsular Malaysia. *Journal of Tropical Forest Science* 5: 97–112.

Medina, E., and E. Cuevas. 1989. Patterns of nutrient accumulation and release in Amazonian forests of the upper Rio Negro basin, pp. 217–240. In J. Proctor, ed., *Mineral Nutrients in Tropical Forest and Savanna Ecosystems*. Blackwell Scientific, Oxford.

Medina, E., V. Garcia, and E. Cuevas. 1990. Sclerophylly and oligotrophic environments: relationships between leaf structure, mineral nutrient content, and drought resistance in tropical rain forests of the upper Rio Negro region. *Biotropica* 22: 51–64.

Medway, Lord. 1972. Phenology of a tropical rain forest in Malaya. *Biological Journal of the Linnean Society* 4: 117–146.

Meinzer, F. C. 1993. Stomatal control of transpiration. *Trends in Ecology and Evolution* 8: 289–294.

Meinzer, F. C., J. L. Andrade, G. Goldstein, N. M. Holbrook, J. Cavelier, and P. Jackson. 1997. Control of transpiration from the upper canopy of a tropical forest: the role of stomatal, boundary layer and hydraulic architecture components. *Plant, Cell and Environment* 20: 1242–1252.

Meinzer, F. C., G. Goldstein, P. Jackson, N. M. Holbrook, M. V. Gutiérrez, and J. Cavelier. 1995. Environmental and physiological regulation of transpiration in tropical forest gap species: the influence of boundary layer and hydraulic properties. *Oecologia* 101: 514–522.

Nadkarni, N. M. 1981. Canopy roots: convergent evolution in rainforest nutrient cycles. *Science* 214: 1023–1024.

Niklas, K. J. 1994. Interspecific allometries of critical buckling height and actual plant height. *American Journal of Botany* 81: 1275–1279.

Oldeman, R. A. A. 1974. L'architecture de la forêt guyanaise *Mémoires O.R.S.T.O.M.* 73: 1–204.

Passioura, J. B., M. C. Ball, and J. H. Knight. 1992. Mangroves may salinize the soil and in so doing limit their transpiration rate. *Functional Ecology* 6: 476–481.

Paulian, R. 1961. La zoogéographie de Madagascar et des îles voisines. *Faune de Madagascar* 13: 1–485.

Peace, W. J. H., and F. D. MacDonald. 1981. An investigation of the leaf anatomy, foliar mineral levels, and water relations of trees of a Sarawak forest. *Biotropica* 13: 100–109.

Pócs, T. 1976. Bioclimatic studies in the Uluguru mountains (Tanzania, East Africa) II. Correlations between orography, climate and vegetation. *Acta Botanica Academiae Scientarum Hungaricae* 22: 163–183.

Putz, F. E. 1984. How trees avoid and shed lianas. *Biotropica* 16: 19–23.

Rada, F., G. Goldstein, A. Orozco, M. Montilla, O. Zabala, and A. Azócar. 1989. Osmotic and turgor relations of three mangrove ecosystem species. *Australian Journal of Plant Physiology* 16: 477–486.

Reich, P. B., B. D. Kloeppel, D. S. Ellsworth, and M. B. Walters. 1995. Different photosynthesis-nitrogen relations in deciduous hardwood and evergreen coniferous tree species. *Oecologia* 104: 24–30.

Richards, P. W. 1952. *The Tropical Rain Forest*. Cambridge University Press, Cambridge.

Richards, P. W. 1996. *The Tropical Rain Forest* (2nd ed). Cambridge University Press, Cambridge.

Richter, W. 1984. A structural approach to the function of buttresses of *Quararibea asterolepis*. *Ecology* 65: 1429–1435.

Riswan, S., and K. Kartawinata. 1991. Species strategy in early stages of secondary succession associated with soil properties status in a lowland mixed dipterocarp forest and kerangas forest in East Kalimantan. *Tropics* 1: 13–34.

Saugier, B., A. Granier, J. Y. Pontailler, E. Dufrâne, and D. D. Baldocchi. 1997. Transpiration of a boreal pine forest measured by branch bag, sap flow and micrometeorological methods. *Tree Physiology* 17: 511–519.

Schimper, A. F. W. 1903. *Plant-Geography upon a Physiological Basis*. Clarendon Press, Oxford.

Schlesinger, W. H., and B. F. Chabot. 1977. The use of water and minerals by evergreen and deciduous shrubs in Okefenokee swamp. *Botanical Gazette* 138: 490–497.

Soepadmo, E. 1973. *IBP-PT—Pasoh Project, Annual Report for 1972*. Unpublished mimeographed report. School of Biological Sciences, University of Malaya, Kuala Lumpur.

Soepadmo, E. 1974. *IBP-PT—Pasoh Project, Annual Report for 1973*. Unpublished mimeographed report. School of Biological Sciences, University of Malaya, Kuala Lumpur.

Steele, S. J., S. T. Gower, J. G. Vogel, and J. M. Norman. 1997. Root mass, net primary production and turnover in aspen, jack pine and black spruce forests in Saskatchewan and Manitoba, Canada. *Tree Physiology* 17: 577–587.

Stocker, G. C., and G. L. Unwin. 1989. The rain forests of northeastern Australia—their environment, evolutionary history and dynamics, pp. 241–259. In H. Lieth and M. J. A. Werger, eds., *Tropical Rain Forest Ecosystems, Biogeographical and Ecological Studies*. Elsevier, Amsterdam.

Sugden, A. M. 1982. The vegetation of the Serranía de Macuira, Guajira, Colombia: a contrast of arid lowlands and an isolated cloud forest. *Journal of the Arnold Arboretum* 63: 1–30.

Sugden, A. M. 1986. The montane vegetation and flora of Margarita Island, Venezuela. *Journal of the Arnold Arboretum* 67: 187–232.

Tanner, E. V. J. 1980a. Studies on the biomass and productivity in a series of montane rain forests in Jamaica. *Journal of Ecology* 68: 573–588.

Tanner, E. V. J. 1980b. Litterfall in montane rain forests of Jamaica and its relation to climate. *Journal of Ecology* 68: 833–848.

Terborgh, J. 1971. Distribution on environmental gradients: theory and a preliminary interpretation of distributional patterns in the avifauna of the Cordillera Vilcabamba, Peru. *Ecology* 52: 23–40.

Terborgh, J. 1992. *Diversity and the Tropical Rain Forest*. Scientific American Library, New York.

Terborgh, J., and J. Mathews. 1993. Partitioning of the light resource via crown architecture in two understory treelets in an Amazonian floodplain forest. Abstracts, 30th Anniversary of the Association for Tropical Biology, San Juan, Puerto Rico.

Tilman, D. 1988. *Plant Strategies and the Dynamics and Structure of Plant Communities*. Princeton University Press, Princeton, NJ.

Tyree, M. T., D. A. Snyderman, T. R. Wilmot, and J.-L. Machado. 1991. Water relations and hydraulic architecture of a tropical tree (*Schefflera morototoni*). *Plant Physiology* 96: 1105–1113.

van Steenis, C. G. G. J. 1972. *The Montane Flora of Java*. E. J. Brill, Leiden.

Waring, R. H., and W. H. Schlesinger. 1985. *Forest Ecosystems: Concepts and Management*. Academic Press, Orlando, FL.

Weaver, P. L. 1990. Succession in the elfin woodland of the Luquillo Mountains of Puerto Rico. *Biotropica* 22: 83–89.

Weaver, P. L., E. Medina, D. Pool, K. Dugger, J. Gonzalez-Liboy, and E. Cuevas. 1986. Ecological observations on the dwarf cloud forest of the Luquillo Mountains in Puerto Rico. *Biotropica* 18: 79–85.

Weaver, P. L., and P. G. Murphy. 1990. Forest structure and productivity in Puerto Rico's Luquillo Mountains. *Biotropica* 22: 69–82.

Whitmore, T. C., and C. P. Burnham. 1969. The altitudinal sequence of forests and soils on granite near Kuala Lumpur. *Malayan Nature Journal* 22: 99–118.

Wilson, B. F., and R. R. Archer. 1979. Tree design: some biological solutions to mechanical problems. *BioScience* 29: 293–298.

Windsor, D. M. 1990. Climate and moisture variability in a tropical forest: long-term records from Barro Colorado Island, Panamá. *Smithsonian Contributions to the Earth Sciences* 29: 1–145.

Wolfe, J. A. 1971. Tertiary climatic fluctuations and methods of analysis of Tertiary floras. *Paleogeography, Paleoclimatology, Paleoecology* 9: 27–57.

Wolfe, J. A. 1978. A paleobotanical interpretation of tertiary climates in the northern hemisphere. *American Scientist* 66: 694–703.

Wolfe, J. A. 1993. A method of obtaining climatic parameters from leaf assemblages. *U. S. Geological Survey Bulletin* 2040: 1–71.

Wolfe, J. A., and G. R. Upchurch. 1987. North American nonmarine climates and vegetation during the late Cretaceous. *Palaeogeography, Palaeoclimatology, Palaeoecology* 61: 33–77.

Zimmermann, M. H. 1983. *Xylem Structure and the Ascent of Sap*. Springer-Verlag, Berlin.

SIX

Biomass and Productivity of Tropical Forest

What factors govern the dry matter content and limit the productivity of tropical forest? Tropical forest, which occupies 4% of this planet's surface, produces 49 billion tons of dry matter per year, containing 22 billion tons of carbon, two-sevenths of the world's primary production (Whittaker 1975). What the factors are that limit the productivity of tropical forest is now a fashionable question. Human activities are pouring 7.1 billion tons of carbon per year into the earth's atmosphere, of which 53% is taken back by its various ecosystems (R. F. Stallard, p. c.). The remaining 3.3 billion tons of carbon (12 billion tons of CO_2) remain in the atmosphere, increasing its CO_2 content by 0.4% and contributing its mite to global warming (Schlesinger 1991). Does increased atmospheric CO_2 increase tropical forest productivity enough to swallow a substantial proportion of the human output of CO_2 (Lloyd et al. 1995, Keller et al. 1996)?

This chapter will stick to simpler questions. How similar are the dry matter content and the productivity of lowland tropical rainforest? What limits the gross production (total amount of CO_2 "fixed") of tropical forest? How is this productivity apportioned? Let us start by considering the "constants of the forest," the ways lowland tropical forests resemble each other.

THE "CONSTANTS OF TROPICAL FOREST"

Lowland tropical forests with > 1.7 m of rain and ≤ 4 dry months per year share a number of features in common, features which I shall call "constants of the forest." In chapter 3, I remarked that such forests have an average annual temperature near 26°C, an average diurnal temperature range near 8°C, and an annual evapotranspiration near 1.4 m/year. What features of structure, production, and dynamics do these forests share?

First, most such forests drop between 6 and 8 tons dry weight of leaves per hectare per year (Table 6.1, Appendix 6.1). In Barro Colorado's Lutz catchment, the forest drops slightly over six tons dry weight of leaves per hectare of ground per year (Leigh and Windsor 1982). Leaf fall is measured by collecting the leaves falling into a set of "litter traps" and weighing them after drying them for 48 hr or more, usually at 60°C. Each litter trap should be 100 × 100 cm, although the ones used in Lutz catchment were circular tubs 1/12 m² in area. Leaf fall is lower in temperate-zone forests, forests with dry seasons of 6 months or more, and in elfin forests of tropical mountaintops than in lowland tropical forests (Table 6.1). Total litterfall (including twigs, flowers, and fruit as well as leaves) is lower in lowland tropical forests on soils poor in phosphorus (Silver 1994). Nevertheless, differences among lowland tropical forests in leaf litter fall may reflect variation in litter fall from year to year and differences in methods of collecting litter (the size and numbers of traps used, the frequency of collection, etc.) as much as they do intrinsic differences among the forests concerned (Proctor 1983). On Barro Colorado's Poacher's Peninsula, 50 × 50 cm traps were used to measure litter fall, and the annual leaf fall was 8 tons/ha (S. J. Wright, p. c.).

Second, tropical forests appear to carry between 6 and 8 ha of leaves (measuring one side only)/ha of ground—that is to say, a leaf area index (LAI) between 6 and 8. Leaf area index is harder to measure than litter fall. In Lutz catchment, I measured the area of leaves fallen into every 10th tub—the leaves from tubs 1, 11, 21, . . . 91 one week, those from tubs 2, 12, 22, . . . the next, and so forth, in rotation (Leigh and Smythe 1978). There, somewhat over 7 ha of leaves fell/ha of ground per year. If the average turnover time for this forest's leaves was 1 year, then, during rainy season, LAI in Lutz catchment is between 7 and 8.

Table 6.1 Leaf area, leaf weight, and leaf litter fall

Site	LAI	Leaf SC	Leaf Fall	Reference
Relation between LAI, leaf standing crop, and weight of leaf litter fall				
Pasoh, Malaysia (lowland)	8.0	8.2	6.4	Kato et al. (1978; LAI, SC); Lim (1978; leaf fall)
Sebulu, Borneo (lowland)	8.0	8.0		Yamakura et al. (1986a)
Banco, Côte d'Ivoire (lowland)		9.0	8.2	Huttel (1975)
N. of Manaus, Brazil (lowland)	5.7	6.3		McWilliam et al. (1993)
New Guinea, 2500 m	5.5	8.9	6.2	Edwards and Grubb (1977; LAI, SC); Edwards (1977; leaf fall)
Elfin forest, P. Rico, 1000 m	2.0	2.9	2.4	Weaver et al. (1986)
Subtropical Australia (NSW)		10.5	5.6	Lowman (1992)
Pseudotsuga, Oregon, 520 m	8.0	12.6	1.8	Marshall and Waring (1986)
Pygmy cypress, CA., lowland	2.1	3.1	0.6*	Westman and Whittaker (1975)
	LAIF	% Eaten	Leaf Fall	Reference
Relation between weight and area of leaf fall				
Barro Colorado, Panama	7.25	7.36	6.10	Leigh and Windsor (1982)
Ducke Res. N. of Manaus	6.1			Roberts et al. (1993)
Igapo near Manaus, Brazil†	3.18	9.4	5.32	Adis et al. (1979)
New Guinea, 2500 m	4.2	−10	6.2	Edwards (1977)
Mt. Glorious, Australia	7.3	12.3	5.5	Hegarty (1991)
Nothofagus, New Zealand 400 m	7.0			Kelliher et al. (1992)
Acer, Michigan	7.6	4.1	4.1	Burton et al. (1991)

LAI, leaf area index; SC, standing crop (tons/ha); leaf fall, total dry weight of falling leaves (tons/ha-year); LAIF, the total area of leaves falling per hectare per year, % eaten, proportion of this fallen leaf surface that is eaten.
*Calculated from proportion of SC grown in current year given in Westman and Whittaker (1975: Table 3).
†Igapo is blackwater swamp forest.

The most reliable way to measure LAI is to fell the vegetation in a plot and measure the areas of all the leaves. Yamakura et al. (1986b) harvested a 1/8-ha plot of dipterocarp forest at Sebulu in east Borneo: these plants carried 1 ha of leaves, so LAI = 8. The plants harvested from an 0.2-ha plot at Pasoh Reserve, Malaysia, had a total leaf area of 1.6 ha: again, LAI = 8.0 (Kato et al. 1978). The LAI for other lowland rainforests and mesic temperate deciduous forests appears to be similar (Table 6.1). On the other hand, where annual rainfall is less than 1600 mm, LAI of woody plants is lower the drier the climate (Harrington et al. 1995).

Third, death rate of trees of all sizes ≥ 10 cm dbh is normally 1–2%/year. For mature forest, death rate is lower in settings with poorer soil (see below). Within a forest, if one considers trees of all species together, death rate is remarkably independent of size (Table 6.2; also see Manokaran and Kochummen 1987: fig. 4; Swaine et al. 1987: fig. 4; Lieberman and Lieberman 1994: fig. 3). This pattern holds even though trees in different forests die in different ways. On Barro Colorado, 57% of a sample of trees whose manner of death was known died from a snapped trunk, and only 15% died standing (Putz and Milton 1982), whereas at Pasoh Reserve in Malaysia, 9% died by trunk snap and 45% died standing (Putz and Appanah 1987). In mature forest on Barro Colorado's plateau, death rate of trees ≥ 20 cm dbh was nearly independent of size class, even between 1982 and 1985 (Table 6.2), which included the disastrous El Niño drought of 1983 (Glynn 1990, Windsor 1990). Since 1985, death rate of trees on BCI's 50-ha plot has averaged about 2% a year for all size classes (Table 6.2), even though death rate varies greatly from species to species (Condit et al. 1995).

Finally, most tropical forests have basal area (total cross-sectional area at 1.3 m above ground of all trees ≥ 10 cm dbh) near 30 m²/ha (Dawkins 1959) and above-ground biomass near 300 tons dry weight/ha (Table 6.3, Appendix 6.1). Old forest on BCI has basal area of 28.5 m²/ha (Leigh et al. 1990).

At this level, even healthy mangrove resembles dry-land rainforest. Why should these four features be constants of tropical forest? To answer, I will begin by asking whether total (gross) productivity is a constant of the forest. The engines of forest production are leaves, which use energy from sunlight to make sugar from carbon dioxide and water. Most leaves belong to trees and shrubs. Tropical forests also contain ground-herbs. Finally, they contain an abundance of lianes (Putz and Mooney 1991) and epiphytes (Benzing 1989, 1990), which use host trees as supports to place their own leaves in the sun. In lowland forests, especially seasonal ones (Gentry 1991), lianes outweigh epiphytes, presumably because it pays to have access to water and soil nutrients. In montane forests, however, especially stunted ones, epiphytes are more prevalent (Table 6.4), as if, on these poor soils, there is no point in being rooted in the ground. Considering all these kinds of leaves together, what governs the total leaf area and the total leaf production of tropical forest?

Table 6.2 Death rates of trees of different diameter classes on 50-ha forest dynamics plots

	Barro Colorado, 1982–1985				Barro Colorado, 1985–1990				Pasoh, Malaysia, 1987–1990			
dbh (mm)	N	D	L	m	N	D	L	m	N	D	L	m
10–49	187934	15925	3.32	2.66	192977	21877	5.27	2.28	264990	9696	2.96	1.26
50–99	26533	2116	3.31	2.51	28398	2881	5.27	2.03	41882	1220	2.94	1.01
100–199	12938	990	3.35	2.38	13031	1285	5.28	1.97	17956	611	2.94	1.18
200–299	3677	392	3.35	3.36	3545	340	5.30	1.90	4618	181	2.91	1.37
300–399	1633	171	3.35	3.30	1786	189	5.38	2.08	1751	56	2.89	1.13
400–599	1468	167	3.36	3.59	1549	178	5.47	2.23	1290	35	2.90	0.95
600+	1166	120	3.46	3.14	809	72	5.55	1.68	717	31	2.88	1.54

Data supplied by Suzanne Loo de Lao on behalf of R. Condit, S. P. Hubbell, R. B. Foster, and J. LaFrankie.
N, number of trees in the size class at the beginning of the census; D, number of these trees that died by the next census; L, average interval (years) between times when a tree is censused; m (%), annual mortality rate = $[1/L] \ln [1 - D/N]$. The BCI trees censused in 1982 are classified by their diameters as measured (or estimated for large, buttressed trees) in 1982. In 1985, diameters of large, buttressed trees were measured above buttresses, and Hubbell and Foster (1990) assigned trees which "shrank" markedly since 1982 their 1985 diameters for 1982 as well. Diameters for trees dying before 1985 could not be corrected, so Hubbell and Foster (1990) report an increased death rate among larger trees, which is an artifact.

LEAF AREA INDEX AND LEAF FALL IN DIFFERENT FOREST TYPES

Why should "normal" lowland tropical forest have LAI between 6 and 8? The first step toward answering this question is to see whether other forests behave the same way. Mesic broadleaf forests of the temperate zone tend to have LAI between 5 and 7, a bit lower than lowland tropical forests. Only as one progresses from the latitude of France to that of southern Sweden does the maximum LAI of beech plantations decrease from 6.9 to 3.2 (Waring and Schlesinger 1985: p. 51). Otherwise, one encounters very low LAI only in nutrient-starved settings, such as the 3-m-tall pygmy cypress forests of California's Mendocino County (Westman and Whittaker 1975) and the low bana of Venezuela's San Carlos de Rio Negro (Bongers et al. 1985), in dry forests with annual rainfall < 1500 mm, or in stunted elfin forests of fog-bound, wind-blown tropical mountaintops such as Puerto Rico's Pico del Este (Weaver et al. 1986).

Colossal leaf area indices (one-sided) of 20 or more have been reported for conifer stands in the Pacific Northwest (Gholz 1982). These values were calculated from species-specific regressions relating the logarithm of the weight of a tree's leaves to the logarithm of its diameter: such regressions exaggerate the weight of leaves on very large trees. Inferring LAI from species-specific regressions of a tree's leaf area on its sapwood area (the cross-sectional area of

Table 6.3 Structure, dynamics, and above-ground biomass in selected forests

	Barro Colorado, Plateau	La Selva, Costa Rica	Luquillos, Puerto Rico	Rosario, Cuba	S Carlos, Venezuela, terra firme	Pasoh, Malaysia	Mangroves, Malaysia
Standing crop (tons/ha)	270	221	379	257	234–261	426	470
Basal area (m²/ha)	29	25	36	30	28	31	33
No. of trees/ha	414	446	710	1183	786	596	500
Annual mortality (%)	1.98	2.03	1.22		1.12	1.19	2.26
No. of deaths counted	2069	1386	167		88	944	33
LAI			6–7	8.5	6.4–7.5	7–8	
Leaf fall (g dry wt/m²)	610	660	494	620	500–757	703	576
Total litter fall (g dry wt/m²)	1152	1090	861	820	625–1025	1110	763

Basal area is of stems ≥ 10 cm dbh; number of deaths counted is total number of deaths of trees ≥ 10 cm dbh recorded on the plot concerned during the study; LAI is leaf area index.
Data for BCI plateau are from the 50-ha plot except for litter fall, which is from Foster (1982). Basal area and standing crop are from R. Condit and S. P. Hubbell (p. c.); stem counts and mortality are from Condit et al. (1995). Data for La Selva are from Lieberman and Lieberman (1994), except for leaf fall, from Ryan et al. (1994), and total litter fall, from Parker (1994). Data for Luquillos, Puerto Rico: standing crop, basal area, and stem count (for 1976) are from Crow (1980: I assumed 46.7% of the stems 8–12 cm dbh were ≥ 10 cm dbh). Death rate for undisturbed plots is from Briscoe and Wadsworth (1970: Table 7); LAI is from Odum (1970: Table 8, excluding plumb-line measurements). Litter fall is for tabonuco forest (Weaver and Murphy 1990: Table 7). Data for Rosario, Cuba: basal area and stem density are from Menendez et al. (1988: Table 8.21); standing crop is from Menendez et al. (1988: Table 8.29), LAI is from Menendez et al. (1988: Table 8.30), and litterfall is from Menendez (1988: Table 9.1). Data for San Carlos de Rio Negro: standing crop is from Medina and Cuevas (1989); stem density and basal area are from Uhl and Murphy (1981); mortality is from Uhl et al. (1988); LAI is from Putz (1983) and Medina and Cuevas (1989); leaf fall is from Jordan and Uhl (1978: p. 396) and from Medina and Cuevas (1989); total litter fall is from Medina et al. (1980: p. 247) and from Cuevas and Medina (1986). Data for Pasoh, Malaya, from Kira (1978), except for stem counts and basal area, which are from T. Kira (p. c. concerning a 1-ha plot), and mortality, from Manokaran et al. (1992: p. 101) based on a 50-ha plot. Data for mangroves from Matang, Malaya, from Putz and Chan (1986).

Table 6.4 Partitioning of biomass (tons dry weight/ha) among trees, lianas and epiphytes

	Sebulu Lowland	Pasoh Lowland	New Guinea 2500 m	French Guiana Lowland	Jamaica, 1500 m	
					Mull Ridge	Mor Ridge
Trees and herbs, stems and branches	853.4	457.6	490	564.6	328	218
Trees and herbs, leaves	7.1	7.5	7.5	8.5	6.7	8.2
Liana stems	10.9	9.1	3.3	7.4	2.3	0.0
Liana leaves	0.78	0.52	0.9	1.0	0.12	0.0
Epiphytes	0.33	0.46	3.4	1.8	0.5	2.8

Data for Sebulu from Yamakura et al. (1986a); for Pasoh, Kato et al. (1978); for New Guinea, Edwards and Grubb (1977); for French Guiana, Lescure et al. (1983); for Jamaica, Tanner (1980b).

its functional xylem), it appears that even in these dense stands of giant conifers, LAI never exceeds 12 (Marshall and Waring 1986). Indeed, LAI can only be so high in coniferous forest because conifer needles scatter more light to lower layers (Waring and Schlesinger 1985).

Curiously, a given area of leaves is supplied by a similar cross-sectional area of sapwood in a western Oregon stand of giant Douglas firs and western hemlocks as in the nutrient-starved tall Amazon caatinga near San Carlos de Rio Negro in Venezuela (Table 6.5). Judging, however, from the data of Klinge and Herrera (1983: Tables 2, 13), the sapwood area on a 10 × 10 m plot in tall caatinga predicts the weight or area of the leaves on that plot's trees no better than does the plot's basal area.

There is a trade-off between competing for light and competing for nutrients (Tilman 1988). In theory, this could reduce LAI in nutrient-poor settings. Where nutrients are in short supply, trees invest more in roots (Table 6.6). Unless the nutrient shortage is extreme, however, plants on poor soil make tougher, longer-lived leaves (Reich et al. 1992: p. 380 and Table 5) without greatly reducing LAI (Table 6.7). The greater investment in roots does reduce the capacity of shoots to compete for light by growing taller. In Jamaica, ridge-top forest on mull humus has faster-growing, straighter trees, a canopy 15 m tall, and a tree-borne LAI of 5.3, whereas ridgetop forest on mor humus, presumably acidified by the antiherbivore defenses in the leaves dropped by the forest, has slower-growing, leaning, twisted trees, a canopy only 6 m tall, and a tree-borne LAI of 4.4 (Tanner 1980b).

Only in the most extreme cases does nutrient shortage greatly reduce LAI. Near San Carlos, Venezuela, the Rio Negro and its tributaries are known as "rivers of hunger" because soils are poor and game so scarce in their catchments: even mosquito populations are low (Sponsel and Loya 1993). Near San Carlos, tall caatinga appears far more starved of nutrients than its terra firme counterpart. Organic matter decays much more slowly in the caatinga, thanks both to the more refractory quality of its fallen leaves and to the frequent waterlogging of its soil (Medina et al. 1990). As Chapin (1980) and Tilman (1988) would expect, the caatinga devotes much more of its biomass to roots (Table 6.6). As Janzen (1974) would expect, caatinga leaves, being harder to replace, are better defended against herbivores. The most decisive deterrent to herbivores is shortage of protein coupled with toughness and abundance of indigestible fiber (Coley 1983, Waterman et al. 1988, Ganzhorn 1992). In tall caatinga, protein content per unit acid detergent fiber (cellulose plus lignin plus cutin) is three-fifths that in the nearby terra firme, and tall caatinga leaves are half again as tough as their terra firme counterparts (Medina et al. 1990). Yet one must penetrate to the yet more nutrient-starved bana to see a major reduction in LAI (Table 6.6).

In both tropical and northern deciduous forest, leaf biomass and leaf area index attain normal values very quickly after succession begins (Table 6.8), provided that the soil and climate are suitable and that sufficient seeds or sprouts are available. The first priority of broadleaf forest seems to be to produce leaves until additional leaves are no longer profitable. Similarly, 4 years after the fall of a large tree

Table 6.5 Basal area (BA), sapwood area (SA), leaf area index (LAI), and leaf weight in two forests

	BA (m^2/ha)	SA (m^2/ha)	LAI (ha/ha)	LA/SA (m^2/m^2)	LW (tons/ha)	LW/SA (tons/m^2)
Oregon						
Douglas fir	66.7	7.3	3.43	4699	4.64	0.64
Western hemlock	24.0	11.8	4.84	4102	6.54	0.55
Total, 1 hectare	90.7	19.1	8.27	4330	11.18	0.59
San Carlos de Rio Negro (thirteen 10 × 10 m plots)						
Tall caatinga	36.9	15.0	5.08	3380	8.0	0.53

Oregon data from Marshall and Waring (1986), tall caatinga data from Klinge and Herrera (1983)

Table 6.6 Gradients of nutrient and drought stress

	San Carlos de Rio Negro, Venezuela					Dry Forest	
	Terra Firme	Tall Caatinga	Tall Bana	Low Bana	Open Bana	Guanica P. Rico	Chamela Mexico
Above-ground SC	248	252	182	40	6	45	51
Below-ground SC	56	105	128	69	42	45	32
BA, stems > 1 cm	34.3	36.9	32.1	15.5	1.7	21.2	23
BA, stems >10 cm	23.1	29.4					
No. stems > 1 cm	11255		22400	28000	600		
No. stems >10 cm	786	999				322	
Leaf area index	6.4–7.5	4.5–5.1	4.8	3.3	1.4	4.3	
Leaf fall, g/m²/year	650	520				434	377

San Carlos de Rio Negro: standing crop (SC) from Medina and Cuevas (1989); basal area (BA) and stem counts ≥ 1 cm and ≥ 10 cm dbh for Terra Firme from Jordan and Uhl (1978: Table 1); Terra Firme leaf area index (LAI) from Putz (1983) and Medina and Cuevas (1989); tall caatinga BA and stem counts from Klinge and Herrera (1983), LAI from Klinge and Herrera 1983 and Medina and Cuevas (1989), bana data from Bongers et al. (1985). Guanica: SC, litter fall from Lugo and Murphy (1986), BA, stem counts, and LAI from Murphy and Lugo (1986). Chamela: SC and BA from Kummerow et al. (1990), litter fall is average of hill and valley site in Martinez-Yrízar and Sarukhán (1990).

opens the forest canopy, light levels at the forest floor are lower than before the tree fell (Smith et al. 1992) as if, by this time, the forest has grown enough foliage to exploit the available light.

What sets the limit on a forest's leaf area? At Pasoh, 1 ha of leaves/ha of ground removes half the remaining light, and 8 ha of leaves/ha of ground intercept $1 - (1/2)^8$, or 99.6%, of the remaining light (Yoda 1974). This rule is deceptively simple. Canopy leaves are heavier per unit weight, and presumably more opaque, than their understory counterparts, but canopy leaves are also more likely to be inclined, thereby spreading the available light over more leaf surface, rather than horizontal (Table 6.9). At Pasoh, these two effects appear to cancel, and branches, which also obstruct light, are distributed in such a way that at Pasoh, each hectare of leaves appears to take up the same amount of light. This is not always true. In tulip poplar forests of different ages near Edgewater, Maryland, Brown and Parker (1994) found no correlation between the total weight or area of a stand's leaves and the amount of light reaching the forest floor. Even though their range of LAI was rather narrow, their result is unsettling.

Table 6.7 Effect of nutrient shortage on leaf weight and leaf lifetime in Montane Forest of Jamaica

	Mor Ridge	Mull Ridge
Leaf area index	5.8	5.8
Standing crop (tons/ha)	8.2 + 1.5	6.8 + 0.3
Leaf fall (tons/ha·year)	4.9	5.3
Leaf lifetime (years)	1.7	1.3

Leaf lifetime is calculated from dividing leaf standing crop (without the bromeliads) by leaf fall.

Leaf area index and standing crop from Tanner (1980b); standing crop is expressed as (tree, sapling, seedling and climber leaves) + bromeliad weights, as given in Tanner (1980b: Table 2); litter fall is from Tanner (1980a).

Do a forest's leaves exploit as much as possible of the incoming light? One way to answer this question is to measure how much light reaches the forest floor and compare this with the level required for leaves to return an adequate profit to their bearers. Measuring the amount of light reaching the forest floor is a subtle business, not without controversies and puzzling discrepancies (Alexandre 1982). Nevertheless, it appears that in mature forests of both the tropics and the temperate zone, about 1% of the incident light reaches the forest floor (Table 6.10). Forests intercept a larger proportion of photosynthetically active radiation, whose wavelengths range from 400 to 700 nm, than total sunlight or total solar radiation (Table 6.10). Forests at drier, or more nutrient-starved sites are less shady. On Mt. Lemmon, near Tucson, Arizona, a Douglas-fir stand (Table 6.10), which is 2600 m above sea level and supports the high basal area of 118 m²/ha, lets nearly 5% of the incoming sunlight reach the forest floor (Whittaker and Niering 1975). Blackwater swampforest, which is considered more nutrient-starved than "whitewater swamp-forest" or upland (terra firme) rainforest (Janzen 1974), also allows more light to reach the forest floor (Table 6.10).

In most rainforests, barely enough light reaches the forest floor to allow its ground herbs to turn a profit. Some shade-tolerant ground-herbs have leaves whose photosynthesis momentarily balances their respiration when light level is as low as 1, or even 0.5, µmol of photosynthetic photons/m² of leaf · sec (Björkman et al. 1972, Mulkey 1986). A leaf's lifetime photosynthesis must, however, do more than balance its daytime respiration. It must compensate for the leaf's nighttime respiration, when no photosynthesis occurs, and pay for that leaf's share of the respiration of roots and stems, and it must accumulate a profit rapidly enough to pay for that leaf's replacement (Givnish 1988). Such ground-herbs may need 2–4 µmol of photosynthetic photons/m²·sec (0.3–0.6% of the density of photosynthetic photons reaching the canopy) just to break

Table 6.8 The relation between forest structure and forest age

Increase of leaf and total biomass (g dry weight/m^2) with age in tropical forest (Snedaker 1970)

Age (years)	1	2	3	4	5	6	7	10
Leaf biomass	706	953	751	845	766	778	1083	744
Total biomass	836	1419	2287	2711	3667	4467	4666	5414

Leaf area, leaf weight, and forest height in forests of different ages and altitude in New Hampshire*

Age, years	3	4	7	11	18	19	22	30	35	40	44	49	57	200
LAI	5.3	5.9	4.8	5.0	7.1	6.2	5.8	6.3	5.6	6.4	7.2	7.0	6.3	5.5
Leaf weight†	2.2	2.6	2.2	2.2	3.0	2.6	2.8	2.9	2.5	2.9	3.1	3.2	3.0	2.5
Height (m)	2	3	6	7	14	13	12	20	23	20	21	23	22	23
Altitude (m)	600	510	495	415	480	570	555	480	510	585	510	560	645	615

Leaf area, basal area, and forest height in forest stands of different ages near Edgewater, Maryland (Brown and Parker 1994)

Age, years	10–14	24–25	35–38	150	340
LAI	4.61	5.08	5.73	6.75	6.14
Leaf weight (g/m^2)	346	418	427	452	421
Basal area	26.0	29.2	31.2	40.0	40.1
Forest height (m)	12.2	13.5	27.2	31.5	38.5

*Leaf weight data from Covington & Aber (1980); other data from Aber 1979).
†Leaf weight (total weight of forest's leaves) in tons dry weight per hectare.

even. On Barro Colorado, many ground-herbs can reproduce only when a fallen tree opens a gap overhead (Smith 1987, Mulkey et al. 1991), as if BCI's forest supports as much leaf area as is profitable.

On Barro Colorado, liane seedlings can grow in deep shade: by reducing their investment in support tissue to a minimum, they become limited by the shortage, not of light, but of suitable supports or "trellises" on which to grow (Putz 1984b). In hardwood forests of the north temperate zone, digging a trench around a shade-tolerant seedling deep enough to protect it from competing roots allows it to grow at light levels which otherwise would not suffice (Perry et al. 1969), presumably because the absence of root competition decreases the energy the seedling must devote to constructing and maintaining roots. Likewise, in drier climates or on poorer soil, plants must devote more of their energy to supporting the growth and maintenance of roots and/or associated mycorrhizae (Janos 1983) to secure the water and/or nutrients they need. Thus, in forests of drier climates or poorer soils, ground-herbs need more light to support growth and reproduction and more light reaches the forest floor unused (Table 6.10).

THE GROSS PRODUCTION (TOTAL PHOTOSYNTHESIS) OF TROPICAL FOREST

All of a forest's activities, plant and animal, depend on the photosynthesis of its leaves. How much energy do its leaves yield? Different tropical forests appear to have similar LAI because so little light reaches the forest floor that additional leaves could not pay for themselves. Are their aggregate rates of "gross photosynthesis" (the total amount of CO_2 fixed, or of sugar synthesized, per hectare of forest per year)

Table 6.9 Leaf weight per unit area (SLW) and leaf angle relative to horizontal in selected forests

	Oak–Hickory E. Tennessee		Nothofagus South Island, NZ		Rainforest	
					Manaus	Pasoh
LAI	4.9		6.7		5.7	8.0
Forest age (years)	45		90		Old	Old
Altitude (m)	335		900		Lowland	Lowland
Stratum	SLW	LA	SLW	LA	SLW	SLW
C	123 (23)	38 (20)	220 (16)	50 (16)	154 (25)	147 (37)
M	80 (15)	20 (12)	145 (9)	25 (12)	118 (16)	105 (22)
U	49 (2)	10 (4)	70 (2)	15 (7)	88 (5)	79 (3)

SLW, specific leaf weight (grams per square meter of leaf) and leaf angle (degrees from horizontal) in the canopy (C), midlevel (M), and understory (U) of selected forests. Numbers in parentheses represent the number of meters above the forest floor at which the parameter was measured. Data for Tennessee, Hutchison et al. (1986); New Zealand, Hollinger (1989); Manaus, McWilliam et al. (1993); Pasoh, Kato et al. (1978).

Table 6.10 Light interception in selected forests

Site	Sensor height (cm)	%	Sensor type, units	Reference
Paleotropics				
Rainforest, Lamington, Queensland, 28° S	0	2.58	S, W/m²	Björkman and Ludlow (1972)
	0	0.44	QS, E	Björkman and Ludlow (1972)
Rainforest, Pasoh, West Malaysia	0	0.3	L, lux	Yoda (1974)
	0	0.4	PAR, W/m²	Yoda (1974)
Rainforest, Danum, Sabah	?	1.58	CP	Whitmore et al. (1993)
Rainforest, Côte d'Ivoire	25	0.7	L	Alexandre (1982)
Rainforest, Gabon	130	3.0	S, 400–1100 nm	Hladik and Blanc (1987)
Amazonian rainforest				
30 km N of Manaus	130	1.2	L	da Conceição (1977)
50 km N of Manaus		1.1	QS, E	McWilliam et al. (1993)
Rio Negro, near Rio Branco				
Upland rainforest	0	1.1	O	Williams et al. (1972)
Whitewater swampforest	0*	1.1–2.2	O	Williams et al. (1972)
Blackwater swampforest	0*	2.9–9.2	O	Williams et al. (1972)
Northern Neotropics				
Monsoon forest, BCI, plateau	61	0.3	CP	Smith et al. (1992)
Rainforest, La Selva, Costa Rica, sites 5–7	130	1.5	QS, E	Rich et al. (1993)
	150	2.2	CP	Rich et al. (1993)
Rainforest, La Selva	60	1.9	CP	Clark et al. (1996)
Rainforest, La Selva	70	0.9–1.5	QS, E	Chazdon and Fetcher (1984), Chazdon (1986)
Rainforest, El Verde, P Rico	0?	0.4	400–700 nm, W/m²	Johnson and Atwood (1970)
Rainforest, El Verde, P Rico	0?	1.5	400–1100 nm, W/m²	Johnson and Atwood (1970)
North America				
Sugar maple, MI, 5 sites	150?	1.3–3.8	SCe	Burton et al. (1991)
Tulip poplar, Edgewater, MD				
10–14 year old	100	1.1	QS, E	Brown and Parker (1994)
45 years old	100	2.8	QS, E	Brown and Parker (1994)
340 years old	100	1.3	QS, E	Brown and Parker (1994)
Tulip poplar, E. TN†	0	1.7	300–3000 nm, W/m²	Hutchison and Matt (1977)
Coast redwood, CA, flat	0	1.0	L	Westman and Whittaker (1975)
Coast redwood, CA, slope	0	1.1	L	Westman and Whittaker (1975)
Hemlock, Great Smokies, TN	0	0.4	L	Westman and Whittaker (1975)
Douglas fir, Mt. Lemmon, AZ	0	4.7	L	Whittaker and Niering (1975)

*Water level.
†Summer and autumn full leaf.
%, percent of light or radiation above the forest which penetrates to the understory sensor; sensor type: S, solarimeter or pyranometer, which measures radiation in W/m²; QS, quantum sensor, which measures PAR, photosynthetically active radiation, photons of wavelength 400–700 nm, in E; L, light meter; O is ozalid paper, which integrates the total radiation received at the forest floor during a specified period, usually one day; CP, canopy photography; and SCe, sunfleck ceptometer.

also similar? Here, we assume that fixing 1 mol (44 g) of CO_2 synthesizes 1/6 mol (30 g) of glucose ($C_6H_{12}O_6$) and that the energy content of glucose is 16.7 kJ/g. A gram dry weight of vegetable matter contains 1/(2.2) g of carbon (Whittaker 1975: p. 225, bottom) and energy content about 18 kJ (Whittaker 1975: frontispiece).

Let us first try to estimate gross photosynthesis for a forest with LAI 8, where each unit of LAI intercepts a fraction e^{-k} of the light that reaches it. At Pasoh, $e^{-k} = 1/2$, so $k = 0.7$. A leaf with A m² of leaves/m² of ground above it receives a fraction e^{-kA} of the light reaching the canopy. When light level is low, photosynthetic rate, P, is proportional to light level I: that is to say, $P = mI$. If P is measured by the number of micromoles of CO_2 fixed and I by the number of micromoles (microEinsteins, µE) of photosynthetic photons received per square meter of leaf per second, then m is the "quantum efficiency" of the leaf, which theoretically cannot exceed 1/8, and actually never exceeds 1/12 (Bugbee and Salisbury 1989). Here, I set $m = 0.05$, its usual value. As I increases, P usually approaches an asymptotic maximum, P_{max}. The simplest formula for photosynthetic rate satisfying these two conditions is

$$P = \frac{mI}{1 + (mI/P_{max})} \quad (6.1)$$

Let us now model how photosynthetic rate, P, should increase with light level, I. In this unspeakably crude model, light combines with "available enzyme" to produce "activated enzyme," which then transforms into ("decomposes to release") available enzyme and photosynthate. Let E be the molar concentration of photosynthetic enzyme, in both available and activated states, and Q the molar concentration of this enzyme which is "activated" by light. Let the rate $P = kQ$ at which activated enzyme transforms into available enzyme and photosynthate equal the rate, $KI(E - Q)$, at which light and available enzyme combine to produce activated enzyme. Then $P = kQ = KI(E - Q) = KI(E - P/k)$. Solving for P, we find

$$P = \frac{KEI}{1 + (KI/k)}. \tag{6.1a}$$

This is the Michaelis-Menten equation (Haldane 1930: pp. 38–39); notice that KE corresponds to m. When light is saturating, $P_{max} = kE$; at saturation, photosynthesis is limited by E, the concentration of the appropriate enzyme. Since the enzymes involved are proteins, which contain nitrogen (indeed, most of a leaf's nitrogen is tied up in photosynthetic machinery; Field and Mooney 1986), this extremely simplistic model does show why P_{max} is higher in leaves with higher nitrogen content (Field and Mooney 1986, Reich et al. 1992).

The total photosynthetic rate, $P_T(t)$, of this forest at time t, given an LAI of 8, is

$$P_T = \int_0^8 \frac{mI(t)\, e^{-kA}\, dA}{1 + mI(t)\, e^{-kA}/P_{max}} = \frac{P_{max}}{k} \int_{1 + mIe^{-8}/P_{max}}^{1 + mI/P_{max}} \frac{du}{u} \tag{6.2}$$

where $u = 1 + mIe^{-kA}/P_{max}$ and $I(t)$ is the photosynthetically active radiation reaching the canopy at time t. If e^{-k8} is so small that mIe^{-k8}/P_{max} is far smaller than 1, as is true if $P \approx mI$ for forest floor herbs, then

$$P_T = \frac{P_{max}}{k} \ln\left[\frac{1 + mI/P_{max}}{1 + mIe^{-8}/P_{max}}\right] \approx \frac{P_{max}}{k} \ln\left[1 + mI/P_{max}\right]. \tag{6.3}$$

Equation 6.3 tells us that, when $mIe^{-kLAI}/P_{max} \ll 1$, adding more leaf layers contributes negligibly to the forest's photosynthesis.

Let us now try to calculate the daily photosynthesis, P_{daily} of a rainforest. Let I_m be the 24-hr average of radiation incident upon the forest canopy. First, let $I(t) = 0$ between 6 P.M. and 6 A.M., and $2I_m$ between 6 A.M. and 6 P.M. Recalling that a 12-hr day contains 12(3600) seconds, then equation 6.3 implies that the forest as a whole fixes

$$P_{daily} = 12(3600)\frac{P_{max}}{k} \ln\left(1 + \frac{2mI_m}{P_{max}}\right) \tag{6.4}$$

µmol, or one-millionth that many moles, of $CO_2/m^2 \cdot$ day. Suppose now that P_{max} is 20 mg CO_2/dm^2 of leaf $\cdot$ hr (Bossel and Krieger 1991), or 2 g CO_2/m^2 of leaf $\cdot$ hr (Kira 1978: p. 579). Since 1 µmol/sec is 0.0036 mol/hr, and 1 mol CO_2 weighs 44 g, 1 µmol CO_2/sec is 0.1584 g CO_2/hr. Thus 2 g $CO_2/m^2 \cdot$ hr is 12.6 µmol $CO_2/m^2 \cdot$ sec, a P_{max} rather typical of those of Neotropical canopy trees (Zotz and Winter 1993). At Pasoh, incident solar radiation averages 164 W/m² (Table 6.13). A watt of solar radiation contains, on average, 1.9 µmol of photosynthetic photons, 1.9 µE of photosynthetically active radiation, which includes roughly 1.2×10^{18} photosynthetic photons. Setting $2I_m = 328$ W/m² = 623 µE/m², equation 6.4 tells us that $P_{daily} = 0.968$ mol CO_2/m. Here, yearly photosynthetic output fixes 353 mol CO_2/m^2 of forest $\cdot$ year, yielding the equivalent of that many moles of sugar in usable energy. This amounts to a yearly gross production of 10.6 kg sugar (about 9.5 kg organic dry matter)/m², or 106 tons of sugar/ha, a power output of 5.6 W/m².

A more refined calculation would let $I(t)$ increase steadily from 0 to $4I_m$ between 6 A.M. and noon and decrease steadily from this peak to 0 between noon and 6 P.M. In this case, P_{daily} is

$$\frac{0.0432 P_{max}^2}{4kmI_m}\left[\left(1 + \frac{4mI_m}{P_{max}}\right)\ln\left(1 + \frac{4mI_m}{P_{max}}\right) - \frac{4mI_m}{P_{max}}\right]. \tag{6.5}$$

This amounts to 0.889 mol $CO_2/m^2 \cdot$ day when $4I_m = 1246$ µE/m², allowing the synthesis of 97 tons of glucose/m² $\cdot$ year, a power output of 5.1 W/m². Tracking $I(t)$ more carefully from hour to hour and day to day would further lower our estimate of gross photosynthesis, but not by much.

Do these equations yield reasonable estimates of total photosynthesis? Zotz and Winter (1993) find that, for canopy leaves of a variety of tree species in central Panama,

$$P_{daily} = 21200 P_{max} - 2100 \tag{6.6}$$

µmol CO_2/m^2. In other words, canopy leaves may be treated, to first approximation, as if they photosynthesize at P_{max} for 6 hr per day and are otherwise inactive. For the ideal forest represented in equation 6.4, at $2I_m = 623$ µE, equation 6.1 suggests that photosynthetic rate in the top layer of leaves is 8.97 µmol $CO_2/m^2 \cdot$ sec, whereas if they photosynthesize at P_{max} for 6 hr a day and are inactive the other six, their average photosynthetic rate would be 6.3 µmol $CO_2/m^2 \cdot$ sec, 70% of the preceding figure. If total daily photosynthesis as calculated from equation 6.4 is scaled down by the same amount, we come up with a gross production of 74 tons/ha. If only the top layer's photosynthesis needs correction in this fashion, we come up with a yearly gross production of 93 tons/ha.

Until recently, empirical estimates of total photosynthesis (usually called gross production) were few and far between. Most of these estimates, moreover, were based on attempts to measure dry matter production and ecosystem respiration: as we shall see, we know remarkably little about ecosystem respiration. Kira (1978: p. 578) estimated that, in the dipterocarp forest at Pasoh, annual gross production was the equivalent of 77.2 tons of organic dry matter/ha. Odum (1970: p. I-217, Table 24.6) estimated that the gross production of the rain forest at El Verde, Puerto Rico was 32.8 g/m² · day, or 120 tons/ha per year. By summing dry matter production and estimates of community respiration, Grier and Logan (1977) calculated a yearly gross production of 161 tons/ha for an old-growth Douglas-fir forest in western Oregon.

Estimates of gross production from total photosynthesis were more reasonable. Based on measures of photosynthesis, Kira (1978) estimated an annual gross production of 86.5 tons/ha for the mixed dipterocarp forest at Pasoh, and Odum (1970: p. I-224) estimated an annual gross production of 61 tons/ha for El Verde.

More recently, gross production has been inferred from the relation of the daily rhythm of a forest's consumption and release of CO_2 to the daily cycle in the solar radiation it receives. Carbon dioxide efflux from the forest is measured by "eddy correlations," the correlation, over successive time intervals, between the vertical component of air movement above the forest and the CO_2 content of this air (Fan et al. 1990). In the Ducke Reserve, about 25 km north of Manaus, Brazil, the net ecosystem consumption of CO_2, P_{net}, in kilograms CO_2 per hectare of forest per hour, depends on the solar radiation incident upon the canopy R (in W/m²) according to the equation

$$P_{net} = \frac{67R}{411 + R} - 15. \qquad (6.7)$$

If we replace R by I, the photosynthetically active radiation incident upon the canopy, assuming that 1 W corresponds to 1.9 µmol of photosynthetic photons, then

$$P_{net} = \frac{67I}{781 + I} - 15. \qquad (6.8)$$

If the negative constant represents forest respiration, then the total forest photosynthesis, P_T, is $67I/(781 + I)$ kg CO_2/hr · ha of forest, which is $42I/(781 + I)$ µmol CO_2/m² of forest · sec. For I between 0 and 2000 µE/m² · sec, $42I/(781 + I)$ is quite nicely approximated (Table 6.11) by $13.3 \ln(1 + 0.00487I)$. If we try to cast this formula into the form of equation 6.3, we find that $13.3 = P_{max}/k$, so if $k = 0.7$, $P_{max} = 9.3$ µmol CO_2/m² · sec; but $m/P_{max} = 0.00487$, so if $m = 0.05$, $P_{max} = 10.3$: not a devastating discrepancy for so crude a calculation.

Now let total photosynthesis, P_T, be $13.3 \ln(1 + 0.00487I)$. What is the forest's daily photosynthesis? Average daytime solar radiation is 322.5 W/m² (Fan et al. 1990: p. 16856) or 622 µE/m² sec. If we assume that solar radiation rises linearly from 0 to 1244 µE from 6 A.M. to noon and declines linearly from noon to 6 P.M., then, according to this formula, a square meter of forest fixes 0.73 mol CO_2 per day, and its annual gross production is 80 tons of sugar/ha, close to Kira's (1978) estimates. Fan et al.'s (1990: p. 16859) more refined calculation suggests that, during the 50-day period of their study, the forest fixed 0.717 mol of CO_2/m² · day, for a gross production of 78.5 tons dry weight of sugar/ha · year. The forest's total respiration was 0.667 mol CO_2/m² · day (2.57 kg C/ha · hr at night, and 4.1 kg C/ha · hr by day)—not readily distinguishable, within the limits of error of this procedure, from its gross production.

The relation between incident solar radiation and total forest photosynthesis may be quite general. A linear regression of total photosynthesis, P_T, on incident solar radiation suggests that, in the Ducke Reserve, hourly P_T increases by 0.015 kg C/ha for each additional watt of solar radiation (Fan et al. 1990). The corresponding regression coefficients for a hardwood forest in Canada and for an oak–hickory forest near Oak Ridge, Tennessee, are both 0.014 (Fan et al. 1990). In a boreal spruce woodland with lichen understory at 54° 50' N in northern Labrador, where the trees, the shrub layer, and the lichens contributed nearly equally to total photosynthesis, Fan et al. (1995) found that when photosynthetically active radiation was lower than 600 µE/m² (solar radiation less than 316 W/m²), hourly P_T increased by 0.0127 kg C/ha for each additional watt of solar radiation before leveling off sharply at a ceiling of 4.5 kg C/ha · hr.

Wofsy et al. (1993) found that that the gross production of Harvard Forest in Petersham, Massachusetts (42° 54' N) represented the fixation of 40.7 tons CO_2/ha · year, representing 28 tons dry weight of vegetable matter. As the growing season here is less than 5 months, its productivity during the growing season is as high as a rainforest's. Between 18 and 27 September, the total photosynthesis, in kilograms CO_2 per hectare of this forest per hour was related to incident photosynthetically active radiation, I (µE/m²), according to the equation

$$(6.9)$$

$$P_{net} = -6.13 + \frac{0.0623I}{1 + 0.0014I} = -6.13 + \frac{45I}{714 + I}.$$

Here, photosynthesis for a given level of radiation was a third lower than at Manaus, perhaps because the leaves were senescing and were photosynthesizing less efficiently.

The similarity in the relation between whole-forest photosynthesis and incident radiation in "mesic" forests is presumably related to the similar LAI of these forests. Incident solar radiation is quite similar for different tropical forests (Tables 6.12, 6.13). As we have seen, most rainforests have enough leaf area to exploit the available

Table 6.11 Comparing two ways of relating total forest photosynthesis to incident photosynthetically active radiation, I

	I					
	100	200	380	760	1330	2000
$42I/(781 + I)$	5.4	8.6	13.8	20.7	26.5	30.2
$13.3 \ln(1 + 0.00487I)$	5.3	9.0	13.9	20.6	26.8	31.6

Table 6.12 Comparison of solar radiation at La Selva and Barro Colorado Island: photosynthetically active radiation (PAR, mol photosynthetic photons/m²·day) and solar radiation (Q, MJ/m²·day)

	Jan.	Feb.	Mar.	Apr.	May	June	July	Aug.	Sep.	Oct.	Nov.	Dec.
La Selva												
PAR, 1987					29.0	29.0	24.0	23.6	28.8	26.8	28.5	26.2
PAR, 1978	28.1	25.5	26.7	31.8	26.2							
Barro Colorado Island												
PAR, 1988	39.9	38.6	41.7	41.4	27.4	26.5	21.2	24.7	25.5	21.6	20.4	23.9
Q, 1988	17.9	16.9	19.5	20.5	14.3	15.3	12.2	14.5	15.1	13.6	13.6	15.8
PAR, 1989	37.6	42.3	46.8	44.3	38.6	28.0	28.0	25.7	28.1	24.0	22.4	27.2
Q, 1989	19.5	20.8	22.9	21.6	18.6	13.5	14.2	14.0	15.6	13.6	12.6	15.6

Data for La Selva from Rich et al. (1993); for BCI, from Windsor (1990: Tables E1, E2).

sunlight. Rainforests of comparable environments should be roughly equally efficient at exploiting solar radiation. Therefore, as long as enough water and the required minimum of nutrients are available, average daily photosynthesis should be much the same for different rainforests. Is gross photosynthesis another constant of the forest?

It seems idiotic to claim that total photosynthesis is independent of soil fertility. Yet a curious comparison by Keyes and Grier (1981) of the productivity of two 40-year-old stands of Douglas fir at nearby sites in western Washington on soils of very different fertility suggests that the dependence of gross production on soil fertility is slight: at the poor-soil site, reduced above-ground productivity was nearly entirely compensated by enhanced below-ground productivity (Table 6.14). Soil fertility appears to affect gross production far less than it does the allocation of the profits therefrom.

Table 6.13 Radiation at different sites: yearly averages of photosynthetically active radiation (PAR, μmol/m²·sec), solar radiation (Q, W/m²), and number of hours of bright sunshine per day (HS)

Site	PAR	Q	HS
La Selva, Costa Rica	317	nd	nd
Barro Colorado Island, Panama	360	187	nd
El Verde, Puerto Rico	nd	185	nd
San Carlos de Rio Negro, Venezuela	nd	165	nd
Manaus, Brazil	nd	186	5.8
Abidjan, Côte d'Ivoire	nd	180	5.7
Bongabo, Zaire	nd	207	5.3
Yangambi, Zaire	nd	199	5.6
Tshibinda, Zaire	nd	216	5.3
Pasoh, Malaysia	nd	164	nd
Singapore, Malaysia	nd	224	5.6
Bukit Belalong, Brunei	nd	181	nd
Danum, Sabah	439	nd	5.7

Data for La Selva from Rich et al. (1993); for BCI, from Windsor (1990); for El Verde, from Odum (1970: table 24); for San Carlos de Rio Negro, from Saldarriaga and Luxmoore (1991: p. 237); for Manaus, Q from Franken and Leopoldo (1984) and HS from Müller (1982); for Abidjan, Q from Bernhard-Reversat et al. (1978) and HS from Müller (1982); Q and HS for Zaire and Singapore from Müller (1982); Q for Pasoh from Soepadmo (1973, 1974); Q for Brunei from Pendry and Proctor (1996: p. 410), and PAR and HS for Danum from Whitmore et al (1993: pp. 135–136).

Gross photosynthesis, however, is lower where leaf area is reduced. A forest at 10° S, 62° W in Rondonia, southwestern Brazil, with an LAI of 4 and an above-ground biomass of about 160 tons dry weight of vegetable matter per hectare, fixes 203 mol $CO_2/m^2 \cdot$ year, an annual gross production of 61 tons of sugar/ha (Grace et al. 1996). Fitting a curve through the middle of the points representing total forest photosynthesis, P_T, as a function of incident radiation, I, in Figure 9 of Lloyd et al. (1995), I find $P_T = 7.65 \ln(1+0.0088I)$. If $0.0088 = m/P_{max}$ and $m = 0.05$, $P_{max} = 5.7$ μmol $CO_2/m^2 \cdot$ sec, while if $7.65 = P_{max}/k$ and $k = 0.7$, $P_{max} = 5.4$. Setting $k = 0.7$, $m = 0.05$, $P_{max} = 5.5$, and $4I_m = 1246$ μE/m² in equation 6.5 gives a daily gross photosynthesis, P_{daily}, of 0.59 mol CO_2/m^2, which, multiplied by 365, gives an annual gross photosynthesis of 215 mol $CO_2/m^2 \cdot$ year, 6% higher than the value from Grace et al.'s (1996) more accurate calculation.

Moreover, net, and presumably gross, production is lower in dry forest. In Chamela, near Mexico's Pacific coast, dry forest with a 7-month dry season, receiving about 700 mm of rain per year, has an LAI of 3.3–3.8; its basal area of plants of > 3.18 cm dbh is 12.7–17.3 m²/ha, and its leaf litter fall is 2.7 tons/ha · year. The net production of this forest is 11 tons of dry matter/ha · year, of which 5 are produced below ground (Martinez-Yrizar et al. 1996). In western Kauai, Hawaii, (tree) LAI and wood production per unit leaf area decreased with rainfall, while nitrogen per unit leaf area, a good index of total daily photosynthesis per unit leaf area

Table 6.14 Biomass and productivity of two Douglas-fir forests on soils of different fertility (data from Keyes and Grier 1981)

	Soil quality	
	Good	Poor
Total biomass (tons/ha)	556	306
Biomass of fine roots (tons/ha)	2.7	8.3
Forest height (m)	33	23
Above-ground production (tons/ha·year)	13.7	7.3
Below-ground production (tons/ha·year)	4.1	8.1
Total production of dry matter (tons/ha·year)	17.8	15.4

(Zotz and Winter 1994) varied by only 10% between sites (Harrington et al. 1995). A koa forest at 500 m with 850 mm of rain/year had a basal area of 8 m²/ha, an LAI of 1.4, and an annual wood production of 0.7 tons/ha, whereas a koa forest at 1100 m with 1750 mm of rain/year had a basal area of 42 m²/ha, an LAI of 5.4, and an annual wood production near 5.7 tons/ha (Harrington et al. 1995). Water, a certain minimum of nutrients, and at least moderate warmth are required for a tropical forest to achieve the standard level of gross production.

The gross production and the LAI of the forests at Ducke and Pasoh Reserves are by no means the maximum that can be achieved under controlled conditions. When interplanetary space travel still seemed to be a real possibility, Bugbee and Salisbury (1988, 1989) explored the limits of wheat productivity under artificial culture conditions. They controlled temperature and light level and maintained CO_2 at 1.2 mmol/mol of air (2900 ppm CO_2 by weight, about eight times the normal atmospheric level). At a light level of 35 mol of photosynthetic photons/m² of substrate · day, absorbing 13.5 mol of photons permits the fixation of 1 mol CO_2. During the 79-day cycle from planting to harvest, the wheat culture absorbed 90% of the incoming photons, allowing the wheat to fix 2.33 mol of CO_2/m² of substrate · day. An astonishing 70% of this was channelled into dry matter production: this culture produced the equivalent of 1.63 mol sugar/m² of substrate · day. During these 79 days, the gross and net production of this wheat culture were equivalent to 273 and 191 tons/ha · year. In these cultures, LAI was 13 by day 24, 17 by day 45, and 11 on day 79. The vertical orientation of wheat leaves allowed a more even sharing of light among leaves than occurs in most forests (Bugbee and Salisbury 1988).

Astonishing productivity also occurs in some natural settings. Out-thrust angles of the rocky, wave-beaten, weather coasts of the northeastern Pacific are often covered by an annual kelp, *Postelsia palmaeformis* (Paine 1979). These kelps begin growing each March. By late June, they form a lush vegetation, which was 43 cm tall by 25 June 1979. On that day, R. T. Paine harvested two 50 × 50 m plots of *Postelsia* from a wave-beaten slope of intertidal rock. He obtained a total of 10.36 kg fresh weight of fronds and 19.73 kg fresh weight of stipes and holdfasts from these two plots. Eleven percent of this is dry matter (Leigh, unpublished data), so 6.6 kg of dry matter grew/m² in 4 months, a production rate which would yield 198 tons of dry matter/ha in a year, if sustained over sufficient time and space. *Postelsia* only contains 11.7 kJ/g dry weight, compared to 17.8 kJ/g dry weight in Bugbee and Salisbury's (1989) wheat. Thus the production rate of this *Postelsia* stand during the 4 months ending 25 June 1979, suitably extrapolated, was equivalent to a dry matter production of 132 tons/ha · year of wheat, or of rainforest plants. This is astonishing production for so high a latitude.

Postelsia can accomplish this because weather coasts collect the wave energy generated by storms all over the ocean, and sea palms are designed to benefit from this energy (Leigh et al. 1987). First, the waves protect *Postelsia* from herbivores, so the plants need not elaborate and maintain defensive compounds. *Postelsia* stands also maintain very high frond area: 8 cm² of Postelsia fronds weigh 1 g when wet (Holbrook et al. 1991: Table III), so the 20.72 kg fresh weight of *Postelsia* fronds/m² on Paine's aforementioned plots imply an LAI of 17. In part, frond area is so high because the fronds hang vertically when not disturbed. A *Postelsia* stand at Garrapata Beach, 10 km south of Point Lobos, California, had an LAI of 14.1, but still let 1.3% of the light reach the substrate (Holbrook et al. 1991: p. 56). However, waves also stir the narrow fronds of *Postelsia*, assuring that light is shared among these fronds much more evenly than it is shared among canopy and understory leaves in a rainforest: most fronds receive a light pulse when a wave strikes (Wing and Patterson 1993). Moreover, if light is pulsed so as to yield flashes at frequencies ranging from 1/2 sec to 1/30 sec, photosynthesis is higher than under constant light of the same average intensity. The most common wave frequency, 1/10 sec, is the frequency of light flecks that yields the highest photosynthetic advantage—1.8–fold—over constant light of the same average intensity (Wing and Patterson 1993).

The laboratory wheat culture and the wave-beaten *Postelsia* stands share three unusual features: freedom from pests, ready access to nutrients, and circumstances (including, but not restricted to, a severe limit on height growth) that allow an extraordinarily equitable sharing of light among leaves, thus permitting uncommonly high LAI. The physiological "constant of the forest," the annual gross productivity of 90 tons sugar/ha in "mesic" tropical forests, reflects the limits imposed by settings where nutrients must be actively sought, all plant parts must be defended against an incredible variety of pests, and trees can grow tall enough that only well-lit leaves can pay their share of the costs of placing them so high, leading to inequitable distribution of light among plants of different height.

HOW IS A FOREST'S ENERGY ALLOCATED?

Let us assume that, in general, a rainforest's gross production supplies the energy content of 90 tons of sugar per hectare of forest per year. Where does it all go?

The Trade-off among Stems, Leaves, and Roots

Let us first consider the nature of the trade-off between above-ground and below-ground production. A forest's gross production, G, depends on its leaf area index, L (called LAI elsewhere) and the effectiveness with which these leaves use the available light. This efficiency depends in turn on the leaves' content of nitrogen and phosphorus, which their roots must supply. In short, a forest needs to intercept light and it needs roots to allow effective use of the light their leaves intercept, which is why leaves and fine roots have first call on a forest's energy (King 1993).

Following King (1993), let us express G as $\varepsilon S(1 - e^{-kL})$, where S is incident solar radiation, $1 - e^{-kL}$ is the proportion of this light intercepted by the forest when it has leaf area index L, and ε is the efficiency with which the forest transforms radiation into sugar. The value of ε increases with nutrient content of the leaves until these leaves contain enough nitrogen and phosphorus to function with full efficiency (King 1993). Formally, ε is a function of U, the total stock of nutrient in the vegetation. Let U be the vegetation's stock of the limiting nutrient, and set $\varepsilon(U) = \varepsilon^*[1 - (1 - U/U^*)^2]$ for $U < U^*$ and ε^* for $U \geq U^*$. This expresses the fact that, beyond a certain point, increased nutrient levels no longer increase photosynthetic efficiency (King 1993). When nutrients are scarce, plants maintain foliar nutrient content by stopping "leaks"—making long-lived leaves to avoid losing nutrients in litterfall (Table 6.7), reducing the amount of litterfall and the phosphorus content per unit weight of litter (Silver 1994), and preserving their own lives rather than producing more offspring—and by investing more in roots to extract nutrients from the soil. All these measures are costly: investing in roots diverts resources away from overtopping neighbors in competition for light; long-lived leaves presuppose growth slow enough that they will not be shaded and rendered useless by later-grown foliage; and the measures that enhance a tree's lifetime, such as stiff, strong wood, slow its growth.

At steady state, a forest's nutrient stock, U, represents the balance between input and loss. Nutrients are supplied at a rate, A, from rainfall and the weathering of bedrock; they are lost by the death of trees, which entails loss of the dead tree's roots, and by leakage of nutrients entering the system from rainfall or bedrock, or recycled by litter fall, through an intact root mat. Formally, the nutrients introduced by rainfall and weathering into the fraction of the forest floor with an intact root mat, $A(1 - m)$, must balance the fraction of the forest's nutrients lost by tree fall and escape through the disrupted mat, mU, and the proportion $(R_v/R)/[1 + R_v/R]$ of nutrients reaching an intact root mat which escape through it. Here, m is the fraction of a forest's root mat that is ineffective thanks to tree mortality, R_v is the root mass per hectare required to prevent half of the nutrients reaching an intact root mat from escaping through it, R is the actual root mass per hectare, while the quantity of nutrients incident upon intact root mat is $(1 - m)$ times $(KU + A)$, where K is the proportion of the forest's nutrient stock recycled as litterfall each year. In sum,

$$A(1 - m) = mU + \frac{(KU + A)[(1 - m)(R_v/R)]}{1 + (R_v/R)}, \quad (6.10)$$

$$U = \frac{A(1 - m)}{m[1 + (R_v/R)] + K(1 - m)(R_v/R)}. \quad (6.11)$$

The nutrient stock, U, can be increased by increasing root biomass, R, decreasing m by enhancing tree longevity, and decreasing K, the proportion of the nutrient stock recycled each year as litterfall.

A study of forests along a gradient of soil quality in Sumatra suggests that infertile soil favors lower fruit production and attracts longer-lived trees. Near Ketambe, Sumatra, litter fall, fruit availability, prevalence of mycorrhizae, soil pH, and availability in the soil of phosphorus was measured along a sequence of terraces of various ages. Leaf fall was lower, fruit more scarce, and mycorrhizae more prevalent on poorer soils (Table 6.15; van Schaik and Mirmanto 1985, van Noordwijk and Hairiah 1986). Forest was much taller on ridges and high terraces, where the soil was poorer. This paradox is explained by the fact that poorer soils attracted longer-lived tree species. On the worst soils in this sequence, more plants had ectomycorrhizae, which are particularly expensive (Allen 1991). In central Amazonia, ectomycorrhizae are not prominent in terra firme forest on oxisols, but they are predominant in forests on white sand (Singer and Araujo 1979), another sign that soil impoverishment calls forth greater investment below ground.

As we have seen, nutrient input, A, must be fairly low before leaf area index decreases notably. Plants compete for light by trying to outgrow each other and compete for nutrients by trying to produce more roots than their neighbors do. A forest's tree trunks, pillars of the green cathedral we so enjoy, and resources on which so many people depend, reflect competition among trees for light (King 1990). The trade-off between competing for light and competing for nutrients is reflected by a trade-off between wood and root production: the more severe the nutrient shortage, the more root production prevails over wood production.

The Carbon Cycle

Now let us consider the "carbon cycle" in selected tropical forests to learn what to look for in analyzing forest production and respiration. The dry matter a forest produces consists of leaves, flowers, fruit, twigs, bark, stemwood, and roots. Dry matter of the first five categories eventually falls as litter (Table 6.16). The respiration associated with the costs of their construction occurs above ground. Their decomposition (excepting twigs, which may begin rotting above ground) only contributes to respiration after they fall to the ground. Similarly, the respiration associated with wood construction occurs above ground, but, by and large, the decomposition of this wood contributes to forest respiration only after the tree falls dead to the ground. In a small forest plot with no dead trees, either standing or fallen, gross production (total photosynthesis) should exceed total respiration by an amount equal to stemwood production. Respiration due to belowground processes is soil respiration less the contribution

Table 6.15 Soil characteristics, mycorrhizal prevalence, leaf fall, and fruit availability in a series of Sumatran forests

	Terrace Number					Mountain Slopes
	1	2	3	4	6	
Soil pH	7.25	6.75	6.00	4.87	4.50	4.63
Percent roots with						
V-A mycorrhizae	4.7	24.6	29.5	19.4	42.3	36.4
Ectomycorrhizae	0	2.0	0	0	1.2	2.0
Basal area (m^2/ha)	18.7	21.3	24.9	26.0	26.6	21.9
Leaf fall (tons/ha·year)	7.97	7.38	6.83	6.00	5.70	5.25
Fruit sources/100 m*	1.56	1.29		1.08	0.82	0.65

Data from van Schaik and Mirmanto (1985) and van Noordwijk and Hairiah (1986).
*Fruit sources per 100 m of trail.

to soil respiration from litter decomposition. At steady state, respiration due to above-ground processes is total photosynthesis less respiration from below-ground processes and consumption by herbivores. With these points in mind, let us now consider how a forest uses the energy supplied by gross photosynthesis.

To assess the allocation of a forest's gross production to above- and below-ground activities, we need to know, at minimum, the forest's gross production, litter fall, and soil respiration. One quickly encounters the problem of whose soil respiration figures to believe. In forest at Paragominas, south of Belém, Trumbore et al. (1995) reported an average soil respiration of 270 mg C/m^2 · hr, while at another Amazonian terra firme forest, San Carlos de Rio Negro, Medina et al. (1980) reported an average soil respiration of 114 mg CO$_2$/m^2 · hr, which is 31 mg C/m^2 · hr. As the latter figure is too low to account even for the decomposition of litter, I use the figures of Kursar (1989) for Barro Colorado, the figures of Fan et al. (1990) for central Amazonia, and the figures of Trumbore et al. (1995) for eastern Amazonia.

In tropical forests, soil respiration is much the same day and night (Fan et al. 1990, Trumbore et al. 1995); it is somewhat lower during the dry season than during the rainy season (Kursar 1989, Trumbore et al. 1995). These characteristics make it feasible to estimate soil respiration from a relatively small series of measurements scattered through the year.

Let us now compare the carbon cycles of three forests: Barro Colorado, Ducke Reserve, near Manaus, and evergreen forest 25–35 m tall near Paragominas, south of Belém, with 360 tons dry weight of above-ground biomass per hectare and an annual rainfall averaging 1750 mm, of which < 250 mm falls during the 5-month dry season. On Barro Colorado, soil respiration averages 2.9 µmol CO$_2$/m^2 · sec during the 4 months of dry season, when the decay of litter is slowed or blocked, and 4.3 µmol CO$_2$/m^2 · sec during the 8 months of rainy season—an annual average of 3.84 µmol CO$_2$/m^2 · sec (Kursar 1989). This represents the release of 1.66 kg C/ha · hr, that is to say, 14.5 tons C/ha · year, or the consumption of 36.3 tons glucose/ha · year. In the Ducke Reserve near Manaus, soil respiration during a few weeks of rainy season (Fan et al. 1990) averaged 2.22 kg C/ha · hr (5.13 µmol CO$_2$/m^2 · sec). This represents the release of 19.4 tons C/ha · year, or the consumption of 48.6 tons glucose/ha · year. At Paragominas, soil respiration averages 2.9 kg C/ha · hr during the 6.5 months of rainy season and 2.4 kg C/ha · hr during the 5.5 months of dry season (Trumbore et al. 1995). This represents the release of 23.7 tons C/ha · year, or the consumption of 59.1 tons glucose/ha · year. It appears that the relatively fertile soil of Barro Colorado allows a higher proportion of gross production to be devoted to above-ground activities than is the case in Amazonia. Perhaps this is why Barro Colorado's forest casts such deep shade (Table 6.10). As we shall see in a later chapter, fruit fall on Barro Colorado is quite high; since conserving nutrients is not a particularly urgent priority, it pays more to invest in offspring. This fruit supports an abundance of vertebrates. When flushing new leaves, many of the plants on Barro Colorado try to "outgrow" their pests rather than poisoning them (Coley 1983, Coley et al. 1985, Aide and Londoño 1989), for if insects do find and destroy one flush of leaves, there are enough resources to produce another. Thus insect populations are probably also higher on Barro Colorado than in less fertile settings.

Table 6.16 Litter fall on Poacher's Peninsula, Barro Colorado, and in the Amazonian Egler Reserve, near Manaus

	Leaves	Fruit	Wood	Other Material
Poachers 1 1988	859	76	150	187
Poachers 2 1988	834	163	195	179
Poachers 3 1988	744	102	176	226
Poachers 4 1988	790	102	176	210
Egler Reserve 1963	640	20	130	
Egler Reserve 1964	480	50	140	

Fall of litter of different categories (g dry weight per square meter per year). Data from Poachers: S. J. Wright (p. c.), based on fifteen 50 × 50 cm litter traps per 2.25-ha plot; data for Egler Reserve from Klinge and Rodrigues (1968), based on ten 50 × 50 cm litter traps. At both sites, traps were collected weekly and their contents sorted and dried in an oven.

The fact that both litter fall and soil respiration excluding litter fall are higher at Paragominas than at Ducke suggests that gross production is also higher at Paragominas: our "constants of the forest" are by no means absolute.

Leaves

As leaves supply the energy on which their plants depend, they have a primary claim, along with roots, on the energy available. Because light level at the forest floor declines exponentially with leaf area index, while even ground-herbs need a certain minimum of light to turn a profit, there is a sharp limit, similar for most "mesic" forests, beyond which additional leaf area is profitless.

This limit probably explains why leaf fall is similar in most "normal" lowland rainforests, about 7 tons dry weight of leaves per hectare per year. Curiously, mesic broadleaf forests within 45° of the equator appear to drop about 600–700 kg dry weight of leaves per hectare per month of growing season; these forests have similar LAI, but the temperate-zone forests have progressively thinner leaves. Leaf production exceeds leaf fall by the amount animals eat, plus the dry matter, if any, plants resorb from their leaves before dropping them. It probably takes 2.25 g of glucose to make 1 g dry weight of leaf (T. Kursar, p. c.). This is much more than the 1.6 g glucose previously thought to be required to produce 1 g dry weight of leaf (cf. Penning de Vries et al. 1974), but is similar to the construction costs of bacteria and fungi. Thus, constructing leaves accounts for > 20% of a tropical forest's energy budget. Nocturnal respiration in tropical leaves consumes about 10% of their net daytime photosynthesis (Zotz and Winter 1993): this amounts to about 9 tons of sugar per hectare per year. At the Ducke Reserve, 25 km northeast of Manaus, nocturnal respiration in the above-ground parts of the forest is reported to release 0.35 kg C/ha of forest · hr (Fan et al. 1990), representing a glucose consumption of 875 g/ha · hr, or 3.8 tons for a year of 12-hr nights. This is less than half the leaf respiration expected from Zotz and Winter's (1993) results. I have no explanation for the discrepancy.

The energy content of a tropical forest's leaves is about one-tenth its gross production. Thus, where average leaf lifetime is 1 year, the average leaf must repay its energy content 10 times over: this amounts to repaying its construction costs nearly 5 times over, or the total costs of its construction and maintenance nearly 3 times over. Averaged over a 24-hr day, 1 m² of leaves must yield 0.53 W— 0 during the night and 1.06 W during the daylight hours. It would take an average of 94 m² of rainforest leaves (the leaves of 12 or 13 m² of forest) to produce the power required to keep a 100-W bulb lit during the day. The burden of fueling the forest, however, is very unequally spread among its leaves. The power output of canopy sun-leaves averages 3.15 W during the daylight hours, and it takes 32 m² of canopy leaves to supply an average of 100 W during the daylight hours.

Roots

Below-ground respiration (the decay of roots and the respiration of roots and mycorrhizae) consumes the equivalent of 20 tons glucose/ha · year on Barro Colorado, 39 tons glucose/ha · year at Ducke, and 49 tons glucose/ha · year at Paragominas (Table 6.17). Unlike leaves, the profit derived from laying on extra roots declines more gradually, so that root expenses, unlike leaf expenses, differ greatly in different forests. Presumably, the extra below-ground expenditure at the Ducke Reserve is a response to the poor soil. At Paragominas, the extra below-ground expenditure reflects the search for water during the dry season; there,

Table 6.17 The carbon cycle in three tropical forests (production and respiration expressed as g C/m²·year)

	Barro Colorado	Ducke	Paragominas
Gross production		3141	
Litter fall			
Leaves	402	280–344	
Total	645	365–390	460
Soil respiration	1440	1960	2365
Below-ground process respiration	800	1560	1900
Fine root production, depth			
0–10 cm	55		60
0–25 cm	72		
0–100 cm			190

Litterfall and roots are assumed to be 50% by weight carbon (Nepstad et al. 1994).
For Barro Colorado, data on litter fall are from S. J. Wright (p. c.); soil respiration is from Kursar (1989), and root production (as measured by fine root biomass) is from Cavelier (1989, 1992) measured on an adjacent mainland point. For Ducke Reserve, litter fall is from Klinge and Rodrigues (1968) for Egler Reserve and from Franken et al. (1979). Figures for respiration are from Fan et al. (1990). For Paragominas, litter fall data are from Nepstad et al. (1994); data on respiration are from Trumbore et al. (1995).

trees sink roots up to 18 m deep, and during one dry season, > 60% of the forest's water was drawn from soil between 2 and 8 m deep (Nepstad et al. 1994). This is why, over the normal range of soil fertility, the primary trade-off is between root and wood production. Leaf production varies much less from place to place within tropical rainforest.

The study of root production is a new occupation. In the tropics, the task has hardly begun. At Paragominas, turnover of fine roots < 1 mm in diameter is about 1 year, so the biomass of fine roots may be equated to their annual production (Trumbore et al. 1995: p. 521): I assume that the same is true for Barro Colorado Island. At Paragominas, annual production of roots > 1 m below the soil surface accounts for 2 tons C/ha, and the respiration of these roots annually accounts for another 2 tons C/ha (Davidson and Trumbore 1995).

Especially where soils are poor, trees need mycorrhizae to help their roots take up nutrients (Janos 1980, 1983). These mycorrhizae can charge handsomely for their services (Allen 1991). Mycorrhizae reduce the dry weight gain per unit leaf area of *Beilschmeidea* seedlings by 23% (Lovelock et al. 1996). The most thorough study of mycorrhizal cost has been in a 180-year-old stand of silver fir (*Abies amabilis*) at 1040 m in western Washington. This stand had a basal area of 74.3 m²/ha, 447 tons/ha of above-ground dry matter, including 21.7 tons dry weight/ha of foliage (LAI = 8.6, one-sided), and 137 tons/ha dry matter in living roots (Grier et al. 1981). This forest's total above-ground production was 4.55 tons/ha, of which only 1.04 tons/ha was needles: this forest's needles must have had an average lifetime of 21 years. This forest's high root biomass and long-lived needles indicate nutrient scarcity. Its annual below-ground dry matter production was 20.3 tons/ha, of which 13.3 tons/ha was fine-root production and 3.7 tons/ha (15% of the forest's total dry matter production) was devoted to mycorrhizae (Vogt et al. 1982). The contribution of mycorrhizae to this forest's respiration could well be greater.

Wood

Trees make wood to compete for light—in particular, to elevate their leaves above their neighbors', so as to get first crack at the incoming sunlight (King 1990). This wood is a long-term investment: the vegetation reoccupying a newly cleared gap quickly produces enough leaves to make full use of the available sunlight, but the wood content of a mature forest represents many years of growth (Table 6.18).

Competition for height invites "cheaters," in the form of lianes that take advantage of the trunks of other trees as supports for their own leaves. The crowns of about half a forest's canopy trees are infested by lianes (Putz 1984b). Lianes slow tree growth (Putz 1984b, Clark and Clark 1990)

Table 6.18 Productivity, standing crop, and leaf area in forests of different ages

Pin cherry stands, New Hampshire (Marks 1974)				
Age (years)	1	4	6	14
Leaf area index	1.1	5.5	6.1	5.4
Biomass, major species (g/m²) (% total)				
Leaves	44 (43%)	228 (10%)	273 (7%)	219 (3%)
Current twigs		151 (7%)	103 (3%)	30 (<1%)
Stem wood and bark	36 (35%)	1378 (59%)	2426 (64%)	4915 (69%)
Live branches		104 (4%)	328 (9%)	550 (8%)
Dead branches		108 (5%)	126 (3%)	422 (6%)
Roots	22 (22%)	348 (15%)	518 (14%)	986 (14%)
Total biomass	102	2318	3774	7126
Production (g/m²·year, including roots)	102	1309	1658	1264
Karri Forest, West Australia (Grove and Malajczuk 1985; litter fall from O'Connell and Menagé (1982)				
Age (years)	4	8	11	36	Mature
Basal area (m²/ha)	12	27	29	38	56
Biomass above ground (tons/ha)	35	66	80	249	~800
Leaf weight (g/m²)	514	770	920	620	—
Age during litter fall measurement	2	6	9	40	Mature
Leaf fall, 1978 (g/m²)	101	335	311	401	381
Total litter fall, 1978 (g/m²)	113	406	494	826	1023
Lowland tropical forests in Venezuela and Colombia (Saldarriaga et al. 1988)					
Stand Age	12	20	35	60	80	Mature
Stems ≥1 cm/ha	5655	7106	5431	2863	3986	3911
Stems ≥ 10 cm/ha	342	461	539	441	656	569
Basal area ≥ 1 cm dbh/ha	12.8	16.9	18.6	24.5	24.0	34.8
Leaves (tons/ha)	6.9	9.6	8.5	6.8	8.2	9.8
Above-ground standing crop (tons/ha)	58.0	76.6	92.3	153.8	144.8	238.0
Roots (tons/ha)	9.8	15.9	19.4	31.8	29.7	58.2

and increase the chances of death for the trees that bear them (Putz 1984b). Lianes rarely die when their host trees do, and pioneer trees, which colonize gaps opened by the fall of particularly large trees, are designed to avoid or shed lianes (Putz 1984a). The abundance of lianes and their contribution to leaf litter fall differs markedly from one forest to another (Table 6.19).

How much wood does a forest make, and how much does wood production depend on environmental conditions? Data from the 50-ha forest dynamics plot in old forest on Barro Colorado (Condit et al. 1995) allows us to estimate wood production in that forest and infer what factors pace it. On this plot, tree mortality is normally about 2% per year, and nearly independent of size class (1985–1995 in Table 6.20). The size composition of the trees on this plot is in balance: despite the additional mortality imposed by the fierce El Niño drought of 1982–1983 (Condit et al. 1995), 1985's size composition resembled 1982's (Table 6.20), suggesting that, within limits, wood production is paced by tree mortality. Thus annual wood production on Barro Colorado must normally be about 2% of its standing crop, about 5 or 6 tons per hectare per year. In forest north of Manaus in central Amazonia, tree mortality is lower, 1.17% per year (Rankin-de-Merona et al. 1990). There, annual wood production must be about 3 tons dry weight per hectare.

What Limits a Forest's Timber Content?

Many broadleaf forests have an above-ground standing crop near 300 tons dry matter per hectare (Table 6.3), most of which is wood. This generalization spans a variety of soil types, from the relatively fertile soils of Barro Colorado to the white sands of Amazonian caatinga and the salty mud flats of tropical estuarine mangrove. Why this similarity? This question cannot be answered completely without assuming too many things that should be deduced or explained. Yet it is useful to explore the issue.

What limits a forest's standing crop? It is logical to think that, as a forest grows, the respiration of its support and supply tissues consumes an ever-growing proportion of its gross production, until only enough energy is left to replace leaves, twigs, and roots as they wear out and to support enough reproduction and growth to replace the trees that die (Waring and Schlesinger 1985). In one of the most appealing of the current crop of forest models, Bossel and Krieger (1991) assert that the forest equilibrates when respiration of the forest's woody tissues consumes enough of the forest's gross production to prevent further dry matter accumulation. They assume, however, that the annual cost of respiration of their woody tissues and the replacement of branches and coarse roots is 6% of the forest's above-ground standing crop. This would cause a rainforest's standing crop to equilibrate at a reasonable value: a forest whose biomass increases by 10 tons dry weight per hectare when young would equilibrate when its standing crop of wood was 333 tons per hectare. On the other hand, such high wood respiration would make nonsense of the budget drawn up for the Ducke Reserve in Table 6.24. What goes on here?

First, wood respiration is much lower than was thought. Ryan and Waring (1992) first saw the problem when they found that higher wood respiration could not explain the diminished wood production of older stands of lodgepole pine at 2800 m near Winter Park, Colorado (Table 6.21). They also found that woody respiration was proportional to sapwood volume, as is also true in other coniferous forests (Ryan et al. 1995).

Table 6.19 Liana abundance, diversity, and leaf production in tropical forest

	Barro Colorado, Panama	Yanamono, Peru	Ituri, Zaire	Analamazaotra, Madagascar (1000 m)	Lambir, Sarawak
Abundance and diversity of lianes (number of species in parentheses)					
Plot size (ha)	1.0	1.0	10.0	0.51	1.0
Number on plot of					
Trees ≥ 10 cm dbh	425	580 (283)	3433 (199)	539 (125)	728
Trees ≥ 5 cm dbh	1022		9382 (268)	1384 (177)	
Lianes ≥ 10 cm dbh	3	26 (17)	38 (22)	1 (1)	2
Lianes ≥ 5 cm dbh	43		427 (75)	39 (12)	27
Lianes ≥ 2 cm dbh	429		3275 (138)		256

	Ulu Segama, Sabah		Mt. Glorious, Queensland		Makokou, Gabon	Barro Colorado, Poacher's Peninsula
	Primary	Logged	Rainforest	Regrowth		
Litter fall						
All leaves (tons/ha·year)	6.53	6.16	6.22	6.31	4.37	7.89
Liane leaves (tons/ha·year)	0.83	1.68	1.47	5.23	1.57	0.93

Data sources: Barro Colorado: tree density provided for 1990 census by Suzanne Loo de Lao on behalf of R. Condit, S. P. Hubbell, and R. B. Foster (p. c.); liane density from Putz (1984b); litter fall from Leigh and Wright (1990). Peru: Gentry et al. (1988). Ituri: Jean Rémy (p. c.) Analamazaotra: Abraham et al. (1996). Lambir: Putz and Chai (1987). Sabah: Berghouts et al. (1994). Queensland: Hegarty (1991). Gabon: Hladik (1974).

Table 6.20 Size composition and tree mortality on Barro Colorado's 50-ha forest dynamics plot

Diameter at breast height (cm)	1982 N	1982–1985 D	m	1985 N	1985–1990 D	m	1990 N	1990–1995 D	m	1995 N
2–≤ 5	65182	5979	2.89	65089	7447	2.31	87704	9959	2.52	83182
5–≤ 10	21382	1647	2.43	24490	2417	1.97	29024	2622	1.98	28906
10–≤ 20	12489	972	2.42	12745	1261	1.97	13154	1153	1.92	13307
20–≤ 30	3698	396	3.36	3506	335	1.89	3594	297	1.80	3654
30–≤ 40	1740	180	3.32	1771	190	2.11	1739	150	1.88	1761
40–≤ 50	984	126	4.06	969	107	2.15	925	85	2.02	934
50–≤ 60	543	53	2.96	563	67	2.30	553	52	2.09	537
60–≤ 70	297	35	3.61	304	30	1.87	327	33	2.26	324
70–≤ 80	172	17	3.08	176	17	1.83	166	15	2.02	173
80–≤ 90	97	10	3.12	90	9	1.89	120	7	1.28	126
90–≤100	53	3	1.44	60	4	1.25	78	4	1.12	71
100+	174	19	3.28	173	12	1.29	170	9	1.15	173

N is the number of trees in each class at the census indicated, D is the number of these which died before the next census, and m is their annual mortality rate, $[(1/T)\ln[N/(N-D)]]$, where T is the average time interval between remeasurements, expressed as a percentage. Figures in this table for 1982–1985 have been revised to correct for inflated diameter estimates in 1982 (Leigh 1996). Data were provided by Suzanne Loo de Lao on behalf of R. Condit, S. P. Hubbell, and R. B. Foster (p. c.); 1982–1985 are mostly as given in Leigh (1996), the other numbers were provided in March 1996.

Ryan et al. (1994) found that respiration of woody tissues was also relatively low at La Selva, Costa Rica. There, too, maintenance respiration of woody tissues appears to be proportional to sapwood volume. In the second-growth tree *Simarouba amara* and the shade-tolerant old forest tree *Minquartia guianensis*, maintenance respiration of woody tissue is 39.6 µmol CO_2/sec · m³ of sapwood when air temperature is 24.6°C (Ryan et al. 1994). If, as they suggest, sapwood volume at La Selva is 23% of wood volume, while a cubic meter of wood weighs 600 kg, on average, when dry, then there is 0.383 m³ of sapwood per ton dry weight of wood. A heroic extrapolation suggests that sapwood respiration releases about 15 µmol CO_2/sec, or 480 mol CO_2/year, per ton of wood, an annual glucose consumption of 14.4 kg glucose/ton of wood · year. Thus sapwood respiration annually amounts to 1.4% of the above-ground standing crop. [Ryan et al. concluded that sapwood respiration accounted for 1.8 tons of carbon (4.5 tons of glucose) in a hectare of forest with an above-ground standing crop of 249 tons of dry matter, a sapwood respiration rate of 1.8% of the forest's above-ground dry matter per year; the discrepancy is unexplained. Thus sapwood respiration appears to be far too low to limit biomass accumulation in forests of either the tropics or the temperate zone (Gower et al. 1996).

What does limit a forest's height and its "wood crop"? Givnish (1988) suggested that a tree stopped growing when a new leaf at its top could no longer pay for its share of the construction and maintenance of the wood and roots required to support the crown and supply it with water and nutrients. King (1990) supposed that the stemwood production of a tall tree, expressed per square meter of crown, declined linearly with the tree's height, and calculated the height at which a tree could no longer increase its wood production compared to competing neighbors by increasing its height. He calculated the optimum height of conifers with conical crowns from stand tables for various plots and found that his predictions were close to, but usually exceeded, observed height.

In fact, a tall tropical tree's growth is slowed because, as it gets taller, it must produce progressively more wood for each cubit it adds to its height. In particular, the higher a tree's leaves, the smaller the leaf area that can be supported by a kilogram of wood for each meter of these leaves' height above the ground (Table 6.22), that is to say, the lower the "leaf support efficiency" in the sense of King (1994). In east Borneo, a 46.5 m tree requires 24.8 tons of wood to carry 1525 m² of leaves. Presumably, a tree that tall would need 767/1525 times as much wood, or 12.5 tons, to support 767 m² of leaves at the same height. A

Table 6.21 Standing crop, wood production, and respiration of woody tissues in lodgepole pine stands of different ages*

Age (years)	Standing Crop (kg C/m²)				Wood production (g C/m² of forest)	Sapwood respiration (g C/m² of forest)
	Leaves	Roots	Sapwood	Total		
40	0.65	0.5	3.6	6.2	210	61
245	0.41	1.1	4.4	9.0	46	79

*Data from Ryan and Waring (1992).

tree 70.7 m tall used 42.7 tons of wood to support 767 m² of leaves, as if (42.7 − 12.5)/(70.7 − 46.5), or 1.25 tons, of wood were needed for each extra meter of height above 46.5 m. Here, 1.63 kg of wood appear needed to lift a square meter of leaves each extra meter above 46.5 m. The tallest trees in the New Guinea cloud forest (Edwards and Grubb 1977) must also lay on extra wood to grow above the main canopy (Table 6.22), although, in general, these montane trees are more stoutly built than their lowland counterparts (partly because their leaves weigh more per unit area). For comparison, a square meter of the stoutly built, slow-growing, elfin forest atop the Luquillos in Puerto Rico uses about 8 kg of wood to support 2 m² of leaves 4 m above the ground—1 kg of wood to raise each m² of leaves another meter above the ground—while on Barro Colorado, a square meter of forest uses 28 kg of wood to support 7 m² of leaves an average of 20 m above the ground—200 g of wood for every meter a square meter of leaves is lifted above the ground. A forest must be quite tall before its growth slows: on Barro Colorado, forest surrounding a tower grew from 27 to 34 m in 20 years, 35 cm/year.

Table 6.22 The cost of supporting leaves at different heights

Height class (m)	N	Wood wt (kg)	Leaf wt (kg)	Leaf area (m²)	Crown/trunk weight ratio	Leaf/wood wt ratio	Avg. ht (m)	LSE (m³/kg)	Basal area (cm²)	Wood density
Sebulu, east Borneo, lowland, 1/8 ha										
70.7	1	42741	112	795	0.294	0.0026	70.7	1.32	13376	0.81
46.5	1	24768	184	1525	0.532	0.0075	46.5	2.86	12668	0.80
40–<45	4	18442	127	1163	0.168	0.0069	42.2	2.66	11411	0.57
35–<40	2	4627	49	478	0.139	0.0106	38.8	3.90	3515	0.57
30–<35	3	2324	30	267	0.270	0.0130	31.8	3.65	1938	0.50
25–<30	6	3190	58	479	0.202	0.0181	26.9	4.04	3110	0.59
20–<25	6	1679	39	443	0.222	0.0233	22.9	6.03	1920	0.49
15–<20	10	1265	43	460	0.279	0.0342	17.1	6.22	1879	0.56
12–<15	10	370	20	233	0.319	0.0540	13.0	8.19	746	0.54
Diameter at breast ht, cm										
100 ≤ dbh	2	67508	297	2321	0.372	0.0044	58.6	2.01	26044	0.80
50 ≤ dbh <100	4	18442	127	1163	0.168	0.0069	42.2	2.66	11411	0.57
30 ≤ dbh <50	4	6551	75	733	0.148	0.0115	33.6	3.76	5245	0.57
25 ≤ dbh <30	6	3595	72	586	0.288	0.0199	28.4	4.64	3242	0.54
20 ≤ dbh <25	5	1446	29	341	0.207	0.0204	23.8	5.60	1857	0.51
15 ≤ dbh <20	5	701	22	206	0.226	0.0308	17.6	5.19	1136	0.46
10 ≤ dbh <15	12	1020	35	406	0.253	0.0344	15.9	6.35	1432	0.56

Height class (m)	N	Wood wt (kg)	Leaf wt (kg)	Leaf area (m²)	Crown/trunk weight ratio	Leaf/wood wt ratio	Avg. ht (m)	LSE (m³/kg)	Basal area (cm²)
Papua New Guinea, 2500 m, 1/25 ha									
37.8	1	1683	10	38	0.484	0.0060	37.8	0.842	3312
≥35*	2	7863	72	262	0.342	0.0093	36.4	1.242	12839
30–<35	3	6737	99	384	0.236	0.0147	32.3	1.843	9700
25–<30	6	6804	91	492	0.306	0.0133	27.1	1.961	9219
20–<25	13	3158	54	289	0.211	0.0170	21.6	1.977	5118
15–<20	10	736	15	77	0.192	0.0209	16.7	1.753	1943
Diameter at breast ht, cm									
dbh ≥ 100*	1	6000	62	225	0.307	0.0103	35.0	1.310	9527
50 ≤ dbh < 100	5	10480	123	535	0.291	0.0117	30.8	1.570	16401
30 ≤ dbh < 50	6	5611	81	405	0.276	0.0145	27.7	1.999	6741
20 ≤ dbh < 30	9	1965	38	196	0.219	0.0193	21.5	2.146	3753
15 ≤ dbh < 20	8	852	25	146	0.236	0.0298	17.2	2.955	1869
10 ≤ dbh < 15	10	440	9	48	0.157	0.0200	15.1	1.645	1257

Data for Sebulu from Yamakura et al. (1986b); data for Papua New Guinea from Edwards and Grubb (1977).
N, the number of trees in this category; Wood wt, the total dry weight of wood of trees in this category, including woody parts of associated climbers and epiphytes; Leaf wt, the total dry weight of leaves of these trees and their associated climbers and epiphytes; Leaf area, the total area (one-sided) of these leaves; Crown/trunk wt ratio, the total dry weight of these trees' branches and leaves (including the total dry weight of associated climbers and epiphytes), divided by the total dry weight of these trees' trunks; Leaf/wood wt ratio, the total dry weight of the leaves of these trees and their climbers and epiphytes, divided by the total dry weight of all their woody parts; Avg. ht, the (unweighted) average height of trees in this category; LSE, Leaf Support Efficiency, the total area of leaves of these trees, and the leaves of their climbers and epiphytes, multiplied by the average height of these trees and divided by the total dry weight of their woody parts; Basal area, the total basal area (sum of cross-sectional area 1.3 m above ground) of these trees' trunks; and wood density, the total dry weight of these trees' trunks and branches, excluding epiphytes, in kg, divided by the fresh volume of these woody parts, in liters.
*Including one 35-m tall tree felled just outside plot.

All this is very ad hoc. We need a precise theory of tree engineering to assess the amount of wood a tree needs to lift its leaves an extra meter upward and to predict the extra risk of mortality thus incurred. Once we know the rates of height growth at various levels of the forest as a function of the light allowed to reach this level by taller trees and the mortality rate of trees at different levels, it should be possible to calculate the equilibrium distribution of tree heights and diameters in a forest (Bossel and Krieger 1991).

Even without such theory it is clear, given the mortality rate of tropical trees, that the slow growth imposed on tall trees by their need to invest disproportionately in wood must limit tree growth. Among other things, taller trees are likely to lack equally tall neighbors, thanks to thinning from mortality. A fully emergent tree does not benefit from further increase of height: it does better to devote its resources to reproducing, improving its roots, or thickening its trunk (King 1990). Where nutrients are in short supply, the advantages accruing to an emergent crown are diminished (Givnish 1984). The canopy in the nutrient-poor heath forests of Malaysia is smoother (Brunig 1983) because it does not pay trees in such nutrient-starved habitats to do more than keep up with their neighbors.

In stunted elfin forests, as in lowland forests on poor soils, little energy is left for growth and reproduction after the costs of leaves and roots have been met. In these elfin forests (Table 6.23), stem density is higher because it is harder for one tree to overtop another, and basal area is higher because trees must build stoutly in order to live long (Lawton 1982, Leigh 1990).

In the north temperate zone, some gymnosperms like redwoods, Douglas fir, or even white pine, create forests of impressive height and enormous basal area and timber volume. How do these forests escape the limits that apply to tropical forests? Leaf production in these forests is so low (Appendix 6.1) that one wonders whether they channel energy differentially into wood production. Moreover, these trees, or at least canopy individuals thereof, are exceedingly long lived. They do not have the high turnover within an intact canopy that is so characteristic of tropical forest: the canopy is dominated, more often than not, by trees that grew up after their site was cleared by a major disturbance, such as a fire or a hurricane (Franklin and DeBell 1988, Stewart 1989). Does the mortality that strikes all size classes evenly in tropical forest play an essential role in structuring that forest?

Trees grow in height to compete for light at the expense of their longevity and reproductive output. In chapter 9, we shall see how, in forests where the seeds of canopy trees disperse farther, the advantage of increased reproduction is enhanced relative to that of height growth, so canopy stature should be lower. There, I consider whether the great height of dipterocarps, and the greater height of redwoods, is related to ineffectively dispersed seeds.

The Energy Budget of Rainforest

In the Ducke Reserve, near Manaus, Fan et al. (1990) found that total forest respiration averaged 3.335 kg/ha · hr: 2.57 by night and 4.1 by day. Extrapolating to a full year, forest respiration at the Ducke Reserve consumed 73 tons of glucose/ha. If the same is true on Barro Colorado, we may draw up energy budgets for both forests (Table 6.24). These budgets are preliminary: we do not know, for example, whether leaf or wood respiration is lower in forests on infertile soil. Nonetheless, Barro Colorado's above-ground respiration budget shows a decided surplus of respiration not accounted for, while the above-ground respiration budget of Ducke Reserve is more or less in balance. Barro Colorado's above-ground respiration surplus probably is taken up in part by the activities of herbivores, mostly insects.

Table 6.23 Altitudinal gradients in tropical rainforest

	Luquillo Mts, Puerto Rico			Tjibodas, Java		
	Tabonuco (440 m)*	Colorado (725 m)	Elfin (1000 m)	(1550 m)	(2400 m)	(3000 m)
Forest height (m)	20–30	8–20	3–5	20–45	10–20	4–9
Basal area, stems > 1 cm					62.0	61.5
Basal area, stems > 4 cm	40	40	55			
Basal area, stems > 10 cm				52.2	56.9	36.4
Stems >10 cm	710		49			
Stems/ha > 4 cm	1750	1850	21900			
Stems/ha > 10 cm	710			427	1516	3828
Leaf area index	6.5	4.49†	3.3			
Leaf fall (g/m²/year)	494	505	245	450		
Litter fall(m²/year)	861	680	310	596		

Luquillo data from Weaver and Murphy (1990), except stems ≥10 cm dbh/ha for Tabonuco forest, which is from Crow (1980). Tjibodas data from Yamada (1975, 1976a,b).
*Altitude in parentheses.
†Excluding bryophytes.

Table 6.24 Annual forest respiration on Barro Colorado Island, Panama, and Ducke Reserve, Amazonia (tons glucose/ha)

	Barro Colorado	Ducke
Soil respiration (total)	36	49
Root and soil processes	23	41
Litter decomposition	13	8
Above-ground respiration	37	24
Leaf maintenance	8	8
Wood construction cost	5–6	3
Sapwood respiration*	4	4
Litter construction cost	13	8
Other (unknown)	6–7	1

*Sapwood respiration is 1.5% of standing crop per year.
Respiration from Fan et al. (1990), except Barro Colorado soil respiration, from Kursar (1989). Construction costs are additional to, and assumed equal to, the dry matter content of the items constructed (T. Kursar, p. c.).

CONCLUDING REMARKS

Most mature tropical rainforests carry about 7 or 8 ha of leaves/ha of ground (i.e., have leaf area index near 7). They drop about 7 tons dry weight of leaves/ha · year, and have basal area about 30 m²/ha. The mortality rate among their trees is between 1 and 2% per year and is roughly the same for trees of all size classes ≥ 10 cm dbh.

When LAI is near 7, about 1% of the light incident upon the canopy reaches the forest floor. The annual average radiation received by most tropical forests is 180 ± 20 W/m². The 2 W/m² reaching the forest floor is too little for additional plants to "break even," therefore the similarity in LAI.

A crude model suggests that a forest with normal LAI and normal levels of incident radiation should fix about 300 mol CO_2/m² · year; a gross production of about 90 tons of sugar/ha · year. The sparse evidence available suggests that gross production is another "constant" of normal forests. On the other hand, a tropical forest in southwest Amazonia with LAI 4 has a gross production of only 60 tons of sugar/ha · year.

Once leaf area is fully developed, a forest's trees can invest their remaining energy either in wood, to grow higher in competition for light, in roots, to compete for nutrients, or in reproductive structures, to make more young. On worse soils, below-ground investment, and therefore soil respiration, is higher.

Systems protected from herbivores, where nutrients are readily available and circumstances allow even distribution of light among leaves, have twice or thrice the gross production of tropical forests: an annual production of 90 tons per hectare per year applies to communities where plants must be defended from herbivores and are free to try to outgrow each other.

In the next chapter, I discuss how much herbivores eat, what they eat, and how plants keep them from eating more.

APPENDIX 6.1

Basal area and litter fall in different forests

Site	Basal area (m²/ha) (stems ≥ 10 cm)	Litter fall (g/m²/year) Leaves	Litter fall (g/m²/year) Total	Reference
Tropical forests				
Las Tuxtlas, Mexico	34.9			Bongers et al. (1988)
Finca Seacté, Guatemala, 900 m, 1974–75	46.0	735	956	Kunkel-Westphal and Kunkel (1979)
1975–76		613	910	Kunkel-Westphal and Kunkel (1979)
Finca Seacté, Guatemala, 1000 m, 1974–75	40.0	763	1012	Kunkel-Westphal and Kunkel (1979)
1975–76		696	992	Kunkel-Westphal and Kunkel (1979)
La Selva, Costa Rica	24.7	660	1090	Ryan et al. (1994), Parker (1994)
Monteverde, Costa Rica, 21 m cloud forest, 1500 m, trees and ground-herbs		480	700	Nadkarni and Matelson (1992a)
Monteverde, 1500 m, epiphytic material		40?	50	Nadkarni and Matelson (1992b)
Volcan Barva, Costa Rica, 100 m, 1985–86	22.7		900	Heaney and Proctor (1989)
1000 m, 1985–86	31.2		660	Heaney and Proctor (1989)
2000 m, 1985–86	28.6		580	Heaney and Proctor (1989)
2600 m, 1985–86	51.2		530	Heaney and Proctor (1989)
Barro Colorado, plateau 1969/70	28.5	578	1089	Foster 1982; (Table 1)
1970–71		643	1215	basal area, R. Condit (p. c.)
Barro Colorado, Lutz Ravine 1972		749	1200	Leigh and Windsor (1982)
1973		689	1096	
1974		635	1034	
1975		536	853	
1976		688	1014	
1977		622	946	
Barro Colorado, Lutz Ravine 1978		572	928	
1979		638	994	

(continued)

Appendix 6.1 (continued)

Site	Basal area (m²/ha) (stems ≥ 10 cm)	Litter fall (g/m²/year) Leaves	Total	Reference
Barro Colorado, Poacher's Peninsula (basal area for all Poachers plots includes only trees ≥ 20 cm dbh).				
Poachers P1				S. J. Wright, (p. c.)
1988	25.6	804 (154, 55?)*	1217	
1989		854 (178, 33?)	1269	
1990		695 (136, 26?)	968	
1991		747 (136, 22?)	1135	
1992		667 (101, 28?)	1054	
P2				
1988	19.3	792 (116, 43?)	1328	
1989		846 (128, 43?)	1306	
1990		734 (88, 23?)	1046	
1991		783 (108, 21?)	1260	
1992		688 (112, 28?)	1125	
P3				
1988	26.9	710 (75, 36?)	1214	
1989		725 (70, 21?)	1268	
1990		714 (89, 16?)	1165	
1991		737 (88, 15?)	1252	
1992		684 (76, 17?)	1111	
Barro Colorado, Poachers P4				
1988	21.2	749 (52, 41?)	1236	
1989		678 (44, 23?)	1118	
1990		618 (52, 11?)	991	
1991		704 (80, 11?)	1120	
Gigante Peninsula, near BCI 1989		712	998	J. Cavelier (p. c.)
Manaus, Brazil 1963	24.8	640	790	Klinge and Rodrigues (1968)
(terra firme) 1964		480	670	Klinge and Rodrigues (1968)
1974		608	780	Franken, Irmler, and Klinge (1979)
1975		680	800	Franken, Irmler, and Klinge (1979)
Bacia Modelo, Manaus (terra firme) 1979–80		441	742	Luizão and Schubart (1987)
Ducke Reserve, riverine forest 1977		430	637	Franken (1979)
Manaus, Igapo forest 1976–77		532	676	Adis, Furch, and Irmler (1979)
Mocambo, Belem, Brazil	27.7	802	990	Klinge (1977)
Maracá, Roraima, Brazil (evergreen)	23.8	630	928	Basal area, Thompson et al. (1992); litter fall, Scott et al. (1992)
St. Elie, Guyane Française 1978	38.2	578	867	Puig and Delobelle (1988); basal
1979		543	746	area is avg. of plots A–D from
1980		549	748	Puig and Lescure (1981)
San Carlos de R. N., Venezuela				
Terra firme 1975–77	27.8	500	620	Basal area, Uhl and Murphy (1981); leaves, Jordan and Uhl (1978);. total litterfall, Medina et al. (1980)
1980–81		757	1025	Cuevas and Medina (1986)
SCRN tall Amazon caatinga 1980–81		399	561	Cuevas and Medina (1986)
Elfin forest, Luquillos, Puerto Rico, 1000 m	49.1	245	310	Weaver et al. (1986)
Blue Mts, Jamaica, 1550 m, 2237 mm rain/year				Tanner (1977, 1980b)
Mor ridge, 1615 m, acid soil	64.7	491	661	
Mull ridge, 1605 m, 1974–75	65.4	532	554	
Mull ridge, 1605 m, 1977–78		563	580	
Wet slope, 1570 m, 1974–75	46.4	436	555	
Gap forest, 1590 m, 1974–75	47.8	550	647	
Colombia, Cordillera Central				
2550 m, 25 m forest, 0.2 ha	30.0	461 (4.27 m²)	703	Veneklaas (1991)
3370 m, 22 m forest, 0.24 ha	38.7	282 (1.52 m²)	431	Veneklaas (1991)
Kade, Ghana	30.9	702	1054	John (1973), from Nye (1961)
1970–2		740	966	John (1973)
Banco, Cote d'Ivoire 1966–67	30	800	1048	Bernhard (1970); basal area from Huttel (1975)
Banco, Côte-d'Ivoire 1967–68		774	1119	Bernhard (1970)
Yapo, Côte d'Ivoire 1967–68	31	618	872	Bernhard (1970); Basal area from Huttel (1975)

Site	Basal area (m²/ha) (stems ≥ 10 cm)	Litter fall (g/m²/year) Leaves	Total	Reference
Yapo, Côte d'Ivoire 1968–69	31	719	995	Bernhard (1970)
Bakundu, Cameroon, plot 1, 1983		872	1356	Songwe et al. (1988)
Bakundu, Cameroon, plot 2, 1983		856	1389	Songwe et al. (1988)
Ipassa, Gabon		650	1330	Hladik (1978)
Bannadpore, W. Ghats, 12° 5' N, 200 m, 1977–78	33.9	318	411	Rai and Proctor (1986a,b)
1978–79		326	402	Rai and Proctor (1986a,b)
Kagneri, W. Ghats, 12° 49' N, 500 m, 1977–78	35.6	298	346	Rai and Proctor (1986a,b)
1978–79		386	450	Rai and Proctor (1986a,b)
Attapadi, Silent Valley, 900 m, 1979–80	59.6	678	866	Pascal (1988)
1980–81		587	811	Pascal (1988)
1981–82		647	883	Pascal (1988)
Pasoh, Malaya, mixed dipterocarp forest	31.4	703	1110	Basal area; Kira (p. c.); litter fall; Kira (1978)
1972		741	1026	Lim (1978)
1973		539	749	Lim (1978)
Ketambe, Sumatra	23.3	614	958	van Schaik and Mirmanto (1985)
Danum Valley, Sabah, uncut 1989	27.4	653	(82.5 lianas)	Burghouts et al. (1994)
			1150	Burghouts et al. (1992)
Danum Valley, Sabah, logged 1989	42.2	620	(168 lianas)	Burghouts et al. (1994)
			1190	Burghouts et al. (1992)
Gunung Mulu, Sarawak, wet, very mild dry season				J. Proctor et al. (1983a, b)
Alluvial forest, 5090 mm rain/year	28	660	1150	
Dipterocarp forest, 5110 mm rain/year	57	540	880	
Heath forest, 5700 mm rain/year	43	560	920	
Limestone forest, 5700 mm rain/year	37	730	1200	
Gunung Silam, Sabah, 280 m, 0.4 ha	38.2	386	651	Proctor et al. (1988, 1989)
G. Silam, 33 m forest, 330 m, 0.4 ha	46.2	450	737	Proctor et al. (1988, 1989)
G. Silam, 27 m forest, 480 m, 0.4 ha	39.8	344	522	Proctor et al. (1988, 1989)
G. Silam, 23 m forest, 610 m, 0.24 ha	43.7	413	560	Proctor et al. (1988, 1989)
G. Silam, 13 m forest, 790 m, 0.04 ha	30.9	366	553	Proctor et al. (1988, 1989)
G. Silam, 11 m forest, 870 m, 0.04 ha	26.7	332	480	Proctor et al. (1988, 1989)
Tjibodas, Java, 1550 m	52.2	450	596	Yamada (1976b)
New Guinea, 2550 m	47	<635	755	Edwards (1977)
Mt Glorious, Australia, 640 m 1985–86 (27° 20' S, forest interior)	69.6	549 (7.3 m²)		Hegarty (1991)
Pin Gin Hill, Australia, 60 m, 1975	35.6	489	961	Spain (1984)
(17° 33' S) 1976			889	Spain (1984)
1977			1053	Spain (1984)
Gadgarra, Atherton, Australia, 680 m, 1975	63.0	492	816	Spain (1984)
(17° 18' S) 1976			886	Spain (1984)
1977			881	Spain (1984)
Atherton Tableland, Australia, 700 m, 1976–77 (Gadgarra State Forest Reserve, 17° 18' S),	61		1030	Brasell et al. (1980)
1977–78			1060	
1978–79			870	
Temperate-zone forests				
Southern Hemisphere				
Eucalyptus diversicolor (karri forest, wet sclerophyll, Pemberton, Western Australia				
1977	56	337	868	O'Connell and Menage (1982)
1978	56	381	1023	O'Connell and Menage (1982)
Northern Hemisphere				
Hardwoods, 590 m, Hubbard Brook, NH	26	285	572	Whittaker et al. (1974); Gosz et al. (1972)
Hardwoods, Coweeta, NC (≥2.5cm)	25.6	451		
20 m sugar maple, NE Minnesota, 47° 52' N		79-year-old overstory		Burton et al. (1991)
1988	31	313 (7.4 m²)		
1989		256 (5.9 m²)		
1990		370 (7.3 m²)		

(*continued*)

Appendix 6.1 (continued)

Site	Basal area (m²/ha) (stems ≥ 10 cm)	Litter fall (g/m²/year) Leaves	Total	Reference
24 m sugar maple, Michigan penisula, 46° 52' N		73-year-old overstory		Burton et al. (1991)
1988	32	382 (6.9 m²)		
1989		275 (4.9 m²)		
1990		314 (6.7 m²)		
28 m sugar maple, Michigan, 44° 23' N		74-year-old overstory		Burton et al. (1991)
1988	30	342 (6.0 m²)		
1989		407 (7.1 m²)		
1990		437 (7.8 m²)		
Virgin oak forest, NJ		467	650	Lang and Forman (1978)
Oak-pine forest, Brookhaven NY	15.6	260	333	Whittaker and Woodwell (1969)
Old Growth (WT), Edgewater MD (≥2.5 cm)	34.8	443		Burnham et al. (1992)
Liriodendron forest, Edgewater (≥2.5 cm)	34.7	383 (5.3 m²)	440	Parker et al. (1989)
Cypress swamp, Battle Cr., MD (≥2.5 cm)	56.3	494		Burnham et al. (1992)
Douglas fir, Andrews EF, Oregon 1977–80	91	184		Marshall and Waring (1986)
		210	466	
Redwoods, Mendocino Co., CA	247		213	Westman & Whittaker (1975)
Giant Sequoia, Sequoia NP, CA 1982–83	159.6	661	923	Stohlgren (1988)
1983–84			197	
1984–85			204	
1985–86			547	
Fir-Pine, Sequoia NP, CA, 1982–83	65.9	264	513	Stohlgren (1988)
1985–86			332	
Cascades, Western Washington, *Abies amabilis*				
23-year stand, 1200 m 1977–78	45.7	104	151	Grier et al. (1981)
180-year stand, 1200 m 1977–78	74.3	103	218	Grier et al. (1981)
Lodgepole pine, Medecine Bows, Wyoming (ca 60 cm rain/year, 2/3 of which is snow)				
70-year, 3050 m, 1280 stems/ha, 1977–79	26	111	147	Fahey (1983)
105-year, 2800 m, 15000 stems/ha, 1977–79	39	100	125	Fahey (1983)
105-year, 2800 m, 1800 stems/ha, 1977–79	40	91	130	Fahey (1983)
105-year, 2800 m, 700 stems/ha, 1977–79	38	90	125	Fahey (1983)
105-year, 2900 m, 1850 stems/ha, 1977–79	64	106	159	Fahey (1983)
240-year, 2950 m, 420 stems/ha, 1977–79	37	87	155	Fahey (1983)

*Liana leaf falls in parentheses, followed by figure with question mark, represents fall of unidentified leaves which might or might not be lianas.

References

Aber, J. D. 1979. Foliage-height profiles and succession in northern hardwood forest. *Ecology* 60: 18–23.

Abraham, J. P., R. Benja, M. Randrianasolo, J. U. Ganzhorn, V. Jeannoda, and E. G. Leigh Jr. 1996. Tree diversity on small plots in Madagascar: a preliminary review. *Revue d'Ecologie (La Terre et la Vie)* 51: 93–116.

Adis, J., K. Furch, and U. Irmler. 1979. Litter production of a central-Amazonian black water inundation forest. *Tropical Ecology* 20: 235–245.

Aide, T. M., and E. C. Londoño. 1989. The effects of rapid leaf expansion on the growth and survival of a lepidopteran herbivore. *Oikos* 55: 66–70.

Alexandre, D. Y. 1982. Étude de l'éclairement du sous-bois d'une forêt dense humide sempervirente (Taï, Côte-d'Ivoire). *Acta Oecologica: Oecologia Generalis* 3: 407–447.

Allen, M. F. 1991. *The Ecology of Mycorrhizae*. Cambridge University Press, Cambridge.

Benzing, D. H. 1989. Vascular epiphytism in America, pp. 133–154. In H. Lieth and M. J. A. Werger, eds., *Tropical Rain Forest Ecosystems: Biogeographical and Ecological Studies*. Elsevier, Amsterdam.

Benzing, D. H. 1990. *Vascular Epiphytes*. Cambridge University Press, New York.

Bernhard, F. 1970. Étude de la litière et de sa contribution au cycle des éléments minéraux en forêt ombrophile de Côte-d'Ivoire. *Oecologia Plantarum* 5: 247–266.

Bernhard-Reversat, F., C. Huttel, and G. Lemée. 1978. Structure and functioning of evergreen rain forest ecosystems of the Ivory Coast, pp. 557–574. In *Tropical Forest Ecosystems*, UNESCO, Paris.

Björkman, O., and M. M. Ludlow. 1972. Characterization of the light climate on the floor of a Queensland rainforest. *Carnegie Institution of Washington Year Book* 71: 85–94.

Björkman, O., M. M. Ludlow, and P. A. Morrow. 1972. Photosynthetic performance of two rainforest species in their native habitat and analysis of their gas exchange. *Carnegie Institution of Washington Year Book* 71: 94–102.

Bongers, F., D. Engelen, and H. Klinge. 1985. Phytomass structure of natural plant communities on spodosols in southern Venezuela: the Bana woodland. *Vegetatio* 63: 13–34.

Bongers, F., J. Popma, J. Meave del Castillo, and J. Carabias. 1988. Structure and floristic composition of the lowland rain forest of Los Tuxtlas, Mexico. *Vegetatio* 74: 55–80.

Bossel, H., and H. Krieger. 1991. Simulation model of natural tropical forest dynamics. *Ecological Modelling* 59: 37–71.

Brasell, H. M., G. L. Unwin, and G. C. Stocker. 1980. The quantity, temporal distribution and mineral-element content of litterfall in two forest types at two sites in tropical Australia. *Journal of Ecology* 68: 123–139.

Briscoe, C. B., and F. H. Wadsworth. 1970. Stand structure and yield in the tabonuco forest of Puerto Rico, pp. B-79–89. In H. T. Odum and R. F. Pigeon, eds., *A Tropical Rain Forest*. Division of Technical Information, U. S. Atomic Energy Commission, Washington, DC.

Brown, M. J., and G. G. Parker. 1994. Canopy light transmittance in a chronosequence of mixed-species deciduous forests. *Canadian Journal of Forest Research* 24: 1694–1703.

Brunig, E. F. 1983. Vegetation structure and growth, pp. 49–75. In F. B. Golley, ed., *Tropical Rain Forest Ecosystems: Structure and Function*. Elsevier, Amsterdam.

Bugbee, B. B., and F. B. Salisbury. 1988. Exploring the limits of crop productivity I. Photosynthetic efficiency of wheat in high irradiance environments. *Plant Physiology* 88: 869–878.

Bugbee, B. B., and F. B. Salisbury. 1989. Current and potential productivity of wheat for a controlled environment life support system. *Advances in Space Research* 9(8): 5–15.

Burghouts, T. B. A., E. J. F. Campbell, and P. J. Kolderman. 1994. Effects of tree species heterogeneity on leaf fall in primary and logged dipterocarp forest in the Ulu Segama Forest Reserve, Sabah, Malaysia. *Journal of Tropical Ecology* 10: 1–26.

Burghouts, T., G. Ernsting, G. Korthals, and T. de Vries. 1992. Litterfall, leaf litter decomposition and litter invertebrates in primary and selectively logged dipterocarp forest in Sabah, Malaysia. *Philosophical Transactions of the Royal Society of London* B 335: 407–416.

Burnham, R. J., S. L. Wing, and G. G. Parker. 1992. The reflection of deciduous forest communities in leaf litter: implications for autochthonous litter assemblages from the fossil record. *Paleobiology* 18: 30–49.

Burton, A. J., K. S. Pregitzer, and D. D. Reed. 1991. Leaf area and foliar biomass relationships in northern hardwood forests located along an 800 km acid deposition gradient. *Forest Science* September 1991: 1041–1059.

Cavelier, J. 1989. *Root Biomass Production and the Effect of Fertilization in Two Tropical Rain Forests*. PhD thesis, Botany School, University of Cambridge.

Cavelier, J. 1992. Fine-root biomass and soil properties in a semi-deciduous and a lower montane rain forest in Panama. *Plant and Soil* 142: 187–201.

Chapin, F. S. III. 1980. The mineral nutrition of wild plants. *Annual Review of Ecology and Systematics* 11: 233–260.

Chazdon, R. L. 1986. Light variation and carbon gain in rain forest understorey palms. *Journal of Ecology* 74: 995–1012.

Chazdon, R. L., and N. Fetcher. 1984. Photosynthetic light environments in a lowland tropical rain forest in Costa Rica. *Journal of Ecology* 72: 553–564.

Clark, D. B., and D. A. Clark. 1990. Distribution and effects on tree growth of lianas and woody hemiepiphytes in a Costa Rican tropical wet forest. *Journal of Tropical Ecology* 6: 321–331.

Clark, D. B., D. A. Clark, P. M. Rich, S. Weiss, and S. F. Oberbauer. 1996. Landscape-style evaluation of understory light and canopy structure: methods and application in a neotropical lowland rain forest. *Canadian Journal of Forest Research* 26: 747–757.

Coley, P. D. 1983. Herbivory and defensive characteristics of tree species in a lowland tropical forest. *Ecological Monographs* 53: 209–233.

Coley, P. D., J. P. Bryant, and F. S. Chapin III. 1985. Resource availability and plant herbivore defense. *Science* 230: 895–899.

Condit, R., S. P. Hubbell, and R. B. Foster. 1995. Mortality rates of 205 Neotropical tree and shrub species and the impact of a severe drought. *Ecological Monographs* 65: 419–439.

Covington, W. W., and J. D. Aber. 1980. Leaf production during secondary succession in northern hardwoods. *Ecology* 61: 200–204.

Crow, T. R. 1980. A rainforest chronicle: a 30-year record of change in structure and composition at El Verde, Puerto Rico. *Biotropica* 12: 42–55.

Cuevas, E., and E. Medina. 1986. Nutrient dynamics within amazonian forest ecosystems I. Nutrient flux in fine litter fall and efficiency of nutrient utilization. *Oecologia* 68: 466–472.

da Conceição, P. N. 1977. Algunos aspectos ecofisiológicos de floresta tropical úmida de terra firme. *Acta Amazônica* 7: 157–178.

Davidson, E. A., and S. E. Trumbore. 1995. Gas diffusivity and production of CO_2 in deep soils of the eastern Amazon. *Tellus* 95B: 550–565.

Dawkins, H. C. 1959. The volume increment of natural tropical high-forest and limitations on its improvements. *Empire Forestry Review* 38: 175–180.

Edwards, P. J. 1977. Studies of mineral cycling in a montane rain forest in New Guinea. *Journal of Ecology* 65: 971–992.

Edwards, P. J., and P. J. Grubb. 1977. Studies of mineral cycling in a montane rain forest in New Guinea I. The distribution of organic matter in the vegetation and soil. *Journal of Ecology* 65: 943–969.

Fahey, T. J. 1983. Nutrient dynamics of above-ground detritus in lodgepole pine (*Pinus contorta* ssp. *latifolia*) ecosystems, southeastern Wyoming. *Ecological Monographs* 53: 51–72.

Fan, S.-M., M. L. Goulden, J. W. Munger, B, C. Daube, P. S. Bakwin, S. C. Wofsy, J. S. Amthor, D. R. Fitzjarrald, K. E. Moore, and T. R. Moore. 1995. Environmental controls on the photosynthesis and respiration of a boreal lichen woodland: a growing season of whole-ecosystem exchange measurements by eddy correlation. *Oecologia* 102: 443–452.

Fan, S.-M., S. C. Wofsy, P. S. Bakwin, and D. J. Jacob. 1990. Atmosphere-biosphere exchange of CO_2 and O_3 in the Central Amazonian forest. *Journal of Geophysical Research* 95: 16851–16864.

Field, C., and H. A. Mooney. 1986. The photosynthesis-nitrogen relationship in wild plants, pp. 25–55. In T. J. Givnish, ed., *On the Economy of Plant Form and Function*. Cambridge University Press, Cambridge.

Foster, R. B. 1982. Famine on Barro Colorado Island, pp. 201–212. In E. G. Leigh, Jr., A. S. Rand, and D. M. Windsor, eds., *The Ecology of a Tropical Forest*. Smithsonian Institution Press, Washington, DC.

Franken, M. 1979. Major nutrient and energy contents of the litterfall of a riverine forest of central Amazonia. *Tropical Ecology* 20: 211–224.

Franken, M., U. Irmler, and H. Klinge. 1979. Litterfall in inundation, riverine and terra firme forests of central Amazonia. *Tropical Ecology* 20: 225–235.

Franken, W., and P. R. Leopoldo. 1984. Hydrology of catchment areas of Central-Amazonian streams, pp. 501–519. In H. Sioli, ed., *The Amazon: Limnology and Landscape Ecology of a Mighty Tropical River and Its Basin*. W. Junk, Dordrecht.

Franklin, J. F., and D. S. DeBell. 1988. Thirty-six years of tree population change in an old-growth *Pseudotsuga-Tsuga* forest. *Canadian Journal of Forest Research* 18: 633–639.

Ganzhorn, J. U. 1992. Leaf chemistry and the biomass of folivorous primates in tropical forests: tests of a hypothesis. *Oecologia* 91: 540–547.

Gentry, A. H. 1988. Tree species richness of upper Amazon forests. *Proceedings of the National Academy of Sciences, USA* 85: 156–159.

Gentry, A. H. 1991. The distribution and evolution of climbing plants, pp. 3–42. In F. E. Putz and H. A. Mooney, eds., 1991. *The Biology of Vines*. Cambridge University Press, Cambridge.

Gholz, H. L. 1982. Environmental limits on above-ground net primary production, leaf area, and biomass in vegetation zones of the Pacific northwest. *Ecology* 63: 469–481.

Givnish, T. J. 1984. Leaf and canopy adaptations in tropical forests, pp. 51–84. In E. Medina, H. A. Mooney, and C. Vasquez-Yanez, eds., *Physiological Ecology of Plants of the Wet Tropics*. W. Junk, The Hague.

Givnish, T. J. 1988. Adaptation to sun and shade: a whole-plant perspective. *Australian Journal of Plant Physiology* 15: 63–92.

Glynn, P. W. (ed). 1990. *Global Ecological Consequences of the 1982–83 El Nino-Southern Oscillation*. Elsevier, Amsterdam.

Gosz, J. R., G. E. Likens, and F. H. Bormann. 1972. Nutrient content of litter fall on the Hubbard Brook Experimental Forest, New Hampshire. *Ecology* 53: 769–784.

Gower, S. T., R. E. McMurtrie, and D. Murty. 1996. Aboveground net primary production decline with stand age: potential cause. *Trends in Ecology and Evolution* 11: 378–383.

Grace, J., Y. Malhi, J. Lloyd, J. McIntyre, A. C. Miranda, P. Meir, and H. S. Miranda. 1996. The use of eddy covariance to infer the net carbon dioxide uptake of Brazilian rain forest. *Global Change Biology* 2: 209–217.

Grier, C. C., and R. S. Logan. 1977. Old-growth *Pseudotsuga menziesii* communities on a western Oregon watershed: biomass distribution and production budgets. *Ecological Monographs* 47: 373–400.

Grier, C. C., K. A. Vogt, M. R. Keyes, and R. L. Edmonds. 1981. Biomass distribution of above- and below-ground production in young and mature *Abies amabilis* zone ecosystems of the Washington Cascades. *Canadian Journal of Forest Research* 11: 155–167.

Grove, T. S., and N. Malajczuk. 1985. Biomass production by trees and understorey shrubs in an age-series of *Eucalyptus diversicolor* F. Muell. stands. *Forest Ecology and Management* 11: 59–74.

Haldane, J. B. S. 1930. *Enzymes*. Longmans Green, London.

Harrington, R. A., J. H. Fownes, F. C. Meinzer, and P. G. Scowcroft. 1995. Forest growth along a rainfall gradient in Hawaii: *Acacia koa* stand structure, productivity, foliar nutrients, and water- and nutrient-use efficiencies. *Oecologia* 102: 277–284.

Heaney, A., and J. Proctor. 1989. Chemical elements in litter in forests on Volcán Barva, Costa Rica, pp. 255–271. In J. Proctor, ed., *Mineral Nutrients in Tropical Forests and Savannas*. Blackwell Scientific Publications, Oxford.

Hegarty, E. E. 1991. Leaf litter production by lianes and trees in a subtropical Australian rain forest. *Journal of Tropical Ecology* 7: 201–204.

Hladik, A. 1974. Importance des lianes dans la production foliare de la forêt équatoriale du Nord-Est du Gabon. *Comptes Rendus de l'Academie des Sciences, Paris* Série D 278: 2527–2530.

Hladik, A. 1978. Phenology of leaf production in rain forest of Gabon: distribution and composition of food for folivores, pp. 51–71. In G. G. Montgomery, ed., *The Ecology of Arboreal Folivores*. Smithsonian Institution Press, Washington, DC.

Hladik, A., and P. Blanc. 1987. Croissance des plantes en sous-bois de forêt dense humide (Makokou, Gabon). *La Terre et la Vie* 42: 209–234.

Holbrook, N. M., M. W. Denny, and M. A. R. Koehl. 1991. Intertidal "trees": consequences of aggregation on the mechanical and photosynthetic properties of sea-palms *Postelsia palmaeformis* Ruprecht. *Journal of Experimental Marine Biology and Ecology* 146: 39–67.

Hollinger, D. Y. 1989. Canopy organization and foliage photosynthetic capacity in a broad-leaved evergreen montane forest. *Functional Ecology* 3: 53–62.

Hubbell, S. P., and R. B. Foster. 1990. Structure, dynamics, and equilibrium status of old-growth forest on Barro Colorado Island, pp. 522–541. In A. H. Gentry, ed., *Four Neotropical Rainforests*. Yale University Press, New Haven, CT.

Hutchison, B. A., and D. R. Matt. 1977. The distribution of solar radiation within a deciduous forest. *Ecological Monographs* 47: 185–207.

Hutchison, B. A., D. R. Matt, R. T. McMillen, L. J. Gross, S. J. Tajchman, and J. M. Norman. 1986. The architecture of a deciduous forest canopy in eastern Tennessee, USA. *Journal of Ecology* 74: 635–646.

Huttel, C. 1975. Recherches sur l'écosystème de la forêt sub-équatoriale de basse Côte-d'Ivoire III. Inventaire et structure de la végétation ligneuse. *La Terre et la Vie* 29: 178–191.

Janos, D. P. 1980. Mycorrhizae influence tropical succession. *Biotropica* 12 (supplement): 56–64.

Janos, D. P. 1983. Tropical mycorrhizas, nutrient cycles and plant growth, pp. 327–345. In S. L. Sutton, T. C. Whitmore, and A. C. Chadwick, eds., *Tropical Rain Forest: Ecology and Management*. Blackwell Scientific Publications, Oxford.

Janzen, D. H. 1974. Tropical blackwater rivers, animals, and mast fruiting by the Dipterocarpaceae. *Biotropica* 6: 69–103.

John, D. M. 1973. Accumulation and decay of litter and net production of forest in tropical West Africa. *Oikos* 24: 430–435.

Johnson, P. L., and D. M. Atwood. 1970. Aerial sensing and photographic study of the El Verde rain forest, pp. B-63–B-78. In H. T. Odum and R. Pigeon, eds., *A Tropical Rain Forest*. Division of Technical Information, U. S. Atomic Energy Commission, Springfield, VA.

Jordan, C. F., and C. Uhl. 1978. Biomass of a "tierra firme" forest of the Amazon basin. *Oecologia Plantarum* 13: 387–400.

Kato, R., Y. Tadaki, and H. Ogawa. 1978. Plant biomass and growth increment studies in Pasoh Forest. *Malayan Nature Journal* 30: 211–224.

Keller, M., D. A. Clark, D. B. Clark, A. M. Weitz, and E. Veldkamp. 1996. If a tree falls in the forest ... *Science* 273: 201.

Kelliher, F. M., B. M. M. Köstner, D. Y. Hollinger, J. N. Byers, J. E. Hunt, T. M. McSeveny, R. Meserth, P. L. Weir, and E.-D. Schulze. 1992. Evaporation, xylem flow, and tree transpiration in a New Zealand broad-leaved forest. *Agricultural and Forest Meteorology* 62: 53–73.

Keyes, M. R., and C. C. Grier. 1981. Above- and below-ground net production in 40-year-old Douglas-fir stands on low and high productivity sites. *Canadian Journal of Forest Research* 11: 599–605.

King, D. A. 1990. The adaptive significance of tree height. *American Naturalist* 135: 809–828.

King, D. A. 1993. A model analysis of the influence of root and foliage allocation on forest production and competition between trees. *Tree Physiology* 12: 119–135.

King, D. A. 1994. Influence of light level on the growth and morphology of saplings in a Panamanian forest. *American Journal of Botany* 81: 948–957.

Kira, T. 1978. Community structure and organic matter dynamics in tropical lowland rain forests of Southeast Asia with special reference to Pasoh Forest, West Malaysia, pp. 561–590. In P. B. Tomlinson and M. H. Zimmermann, eds., *Tropical Trees as Living Systems*. Cambridge University Press, Cambridge.

Klinge, H. 1977. Fine litter production and nutrient return to the soil in three natural forest stands of eastern Amazonia. *Geo-Eco-Trop* 1: 159–167.

Klinge, H., and R. Herrera. 1983. Phytomass structure of natural plant communities on spodosols in southern Venezuela: the tall Amazon Caatinga forest. *Vegetatio* 53: 65–84.

Klinge, H., and W. A. Rodrigues. 1968. Litter production in an area of Amazonian terra firme forest. Part I. Litter-fall, organic carbon and total nitrogen contents of litter. *Amazoniana* 1: 287–302.

Kummerow, J., J. Castillanos, M. Maas, and A. Larigauderie. 1990. Production of fine roots and the seasonality of their growth in a Mexican deciduous dry forest. *Vegetatio* 90: 73–80.

Kunkel-Westphal, I., and P. Kunkel. 1979. Litter fall in a Guatemalan primary forest, with details of leaf-shedding by some common tree species. *Journal of Ecology* 67: 665–686.

Kursar, T. A. 1989. Evaluation of soil respiration and soil CO_2 concentration in a lowland moist forest in Panama. *Plant and Soil* 113: 21–29.

Lang, G. E., and R. T. T. Forman. 1978. Detrital dynamics in a mature oak forest: Hutcheson Memorial Forest, New Jersey. *Ecology* 59: 580–595.

Lawton, R. O. 1982. Wind stress and elfin stature in a montane rain forest tree: an adaptive explanation. *American Journal of Botany* 69: 1224–1230.

Leigh, E. G., Jr. 1990. Tree shape and leaf arrangement: a quantitative comparison of montane forests, with emphasis on Malaysia and south India, pp. 119–174. In J. C. Daniel and J. S. Serrao, eds., *Conservation in Developing Countries: Problems and Prospects*. Bombay Natural History Society and Oxford University Press, Bombay.

Leigh, E. G., Jr. 1996. Epilogue: research on Barro Colorado Island, 1980–94, pp. 469–503. In E. G. Leigh, Jr., A. S. Rand, and D. M. Windsor, eds., *The Ecology of a Tropical Forest* (2nd ed.). Smithsonian Institution Press, Washington, DC.

Leigh, E. G., Jr., R. T. Paine, J. F. Quinn, and T. H. Suchanek. 1987. Wave energy and intertidal productivity. *Proceedings of the National Academy of Sciences, USA* 84: 1314–1318.

Leigh, E. G., Jr., and N. Smythe. 1978. Leaf production, leaf consumption, and the regulation of folivory on Barro Colorado Island, pp. 33–50. In G. G. Montgomery, ed., *The Ecology of Arboreal Folivores*. Smithsonian Institution Press, Washington, DC.

Leigh, E. G., Jr., and D. M. Windsor. 1982. Forest production and regulation of primary consumers on Barro Colorado Island, pp. 111–122. In E. G. Leigh, Jr., A. S. Rand, and D. M. Windsor, eds., *The Ecology of a Tropical Forest*. Smithsonian Institution Press, Washington, DC.

Leigh, E. G., Jr., D. M. Windsor, A. S. Rand, and R. B. Foster. 1990. The impact of the "El Niño" drought of 1982–83 on a Panamanian semideciduous forest, pp. 473–486. In P. W. Glynn, ed., *Global Ecological Consequences of the 1982–83 El Niño-Southern Oscillation*. Elsevier, Amsterdam.

Leigh, E. G., Jr., and S. J. Wright. 1990. Barro Colorado Island and tropical biology, pp. 28–47. In A. H. Gentry, ed., *Four Neotropical Rainforests*. Yale University Press, New Haven, CT.

Lescure, J. P., H. Puig, B. Riera, D. Leclerc, A. Beekman, and A. Beneteau. 1983. La phytomasse épigée d'une forêt dense en Guyane française. *Acta Oecologica: Oecologia Generalis* 4: 237–251.

Lieberman, D., and M. Lieberman. 1987. Forest tree growth and dynamics at La Selva, Costa Rica (1969–1982). *Journal of Tropical Ecology* 3: 347–358.

Lieberman, M., and D. Lieberman. 1994. Patterns of density and dispersion in forest trees, pp. 106–119. In L. A. McDade, K. S. Bawa, H. A. Hespenheide, and G. S. Hartshorn, eds., *La Selva: Ecology and Natural History of a Tropical Rain Forest*. University of Chicago Press, Chicago.

Lim, M. T. 1978. Litterfall and mineral nutrient content of litter in Pasoh Forest Reserve. *Malayan Nature Journal* 30: 375–380.

Lloyd, J., J. Grace, A. C. Miranda, P. Meir, S. C. Wong et al. 1995. A simple calibrated model of Amazon rainforest productivity based on leaf biochemical properties. *Plant, Cell and Environment* 18: 1129–1145.

Lovelock, C. E., D. Kyllo, and K. Winter. 1996. Growth responses to vesicular-arbuscular mycorrhizae and elevated CO_2 in seedlings of a tropical tree, *Beilschmeidia pendula*. *Functional Ecology* 10: 662–667.

Lowman, M. D. 1992. Herbivory in Australian rain forests, with particular reference to the canopies of *Doryphora sassafras* (Monimiaceae). *Biotropica* 24: 263–272.

Lugo, A. E., and P. G. Murphy. 1986. Nutrient dynamics of a Puerto Rican subtropical dry forest. *Journal of Tropical Ecology* 2: 55–72.

Luizão, F. J., and H. O. R. Schubart. 1987. Litter production and decomposition in a terra-firme forest of Central Amazonia. *Experientia* 43: 259–265.

McWilliam, A.-L. C., J. M. Roberts, O. M. R. Cabral, M. V. B. R. Leitao, A. C. L. de Costa, G. T. Maitelli, and C. A. G. P. Zamparoni. 1993. Leaf area index and above-ground biomass of *terra firme* rain forest and adjacent clearings in Amazonia. *Functional Ecology* 7: 310–317.

Manokaran, N., A. R. Kassim, A. Hassan, E. S. Quah, and P. F. Chong. 1992. Short-term population dynamics of dipterocarp trees in a lowland rain forest in peninsular Malaysia. *Journal of Tropical Forest Science* 5: 97–112.

Manokaran, N., and K. M. Kochummen. 1987. Recruitment, growth and mortality of tree species in a lowland dipterocarp forest in Peninsular Malaysia. *Journal of Tropical Ecology* 3: 315–330.

Marks, P. L. 1974. The role of pin cherry (*Prunus pensylvanica* L.) in the maintenance of stability in northern hardwood ecosystems. *Ecological Monographs* 44: 73–88.

Marshall, J. D., and R. H. Waring. 1986. Comparison of methods for estimating leaf-area index in old-growth Douglas-fir. *Ecology* 67: 975–979.

Martínez-Yrízar, A., and J. Sarukhán. 1990. Litterfall patterns in a tropical deciduous forest in Mexico over a five-year period. *Journal of Tropical Ecology* 6: 433–444.

Martínez-Yrízar, A., J. M. Maass, L. A. Perez-Jiminez, and J. Sarukhan. 1996. Net primary productivity of a tropical deciduous forest ecosystem in western Mexico. *Journal of Tropical Ecology* 12: 169–175.

Medina, E., and E. Cuevas. 1989. Patterns of nutrient accumulation and release in Amazonian forests of the upper Rio Negro basin, pp. 217–240. In J. Proctor, ed., *Mineral Nutrients in Tropical Forest and Savanna Ecosystems*. Blackwell Scientific Publications, Oxford.

Medina, E., V. Garcia, and E. Cuevas. 1990. Sclerophylly and oligotrophic environments: relationships between leaf structure, mineral nutrient content, and drought resistance in tropical rain forests of the upper Rio Negro region. *Biotropica* 22: 51–64.

Medina, E., H. Klinge, C. Jordan, and R. Herrera. 1980. Soil respiration in Amazonian rain forests in the Rio Negro basin. *Flora* 170: 240–250.

Menéndez, L. 1988. Dinámica de la producción de hojarasca, pp. 213–242. In R. A. Herrera, L. Menéndez, M. E. Rodríguez, and E. E. García, eds., *Ecologia de los Bosques Siempreverdes de la Sierra del Rosario, Cuba*. Instituto de Ecologia y Sistematica, Academia de Ciencias en Cuba.

Menéndez, L., E. E. García, R. A. Herrera, M. E. Rodríguez, and J. A. Bastart. 1988. Estructura y productividad del bosque siempreverde medio de la Sierra del Rosario, pp. 151–212. In R. A. Herrera, L. Menéndez, M. E. Rodríguez, and E. E. García, eds., *Ecologia de los Bosques Siempreverdes de la Sierra del Rosario, Cuba.* Instituto de Ecologia y Sistematica, Academia de Ciencias en Cuba.

Mulkey, S. S. 1986. Photosynthetic acclimation and water-use efficiency of three species of understory herbaceous bamboo (Gramineae) in Panama. *Oecologia* 70: 514–519.

Mulkey, S. S., A. P. Smith, and S. J. Wright. 1991. Comparative life history and physiology of two understory Neotropical herbs. *Oecologia* 88: 263–273.

Müller, M. J. 1982. *Selected Climate Data for a Global Set of Standard Stations for Vegetation Science.* W. Junk, The Hague.

Murphy, P. G., and A. E. Lugo. 1986. Structure and biomass of a subtropical dry forest in Puerto Rico. *Biotropica* 18: 89–96.

Nadkarni, N. M., and T. J. Matelson. 1992a. Biomass and nutrient dynamics of fine litter of terrestrially rooted material in a neotropical montane forest, Costa Rica. *Biotropica* 24(2a): 113–120.

Nadkarni, N. M., and T. J. Matelson. 1992b. Biomass and nutrient dynamics of epiphytic litterfall in a neotropical montane forest, Costa Rica. *Biotropica* 24: 24–30.

Nepstad, D. C., C. R. de Carvalho, E. A. Davidson, P. H. Jipp, P. A. Lefebvre et al. 1994. The role of deep roots in the hydrological and carbon cycles of Amazonian forests and pastures. *Nature* 372: 666–669.

O'Connell, A. M., and P. M. A. Menagé. 1982. Litter fall and nutrient cycling in karri (*Eucalyptus diversicolor* F. Muell.) forest in relation to stand age. *Australian Journal of Ecology* 7: 49–62.

Odum, H. T. 1970. Summary: an emerging view of the ecological system at El Verde, p. I-191–I-289. In H. T. Odum and R. Pigeon, eds., *A Tropical Rain Forest.* Division of Technical Information, U. S. Atomic Energy Commission, Washington, DC.

Paine, R. T. 1979. Disaster, catastrophe, and local persistence of the sea palm *Postelsia palmaeformis. Science* 205: 685–687.

Parker, G. G. 1994. Soil fertility, nutrient acquisition and nutrient cycling, pp. 54–63. In L. A. McDade, K. S. Bawa, H. A. Hespenheide, and G. S. Hartshorn, eds., *La Selva: Ecology and Natural History of a Neotropical Rain Forest.* University of Chicago Press, Chicago.

Parker, G. G., J. P. O'Neill, and D. Higman. 1989. Vertical profile and canopy organization in a mixed deciduous forest. *Vegetatio* 85: 1–11.

Pascal, J. P. 1988. *Wet Evergreen Forests of the Western Ghats of India.* Institut Français de Pondichery, Pondicherry, India.

Pendry, C. A., and J. Proctor. 1996. The causes of altitudinal zonation of rain forests on Bukit Belalong, Brunei. *Journal of Ecology* 84: 407–418.

Penning de Vries, F. W. T., A. H. M. Brunsting, and H. H. van Laar. 1974. Products, requirements and efficiency of biosynthesis: a quantitative approach. *Journal of Theoretical Biology* 45: 339–377.

Perry, T. O., H. E. Sellers, and C. O. Blanchard. 1969. Estimation of photosynthetically active radiation under a forest canopy with chlorophyll extracts and from basal area measurements. *Ecology* 50: 39–44.

Proctor, J. 1983. Tropical forest litterfall. I. Problems of data comparison, pp. 267–273. In S. L. Sutton, T. C. Whitmore, and A. C. Chadwick, eds., *Tropical Rain Forest: Ecology and Management.* Blackwell Scientific Publications, Oxford.

Proctor, J., J. M. Anderson, P. Chai, and H. W. Vallack. 1983a. Ecological studies in four contrasting lowland rain forests in Gunung Mulu National Park, Sarawak I. Forest environment, structure and floristics. *Journal of Ecology* 71: 237–260.

Proctor, J., J. M. Anderson, S. C. L. Fogden, and H. W. Vallack. 1983b. Ecological studies in four contrasting lowland rain forests in Gunung Mulu National Park, Sarawak II. Litterfall, litter standing crop and preliminary observations on herbivory. *Journal of Ecology* 71: 261–283.

Proctor, J., Y. F. Lee, A. M. Langley, W. R. C. Munro, and T. Nelson. 1988. Ecological studies on Gunung Silam, a small ultrabasic mountain in Sabah, Malaysia. I. Environment, forest structure and floristics. *Journal of Ecology* 76: 320–340.

Proctor, J., C. Phillipps, G. K. Duff, A. Heaney, and F. M. Robertson. 1989. Ecological studies on Gunung Silam, a small ultrabasic mountain in Sabah, Malaysia. II. Some forest processes. *Journal of Ecology* 77: 317–331.

Puig, H., and H.-P. Delobelle. 1988. Production de litière, nécromasse, apports minéraux au sol par la litière en forêt guyanaise. *La Terre et la Vie* 43: 3–22.

Puig, H., and J. P. Lescure. 1981. Étude de la variabilité floristique dans la région de la piste de Saint-Elie. *Bulletin de liaison de groupe de travail sur l'écosystème forestier Guyanais (ECEREX)* 3: 26–29.

Putz, F. E. 1983. Liana biomass and leaf area of a "tierra firme" forest in the Rio Negro basin, Venezuela. *Biotropica* 15: 185–189.

Putz, F. E. 1984a. How trees avoid and shed lianas. *Biotropica* 16: 19–23.

Putz, F. E. 1984b. The natural history of lianas on Barro Colorado Island, Panama. *Ecology* 65: 1713–1724.

Putz, F. E., and S. Appanah. 1987. Buried seeds, newly dispersed seeds, and the dynamics of a lowland forest in Malaysia. *Biotropica* 19: 326–333.

Putz, F. E., and P. Chai. 1987. Ecological studies of lianas in Lambir National Park, Sarawak, Malaysia. *Journal of Ecology* 75: 523–531.

Putz, F. E., and H. T. Chan. 1986. Tree growth, dynamics and productivity in a mature mangrove forest in Malaysia. *Forest Ecology and Management* 17: 211–230.

Putz, F. E., and K. Milton. 1982. Tree mortality rates on Barro Colorado Island, pp. 95–100. In E. G. Leigh, Jr., A. S. Rand, and D. M. Windsor, eds., *The Ecology of a Tropical Forest.* Smithsonian Institution Press, Washington, DC.

Putz, F. E., and H. A. Mooney (eds). 1991. *The Biology of Vines.* Cambridge University Press, Cambridge.

Rai, S. N., and J. Proctor. 1986a. Ecological studies on four rainforests in Karnataka, India I. Environment, structure, floristics and biomass. *Journal of Ecology* 74: 439–454.

Rai, S. N., and J. Proctor. 1986b. Ecological studies on four rainforests in Karnataka, India II. Litterfall. *Journal of Ecology* 74: 455–463.

Rankin-de-Merona, J. M., R. W. Hutchings H., and T. E. Lovejoy. 1990. Tree mortality and recruitment over a five-year period in undisturbed upland rainforest of the Central Amazon, pp. 573–584. In A. H. Gentry, ed., *Four Neotropical Rainforests.* Yale University Press, New Haven, CT.

Reich, P. B., M. B. Walters, and D. S. Ellsworth. 1992. Leaf lifespan in relation to leaf, plant and stand characteristics among diverse ecosystems. *Ecological Monographs* 62: 365–392.

Rich, P. M., D. B. Clark, D. A. Clark, and S. F. Oberbauer. 1993. Long-term study of solar radiation regimes in a tropical wet forest using quantum sensors and hemispherical photography. *Agricultural and Forest Meteorology* 65: 107–127.

Roberts, J., O. M. R. Cabral, G. Fisch, L. C. B. Molion, C. J. Moore, and W. J. Shuttleworth. 1993. Transpiration from an Amazonian rainforest calculated from stomatal conductance measurements. *Agricultural and Forest Meteorology* 62: 175–196.

Ryan, M. G., S. T. Gower, R. M. Hubbard, R. H. Waring, H. L. Gholz, W. P. Cropper Jr., and S. W. Running. 1995. Woody tissue maintenance respiration of four conifers in contrasting climates. *Oecologia* 101: 133–140.

Ryan, M. G., R. M. Hubbard, D. A. Clark, and R. L. Sanford, Jr. 1994. Woody-tissue respiration for *Simarouba amara* and *Minquartia guianensis*, two tropical wet forest trees with different growth habits. *Oecologia* 100: 213–220.

Ryan, M. G., and R. H. Waring. 1992. Maintenance respiration and stand development in a subalpine lodgepole pine forest. *Ecology* 73: 2100–2108.

Saldarriaga, J. G., and R. J. Luxmoore. 1991. Solar energy conversion efficiencies during succession of a tropical rain forest in Amazonia. *Journal of Tropical Ecology* 7: 233–242.

Saldarriaga, J. G., D. C. West, M. L. Tharp, and C. Uhl. 1988. Long-term chronosequence of forest succession in the upper Rio Negro of Colombia and Venezuela. *Journal of Ecology* 76: 938–958.

Schlesinger, W. H. 1991. *Biogeochemistry: An Analysis of Global Change*. Academic Press, San Diego, CA.

Scott, D. A., J. Proctor, and J. Thompson. 1992. Ecological studies on a lowland evergreen rain forest on Maracá Island, Roraima, Brazil. II. Litter and nutrient cycling. *Journal of Ecology* 80: 705–717.

Silver, W. L. 1994. Is nutrient availability related to plant nutrient use in humid tropical forests? *Oecologia* 98: 336–343.

Singer, R., and I. Araujo. 1979. Litter decomposition and ectomycorrhizae in Amazonian forests. *Acta Amazonica* 9: 25–41.

Smith, A. P. 1987. Respuestas de hierbas del sotobosque tropical a claros ocasionados por la caída de árboles. *Revista de Biología Tropical* 35 (supplement): 111–118.

Smith, A. P., K. P. Hogan, and J. R. Idol. 1992. Spatial and temporal patterns of light and canopy structure in a lowland tropical moist forest. *Biotropica* 24: 503–511.

Snedaker, S. C. 1970. Ecological studies on tropical moist forest succession in eastern lowland Guatemala. PhD thesis, University of Florida, Gainesville.

Soepadmo, E. 1972. Progress report on IBP-PT Project at Pasoh, Negri Sembilan, Malaysia (1970–1972). Unpublished mimeographed report, School of Biological Sciences, University of Malaysia, Kuala Lumpur.

Soepadmo, E. 1973. *IBP-PT—Pasoh Project, Negri Sembilan, Malaysia, Annual Report for 1972*. Unpublished mimeographed report, School of Biological Sciences, University of Malaysia, Kuala Lumpur.

Soepadmo, E. 1974. *IBP-PT—Pasoh Project, Negri Sembilan, Malaysia, Annual Report for 1973*. Unpublished mimeographed report, School of Biological Sciences, University of Malaysia, Kuala Lumpur.

Songwe, N. C., F. E. Fasehun, and D. U. U. Okali. 1988. Litterfall and productivity in a tropical rain forest, Southern Bakundu Forest Reserve, Cameroon. *Journal of Tropical Ecology* 4: 25–37.

Spain, A. V. 1984. Litterfall and the standing crop of litter in three tropical Australian rainforests. *Journal of Ecology* 72: 947–961.

Sponsel, L. E., and P. C. Loya. 1993. Rivers of hunger? Indigenous resource management in the oligotrophic ecosystems of the Rio Negro, Venezuela, pp. 435–446. In C. M. Hladik, A. Hladik, O. F. Linares, H. Pagezy, A. Semple, and M. Hadley, eds., *Tropical Forests, People and Food*. UNESCO, Paris.

Stewart, G. H. 1989. The dynamics of old-growth *Pseudotsuga* forests in the western Cascade Range, Oregon, USA. *Vegetatio* 82: 79–94.

Stohlgren, T. J. 1988. Litter dynamics in two Sierran mixed conifer forests. *Canadian Journal of Forest Research* 18: 1127–1135.

Swaine, M. D., J. B. Hall, and I. J. Alexander. 1987. Tree population dynamics at Kade, Ghana (1968–1982). *Journal of Tropical Ecology* 3: 331–345.

Tanner, E. V. J. 1977. Four montane rain forests of Jamaica: a quantitative characterization of the floristics, the soils and the foliar mineral levels, and a discussion of the interrelations. *Journal of Ecology* 65: 883–918.

Tanner, E. V. J. 1980a. Litterfall in montane rain forests of Jamaica and its relation to climate. *Journal of Ecology* 68: 833–848.

Tanner, E. V. J. 1980b. Studies on the biomass and productivity in a series of montane rain forests in Jamaica. 68: 573–588.

Thompson, J., J. Proctor, V. Viana, W. Milliken, J. A. Ratter, and D. A. Scott. 1992. Ecological studies on a lowland evergreen rain forest on Maracá Island, Roraima, Brazil. I. Physical environment, forest structure and leaf chemistry. *Journal of Ecology* 80: 689–703.

Tilman, D. 1988. *Plant Strategies and the Dynamics and Structure of Plant Communities*. Princeton University Press, Princeton, NJ.

Trumbore, S. E., E. A. Davidson, P. B. de Camargo, D. C. Nepstad, and L. A. Martinelli. 1995. Belowground cycling of carbon in forests and pastures of Eastern Amazonia. *Global Biogeochemical Cycles* 9: 515–528.

Uhl, C., K. Clark, N. Dezzeo, and P. Maquirino. 1988. Vegetation dynamics in Amazonian treefall gaps. *Ecology* 69: 751–763.

Uhl, C., and P. G. Murphy. 1981. Composition, structure and regeneration of a tierra firme forest in the Amazon basin of Venezuela. *Tropical Ecology* 22: 219–237.

van Noordwijk, M., and K. Hairiah. 1986. Mycorrhizal infection in relation to soil pH and soil phosphorus content in a rain forest of northern Sumatra. *Plant and Soil* 96: 299–302.

van Schaik, C. P., and E. Mirmanto. 1985. Spatial variation in the structure and litterfall of a Sumatran rain forest. *Biotropica* 17: 196–205.

Veneklaas, E. J. 1991. Litterfall and nutrient fluxes in two montane tropical rain forests, Colombia. *Journal of Tropical Ecology* 7: 319–336.

Vogt, K. A., C. C. Grier, C. E. Meier, and R. L. Edmonds. 1982. Mycorrhizal role in net primary production and nutrient cycling in *Abies amabilis* ecosystems in western Washington. *Ecology* 63: 370–380.

Waring, R. H., and W. H. Schlesinger. 1985. *Forest Ecosystems*. Academic Press, Orlando, FL.

Waterman, P. G., J. A. M. Ross, E. L. Bennett, and A. Glyn Davies. 1988. A comparison of the floristics and leaf chemistry of the tree flora in two Malaysian rain forests and the influence of leaf chemistry on populations of colobine monkeys in the Old World. *Biological Journal of the Linnean Society* 34: 1–32.

Weaver, P. L., E. Medina, D. Pool, K. Dugger, J. Gonzalez-Liboy, and E. Cuevas. 1986. Ecological observations in the dwarf cloud forest of the Luquillo Mountains in Puerto Rico. *Biotropica* 18: 79–85.

Weaver, P. L., and P. G. Murphy. 1990. Forest structure and productivity in Puerto Rico's Luquillo Mountains. *Biotropica* 22: 69–82.

Westman, W. E., and R. H. Whittaker. 1975. The pygmy forest region of northern California: studies on biomass and primary productivity. *Journal of Ecology* 63: 493–520.

Whitmore, T. C., N. D. Brown, M. D. Swaine, D. Kennedy, C. I. Goodwin-Bailey, and W.-K. Gong. 1993. Use of hemispherical photographs in forest ecology: measurement of gap size and radiation totals in a Bornean tropical rain forest. *Journal of Tropical Ecology* 9: 131–151.

Whittaker, R. H. 1975. *Communities and Ecosystems*. Macmillan, New York.

Whittaker, R. H., F. H. Bormann, G. E. Likens, and T. G. Siccama. 1974. The Hubbard Brook ecosystem study: forest biomass and production. *Ecological Monographs* 44: 233–254.

Whittaker, R. H., and W. A. Niering. 1975. Vegetation of the Santa Catalina Mountains, Arizona. V. Biomass, production and diversity along the elevation gradient. *Ecology* 56: 771–790.

Whittaker, R. H., and G. M. Woodwell. 1969. Structure, production and diversity of the oak-pine forest at Brookhaven, New York. *Journal of Ecology* 57: 155–174.

Williams, W. A., R. S. Loomis, and P. de T. Alvim. 1972. Environments of evergreen rain forests on the lower Rio Negro, Brazil. *Tropical Ecology* 13: 65–78.

Windsor, D. M. 1990. Climate and moisture variability in a tropical forest: long-term records from Barro Colorado Island, Panama. *Smithsonian Contributions to the Earth Sciences* 29: 1–145.

Wing, S. R., and M. R. Patterson. 1993. Effects of wave-induced lightflecks in the intertidal zone on photosynthesis in the macroalgae *Postelsia palmaeformis* and *Hedophyllum sessile* (Phaeophyceae). *Marine Biology* 116: 519–525.

Wofsy, S, C., M. L. Goulden, J. W. Munger, S.-M. Fan, P. S. Bakwin, B. C. Daube, S. L. Bassow, and F. A. Bazzaz. 1993. Net exchange of CO_2 in a mid-latitude forest. *Science* 263: 1314–1316.

Yamada, I. 1975. Forest ecological studies of the montane forest of Mt. Pangrango, West Java. I. Stratification and floristic composition of the montane rain forest near Cibodas. *South East Asian Studies* 13: 402–426.

Yamada, I. 1976a. Forest ecological studies of the montane forest of Mt. Pangrango, West Java. II. Stratification and floristic composition of the forest vegetation of the higher part of Mt. Pangrango. *South East Asian Studies* 13: 513–534.

Yamada, I. 1976b. Forest ecological studies of the montane forest of Mt. Pangrango, West Java. III. Litter fall of the tropical montane forest near Cibodas. *South East Asian Studies* 14: 194–229.

Yamakura, T., A. Hagihara, S. Sukardjo, and H. Ogawa. 1986a. Aboveground biomass of tropical rain forest stands in Indonesian Borneo. *Vegetatio* 68: 71–82.

Yamakura, T., A. Hagihara, S. Sukardjo, and H. Ogawa. 1986b. Tree size in a mature dipterocarp forest stand in Sebulu, east Kalimantan, Indonesia. *Southeast Asian Studies* 23: 452–478.

Yoda, K. 1974. Three-dimensional distribution of light intensity in a tropical rain forest of west Malaysia. *Japanese Journal of Ecology* 24: 247–254.

Zotz, G., and K. Winter. 1993. Short-term photosynthesis measurements predict carbon balance in tropical rain-forest canopy plants. *Planta* 191: 409–412.

Zotz, G., and K. Winter. 1994. Predicting annual carbon balance from leaf nitrogen. *Naturwissenschaften* 81: 449.

SEVEN

The Seasonal Rhythm of Fruiting and Leaf Flush and the Regulation of Animal Populations

How much of the forest's production is consumed by herbivores? What prevents herbivores from eating more?

The most conspicuous herbivores are vertebrates. In Panama many of these herbivores eat leaves (especially young leaves), which are sugar factories the plants depend on; other herbivores eat seeds and seedlings. In the Neotropics, consumption of twigs, bark, and the like is less than in the United States, or in Madagascar or New Zealand in the days of elephant birds and moas. To what extent the restricted diets of modern vertebrate herbivores of Neotropical forests is an artifact of the human presence is not yet known.

The principal food of most Neotropical species of vertebrate herbivores is fruit. Most of this fruit is meant to be eaten (in moderation) by animals that will carry the seeds in that fruit away from their parent plant. In Panama some trees have evolved to take advantage of the agoutis that eat their seeds; these trees now depend on agoutis to bury their seeds (thus protecting them from insect pests) and forget enough of them to assure the trees a posterity (Smythe 1978, 1989). Such trees, and many others, need frugivores, but an overabundance of frugivores, hungry enough to eat all the seeds, is counterproductive. Motivated by this belief, the first studies of vertebrate population regulation on Barro Colorado showed how forest trees collectively "control" frugivore populations by seasonal shortage of fruit and new leaves.

How effective a defense against vertebrate frugivores and folivores is this seasonal shortage? Does the forest need the help of predators such as big cats to protect it from vertebrate herbivores? What other defenses does the forest employ against vertebrates?

Vertebrate herbivores are conspicuous. Nonetheless, insects eat more foliage and destroy far more seeds than vertebrates do. To defend young leaves against insects, plants have developed an extraordinary diversity of chemical compounds (Rosenthal and Berenbaum 1991, 1992). Nonetheless, as we shall see, the forest needs insectivorous birds to help shield it from insect attack.

THE SEASONAL RHYTHM OF THE FOREST'S FOOD PRODUCTION

On Barro Colorado, there is an overall seasonal rhythm in the availability of fruit and new leaves (Foster 1982b). Late in the rainy season, when these foods are scarce, animals are hungrier, range farther, take more risks to find food, and are more readily trapped (Smythe 1978). Among vertebrate herbivores, reproductive rates are higher and mortality lower in the season when fruit and new leaves are abundant (Leigh et al. 1982, Adler and Beatty 1997). Do the seasonal rhythms of fruit and new leaves limit animal populations? Is the forest somehow organized to create a seasonal shortage of food that limits its herbivore populations?

These questions cannot be answered in a word. Here, they will be approached in four stages. First, I describe Barro Colorado's seasonal rhythm of leaf flush and fruit fall. This overall rhythm is the composite of the different seasonal rhythms of a whole array of plant species. The second stage is to ask what governs the phenology, that is to say, the timing of leaf flush, flowering, and fruiting, in different plant species. How might natural selection adjust the timing of these activities? To deal with these issues, I consider the effects on different plants of irrigating whole hectares of forest for several successive dry seasons, and ask whether knowledge of the "hydraulic architecture" of different plants will help us understand their responses to irrigation. The third stage is to assess how much food the forest supplies, how many herbivores exploit this bounty, and how much food they need. Finally, I consider the evidence that seasonal shortage of food limits herbivore populations.

Leaf Flush

The forest on Barro Colorado has two major peaks of leaf flush (Leigh and Smythe 1978, Leigh and Windsor 1982). Many evergreen canopy trees flush new leaves near the beginning of the new year, as the dry season begins. Deciduous trees, and many evergreens, flush new leaves at, or somewhat before, the beginning of the rainy season. Sometimes there is a secondary peak of leaf flush in August and September. In 1976, new leaves were scarce in the first week of March, in late October, and in late November (Table 7.1). This rhythm is echoed by understory plants (Table 7.2), which flush most of their leaves either in December or January, after the rains end, in May, when the rains return, and to a lesser extent, in September (Aide 1988, 1993).

Like many other understory plants (Aide 1993), the understory shrub *Psychotria marginata* has two peaks in leaf flush, one at the beginning and one at the end of rainy season. Leaves produced for the dry season have higher weight per unit area than their rainy season counterparts; this presumably allows more mesophyll per unit area, which may explain why the dry season leaves have twice the water use efficiency, that is to say, fix twice as much CO_2 per mole of water transpired, as their rainy season counterparts (Mulkey et al. 1992). Leaves produced for the rainy season have higher stomatal conductance (i.e., exchange CO_2 and water vapor more readily at a given vapor pressure deficit between leaf and atmosphere), so that CO_2 level in the leaf is closer to that in the atmosphere, allowing more efficient photosynthesis when light is low, and more efficient use of sunflecks (Mulkey et al. 1992). Thus this shrub produces leaves suited for each season.

Many canopy trees do likewise. Wright used a canopy crane to follow the seasonal rhythm of leaf flush and leaf fall in nine species of canopy trees and lianas in a park near Panama City (Wright and Colley 1995). Eight of these flushed leaves in April and May, as rainy season began, and in November and December, shortly before the advent of dry season. Kitajima et al. (1997) compared the physiology of May and December leaves for five of these eight species. In two pioneer species, *Cecropia* and *Urera*, with very short-lived leaves, and in one canopy species, *Luehea*, and one midstory species, *Antirrhoea*, with a leaf life span of about 6 months, December leaves had greater dry weight per unit area and therefore higher photosynthetic capacity per unit area, as if to take advantage of the dry season's greater abundance of light. Moreover, December leaves of *Luehea* and *Antirrhoea* had higher water use efficiency, that is to say, fixed more CO_2 per mole of water transpired, than their May counterparts.

Wright's ninth species, *Anacardium excelsum*, replaces all its leaves at the beginning of the dry season. It continues to flush new leaves as the dry season progresses and solar radiation increases, until, by the time the rains return, it has increased its leaf area by two-thirds. *Anacardium* then gradually sheds leaves as the rainy season progresses, decreasing its leaf area by a third before the next dry season leaf exchange. Leaves produced in January are long lived, soon shaded by newer leaves, and do not differ physiologically from April leaves (Kitajima et al. 1997).

Several factors influence the timing of leaf flush. For many species, the increased availability of water at the beginning of the rains is crucial. Plants like *Anacardium excelsum*, whose transpiration rate per unit leaf area is the same in the dry as in the rainy season (Meinzer et al. 1993), and which can be presumed to have reliable access to water all through the year, have maximum leaf area at the end of dry season, presumably because light is most abundant then (cf. Wright and van Schaik 1994). Herbivores also influence the timing of leaf flush. Herbivorous insects are less numerous (Smythe 1982) and inflict less damage (Aide 1992, 1993) during the dry season. Thus many species in seasonal habitats flush leaves before the rains come (Rockwood 1974: p. 143; Murali and Sukumar 1993). Plants can also reduce herbivore damage by flushing leaves when many other species are doing so, thereby "overloading" the available herbivores (Darwin 1859: p. 70; Coley 1982, Aide 1988, 1993). Plants that flush "out of turn" during the rainy season suffer particularly from herbivores (Rockwood 1974, Aide 1992).

On Barro Colorado, it appears least advantageous to flush leaves in October and November: at that season sunlight is decreasing, and cloudiness is increasing, and insect pests are active and hungry.

Flowering and Fruiting

According to Foster (1973, 1982b), Barro Colorado has a single peak season of flowering and a single peak season of seedling germination, both near the beginning of the rainy season, whereas there are two peak seasons of fruit fall. One fruiting peak, involving both animal- and wind-dispersed seeds, lasts from March through June, straddling the beginning of the rainy season. The second peak, involving only animal-dispersed seeds, lasts from August into October.

The single peak in flowering (Foster 1982b) and seed fall (Garwood 1982, 1983) are generally agreed upon. The double peak in fruit fall is more controversial. Whether one sees it or not depends on how one analyzes, and where (and perhaps in what year) one collects data. Foster (1982b) set out thirteen 1/12-m² tubs on each of 24 ha on Barro Colorado's central plateau, collected their contents each week and identified all the seeds among them. The double peak shows most clearly in the weekly record of the average number of species of falling seeds caught per hectare

Table 7.1 The seasonal rhythm of leaf flush on Barro Colorado

Week	No. of deciduous species flushing - Many leaves	No. of deciduous species flushing - Few leaves	Rainfall (mm)	No. of evergreen species flushing - Many leaves	No. of evergreen species flushing - Few leaves	Week	No. of deciduous species flushing - Many leaves	No. of deciduous species flushing - Few leaves	Rainfall (mm)	No. of evergreen species flushing - Many leaves	No. of evergreen species flushing - Few leaves
1 Jan.	0.5	2	5	3	8	27	4	8	20	6	11
2	1	0.5	5	8	9	28	3	8	22	3	15
3	1	1	10	10	9	29	3	6	7	2	13
4	1	0.5	15	7	9	30	3	4	42	3	12
5 Feb.	0.5	0.5	1	4	10	31 Aug.	3	6	95	5	12
6	2	1	0	5	10	32	3	7	2	7	9
7	1	0.5	0	4	8	33	2	4	41	7	8
8	1	1	0	3	11	34	2	7	40	7	11
9 Mar.	0	4	8	1	8	35 Sep.	2	5	54	4	11
10	3	2	0	3	8	36	0	6	40	3	12
11	2	4	5	4	15	37	0.5	4	105	2	13
12	5	3	3	6	9	38	0.5	4	108	5	6
13 Apr.	5	2	0	3	8	39 Oct.	0	3	131	4	6
14	5	2	0	3	15	40	0	2	31	1	12
15	6	1	0	3	10	41	0	2	47	0	8
16	8	6	0	7	11	42	0	2	52	0	9
17 May	9	6	10	9	13	43	0	2	99	0	9
18	8	4	24	7	13	44 Nov.	0	2	95	0	10
19	11	5	19	6	11	45	0	1	116	1	11
20	8	5	27	7	9	46	0	1	50	0	10
21	6	7	45	5	9	47	0.5	0	14	0	6
22 Jun.	3	8	81	6	9	48 Dec.	0.5	0	1	1	6
23	3	9	79	6	8	49	0	0	3	2	7
24	4	9	16	6	10	50	0	0	0	1	10
25	6	6	58	5	14	51	0	0.5	26	6	8
26 Jul.	5	8	97	4	11	52	0.5	0	3	6	7

Number of species flushing a large number or fewer new leaves, week by week, in 1976 (Leigh and Windsor 1982). A species half of whose trees are flushing new leaves that week is counted as 0.5.

(Foster 1982b: fig. 3) for the year beginning August 1969. The double peak also appears in the total number of species of seed caught per week on the plateau as a whole (Table 7.3), although the second peak sometimes (as in 1994) fails to appear in the number of species of seed caught per week from animal-dispersed trees and shrubs. On the other hand, Smythe's (1970) record of the total weight of fruit falling each month into 170 m² of traps near the laboratory clearing for the 17 months beginning September 1966 showed a minor peak in February, a single major peak from May through August, and a prolonged minimum from November through January. The number of species contributing each month to this fruit catch was lowest in November and December and showed a mild peak in May (Smythe 1970). Forget et al. (1994) found that in 1990, the proportion of fallen fruit dropped by the palm *Scheelea zonensis* which was gnawed by mammals progressively increased from almost none on 5 August to > 80% in mid-September. The increased proportion of fruit that was gnawed was thought to reflect the decline in forestwide fruit availability from August onward which Smythe (1970) observed in 1967.

On Barro Colorado, it pays seeds to germinate at the beginning of the rainy season—as long as they are not lured into premature germination by a dry season rain—because seedlings then have the most time possible to root deeply before the next dry season (Garwood 1982, 1983). Although seeds that fall late in the rainy season delay germination until the beginning of the next rainy season (Garwood 1982, 1983), the long wait exposes them to seed-eaters (Schupp

Table 7.2 Proportion of total annual production, P, of new leaves, and percent consumed, C, among understory plants, month by month

	Jan.	Feb.	Mar.	Apr.	May	Jun.	Jul.	Aug.	Sep.	Oct.	Nov.	Dec.
P	0.162	0.074	0.077	0.045	0.221	0.065	0.020	0.022	0.092	0.092	0.050	0.083
C	25.8	20.5	22.0	24.0	29.2	25.4	35.0	25.9	30.3	23.6	33.4	29.9

Data are for 1987 (T. M. Aide, p. c. Figure 1 of Aide (1993) is based on these numbers.

Table 7.3 Average number of species of all kinds, S, and of animal-dispersed trees and shrubs, A, per week dropping seed into traps on Barro Colorado's plateau in successive fortnights

	Jan.																									Dec.
	1	2	3	4	5	6	7	8	9	10	11	12	13	14	15	16	17	18	19	20	21	22	23	24	25	26
S, 1992	27	23	32	47	50	52	57	60	52	42	33	38	34	23	23	24	21	25	32	37	37	38	25	26	25	22
A, 1992	14	11	15	19	17	15	16	20	20	24	17	16	16	8	12	14	13	17	18	20	20	22	16	15	15	12
S, 1993	26	23	32	39	40	43	46	48	53	40	39	29	29	22	16	16	17	17	28	23	23	19	18	16	15	13
A, 1993	14	13	18	19	20	21	18	19	21	20	20	15	17	15	9	11	10	12	17	14	14	11	10	9	8	9
S, 1994	15	21	16	26	32	34	40	49	49	42	43	33	33	34	25	26	30	30	28	27	28	31	24	22	23	22
A, 1994	10	15	11	13	13	10	13	18	18	21	22	18	20	22	16	15	16	16	16	15	15	16	13	14	13	14
S, 1995	25	26	33	37	52	48	58	54	52	49	40	34	34	31	21	23	25	23	25	31	31	31	20	19	21	21
A, 1995	12	17	19	19	25	21	26	25	22	22	17	14	18	18	14	14	13	14	14	17	16	16	9	7	12	10

Data provided by S. Paton on behalf of the Smithsonian Environmental Sciences Program.

1990). Other things being equal, plants do best to ripen their fruits and disperse their seeds soon after the rains begin. Lianes, most of which depend on the wind for seed dispersal, disperse their seeds in the dry season, when many trees are leafless and the wind is still strong (Appendix 7.1).

It also pays many species to flower early in the rainy season because the onset of rainy season is a cue that can ensure a reasonable degree of synchrony in flowering among the members of a species (Foster 1982b). Small insect pollinators are also most abundant at the beginning of the rainy season. Their abundance may in part reflect the flowering peak at that time, but small insects, which are susceptible to dehydration, are less common and less active (Aide 1993) during the dry season.

Seasonal fruiting rhythms help coopt frugivores as dispersers. In central Panama, many large seeds escape insect attack only if buried by agoutis (Smythe 1989, Forget and Milleron 1991, Leigh et al. 1993). Smythe (1970) suggested that trees with large seeds, attractive to agoutis, should fruit in synchrony because when fruit is abundant, agoutis bury surplus seeds, scatter-hoarding them over their territories against future shortage. And indeed, on Barro Colorado, agoutis bury a larger proportion of the seeds of the emergent leguminous tree *Dipteryx panamensis* when these seeds are abundant, and are more likely to forget buried seeds when the forest has an abundant supply of fruit (Forget 1993).

On the other hand, if every species both flowered and fruited at the beginning of the rainy season, there would not be enough pollinators and seed dispersers to meet the demand (Foster 1982b), especially since there would be no food to maintain these animals during the rest of the year. Related plant species sometimes flower (Appanah 1993) or fruit (Snow 1976: pp. 80–81) at different times as if to avoid competing for pollinators or dispersers. Larger pollinators, such as hummingbirds, bats, and bees, are readily available in the dry season. On Barro Colorado, many understory shrubs and treelets, which cannot afford to make energy-rich fruit, bear their fruit during the season of food scarcity (Appendix 7.1), when hungry frugivores are willing to settle for "second best." Many plants, both on Barro Colorado (Foster 1982b) and in Costa Rica (Levey 1988), fruit in September or October, apparently so as to take advantage of the wave of fall migrants passing through at that time from the temperate zone.

THE TIMING OF LEAF FLUSH, FLOWERING, AND FRUITING

On Barro Colorado, the alternation of dry and rainy seasons is a crucial part of life. It is logical to expect that the seasonal activities of many plants would be timed by the onset of the dry or rainy season.

This view was reinforced by the effects on the forest of the aberrant dry season of 1970. That year, the dry season was very wet, and many plants that normally flower in response to the onset of rainy season either failed to flower that year or flowered without bearing fruit. Those of us who watched, stunned, as 28 cm of rain fell in the second week of January 1970, ending that dry season almost before it was born (for the leaf litter of the forest floor was never to crackle under our feet again that year, as it normally does during dry season) had no idea of the horror that was to follow. The extended food shortage resulting from the failure of the September fruiting peak caused large-scale starvation among the island's mammals. The vultures could not cope with the abundance of corpses, and the forest stank (Foster 1982a).

Important species of food trees which normally flower in response to the onset of rainy season, such as *Trichilia cipo*, *Quararibea asterolepis*, *Dipteryx panamensis*, *Chrysophyllum* spp., and *Faramea occidentalis*, have also failed to fruit in other years when the dry season was uncommonly wet or when the onset of rainy season was uncommonly indistinct, such as 1931, 1956, 1958, and 1993: these were also years of hunger in the forest (Foster 1982a, S. J. Wright, p. c.).

On the other hand, a severe dry season, like that of 1992, normally stimulates fruit production. Even the severe El

Niño dry season of 1982–1983 was not very detrimental to vertebrate frugivores (Leigh et al. 1990). Disruption of the normal cycle of the seasons is much more devastating to Barro Colorado's animal community than the exaggeration of the normal seasonal cycle by an El Niño dry season.

In many species, the trees that bear the most fruit one year bear the least in the following year. The tendency of an El Niño year to precede or follow a year with a wet dry season (Windsor 1990) sometimes creates a resonance that can alternate feast and famine in several successive years. This interplay of factors complicates the distinction between cause and effect but does not negate the importance of the cyclic alternation of the seasons for timing flowering and fruit production on Barro Colorado.

How might the alternation between rainy and dry season affect the seasonal rhythms of different plant species? On Barro Colorado, many species flower in response to a dry season rain following a sufficiently long or severe dry spell. For example, the understory shrub *Hybanthus prunifolius* flowers in synchrony a week or 10 days after the first dry season rain of 12 mm or more following a sufficiently long dry spell (Augspurger 1982). The rows of white "snapdragons" hanging from the curved, leaning branches of the many shrubs of this species during a mass flowering is one of the striking sights of this island. Why must they flower in synchrony? A contiguous group of shrubs must come into synchronous flower to attract pollinating bees down from the canopy. The *Hybanthus* population must fruit in synchrony to satiate a predispersal seed predator (Augspurger 1981, 1982). Moreover, *Hybanthus* flushes leaves after flowering; a plant that flushes leaves out of turn during the rainy season suffers particularly from herbivores (Aide 1992). A *Hybanthus* which has not yet flowered in a given dry season can be made to flower by watering the ground at its base if the watering is preceded by a sufficiently long and severe dry spell. Similarly, flowering and leaf flush can be delayed by covering the ground surrounding the stem by plastic, preventing dry season rains from wetting the soil (Aide 1992).

Hybanthus prunifolius is just one of many plants that flower after dry season rains. The most spectacular among these species is *Tabebuia guayacan*. Over the course of a dry season in central Panama, there are several (traditionally, three) waves of flowering, each following a dry season rain. During each wave of flowering, forested slopes are dotted by the bright yellow crowns of "guayacans" in flower.

Tabebuia neochrysantha, a Costa Rican counterpart of the guayacan, flowers and flushes leaves in response to changes in soil moisture content (Reich and Borchert 1982). Reich and Borchert could bring trees into flower by watering them after a sufficiently long dry spell: if sufficient water was supplied, leaf flush followed flowering. They also found that the timing of leaf fall, flowering, and leaf flush was associated with minute changes in trunk diameter, which were assumed (in the short term) to represent gain or loss of water. When dry season began, the soil began to dry out, and trees began to shrink and lose their leaves. At wetter sites, the reduced transpiration occasioned by leaf loss allowed trees to rehydrate enough to trigger flowering, as also happens in *Erythrina poeppigiana* (Borchert 1980). At drier sites, rehydration, trunk expansion, and flowering only occurred after a sufficiently heavy dry season rain. If enough rain fell, trees flushed leaves after flowering. At the driest sites, only the return of rainy season could cause rehydration sufficient to stimulate flowering and leaf flush (Reich and Borchert 1982). Reich and Borchert (1984) extended these principles to interpret the relation of the timing of leaf fall, flowering, leaf flush, and changes in trunk diameter to the seasonal alternation of rainy and dry season and the incidence of occasional dry season rains in different species of tree in a Costa Rican dry forest. Reich and Borchert (1984) ascribed differences among trees and species to differences in their microhabitats and in the depths to which they sank their roots. It seemed as if a full understanding of the factors governing plant phenology was just around the corner.

A surprise was in store, however. To assess the effect of unseasonable rain on the timing of leaf flush and fruit production, Wright (1991) arranged for the watering of two 2.25-ha plots on Barro Colorado's Poacher's Peninsula. These plots were supplied with 6 mm of water per day, 5 days a week, through five successive dry seasons, beginning in 1986. Soil water potential at 25 cm depth never fell below -0.04 MPa in irrigated plots, although by the end of the dry season it was often below -1.0 MPa on control plots (Tissue and Wright 1995). Wright compared the phenological rhythms of plants on irrigated plots with those of conspecifics on nonirrigated control plots nearby. The scientific community of Barro Colorado was shocked when Wright found that, for most species of canopy tree, irrigation had no effect on the timing of leaf fall, leaf flush, flowering, or fruiting: *Tabebuia guayacan* was one of the few exceptions (Wright and Cornejo 1990a,b).

As the experiment continued, however, it became clear that irrigation was affecting the timing, though not the amount, of leaf, flower, and fruit production in understory shrubs, *Psychotria* spp. and *Piper* spp. (Wright 1991, Tissue and Wright 1995). On unirrigated control plots, leaf flush in *Psychotria furcata*, *P. horizontalis*, and *P. marginata* peaked sharply at the beginning of the rainy season. In all these species, irrigation progressively reduced synchrony of leaf flush both among branches on the same plant and among different plants. In *P. furcata*, the species whose leaf production is normally most nearly confined to the beginning of rainy season, peak leaf production on irrigated plots occurred slightly earlier, as well as being slightly more blurred, each year, as if leaf flush were governed by a circannual clock which is reset each year on the control plots by the onset of rainy season (Wright 1991). *P. furcata* also flowers in synchrony at the beginning of each rainy season. Surprisingly, in 1987 (Wright 1991: fig. 5) and 1990 (Tissue and Wright 1995: fig. 3) flowering in this species was as synchronous on irrigated as on control plots, as if flowering and leaf flush were controlled differently.

Why the contrast in effect on canopy versus understory plants? Sternberg et al. (1989) provided a clue. The more CO_2 is available within a leaf, the more strongly the enzymes involved in photosynthesis discriminate against ^{13}C in favor of ^{12}C, and the higher the ratio of ^{12}C to ^{13}C in the leaf's dry matter. When a leaf's stomates are open, the concentration of CO_2 in the leaf is higher (Mulkey et al. 1992), and the discrimination in favor of ^{12}C is stronger. Sternberg et al. (1989) analyzed the isotopic ratio of ^{13}C to ^{12}C in three species of tree on Wright's plots. Irrigation did not significantly affect the ratio of ^{13}C to ^{12}C in canopy leaves of any of these species, but in 1-m saplings from two of them, the ratio of ^{13}C to ^{12}C was significantly higher on control plots (Sternberg et al. 1989: fig. 3). Irrigation appears to have enhanced the availability of water for small saplings, allowing them to keep their stomates open longer, but it seems to have had no effect on the access of canopy trees to water, presumably because the latter are more deeply rooted.

Indeed, many canopy trees behave as if they always have access to adequate water. Leaves of the 47-m tall *Ceiba pentandra* behind the old laboratory had the same water use efficiency in the dry season as in the middle of rainy season, about 3 mmol CO_2 fixed for every mole of water transpired. As daily photosynthesis was higher in the dry season, leaves were actually using more water then (Zotz and Winter 1994b). On the average, these leaves fixed 200 mmol, about 8.8 g, CO_2/m^2 of leaf surface · day, releasing 75 mol (1354 g) of water in the process, transpiring 154 g of water for every gram of CO_2 fixed. During daytime hours, this *Ceiba* thus transpired an average of 31 mg of water/ m^2 of leaf surface · sec. As previously mentioned, an *Anacardium excelsum* nearly doubles its leaf area during the course of the dry season, carrying maximum leaf area at the driest time of year. The average daily transpiration per square meter of its leaves is the same in the dry as in the rainy season (Meinzer et al. 1993), about 18 mg/sec. Why these trees do not drive down the water table, as farmers do in semiarid habitats of the American West, is something of a mystery.

If canopy trees are responding to climate, and the moisture content of the soil above the water table is irrelevant, could the trees be responding to changes in the vapor pressure deficit of the atmosphere? How can a tree be structured to do so? One way to approach this question is to compare the "hydraulic architecture"—the pressure differentials required to draw water through the roots, stems, branches, petioles, and leaves—of different species. This study is the subject of a specific subculture of plant physiology, which will now be outlined.

Studies of the hydraulic architecture of plants are based on one salient fact, one fundamental question, and one central concern. The salient fact is that, when water evaporates from the mesophyll of leaves, it creates a pull that draws more water up from the soil, through the roots and the narrow vessels of the xylem (Zimmermann 1983). This pull can be powerful. In the mesophyll, water evaporates from pores about 2.5 nm (2.5×10^{-9} m) in radius at their narrowest point (Vogel 1988: p. 239; Tyree et al. 1994a: p. 340). Surface tension, T, in water at 25°C is 0.072 newtons/m, or 0.072 Pa · m. If r is the radius of curvature of the meniscus (the surface separating air and water), the force surface tension exerts to "straighten out the meniscus" creates a pull of $2T/r$ Pa when expressed as a force per unit cross-sectional area of meniscus (Thompson 1942: p. 367). Since the minimum value of r in a pore of radius 2.5 nm is likewise 2.5 nm, the maximum pull per unit cross-sectional area is 59 MPa (Tyree et al. 1994a: p. 341). Thus water in the xylem vessels is under tension, which is usually measured as a (negative) water potential or negative pressure. As water must be pulled from the soil, water potential in the plant must be lower than that in the soil surrounding its roots, and water potential in the plant must decline in the direction of flow (Tyree and Ewers 1991).

Zimmermann et al. (1993, 1994) called this "salient fact," the "cohesion theory of sap ascent," into question. They asserted that water as impure as that in xylem cavitates at potentials less negative than −1 MPa and cited measurements with a "xylem pressure probe" of xylem tension near the top of the 35-m *Anacardium excelsum* by the tower crane in Panama (Meinzer et al. 1993) to show that, in that tree at least, xylem tension was too weak to pull water so high against gravity. Holbrook et al. (1995) spun twigs in a centrifuge, with a single leaf at the axis of rotation, to create xylem tension experimentally. They found that xylem sap can support high tension without cavitating. Moreover, measurements of the xylem tension thus created in the leaf, using the Scholander "pressure bomb" to assess the pressure that must be applied to the leaf to counterbalance xylem tension and force water from the leaf through its petiole's cut tip, agreed with the xylem tension predicted from the length of the twig and the speed at which it was spun. Similarly, Pockman et al. (1995) found that the xylem tensions required to produce cavitation as calculated from the size and angular velocity of a spun stem were the same as those inferred from pressure-bomb measurements. Although these experiments have not settled *all* reasonable doubts, I proceed as if the cohesion theory is indeed a salient fact.

The fundamental question is: Suppose a tree is transpiring $E(t)$ mg of water/m^2 of leaves · sec at time t. Given the water potential of the soil surrounding this tree's roots as a function of time, how does the water potential at various heights in the stem, and in branches, petioles and leaves, vary with time (Tyree 1988, Tyree and Ewers 1991)? Consider, for example, a 20-m *Schefflera morototoni*, each square meter of whose leaves are transpiring 60 mg of water/sec. What pressure differential is required to draw water from the roots, or from the root collar, to the leaves, and how is this differential reflected in water potential of various parts of the plant? This question is relevant to understanding the factors that govern the seasonal rhythms of plant phenology if, as Reich and Borchert (1982, 1984) suggest, leaf fall, flowering, and leaf flush are triggered by appropriate changes in the plant's water potential.

The pressure differential (MPa) per meter of stem, dP/dx, required to pull w kg of water/sec through this stem, is governed by the stem's conductance k_h: $k_h = w/(dP/dx)$. The leaf specific conductance, LSC, is the conductance per square meter of leaves supplied:

$$\text{LSC} = k_h/LA = (w/LA)/(dP/dx) = E/(dP/dx). \quad (7.1)$$

This equation applies because the transpiration rate per square meter of leaves is the rate, w, at which water flows through the stem divided by the leaf area, LA, through which this water transpires. The *Schefflera morototoni* studied by Tyree et al. (1991) has uncommonly high LSC, 2–20 times higher than those of the other species of trees and lianes so far studied (Tyree and Ewers 1991, Zotz et al. 1994). Transpiring 60 mg of water/m^2 of leaves · sec creates an above-ground pressure gradient from the root collar to the petioles of this *Schefflera* of 12.5 kPa/m, not much above the 9.8 kPa/m required to counterbalance gravity (Tyree et al. 1991).

Above-ground hydraulic architecture has been studied in detail for selected plants. Roots have proved more of a challenge (Tyree et al. 1994b). Half the resistance to the flow of water from soil to stomates appears to occur below ground and half above ground (Tyree et al. 1995). Above-ground resistance appears to be concentrated in the petioles and leaves (Tyree et al. 1991, Yang and Tyree 1994).

The central concern in the study of hydraulic architecture is the threat of cavitation and the means by which plants palliate or circumvent this threat. If air is sucked into a xylem vessel containing water under tension, the water column in this vessel snaps (cavitates), generating an ultrasonic "acoustic emission" which appropriate equipment allows one to record (Tyree and Sperry 1989a). When the water column cavitates, the vessel fills with air and no longer conducts water unless, and until, water can be pushed into it under positive pressure (Sperry et al. 1987, Cochard et al. 1994). For any given plant, cavitation is more likely the lower its water potential. The sensitivity of a plant to cavitation is measured by the water potential that causes enough cavitation to reduce stem conductance by half.

How do plants avoid cavitation or reduce its impact? This is an urgent question because in most plants xylem tensions are close to levels that would cause "autocalytic" cavitation, where failure of some vessels increases xylem tension in others, creating a "falling domino" effect (Tyree and Sperry 1988, 1989b; Cochard et al. 1994). The impact of cavitation can be reduced either by ensuring that the least essential plant parts fail first—leaves and petioles, which can be readily replaced, or those branches which contribute least to height growth (Zimmermann 1983, Tyree and Ewers 1991)—or that the most easily refilled vessels, those in or just above the roots, cavitate first (Saliendra et al. 1995, Alder et al. 1996). Thus above-ground resistance tends to be concentrated in the petioles and leaves (Tyree et al. 1991, 1993).

Plants employ an extraordinary variety of means to protect vessels in their trunks from cavitation. A water potential of −1.4 MPa suffices to reduce stem conductance of *Schefflera morototoni* by half (Tyree et al. 1991). During the dry season, water potential in the top 25 cm of the soil sometimes falls this low (Wright 1991). *Schefflera* avoids cavitation by rooting deeply, thus securing relatively reliable access to water. Moreover, *Schefflera morototoni* stores extractable water in its stem for emergencies. Tyree et al.'s (1991) study tree stored 70 kg of extractable water in its 462 l of wood, enough so that, with stomates closed, the water lost by its 26 m^2 of leaves during 2 weeks of drought can be replaced without causing cavitation. A water potential of −0.95 MPa suffices to reduce stem conductance of *Pseudobombax septenatum* by half (Machado and Tyree 1994). Machado and Tyree's study tree, however, stored only 2.5 kg of extractable water in its 131 l of wood, not much reserve for its 23 m^2 of leaves. Unlike the evergreen *Schefflera*, *Pseudobombax* drops its leaves about the new year and flushes new leaves in May. Does embolism in the root collar (whose vessels easily refill when the rains return) protect the upper trunk from being embolized by the drying of the soil?

In contrast, the vessels of the shallow-rooted climbing bamboo *Rhipidocladum racemiflorum* are so constructed that a water potential of −4.2 MPa is required to halve its stem conductance (Cochard et al. 1994). This bamboo has many counterparts. Stems of the equally shallow-rooted *Psychotria horizontalis* lose only 30% of their conductance at −5 MPa (Zotz et al. 1994); for all its shallow roots, numbers of this species of shrub increased during the census interval that included the savage El Niño drought of 1982–1983, and its mortality rate was lower during this interval than during the succeeding "normal" census interval (Condit et al. 1995). The mangrove *Rhizophora mangle* must overcome the osmotic potential of seawater (2.2 MPa) to draw pure water from its salt mud: only at −6.2 MPa do stems of this species lose half their conductance (Sperry et al. 1988).

Let us now return to our original question: Can the study of hydraulic architecture provide any clues as to how atmospheric changes can dictate phenological rhythms, regardless of changes in soil moisture availability? Crassulacean acid metabolism (CAM) is a process in which stomates open at night to take up CO_2, when less water is lost per mole of CO_2 acquired, and store the CO_2 in an acid until it is used for daytime photosynthesis (Winter 1985). CAM is considered a water-saving process. Yet, on Barro Colorado, a lakeside *Clusia uvitana* whose roots have year-round access to water fixes nearly as much CO_2 by CAM as does a hemiepiphytic *Clusia* seated at the base of the crown of the 47-m *Ceiba pentandra* behind Barro Colorado's old laboratory (Zotz and Winter 1994a). Moreover, these two *Clusia* show the same seasonal rhythm of CAM use, as if CAM were governed by an *atmospheric* signal. How can this be?

To find out, Zotz et al. (1994) analyzed the hydraulic architecture of these *Clusia*. *Clusia uvitana* cannot support severe xylem tension: cavitation halves stem conductivity at only -1.3 MPa. Like *Schefflera*, these *Clusia* need reliable access to water. On the other hand, the leaf specific conductivity of *Clusia uvitana* is 0.2 (g/sec) · (m stem/MPa)/m^2 leaf = 0.2 g/m · Mpa · sec, 1/15 the value for *Schefflera*. *Clusia* stems are nearly as conductive as *Schefflera* stems of the same diameter, but a *Clusia* branch of basal diameter 4.5 cm carries 17.5 m^2 of leaves, whereas a *Schefflera* branch of the same diameter carries about 3.4 m^2. In both the lakeside and the epiphytic *Clusia*, transpiration peaks at 8 A.M. and 4 P.M., as if the low leaf specific conductivity could not supply enough water to allow transpiration at midday, so that the stomata had to close to prevent cavitation. Apparently, low leaf specific conductivity makes CAM advantageous, even for the lakeside *Clusia*. Thus hydraulic architecture can affect a plant's response to water stress. It provides a potential means by which plants might adjust the timing of flowering and leaf flush. Hydraulic architecture could even "decide" whether a plant responds to atmospheric changes, as do these *Clusia*, or to changes in soil moisture content, as do *Tabebuia* spp. and *Hybanthus prunifolius*. What must yet be discovered is, what other means, if any, do tropical plants use to adjust their phenological rhythms?

WHAT THE FOREST MAKES AND WHO EATS IT

What the Forest Makes

In Barro Colorado's Lutz Catchment, somewhat more than 6 tons dry weight of leaves fell per hectare per year during the 4 years beginning 1 December 1973 (Leigh and Windsor 1982). Burning (or fully metabolizing) 1 g dry weight of leaves (or fruit) releases 19 kJ of energy (Nagy and Milton 1979). Recalling that 1 W is 1 J/sec, this leaf fall represents a power supply to the forest floor averaging 3600 W/ha, 0.36 W/m^2. Counting what animals ate before the leaves fell, this represents a leaf production of at least 7 or 8 tons dry weight per hectare per year. Fruit production is patchier and harder to measure, but forests on relatively fertile soil, such as Barro Colorado, and the Manu, in Amazonian Peru, drop about 2 tons fresh weight of fruit per hectare per year (Smythe 1970, Terborgh 1983), of which about 40% is dry matter (Smythe 1970).

What animals eat these leaves and fruit? How much do they eat? What limits their populations? What, and how much, do the carnivores and insectivores eat? Does the forest need these carnivores and insectivores to protect it from herbivores?

Biomass of the Rainforest and Its Animals

The dry weight of a forest's vegetable matter can be estimated by multiplying the forest's basal area by one half the height of its canopy to obtain the forest's above-ground wood volume, multiplying wood volume by 0.6 to get above-ground dry weight, and multiplying above-ground weight by 1.15 to include roots. If both forests have basal areas of 30 m^2/ha, mature forest 30 m tall on Barro Colorado contains about 300 tons dry weight of vegetable matter/ha, while 38 m tall forest north of Manaus (Fittkau and Klinge 1973) contains about 400 tons/ha.

Fittkau and Klinge (1973) estimated that (1) terra firme forest north of Manaus supports about 200 kg fresh weight of animals/ha (0.02% of the fresh weight of vegetable matter), (2) the total weight of animals living on leaves and fruit is 16 kg/ha (Fittkau and Klinge 1973: Fig. 2B), and (3) 165 kg/ha of animals inhabit the litter and soil, eating detritus or the fungi that decompose it or preying upon the animals that do. The biomasses of some animal groups have been estimated more carefully in other tropical forests. In the light of these other figures, Fittkau and Klinge's (1973) estimates appear rather crude, sometimes no better than guesses (Table 7.4). Nonetheless, there has been no subsequent attempt to estimate the biomass of the whole of a continental rainforest's fauna. However crude their data, Fittkau and Klinge's (1973) conclusions that far more animal activity is connected with the decomposition of dead plant matter than with the consumption of (pieces of) live plants and that the overwhelming majority of a tropical forest's animal biomass is contributed by insects and other arthropods must be correct. May we conclude that the forest is not limited by its herbivores?

REGULATION OF POPULATIONS OF VERTEBRATE FOLIVORES AND FRUGIVORES

The consumption of vegetable matter by vertebrate folivores can be assessed by estimating the population density of the different species of herbivores and multiplying the density of each species by its per capita consumption of vegetable matter. The latter can either be measured directly (perhaps by the "doubly-labeled water" technique of Nagy 1987), or calculated from standard regressions of feeding rate on body size (Nagy 1987).

Most of the mammals censused on Barro Colorado Island and at other tropical sites have been counted by some form of strip census (Glanz 1982). King's method is the most usual technique (Wright et al. 1994). King's estimate of the density, D, of a species of solitary mammal is $D = N/2LR$, where N is the number encountered, L is the total length of the census walk, and R is the average (species-specific) distance at which the animal is first sighted. For social animals, N is the number of *groups* encountered. For many species, such as howler monkeys, white-faced mon-

Table 7.4 Animal biomass (kg/ha) in selected tropical forests

Category	Manaus	Manu	BCI
All animals	200		
Mammals	8.4	13.8	48.0
Marsupials	2.0	0.7	0.6
Armadillos	0.5	?	2.0
Sloths	0.8	0	24.0
Anteaters	0.3	?	0.3
Monkeys	0.4	6.5	5.1
Tapir, deer, peccaries	0.4	4.4	4.1
Rodents	3.1	1.7	11.3
Carnivores	0.9	0.7	1.3
Birds	3.4	1.8	2.3
Amphibians and reptiles	3.4		
Soil Fauna	165		
Ants	34		
Termites	28		
Earthworms	10		
Others	84		
Other insects	27		
Stingless bees	4.7	?	0.5
Wasps (Vespidae)	4.7		
Lepidoptera	3.4		
Orthoptera	1.1		
Hemiptera	2.3		
Coleoptera	10.6		
Spiders	2.1		
Consumers of leaves and fruit	16.0		
Mammalian	6.7	13.1	42.0
Avian	1.8	1.3	0.8
Arthropod	7.5		

Data for site north of Manaus from Fittkau and Klinge (1973). Mammal data for Manu from Janson and Emmons (1990); bird data from Terborgh et al. (1990). Mammal data for BCI from Appendix 7.1; bird data from Willis (1980). For comparison with Fittkau and Klinge's insect figures, canopy fogging by John Tobin (p. c.) near BCI found 800 g dry weight of ants/ha of ground and 2 kg dry weight of other kinds of insects/ha of ground in tree crowns.

keys, and coati bands, population estimates from King's method square nicely with those from more thorough studies. However, King's method is unreliable for cryptic animals like sloths. It can also underestimate populations of animals that are not habituated to human activity. Using this method, Wright et al. (1994) estimated 56 pacas and 203 agoutis/km^2 near the laboratory clearing, and 8 pacas and 77 agoutis/km^2 in comparable habitat farther away. Squirrels and lone male coatis are also more readily seen near the laboratory clearing. Presumably, King's method also underestimates population sizes at other sites where animals are even less habituated (Wright et al. 1994).

Barro Colorado supports about 4.8 tons of nonflying mammals/km^2, of which 2.63 tons are edentates (Table 7.4). These mammals eat about 600 kg dry weight of food/ha · year (60 tons/km^2 · year, > 12 times the total fresh weight of their bodies), of which edentates account for 177 kg (Table 7.5). Barro Colorado's nearest well-studied counterpart is Peru's Parque Manu, in western Amazonia. The Manu also has a severe dry season (though less severe than BCI's), and a flora characteristic of fertile soil (Foster and Brokaw 1982), but it has many habitats BCI lacks (Terborgh 1983). Published censuses, which may underestimate actual populations of some species (Wright et al. 1994) indicate that the Manu supports only 1.36 tons of nonflying mammals/km^2, almost none of which are edentates. Based on these censuses and published regressions of feeding rate on mammalian body weight (Nagy 1987), the Manu's nonedentates eat 273 kg dry weight of food/ha · year, 20 times their body weight. Considering the errors involved in these estimates, their aggregate consumption is roughly the same as the 423 kg/ha · year consumed by BCI's nonedentates. Mammals weighing < 1 kg account for less of BCI's than of the Manu's nonedentate consumption: on BCI, mammals weighing > 1 kg, exclusive of cats and edentates, eat 380 kg dry weight of food/ha · year, compared to 200 at the Manu. Let us consider in detail what, and how much, these vertebrates eat.

How Much Do Different Mammals Eat?

Folivores

Based on information in chapter 2 and mammal censuses by Glanz (1982) and Wright et al. (1994), vertebrates on Barro Colorado eat less than 300 kg dry weight of leaves/ha · year (Table 7.5). There are two major uncertainties in Table 7.5. The first is the brocket deer, *Mazama americana*. This deer eats fruit and fallen leaves as well as fresh foliage, but in unknown proportions (J. Giacalone, p. c.). The second is the coendou, *Coendou rothschildii*. Glanz (1982) estimated a population density of 10 coendous/km^2, but Enders (1935) reported that several coendous were found in the crowns of the trees felled for Barro Colorado's laboratory clearing, suggesting that there could have been as many as 1 coendou/ha.

Barro Colorado's consumption of leaves by vertebrates, < 300 kg/ha · year, may be typical of sites in the northern neotropics, but leaf consumption by vertebrates appears far lower in the Manu (Table 7.6). Sloths and iguanas are rare in the Manu, and howler monkeys are much less common there (Janson and Emmons 1990).

Frugivores

Even on BCI, with its abundance of sloths, vertebrates eat more fruit than leaves, even though leaf production far surpasses fruit production. Judging from the numbers and weights of the animals involved and Nagy's (1987) regressions of feeding rate on body weight, fruit consumption by nonflying mammals (Table 7.6) and the numbers and kinds of these frugivores (Appendix 7.2) are far more similar on BCI and Manu than are leaf consumption and mammalian folivore populations.

Unless we are misled by the problems of censusing unhabituated animals in truly pristine areas (Wright et al. 1994), terrestrial mammals, especially agoutis and pacas,

Table 7.5 Abundance and food consumption by folivores on Barro Colorado Island

Species	N/ha	Wt (kg)	Wt/ha	F (kg/year)	F/ha	L/ha
Alouatta palliatta	0.8	5.5	4.4	110	88	35
Bradypus variegatus	5.0	3.0	20.0	22	110	110
Choloepus hoffmannii	1.0	4.0	4.0	22	22	22*
Mazama americana	0.03–0.2	20.0	≤4.0	282	≤60	?†
Tapirus bairdii	0.007	300.0	2.1	2020	14	≤14†
Coendou rothschildii	0.1–1.0	4.5	0.5–4.5	95	9–95	9–95†
Iguana iguana	0.7	2.0	1.4	5	7	7

Number per hectare (N/ha); weight (wt) per animal and weight per hectare; yearly food consumption per animal (F); food consumption by animals of that species per hectare (F/ha); and foliage consumption per hectare (L/ha). Information is from chapter 2, except *C. hoffmannii*, population density of which is given by Glanz (1982); *M. americana*, population density of which is from Wright et al. (1994); *Tapirus bairdii*, population density of which is from Terwilliger (1978); and *C. rothschildii*, population estimates of which are from Glanz (1982) and Enders (1935), respectively.
*Feeding rate assumed equal to *Bradypus*, as given by Nagy and Montgomery (1980).
†Feeding rate f (g dry weight/day) is calculated from the animal's fresh weight, x, by the regression $f = 0.577x^{0.727}$ (Nagy 1987: Table 3, regression for eutherian herbivores). $F = 0.365f$ kg/year.

the animals most preyed upon by big cats (Emmons 1987) eat far less fruit in Manu than on Barro Colorado (Table 7.6, Appendix 7.2). In contrast, arboreal mammals appear to eat somewhat more fruit in Manu (Table 7.6). Terrestrial birds eat more fruit in the Manu than on Barro Colorado, while in both places birds in tree crowns eat about 20 kg dry weight of fruit/ha · year (Table 7.6).

We know much less about how much bats eat. We are well informed, however, about *Artibeus jamaicensis*, the Jamaican fruit bat, which is overwhelmingly the most common fruit-eating bat on Barro Colorado Island (Bonaccorso 1979, Handley et al. 1991). Barro Colorado averages two *Artibeus jamaicensis* per hectare (Leigh and Handley 1991). A 50-g *Artibeus jamaicensis* eats more than its weight in fruit, mostly figs, each night: 70 g fresh weight, or 13 g dry weight, of fruit per night (Morrison 1980), or about 5 kg dry weight of fruit per year. If the two *Artibeus jamaicensis* eat half of all the fruit bats eat on Barro Colorado, that island's bats eat 20 kg dry weight of fruit/ha · year.

Restraints on Vertebrate Herbivory

What limits the populations of vertebrate folivores and frugivores? Several explanations have been suggested: predators, seasonal shortage of suitable food, and quality of the food available.

Do Predators Control Herbivore Populations?

As a whole, rainforest seems to be quite luxuriant, and one would hardly think it was herbivore limited. How could the forest be any greener and shadier in the absence of herbivores when, as things stand now, barely enough light reaches the ground for herbs of the forest floor to survive and reproduce (Smith 1987, Mulkey et al. 1991)? To be sure, Kitajima (1994) shed a new and curious light on this observation by showing that it is herbivores and pathogens, not lack of light per se, that kills light-demanding seedlings. Nonetheless, vertebrate herbivores consume only a small proportion of the foliage produced on Barro Colorado.

Should we conclude, with Hairston, Smith, and Slobodkin (1960), that predators prevent herbivores from limiting plant populations? On Isle Royale in Lake Superior, balsam firs (*Abies balsamea*) are slowly being grazed down by moose: they can only respond to favorable growing seasons when wolves depress the moose population (McLaren and Peterson 1994). Terborgh (1986, 1988) suggested that mammal populations are far higher on Barro Colorado than in Manu because Barro Colorado lacks the big cats—pumas and jaguars—and other large predators such as harpy eagles, which supposedly limit mammal populations in Manu. Terborgh inferred that Barro Colorado's swollen populations of mammalian herbivores exert a great effect on the composition of its vegetation.

It is of the utmost importance to examine Terborgh's argument. If predators defend trees from herbivores, trees can invest less in chemical and structural defenses of their own. Free defense from natural enemies is a precious gift: on the most exposed rocky shores of the northeastern Pacific, fast-growing sea palms maintain extraordinary levels of productivity where waves knock away their consumers and predators (Leigh et al. 1987). Before they were extirpated from this same region, sea otters protected subtidal kelps by eating the sea urchins and other herbivores that would otherwise destroy them. Australasia lacks a comparable top predator. Thus subtidal Australasian kelps invest much more in chemical defense. This has sparked an arms race: Australian herbivores are much more tolerant of antiherbivore chemicals in their food. Australasian kelps are more heavily eaten than subtidal kelps at northeastern Pacific sites with sea otters, but they are eaten far less rapidly than subtidal kelps of northeastern Pacific sites without sea otters (Steinberg et al. 1995).

Table 7.6 Consumption of leaves, fruit, and insects on Barro Colorado Island, Panama, and in the Parque Manu of Amazonian Peru (kg dry weight/ha · year)

	Barro Colorado	Manu
Fruit eaten in trees by		
Mammals	95	107
Birds	21	21
Bats	20?	?
Fruit eaten on ground by		
Mammals	260	80
Birds	6	15
Leaves eaten by mammals	250	35
Insects (excluding ants and termites) eaten in trees by		
Mammals	5	16
Birds	32	25
Insects (excluding ants and termites) eaten on ground by		
Mammals	10	1
Birds	1	5

Mammal data from Glanz (1982) and Wright et al. (1994) for BCI and Janson and Emmons (1990) for Manu; bird data from Willis (1980) for BCI and Terborgh et al. (1990) for Manu; feeding rates from Nagy (1987).

Terborgh's argument has two weaknesses. First, he asserts that populations of mammalian herbivores are much higher on Barro Colorado than at other sites where big cats are common. This proposition is doubtful (Glanz 1990, Wright et al. 1994). Terborgh also assumes that in Manu, big cats eat enough to limit populations of mammalian herbivores. Is this second assumption reasonable?

Appendix 7.2 suggests that large mammals, mammals weighing ≥ 1 kg, excluding edentates and cats, annually eat about 400 kg dry weight of fruit, leaves, and insects/ha on Barro Colorado, and about 200 kg/ha in Manu. If the proverb that it takes 10 pounds of feed to make a pound of beef applies to wild animals, then each year, about 40 kg dry weight of mammals/ha become available for harvest by BCI's carnivores, about 20 for Manu's.

Emmons (1987) estimated that a 7.5 km² area in Manu supported the equivalent of 1/5 of a 34-kg jaguar and 1/7 of an equal-sized puma. She believed that each animal ate about 550 kg of prey a year, so that jaguars and pumas, respectively, harvested 110 and 77 kg of vertebrates from the study area each year, a yearly total of 24 kg/km². Most of these prey were animals weighing >1 kg, but not all were mammals.

Emmons's study area also supported three ocelots weighing 8 kg each per 4 km². She attributed a daily feeding rate of 60–90 g/kg body weight to these ocelots and concluded that each year Manu's ocelots ate at least 131 kg/km², perhaps as much as 200 kg/km², of prey. Slightly more than half the food they ate consisted of animals weighing <1 kg (Emmons 1987: Fig. 2B). Thus Emmons (1987) suggested that Manu's cats annually eat 1–1.2 kg of prey weighing ≥ 1 kg each/ha, of which only 240 g is consumed by jaguars and pumas.

If, instead, we apply Nagy's (1987) formulae to the census figures in Janson and Emmons (1990), we would find that each year, Manu's big cats eat 26 kg/km² and its ocelots eat 130 kg/km dry weight of mammals. If half the ocelots' intake consists of animals weighing > 1 kg each, then Manu's cats annually eat about 1 kg of animals weighing ≥ 1 kg each/ha, about 1/20 of the 20 kg/ha theoretically available for harvest. All but 250 g of this is consumed by ocelots, which are just as abundant on Barro Colorado as at Manu (Glanz 1982). It seems hardly reasonable to assume that, eating as little as they do, jaguars and pumas can explain the rarity of agoutis, pacas, and coatis in Manu.

Have Manu's big cats "pushed its frugivores into the trees"? Even this appears doubtful. In the 1930s, Barro Colorado was being visited by at least four distinct puma, yet, then as now, agoutis and coatis were conspicuously common there (Enders 1935).

The ratios of the aggregate weight of Manu's big cats to that of their prey falls within the range of values published for the great African savannah ecosystems (Emmons 1987). Are lions equally unable to regulate their prey populations? It appears that these predators merely weed out the sick, the old, and the surplus young, thereby stabilizing the populations of their prey.

Are Herbivores Limited by Episodic Shortage of Suitable Food?

Sinclair (1975) showed that herbivores could be food-limited without limiting plant populations if plants produced edible materials, such as fruit and new leaves, only at certain seasons. And indeed, on Barro Colorado and in many, if not most, other rainforests, herbivore populations appear limited by seasonal shortage of fruit and new leaves. Sometimes predators or parasites apply the coup de grâce, as when an ocelot eats a young agouti which food shortage has forced to venture dangerously far out (Smythe

1978), or when screwworms fatally infect botfly wounds in howler monkeys weakened from lack of suitable food (Milton 1982); still, food shortage set the stage for these deaths.

Impact of Seasonal Food Shortage upon Herbivores On Barro Colorado, seasonal rhythms of fruiting and leaf flush play a crucial role in defending tropical forests against herbivores, as we have already seen. Many species of plant flush new leaves in synchrony at the beginning of the rainy season (Leigh and Smythe 1978, Coley 1982). Leaves produced later on, out of synchrony, by these same species, are far more heavily eaten (Coley 1982, Aide 1993). Similarly, fallen seeds face a higher risk of being eaten when fruit is in short supply. Seeds of the common understory treelet *Faramea occidentalis* fall in November and December, during the major season of food shortage. These seeds disappear rapidly until fruit abundance increases in March and April. Then disappearance rate drops sharply, rising again during the minor food shortage in June and July (Schupp 1990: pp. 507–508 and Fig. 5). It appears that selection by herbivores is strong enough to influence the seasonal rhythm of leaf flush and fruit fall on Barro Colorado. Driven by the advantage to their individuals, BCI's tree species have "conspired" to create a seasonal shortage of food that limits vertebrate herbivore populations.

Selection has also helped to shape other fruiting rhythms. In the lowland rainforests of Malaysia and Indonesia, where the distinction between dry and rainy season is less clearcut than on Barro Colorado, dipterocarps and many other species fruit gregariously in response to the end of a long drought or sunny spell. The fruiting is more abundant the longer and more intense the sunny spell and the longer the interval since the last gregarious fruiting (van Schaik 1986). The signal that initiates the flowering that leads to gregarious fruiting may be the nocturnal cooling that occurs when the skies are clear night after night (Ashton et al. 1988). Nevertheless, trees of different species, which flower in series, one building up pollinator populations for the next, fruit in synchrony (Chan 1980, Ashton et al. 1988), as if synchronous fruiting were favored by some factor other than the common flowering signal. And indeed, fruit that appears "out of turn" in these dipterocarp forests suffers far greater damage (Janzen 1974, Chan 1980).

The influence of herbivores on rhythms of fruiting and leaf flush reflects the obvious fact that herbivores are hungrier when less food is available. Squirrels and terrestrial frugivores of central Panama, like the prosimians of Gabon, are far harder to trap when fruit is abundant (Charles-Dominique 1971: p. 212, Smythe et al. 1982, Gliwicz 1984, Adler and Lambert 1996): bait loses its attractiveness when the forest is full of ripe fruit. Similarly, food placed in the forest when food is scarce is eagerly consumed, while food put out during a season of abundance is eaten far more sparingly, or utterly disdained (Charles-Dominique 1971: p. 212; Smythe 1978: p. 9). Terrestrial frugivores, especially agoutis (Smythe 1978) and coatis (Russell 1982) range farther and take more risks to find food in the lean season. Agoutis abandon their young earlier (Smythe 1978: p. 36), and are far less tolerant of conspecifics (Smythe 1978: pp. 10–11) during this time.

On Barro Colorado, too little fruit falls during the season of greatest food shortage to feed the terrestrial frugivores. The terrestrial frugivores of BCI, mammals and birds, require 265 kg dry weight of fruits and seeds/ha · year (Table 7.6), 22 kg dry weight/ha · month. In Lutz catchment, near where Smythe measured fruit fall, there is one adult 3-kg agouti/ha, and two adult 9-kg pacas/3 ha (Smythe et al. 1982). If there are half as many young as adults, and if a young weighs half as much as an adult, the agoutis and pacas of Lutz catchment alone require 17.5 kg dry weight, or 44 kg fresh weight, of food/ha · month. Remembering that much of what falls is inedible (husks and the like), too little fruit fell in December 1966 and December 1967 to feed the terrestrial frugivores of Lutz catchment (Table 7.7). Indeed, from November through January of 1966–1967 and 1967–1968, the supply of falling fruit must have been inadequate, and the terrestrial frugivores presumably fell back on other resources.

In Manu, terrestrial frugivores, mammals and birds, require at least 94 kg dry weight of fruits and seeds/ha · year (Table 7.6); this requirement is an underestimate if census methods have underestimated some populations of terrestrial mammals (Wright et al. 1994). Terrestrial mammals of Manu thus need at least 8 kg dry weight, or perhaps 20 kg fresh weight, per month. All which falls is not edible, so the May 1977 fruit fall may have been too little to feed Manu's terrestrial frugivores (Table 7.7). Manu's variety of habitats undoubtedly reduces the impact of the shortage for those animals capable of moving about, but it does not eliminate this impact (Terborgh 1983).

The terrestrial frugivores of Gabon have not been completely censused, so the quantitative insufficiency of the fruit production during the season of food shortage for its frugivore populations cannot yet be demonstrated. Nonetheless, Charles-Dominique (1971: p. 209), Gautier-Hion and Gautier (1979: p. 501, 507), Emmons (1980: p. 42) and Gautier-Hion and Michaloud (1989: p. 1827) have identified the major dry season of July and August, when fruit is in shortest supply, as a "critical period" for frugivores of Gabon.

Year-to-year variation in food supply also affects herbivore populations, as if they are indeed food limited. On Barro Colorado, in years when the September fruit crop is low, female coatis that have not yet bred delay breeding an extra year (Russell 1982). We have already noticed how disastrous the failure in 1970 of the September fruiting peak was for Barro Colorado's mammals.

Seasonal food shortage has shaped the demographic patterns of many of BCI's mammal populations. During the season of fruit shortage, death rates, especially of young, are higher among many animals, such as agoutis

Table 7.7 Fruit falling in successive months at different sites (kg fresh weight/ha)

Site	Mar.	Apr.	May	Jun.	Jul.	Aug.	Sep.	Oct.	Nov.	Dec.	Jan.	Feb.
Barro Colorado 1966–67							100	125	63	20	50	225
Barro Colorado 1967–68	50	100	330	580	312	200	125	100	50	40	50	
Cocha Cashu 1976–77												
Forest	115	68	107	236	72	130	119	76	29	53	65	75
Levee	154	147	175	383	182	287	239	186	8	149	187	66
Riverside	106	185	329	242	187	1021	276	128	69	37	39	44
Average	125	133	204	287	147	479	211	130	35	80	97	62

BCI measurements began September 1966 and represent the fall into 75 traps totaling nearly 175 m² of trap surface (Smythe 1970); Manu measurements began September 1976, but are listed as beginning in March (i.e., September is labeled March, October is labeled April, etc.) to harmonize seasons of BCI and Manu. Manu data are based on fifty 0.08 = m² traps in each forest type: 12 m² of trap surface in all (Terborgh 1983).

(Smythe 1978), pacas (Smythe et al. 1982), howler monkeys (Milton 1982, 1990), and sloths (Montgomery and Sunquist 1978: p. 344; Milton 1990). Births of some animals, such as coatis (Russell 1982), frugivorous bats (Morrison 1978, Bonaccorso 1979), and iguanas (Burghardt et al. 1977) are timed to occur when food is abundant. Birth rates of other animals, such as white-faced monkeys (Mitchell 1989), howler monkeys (Milton 1982), spiny rats (Gliwicz 1984), and squirrels (Glanz et al. 1982) are lower during the lean season.

Seasonal Niche Differentiation when Food Is Scarce Niches become more distinct during the season of food shortage. Terrestrial frugivores of Barro Colorado (Smythe 1978: pp. 45–46), which share in the abundance of fruit during the season of fatness, each specialize to a different diet during the lean season, as if to avoid competitive displacement when competition for food is most intense (Zaret and Rand 1971). When food is scarce, peccaries root and browse more, agoutis fall back on seeds which they buried on their territory in the season of abundance, pacas browse seedlings, and coatis search the leaf litter for arthropods, snails, and small vertebrates (Smythe 1978, Russell 1982).

Zaret and Rand (1971) reported decreased dietary overlap among stream fishes in central Panama in response to seasonal shortage of food. Since niches were most distinct when competition was most intense, this seasonal decrease in dietary overlap, this seasonal "behavioral character displacement," suggested to them that niche differentiation minimizes the likelihood of the competitive displacement of one species by another, as Hutchinson (1959) thought. The decrease of dietary overlap during the lean season has been a recurrent theme in tropical biology. Charles-Dominique (1971: p. 225) found that the diets of five species of prosimians of the rainforests near Makokou, Gabon, overlapped least during the major dry season, when fruit and insects were most scarce. Gautier-Hion and Gautier (1979) found that the diets of three species of guenon (*Cercopithecus* spp.) living in this same forest overlapped least in the lean season. Emmons (1980: p. 42), working with seven species of squirrel in this same Gabonais forest, provided the most thorough demonstration that dietary overlap was minimal when food was most scarce. Similarly, the monkeys of the Manu eat fruit when it is abundant, but each species falls back on a different specialty when food is scarce (Terborgh 1983, 1986).

Meanwhile, the role of competition in "structuring" ecological communities was being questioned. Connell (1980) argued that the role of niche differentiation in preventing competitive displacement is untestable because present niche differentiation is at best a response to competition that has occurred in the past and is no longer observable.

Smith et al. (1978), however, found that diets of coexisting species of ground finches on the Galapagos Islands were most distinct during a season of food shortage harsh enough to cause sharp declines in their populations. Smith et al. (1978) cited other examples, most of them far less well documented, of decreased competitive overlap in a time of food shortage. Unlike Zaret and Rand, they asked why animals should restrict their diets most narrowly when food was scarcest: it appears that, during times of abundance, types of food were available which all species could easily eat, whereas the types of food available in the lean season were most readily found and/or attacked or chewed by appropriate specialists. Schoener (1982) amplified and confirmed these arguments. Later, Rosenzweig and Abramsky (1986), inspired by some gerbils in the Negev of southern Israel, developed a mathematical theory of an imaginary "centrifugal community organization" where several coexisting species share a common first preference, but have distinct secondary preferences. Their theory predicted that when food was short, diets of these species would overlap less than when food was abundant, a contrast with the common sense that says animals will be least choosy when food is hardest to find. It turns out that this bit of common sense presupposes that an animal's specialty coincides with its preference. Rosenzweig and Abramsky also remarked that, for animals to which their theory applied, niche differentiation was related to a competition which was *not* lost in the unknowable past. They had happened quite inadvertently upon a major feature of community organization among tropical mammalian frugivores.

What if Shortages of Fruit and New Leaves Coincide? Terborgh and van Schaik (1987) suggested that in the Neotropics, seasonal shortages of fruit and new leaves coincide, while in Africa and Asia, they do not. Because fruit and new leaves are somewhat interchangeable resources, they suggested that the simultaneity in the Neotropics of the shortages of fruit and new leaves keep the total biomass of primates lower than in Africa and Asia. Primate biomasses do appear lower in undisturbed Neotropical forests than in their Paleotropical counterparts (Table 7.8). Is it fair to ascribe this contrast to the contrasting rhythms of fruit and leaf production between the New World and the Old?

In the submontane rainforest at Ranomafana, in Madagascar, seasonal shortages of fruit and new leaves do not coincide (Overdorff 1992). At Makokou, Gabon, shortages of fruit and new leaves appeared briefer and less coincident in 1970–1971 than is the rule on Barro Colorado (Table 7.9). The season of fruit shortage at Gabon, however, is somewhat unpredictable. Gautier-Hion et al. (1985: p. 420) declare that in Gabon, fruit is normally scarcest during the major dry season in June and July. In 1968, however, the shortage fell in August and September (Charles-Dominique 1971: p. 210); in 1970 it persisted into August, in 1971 it was most severe in May (Table 7.9), and in 1981, admittedly an abnormal year, fruit was scarcest in April and May (Gautier-Hion et al. 1985: p. 409). Is Gabon's leaf shortage equally unpredictable? In 1968, insects were scarce in July and August (Charles-Dominique 1971: p. 210); did this low reflect a shortage of new leaves? The one conclusion we can draw is that the evidence from Gabon is not sufficient to decide whether the reason primate biomasses in Africa are high is because in Africa, shortages of fruit and new leaves do not coincide. As Terborgh and van Schaik (1987) recognize, and as relevant funding agencies so far have not, this question is in urgent need of further attention.

Are Herbivore Numbers Limited by the Quality of Their Food?

Like insects (Coley 1983), primates prefer their leaves tender and rich in protein (Milton 1979, Davies et al. 1988). The biomass of folivorous primates does indeed appear to be higher the higher the ratio of the protein (or nitrogen) content to the content of lignins and cell walls (as measured by acid detergent fiber) in the mature leaves of their forest (Waterman et al. 1988, Oates et al. 1990, Ganzhorn 1992). If we separate the frugivores of Asia and Africa from those of Madagascar, we find that total biomass of folivorous primates increases with the ratio of protein (or nitrogen) to acid detergent fiber in both places (Table 7.10), as if food quality were involved in limiting primate populations.

Leaf-eating primates of deciduous dry forests in Madagascar such as Morondava N5, which averages 750 mm of rain a year, and Ampijoroa, which averages 1250 mm of rain a year, eat far more foliage per hectare than primates of rainforests such as Douala-Édéa, Analamazaotra, and Sepilok (Table 7.10). Why? Leaves which are tougher and less rich in protein tend to be longer lived (Coley 1987a,b, 1988, Reich et al. 1991, 1992). The longer-lived a leaf and

Table 7.8 Primate biomass and consumption in different tropical forests

Site	Biomass (kg/km^2)	Consumption (kg dry wt/ha·yr)	Reference
New World			
Barro Colorado, Panama	504	95	Wright et al. (1994)
Manu, Peru	646	130	Janson and Emmons (1990)
Africa			
Kibale, Uganda	2317	416	Struhsaker (1975)
Tiwai, Sierra Leone	1375	248	Oates et al. (1990)
Ituri, Zaire	709	127	Thomas (1991)
Lomako, Zaire	1034	184	McGraw (1994)
Ipassa, Gabon	646	131	Bourlière (1985)
Tai, Côte d'Ivoire*	1074	198	Galat and Galat-Luong (1985)
Madagascar			
Analamazaotra (1000 m)	375	89	Ganzhorn (1992)
Ampijoroa (dry forest)	771	168	Ganzhorn (1992)
Morondava N5 (dry forest)	685	178	Ganzhorn (1992)
Morondava CS7 (dry forest)	583	142	Ganzhorn (1992)
Asia			
Ceylon, dry forest	2750	547	Hladik and Hladik (1972)
Kuala Lompat, Malaysia	834	159	Raemaekers and Chivers (1980)

Consumption, F (kg dry weight of food/year) of an animal of average weight, x (g), is calculated from the regression $F = 0.085775 x^{0.822}$ (Nagy 1987).
*Tai figures calculated using weights from Oates et al. (1990) and multiplying densities and feeding rates by 6/5 to correct for overlap in home ranges; about a third of the average troop's home range is shared with another (Whitesides et al. 1988). Chimpanzees added as in Bourlière (1985).

Table 7.9 Rhythms of fruit and leaf production in Gabon and Panama

		Jul.	Aug.	Sep.	Oct.	Nov.	Dec.	Jan.	Feb.	Mar.	Apr.	May	Jun.	Jul.
Makokou, Gabon														
Fruit	1970–71		13/14	16/22	20/19	18/16	18/27	25/25	18/25	33/26	18/13	9/5	13/14	12/12
	1975–76A	23	24	29	33	39	37	43	50	37	30	34	16	
	1975–76B	60	70	110	240	150	130	180	170	120	80	75	75	
	1981	38	52	84	74	56	70	48	38	24	14	13	22	
New Leaves	1970–71		3	18	72	62	58	49	56	54	73	48	40	28

		Jan.	Feb.	Mar.	Apr.	May	Jun.	Jul.	Aug.	Sep.	Oct.	Nov.	Dec.
Barro Colorado, Panama													
Fruit	1967	15	11	13	25	21	24	15	17	20	21	15	12
New Leaves	1978	9	4	6	16	14	2	3	4	6	0.4	1	6

Gabon, fruit: 1971–72, number of species near Makokou bearing fruit eaten by chimpanzees, successive half-months, from Hladik (1973); 1975–76A, number of species of fruit and seed on the path along a 2-km forest transect, and 1976B, number of fruits and seeds found on this trail each month, read off from Gautier-Hion (1980), Figs 7d and 7c; 1981, number of fruiting sites per 6 km, from Gautier-Hion et al. (1985); Gabon leaves, number of trees near Makokou (out of 300, representing 150 species) flushing new leaves in successive months, from Hladik (1978); monthly number of species of trees near the laboratory on Barro Colorado Island bearing ripe fruit eaten by monkeys, month by month, from Hladik and Hladik (1969), and monthly percentage of > 300 trees, representing about 150 species, near BCI's laboratory, flushing new leaves, from Wolda (1982).

the lower its nitrogen content, the less energy per day each gram of that leaf produces from photosynthesis (Reich et al. 1992). It pays plants of habitats that alternate reliably wet rainy seasons with long, severe dry seasons to have deciduous leaves (Reich 1995). Short-lived leaves, however, cannot afford a low ratio of nitrogen to fiber: the low nitrogen level limits the rate of photosynthesis, keeping it too low to pay for all that fiber within the leaf's short life. Such plants can afford to defend their leaves with ephemeral poisons, such as alkaloids (Coley et al. 1985). Large herbivores, however, can cope with such poisons by eating mixtures of many different kinds of leaves (Milton 1979). Do the short lives of deciduous leaves condemn them to be good fodder for vertebrate herbivores? The deciduous forest on the northern flank of India's Nilgiri Hills is adorned by refulgent peacocks and teeming with elephants, deer, and other large animals. Is the short lives of its leaves what allows this forest, with its 1300 mm of rain a year, to support 15 tons of mammals per square kilometer (Karanth and Sunquist 1992), 10 times as much as the pristine Manu (Table 7.11)? Was the Neotropical "megafauna," which disappeared more than 10,000 years ago at the hands of human hunters who had crossed over from Asia, restricted to dry forest for similar reasons?

INSECTS OF TREE CROWNS

Much research has been devoted to the study of the vertebrate herbivores in tropical rainforests and the factors that limit their populations. Yet, in these forests, insects consume much more foliage than vertebrates do. The reverse may be true in the grasslands and savannas of tropical Africa and even (perhaps) in the dry forests of south India, but that is another story. Do seasonal shortages of new leaves also limit insect populations? Do forests need help from insect-eating predators in controlling their insect pests?

Table 7.10 Relation of leaf quality, R, to biomass of leaf-eating primates (kg/km^2) and their consumption (kg dry weight of food eaten/ha·year) in selected forests

Site	R	Biomass	Consumption	Reference
Uganda, moist forest (Kibale)	0.510	1875	377	Waterman et al. (1988)
Sierra Leone, moist forest (Tiwai)	0.345	786	169	Oates et al. (1990)
Malaya, rainforest (Kuala Lompat)	0.242	876	168	Waterman et al. (1988)
Madagascar, dry forest (Morondava N5)	0.479	438	105	Ganzhorn (1992)
South India (Kakuchi)	0.242	532	93	Waterman et al. (1988)
Madagascar, dry forest, Ampijoroa	0.316	327	71	Ganzhorn (1992)
Madagascar, dry forest (Morondava CS7)	0.449	226	56	Ganzhorn (1992)
Madagascar, moist forest, 1000 m (Analamazaotra)	0.229	167	37	Ganzhorn (1992)
Cameroun, moist forest (Douala-Édéa)	0.202	189	34	Waterman et al. (1988)
Sabah, Borneo, rainforest (Sepilok)	0.167	64	12	Waterman et al. (1988)

R is the average of the ratio of protein content to the content of acid detergent fiber in mature leaves, averaged over the different species of canopy tree in the forest. Waterman et al. (1988) weighted their averages by the relative abundances of the tree species concerned; Ganzhorn (1992) did not.

Table 7.11 Biomass and total consumption by different sizes of mammals (excluding cats, dogs, and edentates) in the rainforest of Peru's Manu and in deciduous forest at Nagarahole, South India

	Manu		Nagarahole	
Size Class (kg)	Biomass (kg/km^2)	Consumption (kg/ha·year)	Biomass (kg/km^2)	Consumption (kg/ha·year)
> 1000	0	0	6870*	272*
100–1000	0	0	5047	322
10–100	405	54	2593	294
1–10				
Primates	543	107	235	42
Others	173	35	?	?
< 1 kg	210	61	?	?
Totals	1331	257	14745+	930+

Data for Manu from Janson and Emmons (1990); data for Nagarahole from Karanth and Sunquist (1992). Feeding rate, F, (kg dry matter/year) of an animal weighing x grams is assumed to be $F = 0.365(0.577)x^{0.727}$ (Nagy 1987: regression for eutherian herbivores in Table 3).
*Category consists entirely of elephants.

Numbers and Biomass of Arboreal Insects

We estimated consumption by tropical vertebrates by counting them and assessing how much each one ate. Can the same be done with insects?

Counting and weighing the insects of tree crowns is a major problem facing the new science of canopy biology (Moffett 1993, Lowman and Nadkarni 1995). The technique of filling a tree's crown, or a column of forest, with insecticidal fog has gone far to bring this problem within human reach (Erwin and Scott 1980). Nonetheless, the primary focus of the vast amount of attention now devoted to canopy insects is on counting the species (Erwin 1982, Stork 1987, May 1990). Even though 74% of the animal species so far described are insects (May 1990), it is possible that no more than 3% of the world's insect species have been described (Erwin 1982).

So far, few insect-fogging studies tell us the number of arthropods per square meter of forest canopy and still fewer tell us the insect biomass. The data so far gathered suggest that upland tropical rainforest supports a few hundred insects per square meter of ground, or from one to several kilograms (dry weight) of insects per hectare (Table 7.12). A forest's arboreal vertebrates may well weigh more *in toto* than its arboreal insects, contrary to the estimates of Fittkau and Klinge (1973). It is clear, first, that a high proportion of these insects are ants, and second, that insect numbers and biomass are highest early in the rainy season (Table 7.12). But we do not know the annual average of the biomass of any rainforest's phytophagous insects or the seasonal variation in this biomass.

How do the numbers and biomass of arboreal insects in tropical forest compare with those in temperate-zone trees? This is a bit of a mystery. It is clear that ants are remarkably rare in the crowns of temperate-zone trees (Moran and Southwood 1982, Schowalter and Crossley 1988). The fogging data of Moran and Southwood (1982) suggest that the numbers and biomass of phytophagous insects are considerably higher in the temperate zone than in the tropics, which conflicts with the conclusion of Coley and Aide (1991) and Coley and Kursar (1996) that insect herbivory is more intense in the tropics. The branch-bagging data of Schowalter and Crossley (1988) yield figures for insect biomass more in line with tropical values, but how many of the relevant insects escape while the branch is being bagged?

How Much Foliage Do Insects Eat?

Inference from Insect Biomass

Schowalter et al. (1981) estimate from the holes and gaps of litter leaves that in 1977 the chewing herbivores of an undisturbed catchment in Coweeta, North Carolina, ate 1.9% of that forest's leaf area, or 62 kg dry weight of foliage/ha · year. Schowalter and Crossley (1988) report that the biomass of chewing folivorous insects averaged 715 g dry weight/ha during the 1977 growing season. If the growing season lasts 200 days, 715 g of folivores ate an average of 62,000/200, or 310 g dry weight of foliage/day, a daily intake of 0.43 g dry weight of foliage/g dry weight of chewing insect. Schowalter et al.'s daily feeding rate of 0.68 g of foliage/g chewing insect does not square with their estimates of herbivore biomass and total consumption and the length of the growing season. Schowalter et al. (1981: p. 1015) report that chewing herbivores elsewhere eat between half and 1.5 times their body weight per day. If Barro Colorado has an average of 2.3 kg dry weight of arboreal insects other than ants per hectare (Table 7.12), and 1 kg of these are chewing folivores, chewing folivores eat between 160 and 550 kg dry weight of foliage/ha · year.

Table 7.12 Numbers and biomass (dry weight) of arboreal insects of different categories

Site		Total kg/ha	Total N/m²	Ants kg/ha	Ants N/m²	Homoptera kg/ha	Homoptera N/m²	Coleoptera kg/ha	Coleoptera N/m²
Pipeline Road, Panama,	July 1976				240	0.2	11.8		42
	October 1975				35	0.06	4.8		21
	March/April 1976				71	0.06	3.5		6
Gigante, BCNM, Panama,	just before rains 1992	1.7		0.7					
	just after rains 1992	4.4		1.0					
Near Manaus, Brazil									
Varzea		0.2	32	0.11	17	0.001	0.3	0.55	4
Igapó		0.7	62	0.23	27	0.014	1.2	0.12	3
Terra firme		1.8	161	0.44	86	0.16	17.2	0.15	11
Brunei, Borneo			117		22		13*		20
Britain									
Betula crown		14	1290	0.06	3				
Quercus robur crown		6.0	591	0.02	0.7				
Salix cinerea crown		2.8	140	0.0	0.1				
Salix alba crown		1.4	83	0.0	0.0				
Coweeta, North Carolina, 1977		1.6		0.02		0.28		0.34	

Insect numbers, N, per square meter. Pipeline Road ant data from Montgomery (1985); Homoptera data from Wolda (1979); and Coleoptera data from Erwin and Scott (1980). Data from Gigante, a mainland peninsula near Barro Colorado Island, Panama, from John Tobin (p. c.). Manaus data from Adis et al. (1984); Brunei data from Stork (1991); British tree data from Moran and Southwood (1982); Coweeta data from Schowalter and Crossley (1988).
*Includes Hemiptera.

Perhaps the most important lesson of Schowalter et al. (1981) is their remark (p. 1016) that, according to the literature, sucking insects consume an average of 2.5 times their own dry weight in dry matter, or 12.5 times their dry weight in liquid from their hosts' phloem, per day. Ant-tended aphids consume 5 times their dry weight in dry matter per day, which happens to be 5 times their wet weight in phloem liquid per day (Schowalter et al. 1981: p. 1016). Even though the biomass of sucking herbivores at Coweeta was less than half that of chewing herbivorous insects, Schowalter et al. (1981: p. 1015) thought the forest lost more dry matter to sucking than to chewing herbivores. In those tropical forests sampled to date, suckers outnumber chewers (Table 7.13). If feeding rates of sucking herbivores is as much higher than those of chewing herbivores in tropical forest as in the fields and forests of the temperate zone, well over half the consumption of tropical insect herbivores may leave no visible mark. The African ant–defended tree *Leonardoxa africana* carries more sucking than chewing insects, and sucking insects reduce the plant's leaf area as much as chewing ones do (Gaume et al. 1997). Chewing insects have hogged our attention, because their outbreaks cause visible, sometimes spectacular, defoliations (Janzen 1981). Are plant "lice" a bigger overall drain on the forest?

Measuring Herbivory from the Holes and Gaps in Fallen Leaves

To measure forestwide herbivory by chewing insects in Barro Colorado's Lutz ravine, Leigh collected fallen leaves once a week from every 10th of a series of one hundred 0.0792-m² tubs set in a circle around Barro Colorado's Lutz catchment. Leaves from tubs 1, 11, 21, ... were collected one week, leaves from tubs 2, 12, 22, ... the next, and so forth (Leigh and Smythe 1978). For the first 2 years, Leigh

Table 7.13 Proportions of chewing and sucking herbivores among the insects in different tree crowns

	Brunei		Britain			South Africa		
	S	N	S	N	B	S	N	B
Chewing insects	0.092	0.104	0.123	0.061	0.136	0.155	0.102	0.418
Sucking insects	0.085	0.120	0.156	0.554	0.439	0.090	0.262	0.104
All phytophages	0.176	0.223	0.279	0.615	0.575	0.245	0.364	0.522

Proportion of chewing herbivores, sucking herbivores, and total phytophages among the species (S), individuals (N), and biomass (B) of insects collected by fogging tree crowns. Brunei data from Stork (1987); data for Britain and South Africa from Moran and Southwood (1982). Guilds are classified according to the schema of Moran and Southwood (1982).

used a planimeter to measure the area of the fallen leaves, including eaten parts, and the proportion of this area consisting of holes and gaps. Afterwards, Windsor computerized the process. Of the area of the leaves falling during the 4 years beginning 1 December 1973, 7.24% consisted of holes and gaps (Leigh and Windsor 1982). If leaves live a year and are eaten after they had expanded and reached full weight, this represents a consumption of 476 kg dry weight of foliage/ha · year.

This calculation involves several shaky assumptions. First, on Barro Colorado's Poacher's Peninsula, where larger traps have been used to catch falling litter and where this litter has been collected by better supervised assistants, leaf fall consistently averages 8 tons dry weight per hectare per year (S. J. Wright, p. c.). If the true litter fall in Lutz catchment is also 8 tons/ha · year, holes and gaps account for $(0.0724)(8000/0.9276) = 624$ kg dry matter/ha · year, rather than 476.

Second, herbivores prefer young, growing leaves (Aide and Londoño 1989), and the holes they make grow with the leaf (Reichle et al. 1973). Reichle et al. (1973) measured rates of herbivory (hole formation) on leaves of different sizes and stages of growth in a stand of tulip poplars, *Liriodendron tulipifera*. They found that, on the average, holes created by herbivores expanded nearly threefold with the growth of their leaves (Table 7.14).

Third, is it fair to assume that average leaf lifetime is a year? On Barro Colorado, understory leaves live much longer (Coley 1988). In some tropical forests where both leaf biomass and annual litterfall have been measured, the match between the two suggests an average leaf lifetime of 1 year (Table 7.15). Average leaf lifetime tends to be longer the poorer the soil, however, as shown by the figures for San Carlos de Rio Negro. Average leaf lifetime is also longer than a year for some Australian broadleaf forests (Lowman 1985, 1992), and much longer than a year in many coniferous forests (Gower et al. 1993, Grier et al. 1981). Even where average leaf lifetime is a year, however, leaf lifetime varies markedly from species to species (Coley 1988, Reich et al. 1991). Because shorter-lived leaves are more heavily eaten (Coley 1987a,b), our calculation underestimates herbivory rate. There are other problems. One plant species withdraws 19% of the dry matter from its leaves before dropping them (Bray and Gorham 1964). If this were more generally true (which I doubt), this would aggravate our underestimate of leaf consumption.

Perhaps the greatest problem with the calculation is its neglect of leaves eaten whole. Coley (1982) found that 38% of the herbivory on mature leaves of old forest plants consisted of leaves eaten whole. How these various errors balance, I cannot say.

Following Herbivory on Marked Leaves

The only reliable way to measure herbivory is to follow the fate of marked leaves. Coley (1983) measured rates of consumption of marked young and mature leaves she could reach on saplings in gaps. On mature leaves of species of old forest ("persistents"), insect herbivores cropped 0.04% of the available leaf surface per day—1% per 25 days or 14.6% per year. Coley (1983) found that herbivores devoured young leaves 24 times faster. If leaves are young for a month and their average lifetime is a year, then total herbivory on these saplings amounts to 43% of leaf production—which agrees with Coley and Kursar's (1996: p. 306) estimate that leaves suffer 70% of their lifetime damage before they mature. This estimate is high, however: holes and gaps made in young leaves grow with the leaf. If holes and gaps in young leaves expand threefold, on the average, before the leaves mature, as in tulip poplar forests of Tennessee (Reichle et al. 1973), annual consumption is about 24% of these saplings' leaf production. Is this figure representative?

Wright measured herbivory on marked leaves of the nine most common species of canopy tree and liane whose sun-leaves he could reach from a tower crane in the Parque Metropolitano of Panama City. This park averages 1740 mm of rain per year, a third less than Barro Colorado. Yet the loss of area during a leaf's lifetime, averaged over the nine species, was only 8.3% (Wright and Colley 1996: p. 17). It appears that lifetime damage of leaves is much lower in canopy than among the small saplings of treefall gaps studied by Coley (1983). In a deciduous dry forest in Mexico, with one-third Barro Colorado's rainfall and an 8-month dry season, a leaf's lifetime loss of area, averaged over 12 common species, was 17.1% (Filip et al. 1995). Like mammals, insects also find dry forest leaves more edible than those of rainforest.

Judging by Table 11.1 of Coley and Kursar (1996), young leaves of "persistent" species in forest understories lose 0.554% of their area per day to herbivores, whereas mature leaves lose 0.025% per day. On Barro Colorado, the corresponding figures are 0.89% and 0.037% per day (Coley 1983: Table 1). Does Barro Colorado's higher consumption rates reflect an unusually fertile soil, which favors outgrowing rather than poisoning herbivores (cf. Coley et al. 1985)?

When Lowman (1985, 1992) measured herbivory on marked leaves at all levels of temperate and subtropical forest of New South Wales in Australia, she found extraor-

Table 7.14 Leaf surface eaten by herbivores and area of the resulting holes when full-grown, expressed as percentage of total leaf area (LA) at full expansion, in a tulip poplar stand at Oak Ridge Tennessee (data from Reichle et. al. 1973)

Year	% LA Consumed	% Area of Fallen Leaves Consisting of Holes and Gaps
1965	1.9	5.6
1966	3.4	10.1
1967	2.5	7.3

Table 7.15 Leaf biomass and annual leaf fall in selected forests

	Banco, Plateau, Côte d'Ivoire	Pasoh, Malaysia	Manaus, Brazil*	S. Carlos de R. Negro, Venezuela		New South Wales, Australia			Medicine Bow, Wyoming
				Oxisol	Tall caatinga	Cool temp.	Warm temp.	Subtropical	
Biomass	9.0	7.7	9.0	9.2	6.9	7.06	8.91	10.5	9.2
Leaffall	8.2	7.0	8.3	7.6	4.8	3.5/6.2	4.1/6.6	5.6/7.4	0.9

Biomass in tons dry weight per hectare; leaf litterfall in tons dry weight per hectare per year. Leaf fall/leaf production is tabulated for Australian sites. Data for Banco from Bernhard-Reversat et al. (1978); for Pasoh, Kira (1978); for Manaus, Roberts et al. (1993); for Venezuela, litterfall is from Cuevas and Medina (1986), biomass from Cuevas and Medina (1989); data for Australia, Lowman (1992); coniferous forest in Wyoming, Fahey (1983).
*Leaf area (hectares per hectare of ground), not leaf weight.

dinarily high rates of herbivory. In subtropical forest at 34° S latitude, insects ate 1.76 tons dry weight of foliage/ha · year. Annual leaf fall there was 5.59 tons/ha; thus, insects consumed 24% of the leaf production. In nearby temperate forests, leaf litterfall was lower, while the herbivores ate more, consuming 40% of the leaf production (Table 7.16). Herbivory rates in Australia's *Eucalyptus* forests (Fox and Morrow 1983) are far higher than those characteristic of temperate-zone forests elsewhere, except during major insect outbreaks. The "arms race" between plant and insect seems more intense in Australia than elsewhere.

Do Birds Protect the Forest from Insect Herbivores?

There are several reasons for suspecting that forests need the help of insect-eating predators to protect them from insect pests. Many plants defend their leaves by compounds that slow the growth of the insects that eat them: such compounds would not be useful unless slowing an insect's growth rate increased the chance that it is eaten (Feeny 1992: p. 21). Coley et al. (1985) assume that fertile soils allow poorly defended plants to outrun their herbivores, a proposition that makes sense only if predators limit the ability of insects to flock to the feast thus offered. Do birds eat enough folivorous insects to limit these insect populations?

How Much Insect Matter Do Birds Eat?

One can estimate a bird's feeding rate from its diet using regressions published by Nagy (1987). Thus, if we know the numbers, weights, and diets of birds on Barro Colorado, we can calculate the dry weight of folivorous insects its birds eat per hectare per year. Willis (1980) estimated population densities of the species of bird on Barro Colorado for the decade ending in 1970. He carefully measured the populations of various species of antbird and estimated densities of other birds by their ratios to the density of some antbird. His data suggest that, excluding woodpeckers, barkgleaners, and the like, Barro Colorado Island carries 211 g/ha of birds that live almost entirely on arboreal insects, and 199 g/ha of omnivorous birds, roughly half of whose diet is arboreal insects (Table 7.17). These birds eat 24 g dry weight of arboreal, presumably plant-eating, insects per hectare per year (Table 7.18). Douglas Robinson (p. c.) conducted a much more systematic census of Barro Colorado's birds in 1995, with similar results (Tables 7.17, 7.18).

Karr (1971) censused the birds of a small plot on the mainland, 12 km east of Barro Colorado Island, near the former Limbo Hunt Club. He found 246 g/ha of birds feeding primarily on plant-eating (leaf-eating) insects, and 234 g/ha of omnivores: together, these birds ate 28.6 kg dry weight of plant-eating insects/ha · yr (Tables 7.17, 7.18). The small size of his plot was sharply criticized by Terborgh et al. (1990). Nonetheless, when Douglas Robinson and Scott Robinson (p. c.), using the same techniques by which Terborgh et al. (1990) censused the birds of Manu, censused birds on a 104-ha plot near Limbo Hunt Club, which included Karr's plot, their data suggest a similar rate of consumption by birds of plant-eating insects (Tables 7.17, 7.18). Despite very different census techniques, these censuses reveal certain basic differences between Barro Colorado Island and Limbo Hunt Club, such as the greater abundance of ground-feeding insectivores at the latter site.

Just as annual leaf litter fall is similar in different tropical forests (chapter 6), so different forests support a similar abundance and biomass of insect-eating birds, and they eat similar quantities of folivorous insects (Tables 7.17, 7.18). Perhaps the birds of Manu eat less folivorous insect matter because of heavier consumption of arboreal insects there by arboreal mammals (Table 7.6).

Table 7.16 Leaf production and leaf consumption (tons/ha) in three Australian forests (data from Lowman 1985, 1992)

	Leaf Fall	Leaf Consumption	Leaf Production	% Eaten
Cool temperate	3.53	2.63	6.16	43
Warm temperate	4.05	2.50	6.55	38
Subtropical	5.59	1.76	7.35	24

Table 7.17 Bird populations (numbers and biomass) in different rainforests

No. (wt, kg) birds/ha	Manu	Limbo		Barro Colorado		Guyane Fr.	Semengoh
		1970	1994	1965	1995		
Total	18 (1.70)	35 (1.32)	32 (1.76)	23 (1.26)	26 (0.96)	15 (1.5)	30 (1.25)
Arboreal insectivores	7 (0.18)	15 (0.25)	15 (0.25)	10 (0.21)	14 (0.21)	9	10 (0.29)
Other insectivores	4 (0.14)	9 (0.27)	7 (0.20)	3 (0.09)	5 (0.09)	1.5	7 (0.30)
Omnivores	3 (0.23)	4 (0.23)	7 (0.34)	6 (0.20)	5 (0.15)	2.4	8 (0.23)
Frugivores/granivores	4 (1.15)	7 (0.57)	4 (0.97)	3 (0.76)	2 (0.51)	2.0	9 (0.43)

Data for Manu from Terborgh et al. (1990); 1970 data for Limbo Hunt Club, central Panama from Karr (1971), 1994 data for Limbo from Douglas Robinson (p. c.); 1965 data for BCI from Willis (1980); 1995 data for BCI from Douglas Robinson (p. c.); data for Guyane Française from Thiollay (1986); and data for Semengoh, Sarawak from Fogden (1976).

Can Birds Protect Their Forest from Insects?

If it takes 10 pounds of foliage to make a pound of insects (as the saying claimed that it took 10 pounds of feed to make a pound of cow; Leigh 1975), then it takes 260 kg dry weight of foliage to feed the folivorous insects birds eat. Perhaps as much as half the foliage insects other than leafcutter ants eat is destined to fatten prey for insect-eating birds. At any rate, birds eat a far higher proportion of insect production than jaguars and pumas do of mammal production. Do these birds control the populations of folivorous insects?

On Barro Colorado Island, excluding birds from tangles of dead leaves nearly doubles the total number of crickets, katydids, spiders, and roaches found therein (Gradwohl and Greenberg 1982). In northern hardwood forest, excluding birds from understory plants leads to a substantial increase in the numbers of caterpillars on them (Holmes et al. 1979). In oak-hickory forest of Missouri, excluding birds from white oak saplings by placing cages over the saplings doubles the number of insects per sapling, nearly doubles the rate of leaf consumption by insects, and reduces above-ground production of caged plants by a fifth compared to uncaged controls (Marquis and Whelan 1994).

The best evidence that a forest depends on birds for defense against insects comes, however, from Australia. Australian bell miners sometimes join together to defend psyllid-laden eucalypts from other insectivorous birds, thereby allowing the psyllids on these eucalypts to maintain high population levels (Loyn et al. 1983, Loyn 1987). Where the bell miners succeed in excluding other insectivorous birds from a patch of eucalypts, the eucalypts in question decline and die, as if they needed vigorous insectivorous birds to protect them from their psyllids. In normal forests, a few hundred grams of birds per hectare may preserve the life of a few hundred tons of vegetation per hectare. It would be an extraordinary paradox if so many owe their ways of life to so few, but tropical biology is prodigal of such paradoxes.

Table 7.18 Dry weight (kg/ha·year) of food eaten by different categories of birds

	Semengoh	Limbo		Barro Colorado		Manu
		1970	1994	1965	1995	
Folivorous insects	28	29	31	24	24	20
Other insects	16	22	16	7	9	11
Fruit	30	26	36	28	20	38

Feeding rates calculated applying regressions of Nagy (1987) to data for Semengoh from Fogden (1976), 1970 data for Limbo from Karr (1971), 1994 data for Limbo from Douglas Robinson (p. c.), 1965 data for Barro Colorado from Willis (1980), 1995 data for Barro Colorado from Douglas Robinson (p. c.), and Manu data from Terborgh et al. (1990).

Feeding rate f, in grams dry weight eaten per day, for a nonpasserine weighing w grams, is assumed to obey the formula $\ln f = \ln 0.648 + 0.651 \ln w$, while feeding rate for a passerine weighing w grams is assumed to obey the formula $\ln f = \ln (0.398) + 0.850 \ln w$ (Nagy 1987).

Consumption of folivorous insects represents consumption by arboreal insectivores (excluding woodpeckers and bark-gleaners) plus half the consumption by omnivores. Consumption by other insectivores represents consumption by woodpeckers, bark-gleaners, and foragers in tangles of dead leaves, plus consumption by terrestrial (ground-feeding) insectivores. Consumption of fruit represents consumption by frugivorous birds, arboreal and terrestrial, plus half the consumption by omnivores.

CONCLUDING REMARKS

Plants can compete with each other in two basic ways. They can support herbivore populations (and/or fires) sufficient to ruin their competitors, as prairie grasses do. Or they can try to outgrow their competitors by reducing losses to herbivores, as most trees do: this is the way of the rainforest.

Tropical trees defend themselves against vertebrate folivores by making mature leaves that are too tough and poor in protein for these animals to eat. These vertebrates must therefore subsist on fruit and/or young leaves. What cues different plants use to time fruit fall and leaf flush is still a mystery, although some clues are beginning to appear. What is clear is that, on Barro Colorado, there is a season when new leaves are scarce and too little fruit falls to feed the vertebrate frugivores. In all tropical forests, populations of vertebrate herbivores are limited by such times of shortage. Rainforests do not need big cats or other predators to protect them from vertebrate herbivores.

Insects consume far more foliage and destroy much more fruit than vertebrates do. Most insects also find mature leaves unfit to eat: in rainforest, leaves suffer the majority of their damage before they toughen and mature. Nonetheless, forests need insectivorous birds and other consumers to protect them from insect pests. How tree diversity helps defend plants against insects is the subject of the next chapter.

APPENDIX 7.1

Number of species of plants of different categories dropping fruit into traps each week on the central

	Week Number																					
	1	2	3	4	5	6	7	8	9	10	11	12	13	14	15	16	17	18	19	20	21	22
1992																						
Overstory trees: All	9	9	7	8	13	12	15	13	19	14	20	14	12	16	19	25	20	18	17	10	6	15
Animal dispersed	7	7	6	7	11	10	13	10	12	8	13	7	7	8	10	15	12	10	13	7	4	11
Midstory trees: All	3	5	1	2	4	2	6	7	5	5	5	7	7	6	8	6	9	7	14	7	7	6
Animal dispersed	3	4	1	2	3	2	6	6	5	5	4	5	5	5	7	5	8	6	13	7	7	6
Understory trees and shrubs: All	4	3	2	4	1	3	3	2	0	5	2	2	2	5	2	1	3	1	3	4	4	4
Animal-dispersed	4	3	2	4	1	3	2	1	0	4	0	1	2	4	2	1	3	1	3	4	3	2
Lianes (all)	6	8	9	10	11	14	21	22	21	25	24	23	24	30	29	20	20	12	13	8	6	10
Vines (all)	2	1	1	0	0	0	1	0	0	1	1	1	3	2	1	2	3	2	2	0	1	3
Epiphytes (all)	1	1	0	0	0	1	1	1	2	1	2	1	4	3	3	3	4	4	3	3	3	2
Ficus spp.	0	1	1	1	1	1	1	0	1	1	1	0	0	0	0	0	0	0	0	0	0	1
Total	25	28	21	25	30	33	48	45	48	52	55	48	52	62	62	57	59	44	52	32	26	40
1993																						
Overstory trees: All	13	9	11	11	15	12	17	14	17	15	16	22	17	16	17	19	19	22	18	16	18	17
Animal dispersed	10	6	7	6	10	9	11	11	12	10	11	14	10	8	11	10	10	14	12	12	12	13
Midstory trees: All	2	3	4	4	5	6	5	5	5	7	7	3	6	6	7	5	5	5	5	5	5	5
Animal dispersed	2	3	4	4	5	6	4	5	5	7	7	3	6	6	7	5	5	5	5	5	5	5
Understory trees and shrubs: All	4	2	2	2	3	3	5	3	5	1	3	4	3	5	5	4	5	8	5	2	3	3
Animal dispersed	4	2	2	2	2	3	4	3	4	1	3	4	2	4	3	1	3	5	4	1	2	2
Lianes (all)	5	5	6	3	7	8	11	13	11	15	12	16	14	18	15	17	16	15	13	8	9	10
Vines (all)	1	0	0	0	0	1	0	0	0	1	0	0	1	2	3	1	1	1	2	2	2	2
Epiphytes (all)	3	2	2	1	2	1	2	2	1	1	1	1	1	2	1	2	2	2	1	1	2	2
Ficus spp.	1	0	0	0	0	0	0	0	0	0	0	0	0	0	0	0	2	2	2	0	0	0
Total	29	22	25	21	32	31	40	37	39	40	39	46	42	49	48	47	50	55	46	34	39	39
1994																						
Overstory trees: All	7	4	10	8	6	8	9	7	12	8	7	15	12	14	14	18	14	21	17	21	17	18
Animal dispersed	7	4	9	7	6	5	7	5	8	5	3	9	5	7	7	11	7	12	11	14	12	12
Midstory trees: All	4	2	3	1	3	2	3	4	3	4	3	3	5	7	8	8	7	7	7	9	10	7
Animal dispersed	4	2	3	1	3	2	3	4	3	4	3	3	5	7	7	6	5	6	6	8	9	6
Understory trees and shrubs: All	1	2	4	5	4	2	5	3	4	2	1	2	0	2	0	5	6	3	4	1	1	3
Animal dispersed	1	2	4	5	4	2	5	2	3	2	1	1	0	1	0	4	4	2	3	0	1	3
Lianes	3	2	4	4	3	2	11	9	11	16	14	15	16	21	21	19	16	16	12	9	10	14
Vines	1	0	1	1	1	0	0	0	0	0	1	1	0	0	0	1	1	3	0	2	3	2
Epiphytes	1	1	0	0	0	0	0	0	2	2	3	3	2	1	3	1	1	2	1	1	1	0
Ficus spp.	1	0	0	0	0	1	0	0	0	0	0	0	0	0	0	0	0	0	0	0	0	0
Total	18	11	22	19	17	15	28	23	32	32	29	39	35	45	46	52	45	52	41	43	42	44
1995																						
Overstory trees: All	11	9	13	11	15	16	15	18	21	16	15	22	21	23	19	24	22	20	19	20	12	19
Animal dispersed	8	6	10	7	11	12	10	14	15	10	10	14	12	14	11	17	14	12	10	14	8	12
Midstory trees: All	2	2	6	4	5	6	7	6	11	9	7	9	11	10	7	10	7	10	8	5	6	7
Animal dispersed	2	2	6	4	5	5	6	5	10	8	7	9	10	10	7	9	7	10	8	5	6	6
Understory trees and shrubs: All	3	4	5	3	3	3	2	4	6	5	5	2	2	4	4	3	1	4	5	4	2	3
Animal dispersed	2	3	4	2	2	2	1	2	3	4	2	0	1	4	3	3	0	1	3	3	0	1
Lianes	6	5	4	3	6	7	6	13	15	17	19	11	20	16	13	15	13	13	14	11	8	11
Vines	2	2	1	1	3	0	0	1	0	2	0	2	0	2	1	3	1	3	1	1	3	3
Epiphytes	2	1	0	0	1	0	1	0	0	1	2	2	3	3	2	7	4	4	4	3	3	1
Ficus spp.	0	0	0	0	0	0	0	1	0	0	0	0	0	0	0	0	1	1	1	1	1	0
Total	26	23	29	22	33	32	31	42	54	50	48	48	57	58	46	62	48	55	52	45	35	44

Data provided by S. Paton on behalf of the Smithsonian Environmental Sciences Program.

plateau of Barro Colorado, 1992–1995

	Week Number																													
23	24	25	26	27	28	29	30	31	32	33	34	35	36	37	38	39	40	41	42	43	44	45	46	47	48	49	50	51	52
15	14	16	12	6	6	9	7	13	7	9	12	15	11	14	17	14	15	14	14	14	12	7	10	7	9	11	8	7	10
10	10	11	7	3	2	6	4	10	6	6	10	13	8	10	12	9	10	9	10	10	9	5	6	5	7	8	4	3	6
4	4	5	5	3	3	6	4	5	4	6	2	3	5	4	4	5	6	4	4	5	4	4	5	3	3	5	4	2	2
4	4	5	5	3	3	4	4	5	4	6	2	3	5	4	4	5	6	4	4	5	4	4	5	3	3	5	4	2	2
5	3	4	1	1	5	1	7	2	1	1	1	2	3	2	6	4	6	7	6	8	8	7	5	6	6	5	4	6	4
3	1	2	1	1	4	0	6	1	1	0	1	2	2	1	5	3	6	7	6	8	8	7	5	6	6	5	4	6	4
11	8	9	9	6	7	2	4	6	2	2	3	2	6	6	5	7	9	8	7	8	6	4	3	5	7	3	4	5	3
3	2	1	1	1	1	2	2	2	1	0	0	0	1	1	2	1	3	2	2	4	1	2	2	2	2	2	2	1
3	2	1	3	2	3	2	0	0	1	2	2	1	1	1	2	2	2	1	2	2	1	0	0	1	0	0	1	1	1
1	0	0	1	1	1	1	0	2	0	1	0	1	0	0	0	0	0	0	1	1	1	1	1	0	1	1	0	0	0
42	33	36	32	20	26	22	24	30	17	22	20	24	26	28	35	34	39	37	36	40	36	24	26	24	28	27	23	23	21
14	11	12	12	7	6	5	5	7	10	7	5	4	9	7	8	6	9	7	4	7	3	4	6	2	4	6	5	7	4
10	8	9	9	5	6	4	4	7	8	6	4	3	9	7	7	5	7	6	4	6	2	3	4	1	4	5	4	6	4
6	3	6	7	7	5	1	3	2	1	3	1	1	4	4	7	4	3	4	2	2	2	1	1	5	1	1	2	1	3
6	3	6	7	7	5	1	3	2	1	3	1	1	4	4	7	4	3	4	2	2	2	1	1	5	1	1	2	1	3
1	2	1	1	4	2	5	1	2	1	3	2	3	3	4	5	4	5	6	6	5	5	5	6	4	2	1	3	2	1
1	1	1	1	4	2	4	1	2	1	3	2	3	3	4	5	3	5	6	6	5	5	5	6	4	2	1	3	2	1
7	7	8	3	4	3	5	3	3	3	4	1	1	3	5	5	3	2	4	4	2	2	4	3	3	3	2	2
2	1	2	2	1	0	1	0	0	0	1	0	1	1	1	2	1	3	4	3	1	2	3	3	2	2	2	2	2	0
2	2	2	2	1	3	0	1	1	1	1	2	2	3	3	2	2	1	2	1	1	1	1	0	0	0	0	0	0
0	0	0	0	0	1	1	1	0	1	1	0	0	1	0	1	0	0	1	0	0	1	1	1	1	1	1	1	1	1
32	26	31	27	23	21	16	16	15	17	19	14	12	22	23	30	22	24	26	20	20	18	17	19	18	13	14	16	15	11
13	12	15	13	20	15	15	12	15	11	14	9	11	11	10	11	10	9	8	8	8	9	8	7	7	9	6	8	9	12
10	10	12	11	17	13	13	9	13	10	12	7	9	9	8	8	7	7	5	5	5	6	4	5	5	7	4	6	7	9
8	7	7	7	6	3	2	3	3	3	3	4	5	5	4	5	4	5	4	6	3	4	4	2	3	2	4	6	4	3
7	6	6	6	5	2	2	3	3	3	3	4	5	5	4	5	4	5	4	6	3	4	4	2	3	2	4	6	4	3
2	2	2	2	3	3	4	1	0	1	4	3	2	2	1	6	3	4	5	5	8	7	5	5	6	5	2	3	2	3
2	1	2	2	3	3	4	1	0	1	3	3	2	2	1	5	3	4	4	5	7	7	5	5	6	5	2	3	2	3
8	10	4	6	5	6	3	4	4	4	6	7	6	6	4	5	4	4	6	3	6	5	5	2	1	2	5	5	3	2
2	2	1	3	0	3	2	2	2	1	1	1	1	2	1	2	2	1	3	3	3	4	5	3	4	2	2	3	2	2
0	0	2	3	2	1	1	0	2	3	3	2	2	2	3	2	2	3	2	2	2	2	0	2	1	1	1	1	0	1
0	0	0	0	0	0	1	0	2	1	1	1	2	2	1	0	2	1	0	0	0	0	0	1	0	0	0	0
33	33	31	34	36	31	28	22	28	24	32	27	29	30	24	31	27	27	28	27	30	31	27	21	23	21	20	26	20	23
15	12	11	13	11	10	9	8	13	11	9	10	9	9	12	8	12	8	8	8	10	8	7	4	5	5	7	6	6	8
9	8	9	9	9	9	8	7	11	8	6	7	7	8	9	7	10	5	6	6	7	6	4	2	3	2	5	4	3	6
8	4	6	6	7	4	6	2	2	3	2	5	3	4	3	6	4	4	3	4	4	2	2	4	1	1	4	1	1	2
7	4	6	6	7	4	6	2	2	3	2	5	3	4	3	6	4	4	3	4	4	2	2	4	1	1	3	1	1	2
0	2	4	2	4	3	2	2	2	1	3	3	4	3	1	1	4	6	6	6	7	6	2	4	4	3	6	4	4	3
0	0	4	2	4	3	2	2	2	1	3	2	3	2	1	1	4	6	6	6	7	6	2	4	4	3	6	4	4	3
9	8	6	10	7	6	3	2	6	2	6	5	3	6	2	7	4	7	7	8	7	3	3	2	4	6	2	4	6	3
1	2	2	3	2	2	1	0	1	1	1	1	0	2	2	4	5	4	5	3	4	4	3	2	3	2	3	3	2	4
3	2	2	1	3	3	3	2	1	1	2	2	2	1	2	2	1	2	2	1	2	3	2	0	1	1	1	1	0	1
0	1	0	1	0	0	0	0	0	1	0	0	0	0	0	0	0	0	1	0	1	1	1	1	1	0	0	1	1
36	31	31	36	34	28	25	16	25	20	23	26	21	25	22	28	30	31	32	30	35	27	20	19	19	19	23	19	20	22

APPENDIX 7.2 The mammals of BCI and Manú

Animal	Habitat	Diet	Density (no./km²) BCI	Density (no./km²) Manu	Weight/Animal (g) BCI	Weight/Animal (g) Manu	Biomass (kg/km²) BCI	Biomass (kg/km²) Manu	Feeding Rate (g/day) BCI	Feeding Rate (g/day) Manu	Energy Use (kg/ha·year) BCI	Energy Use (kg/ha·year) Manu
Metachirus nudicaudatus	T	I, F	scarce	12		420		5.0		29		1.3
Micoureus cinerea	A	I, F, N		10		150		1.5		14		0.5
Caluromys derbianus	A	75%F, 25%I	20		300		6.0		23		1.7	
Caluromysiops irrupta	A	F		10		250		2.5		20		0.7
Marmosa robinsoni	A	50%F, 50%I	27		60		1.6		8		0.8	
Marmosops noctivaga	A	50%F, 50%I		15		80		1.2		9		0.5
Philander opossum	T/A	50%F, 50%I	13	25	270	270	3.5	6.8	21	21	1.0	1.9
Didelphis marsupialis	T/A	50%F, 33%I	47	55	1000	1000	47.0	55.0	52	52	9.0	10.3
Aotus trivirgatus	A	65%F, 20%I, 15%L	3	40	700	700	2.1	28.0	51	51	0.6	7.5
Callicebus moloch	A	F, U, L, I		24		700		16.8		51		4.5
Alouatta spp.	A	55%F, 45%L	80	30	5500	6000	440.0	180.0	303	326	88.5	35.7
Cebus spp.	A	65%F, 20%I, 15%L	20	75	2600	2500	52.0	187.5	151	146	11.0	40.0
Saimiri sciureus	A	F, I, N		60		800		48.0		57		12.5
Ateles spp.	A	80%F, 20%L	1	25	5000	7000	5.0	175.0	258	340	0.9	31.0
Saguinus spp.	A	60%F, 30%I, 10%L	3	28	800	370	2.4	10.4	57	30	0.6	3.1
Tamandua spp.	T/A	I*	5	scarce	5000	5000	25.0		246		4.5	
Cyclopes didactylus	A	I*	50		300		15.0		25		4.4	
Bradypus spp.	A	L	500	scarce	2980		1490.0		45		81.7	
Choloepus hoffmanni	A	L, F	50	scarce	3500		350.0		80?		29?	
Dasypus novemcinctus	T	I	50	scarce	4000		200.0		146		26.6	
Tapirus spp.	T	L, F	0.5	0.5	300000	150000	150.0	80.0	5534	3504	10.1	6.4
Mazama americana	T	L, F	3	3	15000	30000	45.0	78.0	636	1000	7.0	11.0
Tayassu tajacu	T	S, F, L	10	6	23000	25000	230.0	150.0	904	969	33.0	21.2
Tayassu pecari	T	S, F, L		3		35000		105.0		1277		14.0
Coendou spp.	A	L, U	10	scarce	4500	4500	45.0		237	237	8.6	
Hydrochaeris hydrochaeris	T	L	present	1.6	45000	45000	?	72	1393	1393	?	8.1
Sciurus granatensis	A	S, F	180		250		45.0		22		14.4	
Sciurus spadiceus	T/A	S, F		25		600		15.0		45		4.1
Oryzomys spp.	T	S, F, I	133	180	70	70	9.3	12.6	8	8	3.7	5.1
Agouti paca	T	F, S, L	40	4	8000	8000	320.0	28.0	380	380	55.5	4.9
Dasyprocta spp.	T	S, F	100	5	2800	4000	280.0	20.0	160	215	58.5	3.9
Myoprocta pratti	T	S, F		5		1500		7.5		96		1.8
Proechimys spp.	T	S, F	180	230	350	270	65.3	62.1	29	23	19.8	19.7
Nasua spp.	T/A	F, I	24	scarce	3000	2200	72.0		170	131	14.9	12.0
Potos flavus	A	F	20	25	2200	2200	44.0	55.0	131	131	9.6	12.0
Bassaricyon gabbii	A	F, N, I	scarce	7	1000	1000		7.0	69	69		1.7
Eira barbara	T/A	M, F	2	scarce	4000	4000	8.0		215	215	1.6	
Felis pardalis	T	M	0.8	0.8	12000	9300	9.6	7.4	530	430	1.5	1.3
Felis concolor	T	M		0.024		29000		0.7		1094		0.1
Panthera onca	T	M		0.035		35000		1.2		1277		0.2

The BCI census is from Glanz (1982); the Manu census is from Janson and Emmons (1990). Body weights for BCI are from Eisenberg and Thorington (1973) except for *Tamandua* and *Cyclopes*, which are from Lubin (1983), and *Bradypus* and *Choloepus*, which are from Montgomery and Sunquist (1975). Weight records for Manu are from Janson and Emmons (1990). Diets are from Janson and Emmons (1990), except for *Caluromys, Philander*, and *Didelphis*, which are from Julien-Laferrière and Atramentowicz (1990); *Aotus, Alouatta, Cebus, Ateles*, and *Saguinus*, which are from Hladik and Hladik (1969); *Tamandua* and *Cyclopes*, which are from Lubin (1983); *Choloepus*, which is from Montgomery and Sunquist 1975; *Coendou*, which is from Charles-Dominique et al. (1981) and Glanz (1982); and *Proechimys*, from Adler 1995).

In several cases we have assigned the same diet to comparable species of a genus at the two sites: *Alouatta palliatta* at BCI and *A. seniculus* at Manu; *Cebus capucinus* at BCI, *C. apella* and *C. albifrons* at BCI; *Ateles geoffroyi* at BCI, *A. paniscus* at Manu; *Tamandua mexicana* at BCI, *T. tetradactyla* at Manu; *Tapirus bairdii* at BCI, *T. terrestris* at Manu; *Coendou rothschildi* at BCI, *C. prehensilis* at Manu; *Dasyprocta punctata* at BCI, *D. variegata* at Manu.

Feeding rates (grams dry weight per day) are from the formulae of Nagy (1987). For marsupials, feeding rate is $0.492x^{0.673}$, where x is the animal's fresh weight in grams, whereas for placental mammals, feeding rate is $0.235x^{0.822}$, except for tapirs and capybaras, whose feeding rate is $0.577x^{0.727}$.

Habitat: T, terrestrial; A, arboreal. Diet: L, leaves, S, seeds; F, ripe fruit; U, unripe fruit; I, insects; N, nectar; M, mammals (and other vertebrates).

*The diet of *Tamandua* consists of ants and termites, while *Cyclopes* eat only ants.

References

Adis, J., Y. D. Lubin, and G. G. Montgomery. 1984. Arthropods from the canopy of inundated and terra firme forests near Manaus, Brazil, with critical considerations on the pyrethrum-fogging technique. *Studies on Neotropical Fauna and Environment* 19: 223–236.

Adler, G. H. 1995. Fruit and seed exploitation by Central American spiny rats, *Proechimys semispinosus*. *Studies on Neotropical Fauna and Environment* 30: 237–244.

Adler, G. H., and R. P. Beatty. 1997. Changing reproductive rates in a Neotropical forest rodent, *Proechimys semispinosus*. *Journal of Animal Ecology* 66: 472–480.

Adler, G. H., and T. D. Lambert. 1996. Ecological correlates of trap response of a Neotropical forest rodent, *Proechimys semispinosus*. *Journal of Tropical Ecology* 13: 59–68.

Aide, T. M. 1988. Herbivory as a selective agent on the timing of leaf development in a tropical understory community. *Nature* 336: 574–575.

Aide, T. M. 1992. Dry season leaf production: an escape from herbivory. *Biotropica* 24: 532–537.

Aide, T. M. 1993. Patterns of leaf development and herbivory in a tropical understory community. *Ecology* 74: 455–466.

Aide, T. M., and E. C. Londoño. 1989. The effects of rapid leaf expansion on the growth and survival of a lepidopteran herbivore. *Oikos* 55: 66–70.

Alder, N. N., J. S. Sperry, and W. T. Pockman. 1996. Root and stem xylem embolism, stomatal conductance, and leaf turgor in *Acer grandidentatum* populations along a soil moisture gradient. *Oecologia* 105: 293–301.

Appanah, S. 1993. Mass flowering of dipterocarp forests in the aseasonal tropics. *Journal of Biosciences* 18: 457–474.

Ashton, P. S., T. J. Givnish, and S. Appanah. Staggered flowering in the Dipterocarpaceae: new insights into floral induction and the evolution of mast fruiting in the aseasonal tropics. *American Naturalist* 132: 44–66.

Augspurger, C. K. 1981. Reproductive synchrony of a tropical shrub: experimental studies on effects of pollinators and seed predators on *Hybanthus prunifolius* (Violaceae). *Ecology* 62: 775–788.

Augspurger, C. K. 1982. A cue for synchronous flowering, pp. 133–150. In E. G. Leigh, Jr., A. S. Rand, and D. M. Windsor, eds., *The Ecology of a Tropical Forest*. Smithsonian Institution Press, Washington, DC.

Bernhard-Reversat, F., C. Huttel, and G. Lemée. 1978. Structure and functioning of evergreen rain forest ecosystems of the Ivory Coast, pp. 557–574. In *Tropical Forest Ecosystems*, UNESCO, Paris.

Bonaccorso, F. J. 1979. Foraging and reproductive ecology in a Panamanian bat community. *Bulletin of the Florida State Museum, Biological Sciences* 24: 359–408.

Borchert, R. 1980. Phenology and ecophysiology of tropical trees: *Erythrina poeppigiana* O. F. Cook. *Ecology* 61: 1065–1074.

Bourlière, F. 1985. Primate communities: their structure and role in tropical ecosystems. *International Journal of Primatology* 6: 1–27.

Bray, J. R., and E. Gorham. 1964. Litter production in forests of the world. *Advances in Ecological Research* 2: 101–157.

Burghardt, G. M., H. W. Greene, and A. S. Rand. 1977. Social behavior in hatchling green iguanas: life in a reptile rookery. *Science* 195: 689–691.

Chan, H. T. 1980. Reproductive biology of some Malaysian dipterocarps II. Fruiting biology and seedling studies. *Malaysian Forester* 43: 438–451.

Charles-Dominique, P. 1971. Éco-éthologie des prosimiens du Gabon. *Biologia Gabonica* 7: 121–228.

Charles-Dominique, P., M. Atramentowicz, M. Charles-Dominique, H. Gérard, A. Hladik, C. M. Hladik, and M. F. Prévost. 1981. Les mammifères frugivores arboricoles nocturnes d'une forêt guyanaise: inter-relations plantes-animaux. *Revue d'Écologie (LaTerre et la Vie)* 35: 341–435.

Cochard, H., F. W. Ewers, and M. T. Tyree. 1994. Water relations of a tropical vine-like bamboo (*Rhipidocladum racemiflorum*): root pressures, vulnerability to cavitation and seasonal changes in embolism. *Journal of Experimental Botany* 45: 1085–1089.

Coley, P. D. 1982. Rates of herbivory on different tropical trees, pp. 123–132. In E. G. Leigh, Jr., A. S. Rand, and D. M. Windsor, eds., *The Ecology of a Tropical Forest*. Smithsonian Institution Press, Washington, DC.

Coley, P. D. 1983. Herbivory and defensive characteristics of tree species in a lowland tropical forest. *Ecological Monographs* 53: 209–233.

Coley, P. D. 1987a. Interspecific variation in plant anti-herbivore properties: the role of habitat quality and rate of disturbance. *New Phytologist* 106 (supplement): 251–263.

Coley, P. D. 1987b. Patrones en las defensas de las plantas: porqué los herbivoros prefieren ciertas especies? *Revista de Biologia Tropical* 35 (supplement 1): 151–164.

Coley, P. D. 1988. Effects of plant growth rate and leaf lifetime on the amount and type of anti-herbivore defense. *Oecologia* 74: 531–536.

Coley, P. D., and T. M. Aide. 1991. Comparison of herbivory and plant defenses in temperate and tropical broad-leaved forests, pp. 25–49. In P. W. Price, T. M. Lewinsohn, G. W. Fernandes, and W. W. Benson, eds., *Plant-Animal Interactions: Evolutionary Ecology in Tropical and Temperate Regions*. John Wiley and Sons, New York.

Coley, P. D., J. P. Bryant, and F. Stuart Chapin III. 1985. Resource availability and plant herbivore defense. *Science* 230: 895–899.

Coley, P. D., and T. Kursar. 1996. Anti-herbivore defenses of young tropical leaves: physiological constraints and ecological tradeoffs, pp. 305–336. In S. S. Mulkey, R. L. Chazdon, and A. P. Smith, eds., *Tropical Forest Plant Physiology*. Chapman and Hall, New York.

Condit, R., S. P. Hubbell, and R. B. Foster. 1995. Mortality rates of 205 neotropical tree and shrub species and the impact of a severe drought. *Ecological Monographs* 65: 419–439.

Connell, J. H. 1980. Diversity and the coevolution of competitors, or the ghost of competition past. *Oikos* 35: 131–138.

Cuevas, E., and E. Medina. 1986. Nutrient dynamics within amazonian forest ecosystems. I. Nutrient flux in fine litter fall and efficiency of nutrient utilization. *Oecologia* 68: 466–472.

Darwin, C. R. 1859. *On the Origin of Species by Means of Natural Selection*. John Murray, London.

Davies, A. G., E. L. Bennett, and P. G. Waterman. 1988. Food selection by two South-east Asian colobine monkeys (*Presbytis rubicunda* and *Presbytis melalophos*) in relation to plant chemistry. *Biological Journal of the Linnean Society* 34: 33–56.

Eisenberg, J. F., and R. W. Thorington, Jr. 1973. A preliminary analysis of a neotropical mammal fauna. *Biotropica* 5: 150–161.

Emmons, L. H. 1980. Ecology and resource partitioning among nine species of African rain forest squirrels. *Ecological Monographs* 50: 31–54.

Emmons, L. H. 1987. Comparative feeding ecology of felids in a neotropical rainforest. *Behavioral Ecology and Sociobiology* 20: 271–283.

Enders, R. K. 1935. Mammalian life histories from Barro Colorado Island, Panama. *Bulletin of the Museum of Comparative Zoology* 78: 385–502.

Erwin, T. L. 1982. Tropical forests: their richness in Coleoptera and other arthropod species. *Coleopterists Bulletin* 36: 74–75.

Erwin, T. L., and J. C. Scott. 1980. Seasonal and size patterns, trophic structure, and richness of Coleoptera in the tropical arboreal ecosystem: the fauna of the tree *Luehea seemannii* Triana and Planch in the Canal Zone of Panama. *Coleopterists Bulletin* 34: 305–322.

Fahey, T. J. 1983. Nutrient dynamics of above-ground detritus in lodgepole pine (*Pinus contorta* ssp. *latifolia*) ecosystems, southeastern Wyoming. *Ecological Monographs* 53: 51–72.

Feeny, P. 1992. The evolution of chemical ecology: contributions from the study of herbivorous insects, pp. 1–44. In G. A. Rosenthal and May R. Berenbaum, eds., *Herbivores: Their Interactions with Secondary Plant Metabolites*. vol. 2, *Ecological and Evolutionary Processes*. Academic Press, San Diego, CA.

Filip, V., R. Dirzo, J. M. Maass, and J. Sarukhán. 1995. Within- and among-year variation in the levels of herbivory on the foliage of trees from a Mexican tropical deciduous forest. *Biotropica* 27: 78–86.

Fittkau, E. J., and H. Klinge. 1973. On biomass and trophic structure of the Central Amazonian rainforest ecosystem. *Biotropica* 5: 2–14.

Fogden, M. 1976. A census of a bird community in tropical rain forest in Sarawak. *Sarawak Museum Journal* 24(45) (n. s.): 251–267.

Forget, P.-M. 1993. Post-dispersal predation and scatterhoarding of *Dipteryx panamensis* (Papilionaceae) seeds by rodents in Panama. *Oecologia* 94: 255–261.

Forget, P.-M., and T. Milleron. 1991. Evidence for secondary seed dispersal by rodents in Panama. *Oecologia* 87: 596–599.

Forget, P.-M., E. Munoz, and E. G. Leigh Jr. 1994. Predation by rodents and bruchid beetles on seeds of *Scheelea* palms on Barro Colorado Island, Panama. *Biotropica* 26: 420–426.

Foster, R. B. 1973. *Seasonality of Fruit Production and Seed Fall in a Tropical Forest Ecosystem in Panama*. Ph.D. dissertation, Department of Botany, Duke University, Durham, NC.

Foster, R. B. 1982a. Famine on Barro Colorado Island, pp. 201–212. In E. G. Leigh, Jr., A. S. Rand, and D. M. Windsor, eds., *The Ecology of a Tropical Forest*. Smithsonian Institution Press, Washington, DC.

Foster, R. B. 1982b. The seasonal rhythm of fruitfall on Barro Colorado Island, pp. 151–172. In E. G. Leigh, Jr., A. S. Rand, and D. M. Windsor, eds., *The Ecology of a Tropical Forest*. Smithsonian Institution Press, Washington, DC.

Foster, R. B., and N. V. L. Brokaw. 1982. Structure and history of the vegetation of Barro Colorado Island, pp. 67–81. In E. G. Leigh, Jr., A. S. Rand, and D. M. Windsor, eds., *The Ecology of a Tropical Forest*. Smithsonian Institution Press, Washington, DC.

Fox, L. R., and P. A. Morrow. 1983. Estimates of damage by herbivorous insects on *Eucalyptus* trees. *Australian Journal of Ecology* 8: 139–147.

Galat, G., and A. Galat-Luong. 1985. La communauté de Primates diurnes de la forêt de Taï, Côte-d'Ivoire. *Revue d'Écologie (La Terre et la Vie)* 40: 3–32.

Ganzhorn, J. U. 1992. Leaf chemistry and the biomass of folivorous primates in tropical forests: tests of a hypothesis. *Oecologia* 91: 540–547.

Garwood, N. C. 1982. Seasonal rhythm of seed germination in a semideciduous tropical forest, pp. 173–185. In E. G. Leigh, Jr., A. S. Rand, and D. M. Windsor, eds., *The Ecology of a Tropical Forest*. Smithsonian Institution Press, Washington, DC.

Garwood, N. C. 1983. Seed germination in a seasonal tropical forest in Panama: a community study. *Ecological Monographs* 53: 159–181.

Gaume, L., D. McKey, and M.-C. Anstett. 1997. Benefits conferred by "timid" ants: active anti-herbivore protection of the rainforest tree *Leonardoxa africana* by the minute ant *Petalomyrmex phylax*. *Oecologia* 112: 209–216.

Gautier-Hion, A. 1980. Seasonal variations of diet related to species and sex in a community of *Cercopithecus* monkeys. *Journal of Animal Ecology* 49: 237–269.

Gautier-Hion, A., J. M. Duplantier, L. Emmons, F. Feer, P. Heckestweiler, A. Moungazi, R. Quris, and C. Sourd. 1985. Coadaptation entre rythmes de fructification et frugivorie en forêt tropicale humide du Gabon: mythe ou réalité? *Revue d'Écologie (La Terre et la Vie)* 40: 405–434.

Gautier-Hion, A., and J. P. Gautier. 1979. Niche écologique et diversité des espèces sympatriques dans le genre *Cercopithecus*. *La Terre et la Vie* 33: 493–507.

Gautier-Hion, A., and G. Michaloud. 1989. Are figs always keystone resources for tropical frugivorous vertebrates? A test in Gabon. *Ecology* 70: 1826–1833.

Glanz, W. E. 1982. The terrestrial mammal fauna of Barro Colorado Island: censuses and long-term changes, pp. 455–468. In E. G. Leigh, Jr., A. S. Rand, and D. M. Windsor, eds., *The Ecology of a Tropical Forest*. Smithsonian Institution Press, Washington, DC.

Glanz, W. E. 1990. Neotropical mammal densities: how unusual is the community on Barro Colorado Island, Panama?, pp. 287–313. In A. H. Gentry, ed. *Four Neotropical Rainforests*. Yale University Press, New Haven, CT.

Glanz, W. E., R. W. Thorington Jr., J. Giacalone-Madden, and L. R. Heaney. 1982. Seasonal food use and demographic trends in *Sciurus granatensis*, pp. 239–252. In E. G. Leigh, Jr., A. S. Rand, and D. M. Windsor, eds., *The Ecology of a Tropical Forest*. Smithsonian Institution Press, Washington, DC.

Gliwicz, J. 1984. Population dynamics of the spiny rat *Proechimys semispinosus* on Orchid Island (Panama). *Biotropica* 16: 73–78.

Gower, S. T., P. B. Reich, and Y. Son. 1993. Canopy dynamics and aboveground production of five tree species with different leaf longevities. *Tree Physiology* 12: 327–345.

Gradwohl, J., and R. Greenberg. 1982. The effect of a single species of avian predator on the arthropods of aerial leaf litter. *Ecology* 63: 581–583.

Grier, C. C., K. A. Vogt, M. R. Keyes, and R. L. Edmonds. 1981. Biomass distribution and above- and below-ground production in young and mature *Abies amabilis* zone ecosystems of the Washington Cascades. *Canadian Journal of Forest Research* 11: 155–167.

Hairston, N. G., F. E. Smith, and L. B. Slobodkin. 1960. Community structure, population control, and competition. *American Naturalist* 94: 421–425.

Handley, C. O. Jr., D. E. Wilson, and A. L. Gardner (eds). 1991. *Demography and Natural History of the Common Fruit Bat, Artibeus jamaicensis, on Barro Colorado Island, Panamá*. Smithsonian Institution Press, Washington, DC.

Hladik, A., and C. M. Hladik. 1969. Rapports trophiques entre végétation et primates dans la forêt de Barro Colorado (Panama), *La Terre et la Vie* 23: 25–117.

Hladik, A. 1978. Phenology of leaf production in rain forest of Gabon: distribution and composition of food for folivores, pp. 51–71. In G. G. Montgomery, ed., *The Ecology of Arboreal Folivores*. Smithsonian Institution Press, Washington, DC.

Hladik, C. M. 1973. Alimentation et activité d'un groupe de chimpanzés reintroduit en forêt gabonaise. *La Terre et la Vie* 27: 343–413.

Hladik, C. M., and A. Hladik. 1972. Disponibilités alimentaires et domaines vitaux des Primates à Ceylan. *La Terre et la Vie* 26: 149–215.

Holbrook, N. M., M. J. Burns, and C. B. Field. 1995. Negative xylem pressures in plants: a test of the balancing pressure technique. *Science* 270: 1193–1195.

Holmes, R. T., J. C. Shultz, and P. Nothnagle. 1979. Bird predation on forest insects: an exclosure experiment. *Science* 206: 462–463.

Hutchinson, G. E. 1959. Homage to Santa Rosalia, or, why are there so many kinds of animals? *American Naturalist* 93: 145–159.

Janson, C. H., and L. H. Emmons. 1990. Ecological structure of the nonflying mammal community at Cocha Cashu Biological Station, Manu National Park, Peru, pp. 314–338. In A. H. Gentry, ed. *Four Neotropical Rainforests*. Yale University Press, New Haven, CT.

Janzen, D. H. 1974. Tropical blackwater rivers, animals, and mast fruiting by the Dipterocarpaceae. *Biotropica* 6: 69–103.

Janzen, D. H. 1981. Patterns of herbivory in a tropical deciduous forest. *Biotropica* 13: 271–282.

Julien-Laferrière, D., and M. Atramentowicz. 1990. Feeding and reproduction of three didelphid marsupials in two Neotropical forests (French Guiana). *Biotropica* 22: 404–415.

Karanth, K. U., and M. E. Sunquist. 1992. Population structure, density and biomass of large herbivores in the tropical forests of Nagarahole, India. *Journal of Tropical Ecology* 8: 21–35.

Karr, J. R. 1971. Structure of avian communities in selected Panama and Illinois habitats. *Ecological Monographs* 41: 207–234.

Kira, T. 1978. Community archicture and organic matter dynamics in tropical lowland rain forests of Southeast Asia with special reference to Pasoh Forest, West Malaysia, pp. 561–590. In P. B. Tomlinson and M. H. Zimmerman, eds., *Tropical Trees as Living Systems*. Cambridge University Press, Cambridge.

Kitajima, K. 1994. Relative importance of photosynthetic traits and allocation patterns as correlates of seedling shade tolerance of 13 tropical trees. *Oecologia* 98: 419–428.

Kitajima, K., S. S. Mulkey, and S. J. Wright. 1997. Seasonal leaf phenotypes in the canopy of a tropical dry forest: photosynthetic characteristics and associated traits. *Oecologia* 109: 490–498.

Leigh, E. G. Jr. 1975. Structure and climate in tropical rainforest. *Annual Review of Ecology and Systematics* 6: 67–86.

Leigh, E. G. Jr., and Handley, C. O. Jr. 1991. Population estimates, pp. 77–87. In C. O. Handley, Jr., D. E. Wilson, and A. L. Gardner, eds., *Demography and Natural History of the Common Fruit Bat, Artibeus jamaicensis, on Barro Colorado Island, Panamá*. Smithsonian Institution Press, Washington, DC.

Leigh, E. G. Jr., R. T. Paine, J. F. Quinn, and T. H. Suchanek. 1987. Wave energy and intertidal productivity. *Proceedings of the National Academy of Sciences, USA* 84: 1314–1318.

Leigh, E. G. Jr., A. S. Rand, and D. M. Windsor (eds). 1982. *The Ecology of a Tropical Forest*. Smithsonian Institution Press, Washington, DC.

Leigh, E. G. Jr., and N. Smythe. 1978. Leaf production, leaf consumption, and the regulation of folivory on Barro Colorado Island, pp. 33–50. In G. G. Montgomery, ed., *The Ecology of Arboreal Folivores*. Smithsonian Institution Press, Washington DC.

Leigh, E. G. Jr., and D. M. Windsor. 1982. Forest production and regulation of primary consumers on Barro Colorado Island, pp. 111–122. In E. G. Leigh, Jr., A. S. Rand, and D. M. Windsor, eds., *The Ecology of a Tropical Forest*. Smithsonian Institution Press, Washington, DC.

Leigh, E. G. Jr., D. M. Windsor, A. S. Rand, and R. B. Foster. 1990. The impact of the "El Niño" drought of 1982–83 on a Panamanian semideciduous forest, pp. 473–486. In P. W. Glynn, ed., *Global Ecological Consequences of the 1982–83 El Niño-Southern Oscillation*. Elsevier, Amsterdam.

Leigh, E. G. Jr., S. J. Wright, F. E. Putz, and E. A. Herre. 1993. The decline of tree diversity on newly isolated tropical islands: a test of a null hypothesis and some implications. *Evolutionary Ecology* 7: 76–102.

Levey, D. J. 1988. Spatial and temporal variation in Costa Rican fruit and fruit-eating bird abundance. *Ecological Monographs* 58: 251–269.

Lim, M. T. 1978. Litterfall and mineral content of litter in Pasoh forest. *Malayan Nature Journal* 30: 375–380.

Lowman, M. D. 1985. Temporal and spatial variability in insect grazing of the canopies of five Australian rainforest tree species. *Australian Journal of Ecology* 10: 7–24.

Lowman, M. D. 1992. Herbivory in Australian rain forests, with particular reference to the canopies of *Doryphora sassafras* (Monimiaceae). *Biotropica* 24: 263–272.

Lowman, M. D., and N. M. Nadkarni (eds). 1995. *Forest Canopies*. Academic Press, San Diego, CA.

Loyn, R. H. 1987. The bird that farms the dell. *Natural History* 6/87: 54–60.

Loyn, R. H., R. G. Runnalls, G. Y. Forward, and J. Tyers. 1983. Territorial bell miners and other birds affecting populations of insect prey. *Science* 221: 1411–1413.

Lubin, Y. D. 1983. *Tamandua mexicana* (oso jaceta, hormiguero, tamandua, banded anteater, lesser anteater), pp. 494–496. In D. H. Janzen, ed., *Costa Rican Natural History*. University of Chicago Press, Chicago.

McGraw, S. 1994. Census, habitat preference, and polyspecific associations of six monkeys in the Lomako Forest, Zaire. *American Journal of Primatology* 34: 295–307.

Machado, J.-L., and M. T. Tyree. 1994. Patterns of hydraulic architecture and water relations of two tropical canopy trees with contrasting leaf phenologies: *Ochroma pyramidale* and *Pseudobombax septenatum*. *Tree Physiology* 14: 219–240.

McLaren, B. E., and R. O. Peterson. 1994. Wolves, moose, and tree rings on Isle Royale. *Science* 266: 155–158.

Marquis, R. J., and C. J. Whelan. 1994. Insectivorous birds increase growth of white oak through consumption of leaf-chewing insects. *Ecology* 75: 2007–2014.

May, R. M. 1990. How many species? *Philosophical Transactions of the Royal Society* B330: 293–304.

Meinzer, F. C., G. Goldstein, N. M. Holbrook, P. Jackson, and J. Cavelier. 1993. Stomatal and environmental control of transpiration in a lowland tropical forest tree. *Plant, Cell and Environment* 16: 429–436.

Milton, K. 1979. Factors influencing leaf choice by howler monkeys: a test of some hypotheses of food selection by generalist herbivores. *American Naturalist* 114: 362–378.

Milton, K. 1982. Dietary quality and demographic regulation in a howler monkey population, pp. 273–289. In E. G. Leigh, Jr., A. S. Rand, and D. M. Windsor, eds., *The Ecology of a Tropical Forest*. Smithsonian Institution Press, Washington, DC.

Milton, K. 1990. Annual mortality patterns of a mammal community in central Panama. *Journal of Tropical Ecology* 6: 493–499.

Mitchell, B. J. 1989. Resources, Group Behavior and Infant Development in White-faced Capuchin Monkeys, *Cebus capucinus*. PhD dissertation, Department of Zoology, University of California, Berkeley, CA.

Moffett, M. W. 1993. *The High Frontier: Exploring the Rainforest Canopy*. Harvard University Press, Cambridge, MA.

Montgomery, G. G. 1985. Impact of vermilinguas (*Cyclopes, Tamandua*: Xenarthra = Edentata) on arboreal ant populations, pp. 351–363. In G. G. Montgomery, ed., *The Evolution* and *Ecology of Armadillos, Sloths and Vermilinguas*. Smithsonian Institution Press, Washington, DC.

Montgomery, G. G., and M. E. Sunquist. 1975. Impact of sloths on neotropical forest energy flow and nutrient cycling, pp. 69–98. In F. B. Golley and E. Medina, eds., *Tropical Ecological Systems: Trends in Terrestrial and Aquatic Research*. Springer-Verlag, New York.

Montgomery, G. G., and M. E. Sunquist. 1978. Habitat selection and use by two-toed and three-toed sloths, pp. 329–359. In G. G. Montgromery, ed., *The Ecology of Arboreal Folivores*. Smithsonian Institution Press, Washington, DC.

Moran, V. C., and T. R. E. Southwood. 1982. The guild composition of arthropod communities in trees. *Journal of Animal Ecology* 51: 289–306.

Morrison, D. W. 1978. Foraging ecology and energetics of the frugivorous bat *Artibeus jamaicensis*. *Ecology* 59: 716–723.

Morrison, D. W. 1980. Efficiency of food utilization by fruit bats. *Oecologia* 45: 270–273.

Mulkey, S. S., A. P. Smith, and S. J. Wright. 1991. Comparative life history and physiology of two understory Neotropical herbs. *Oecologia* 88: 263–273.

Mulkey, S. S., A. P. Smith, S. J. Wright, J. L. Machado, and R. Dudley. 1992. Contrasting leaf phenotypes control seasonal variation in water loss in a tropical forest shrub. *Proceedings of the National Academy of Sciences, USA* 89: 9084–9088.

Murali, K. S., and R. Sukumar. 1993. Leaf flushing phenology and herbivory in a tropical dry deciduous forest, southern India. *Oecologia* 94: 114–119.

Nagy, K. A. 1987. Field metabolic rate and food requirement scaling in mammals and birds. *Ecological Monographs* 57: 111–128.

Nagy, K. A., and K. Milton. 1979. Energy metabolism and food consumption by wild howler monkeys (*Alouatta palliatta*). *Ecology* 60: 475–480.

Nagy, K. A., and G. G. Montgomery. 1980. Field metabolic rate, water flux and food consumption in three-toed sloths (*Bradypus variegatus*). *Journal of Mammalogy* 61: 465–472.

Oates, J. F., G. H. Whitesides, A. G. Davies, P. G. Waterman, S. M. Green, G. L. Dasilva,, and S. Mole. 1990. Determinants of variation in tropical forest primate biomass: new evidence from West Africa. *Ecology* 71: 328–343.

Overdorff, D. J. 1992. Differential patterns in flower feeding by *Eulemur fulvus* and *Eulemur rubriventer* in Madagascar. *American Journal of Primatology* 28: 191–203.

Pockman, W. T., J. S. Sperry, and J. W. O'Leary. 1995. Sustained and significant negative water pressure in xylem. *Nature* 378: 715–716.

Raemaekers, J. J., and D. J. Chivers. 1980. Socio-ecology of Malayan forest primates, pp. 279–316. In D. J. Chivers, ed., *Malayan Forest Primates*. Plenum Press, New York.

Reich, P. B. 1995. Phenology of tropical forests: patterns, causes, and consequences. *Canadian Journal of Botany* 73: 164–174.

Reich, P. B., and R. Borchert. 1982. Phenology and ecophysiology of the tropical tree, *Tabebuia neochrysantha* (Bignoniaceae). *Ecology* 63: 294–299.

Reich, P. B., and R. Borchert. 1984. Water stress and tree phenology in a tropical dry forest in the lowlands of Costa Rica. *Journal of Ecology* 72: 61–74.

Reich, P. B., C. Uhl, M. B. Walters, and D. S. Ellsworth. 1991. Leaf lifespan as a determinant of leaf structure and function among 23 amazonian tree species. *Oecologia* 86: 16–24.

Reich, P. B., M. B. Walters, and D. S. Ellsworth. 1992. Leaf lifespan in relation to leaf, plant and stand characteristics among diverse ecosystems. *Ecological Monographs* 62: 365–392.

Reichle, D. E., R. A. Goldstein, R. I. van Hook, Jr., and G. J. Dodson. 1973. Analysis of insect consumption in a forest canopy. *Ecology* 54: 1076–1084.

Roberts, J., O. M. R. Cabral, G. Fisch, L. C. B. Molion, C. J. Moore, and W. J. Shuttleworth. 1993. Transpiration from an Amazonian rainforest calculated from stomatal conductance measurements. *Agricultural and Forest Meteorology* 65: 175–196.

Rockwood, L. L. 1974. Seasonal changes in the susceptibility of *Crescentia alata* leaves to the flea beetle, *Oedionychus* sp. *Ecology* 55: 142–148.

Rosenthal, G. A., and M. R. Berenbaum (eds). 1991. *Herbivores: Their Interactions with Secondary Plant Metabolites*, vol. 1, *The Chemical Participants*. Academic Press, San Diego, CA.

Rosenthal, G. A., and M. R. Berenbaum (eds). 1992. *Herbivores: Their Interactions with Secondary Plant Metabolites*, vol. 2, *Ecological and Evolutionary Processes*. Academic Press, San Diego, CA.

Rosenzweig, M. L., and Z. Abramsky. 1986. Centrifugal community organization. *Oikos* 46: 339–348.

Russell, J. K. 1982. Timing of reproduction by coatis (*Nasua narica*) in relation to fluctuations in food resources, pp. 413–431. In E. G. Leigh, Jr., A. S. Rand, and D. M. Windsor, eds., *The Ecology of a Tropical Forest*. Smithsonian Institution Press, Washington, DC.

Saliendra, N. Z., J. S. Sperry, and J. P. Comstock. 1995. Influence of leaf water status on stomatal response to humidity, hydraulic conductance, and soil drought in *Betula occidentalis*. *Planta* 196: 357–366.

Schoener, T. W. 1982. The controversy over interspecific competition. *American Scientist* 70: 586–595.

Schowalter, T. D., and D. A. Crossley, Jr. 1988. Canopy arthropods and their response to forest disturbance, pp. 207–218. In W. T. Swank and D. A. Crossley, Jr., eds., *Forest Hydrology and Ecology at Coweeta*. Springer-Verlag, New York.

Schowalter, T. D., J. W. Webb, and D. A. Crossley. 1981. Community structure and nutrient content of canopy arthropods in clearcut and uncut forest ecosystems. *Ecology* 62: 1010–1019.

Schupp, E. W. 1990. Annual variation in seedfall, postdispersal predation, and recruitment of a Neotropical tree. *Ecology* 71: 504–515.

Sinclair, A. R. E. 1975. The resource limitations of trophic levels in tropical grassland ecosystems. *Journal of Animal Ecology* 44: 497–520.

Smith, A. P. 1987. Respuestas de hierbas del sotobosque tropical a claros ocasionados por la caída de árboles. *Revista de Biología Tropical* 35 (supplement 1): 111–118.

Smith, J. N. M., P. R. Grant, B. R. Grant, I. J. Abbott, and L. K. Abbott. 1978. Seasonal variation in feeding habits of Darwin's ground finches. *Ecology* 59: 1137–1150.

Smythe, N. 1970. Relationships between fruiting seasons and seed dispersal methods in a Neotropical forest. *American Naturalist* 104: 25–35.

Smythe, N. 1978. The natural history of the Central American agouti (*Dasyprocta punctata*). *Smithsonian Contributions to Zoology* 257: 1–52.

Smythe, N. 1982. The seasonal abundance of night-flying insects in a neotropical forest, pp. 309–318. In E. G. Leigh, Jr., A. S. Rand, and D. M. Windsor, eds., *The Ecology of a Tropical Forest*. Smithsonian Institution Press, Washington, DC.

Smythe, N. 1989. Seed survival in the palm *Astrocaryum standleyanum*: evidence for dependence upon its seed dispersers. *Biotropica* 21: 50–56.

Smythe, N., W. E. Glanz, and E. G. Leigh, Jr. 1982. Population regulation in some terrestrial frugivores, pp. 227–238. In E. G. Leigh, Jr., A. S. Rand, and D. M. Windsor, eds., *The*

Ecology of a Tropical Forest. Smithsonian Institution Press, Washington, DC.

Snow, D. W. 1976. *The Web of Adaptation.* Cornell University Press, Ithaca, NY.

Sperry, J. S., N. M. Holbrook, M. H. Zimmermann, and M. T. Tyree. 1987. Spring filling of xylem vessels in wild grapevine. *Plant Physiology* 83: 414–417.

Sperry, J. S., M. T. Tyree, and J. R. Donnelly. 1988. Vulnerability of xylem to embolism in a mangrove vs an inland species of Rhizophoraceae. *Physiologia Plantarum* 74: 276–283.

Steinberg, P. D., J. A. Estes, and F. C. Winter. 1995. Evolutionary consequences of food chain length in kelp forest communities. *Proceedings of the National Academy of Sciences, USA* 92: 8145–8148.

Sternberg, L. S. L., S. S. Mulkey, and S. J. Wright. 1989. Ecological interpretation of leaf carbon isotope ratios: influence of respired carbon dioxide. *Ecology* 70: 1317–1324.

Stork, N. E. 1987. Guild structure of arthropods from Bornean rain forest trees. *Ecological Entomology* 12: 69–80.

Stork, N. E. 1991. The composition of the arthropod fauna of Bornean lowland rain forest trees. *Journal of Tropical Ecology* 7: 161–180.

Struhsaker, T. T. 1975. *The Red Colobus Monkey.* University of Chicago Press, Chicago.

Terborgh, J. 1983. *Five New World Primates.* Princeton University Press, Princeton, NJ.

Terborgh, J. 1986. Community aspects of frugivory in tropical forests, pp. 372–384. In A. Estrada and T. H. Fleming, eds., *Frugivores and Seed Dispersal.* W. Junk, Dordrecht.

Terborgh, J. 1988. The big things that run the world—a sequel to E. O. Wilson. *Conservation Biology* 2: 402–403.

Terborgh, J., S. K. Robinson, T. A. Parker III, C. A. Munn, and N. Pierpont. 1990. Structure and organization of an Amazonian forest bird community. *Ecological Monographs* 60: 213–238.

Terborgh, J., and C. P. van Schaik. 1987. Convergence vs. nonconvergence in primate communities, pp. 205–226. In J. H. R. Gee and P. S. Giller, eds., *Organization of Communities.* Blackwell Scientific Publications, Oxford.

Terwilliger, V. J. 1978. Natural history of Baird's tapir on Barro Colorado Island, Panama Canal Zone. *Biotropica* 10: 211–220.

Thiollay, J. M. 1986. Structure comparée du peuplement avien dans trois sites de forêt primaire en Guyane. *Revue d'Écologie (La Terre et la Vie)* 41: 59–105.

Thomas, S. C. 1991. Population densities and patterns of habitat use among anthropoid primates of the Ituri Forest, Zaire. *Biotropica* 23: 68–83.

Thompson, D. W. 1942. *On Growth and Form.* Cambridge University Press, Cambridge.

Tissue, D. T., and S. J. Wright. 1995. Effect of seasonal water availability on phenology and the annual shoot carbohydrate cycle of tropical forest shrubs. *Functional Ecology* 9: 518–527.

Tyree, M. T. 1988. A dynamic model for water flow in a single tree: evidence that models must account for hydraulic architecture. *Tree Physiology* 4: 195–217.

Tyree, M. T., H. Cochard, P. Cruiziat, B. Sinclair, and T. Ameglio. 1993. Drought-induced leaf shedding in walnut: evidence for vulnerability segmentation. *Plant, Cell and Environment* 16: 879–882.

Tyree, M. T., S. D. Davis, and H. Cochard. 1994a. Biophysical perspectives of xylem evolution: is there a tradeoff of hydraulic efficiency for vulnerability to dysfunction? *IAWA Journal* 15: 335–360.

Tyree, M. T., and F. W. Ewers. 1991. Tansley review no. 34: the hydraulic architecture of trees and other woody plants. *New Phytologist* 119: 345–360.

Tyree, M. T., S. Patiño, J. Bennink, and J. Alexander. 1995. Dynamic measurements of root hydraulic conductance using a high-pressure flowmeter in the laboratory and field. *Journal of Experimental Botany* 46: 83–94.

Tyree, M. T., D. A. Snyderman, T. R. Wilmot, and J.-L. Machado. 1991. Water relations and hydraulic architecture of a tropical tree (*Schefflera morototoni*). *Plant Physiology* 96: 1105–1113.

Tyree, M. T., and J. S. Sperry. 1988. Do woody plants operate near the point of catastrophic xylem dysfunction caused by dynamic water stress? *Plant Physiology* 88: 574–580.

Tyree, M. T., and J. S. Sperry. 1989a. Characterization and propagation of acoustic emission signals in woody plants: towards an improved acoustic emission counter. *Plant, Cell and Environment* 12: 371–382.

Tyree, M. T., and J. S. Sperry. 1989b. Vulnerability of xylem to cavitation and embolism. *Annual Reviews of Plant Physiology and Molecular Biology* 40: 19–38.

Tyree, M. T., S. Yang, P. Cruiziat, and B. Sinclair. 1994b. Novel methods of measuring hydraulic conductivity of tree root systems and interpretation using AMAIZED. *Plant Physiology* 104: 189–199.

van Schaik, C. P. 1986. Phenological changes in a Sumatran rain forest. *Journal of Tropical Ecology* 2: 327–347.

Vogel, S. 1988. *Life's Devices.* Princeton University Press, Princeton, NJ.

Waterman, P. G., J. A. M. Ross, E. L. Bennett, and A. G. Davies. 1988. A comparison of the floristics and leaf chemistry of the tree flora in two Malaysian rain forests and the influence of leaf chemistry on populations of colobine monkeys in the Old World. *Biological Journal of the Linnean Society* 34: 1–32.

Whitesides, G. H., J. F. Oates, S. M. Green, and R. P. Kluberdanz. 1988. Estimating primate densities from transects in a West African rain forest: a comparison of techniques. *Journal of Animal Ecology* 57: 345–367.

Willis, E. O. 1980. Ecological roles of migrant and resident birds on Barro Colorado Island, Panama, pp. 205–225. In A. Keast and E. M. Morton, eds., *Migrant Birds in the Neotropics.* Smithsonian Institution Press, Washington, DC.

Windsor, D. M. 1990. Climate and moisture variability in a tropical forest: long-term records from Barro Colorado Island, Panama. *Smithsonian Contributions to the Earth Sciences* 29: 1–145.

Winter, K. 1985. Crassulacean acid metabolism, pp. 329–387. In J. Barber and N. R. Baker, eds., *Photosynthetic Mechanisms and the Environment.* Elsevier, Amsterdam.

Wolda, H. 1979. Abundance and diversity of Homoptera in the canopy of a tropical forest. *Ecological Entomology* 4: 181–190.

Wolda, H. 1982. Seasonality of Homoptera on Barro Colorado Island, pp. 319–330. In E. G. Leigh, Jr., A. S. Rand, and D. M. Windsor, eds., *The Ecology of a Tropical Forest.* Smithsonian Institution Press, Washington, DC.

Wright, S. J. 1991. Seasonal drought and the phenology of understory shrubs in a tropical moist forest. *Ecology* 72: 1643–1657.

Wright, S. J., and M. Colley. 1996. *Accessing the Canopy.* Smithsonian Tropical Research Institute, Balboa, Panama.

Wright, S. J., and F. J. Cornejo. 1990a. Seasonal drought and leaf fall in a tropical forest. *Ecology* 71: 1165–1175.

Wright, S. J., and F. H. Cornejo. 1990b. Seasonal drought and the timing of flowering and leaf fall in a neotropical forest, pp. 49–61. In K. S. Bawa and M. Hadley, eds., *Reproductive Ecology of Tropical Forest Plants.* UNESCO, Paris, and Parthenon Publishing, Park Ridge, NJ.

Wright, S. J., M. E. Gompper, and B. De Leon. 1994. Are large predators keystone species in Neotropical forests? The evidence from Barro Colorado Island. *Oikos* 71: 279–294.

Wright, S. J., and C. P. van Schaik. 1994. Light and the phenology of tropical trees. *American Naturalist* 143: 192–199.

Yang, S., and M. T. Tyree. 1994. Hydraulic architecture of *Acer saccharum* and *A. rubrum*: comparison of branches to whole trees and the contribution of leaves to hydraulic resistance. *Journal of Experimental Botany* 45: 179–186.

Zaret, T. M., and A. S. Rand. 1971. Competition in tropical stream fishes: support for the competitive exclusion principle. *Ecology* 52: 336–342.

Zimmermann, M. H. 1983. *Xylem Structure and the Ascent of Sap*. Springer, Berlin.

Zimmermann, U., A. Haase, D. Langbein, and F. Meinzer. 1993. Mechanisms of long-distance water transport in plants: a re-examination of some paradigms in the light of new evidence. *Philosophical Transactions of the Royal Society of London* B 341: 19–31.

Zimmermann, U., F. C. Meinzer, R. Benkert, J. J. Zhu, H. Schneider, G. Goldstein, E. Kuchenbrod, and A. Haase. 1994. Xylem water transport: is the available evidence consistent with the cohesion theory? *Plant, Cell and Environment* 17: 1169–1181.

Zotz, G., M. T. Tyree, and H. Cochard. 1994. Hydraulic architecture, water relations and vulnerability to cavitation of *Clusia uvitana* Pittier: a C_3-CAM tropical hemiepiphyte. *New Phytologist* 127: 287–295.

Zotz, G., and K. Winter. 1994a. A one-year study on carbon, water and nutrient relationships in a tropical C_3-CAM hemi-epiphyte: *Clusia uvitana* Pittier. *New Phytologist* 127: 45–60.

Zotz, G., and K. Winter. 1994b. Photosynthesis of a tropical canopy tree, *Ceiba pentandra*, in a lowland forest in Panama. *Tree Physiology* 14: 1291–1301.

EIGHT

Tropical Diversity

One of the most striking features of tropical forest is the extraordinary diversity of its inhabitants. Diversity of plants, insects, bats, and birds is conspicuously higher in tropical forests than in their temperate zone counterparts (Karr 1971, MacArthur 1972, Erwin 1982, Findley 1993), just as in marine settings the diversity of molluscs is higher at lower latitudes (Fischer 1960). As the diversity of tropical plants provides the foundation for the diversity of their consumers and the carnivores and insectivores that prey upon those consumers (Hutchinson 1959), I shall focus on why there are so many kinds of tropical trees. The diversity of tropical trees is one of the most extraordinary mysteries of tropical forest; it has yet to receive a definitive explanation.

DOCUMENTING TREE DIVERSITY

How Can We Measure Tree Diversity?

Before discussing the diversity of tropical trees, let us consider how to measure diversity in a manner that does not depend unduly on the number of trees in the sample. The fewer words on this subject, the better, for we can diminish the effects of sample size, but we cannot eliminate them entirely.

Fisher measured diversity by a quantity α, which appears as a constant in the distribution of Fisher et al. (1943), according to which the number of species, $p(m)$, represented by m individuals apiece in a sample of total size, N, is

$$p(m) = \frac{\alpha x^m}{m}. \quad (8.1)$$

Where this is true, the total number N of individuals in the sample and the total number, S, of species among them are related to α and x by the equations

$$N = \frac{\alpha x}{1-x}, \quad S = \alpha \ln\left(\frac{1}{1-x}\right). \quad (8.2)$$

If we invert the equation for N to set $x = N/(N + \alpha)$ and substitute this for x in the equation for S, we obtain

$$S = \alpha \ln\left(1 + \frac{N}{\alpha}\right). \quad (8.3)$$

Where α is independent of sample size, S increases linearly with $\ln(N)$ when $N/\alpha >> 1$ (Williams 1964).

The log series distribution was invented to fit the distribution of individuals among species in various insect samples (Fisher et al. 1943). Fisher's α is still a favored measure of insect diversity. Even where the log-series distribution does not apply, α as calculated from equations 8.2 has proved useful (Taylor et al. 1976, Wolda 1987).

As a measure of diversity, α varies less with sample size than does the number of species (Table 8.1). Square plots of 0.1 ha are insufficient to sample forest diversity, but 2 × 500 m plots often provide unbiased estimates of this diversity (Table 8.2). A square hectare plot often provides an unbiased estimate of the diversity of stems ≥ 1 and ≥ 10 cm dbh on a much larger plot, whereas narrow 1-ha rectangles overestimate it (Table 8.2). Nevertheless, the dependence of α on plot size and on the sizes of plants included in the sample differs greatly in different biogeographic regions. At Makokou in Gabon, diversity is highest when small stems are included, which is not true elsewhere, whereas α depends more strikingly on plot size at Mudumalai than in other areas (Appendix 8.1).

The log series is a "null model" for the distribution of abundance of tree species in communities where a tree's prospects of death or reproduction are not affected by what species it belongs to. To be specific, consider a forest with a large but constant number, M, of reproductive trees, and suppose that (1) trees die at random, one by one, without

Table 8.1 Number, N, of stems, number, S, of species, and diversity, α, of stems on square plots of different sizes on Barro Colorado Island, Panama, and Pasoh, Malaysia (data from Condit et al. 1996b and R. Condit, S. P. Hubbell and R. B. Foster, p. c.)

Plot size (ha)	Stems ≥ 1 cm dbh			Stems ≥ 5 cm dbh			Stems ≥ 10 cm dbh		
	N	S	α	N	S	α	N	S	α
Barro Colorado Island (1990 census)									
0.1	488	83	29.3	102	44	30.1	41	25	28.4
0.25	1220	118	32.4	256	73	34.6	106	45	30.4
1.0	4881	173	35.0	1022	129	39.2	425	91	36.0
6.25	30509	234	34.5	6387	201	39.4	2654	162	38.1
25.0	122036	280	34.2	25550	243	37.2	10617	207	36.5
50.0	244072	303	34.1	51100	270	37.4	21233	229	35.9
Pasoh (1986–89 census)									
0.1	670	212	106.4				53	41	81.2
0.25	1676	323	119.0	343	163	121.6	132	86	107.0
1.0	6702	495	123.3	1372	327	135.9	529	206	123.8
6.25	41889	681	115.5	8573	558	133.6	3307	441	136.7
25.0	167554	782	106.2	34290	695	123.4	13226	608	131.6
50.0	335108	817	100.7	68580	752	118.1	26452	683	127.9

regard to what species they belong to, (2) each dead tree is immediately replaced, (3) this replacement has probability u of being a new species, and (4) if the replacement is not a new species, it is the young of a tree chosen at random from those alive the instant before its predecessor died [this is model 2 of Watterson (1974)]. If these assumptions are met, then $\alpha = Mu/(1-u)$ and $x = 1-u$, and a random sample from this forest also follows the log series distribution with the same value of α (Watterson 1974; for details, see Appendix 8.2).

When the log series applies, Fisher's α is closely related to Simpson's (1949) measure of diversity. The latter may be expressed as $1/F$, where F, the "relative dominance," the probability that two stems sampled at random from an infinite stand are the same species, or

$$F = \sum_{i=1} p_i^2. \quad (8.4)$$

Here, p_i is the proportion of individuals in the stand that belong to species i. In a finite sample of N individuals, F is best estimated by F', the probability that two individuals randomly chosen from this sample are the same species (Watterson 1974). Here,

$$F' = \sum_{i=1}^{s} \frac{N_i(N_i - 1)}{N(N - 1)}, \quad (8.5)$$

where S is the number of species in the sample, and N_i is the number of individuals in the sample which belong to species i. If we let F_S be the relative dominance in our sample, treated as a whole stand, then

$$F_S = \sum_{i=1}^{s} N_i^2/N^2; \quad F' = \frac{NF_S - 1}{N - 1} \quad (8.6)$$

$$1 - F_S = \left(\frac{N-1}{N}\right)(1 - F'); \quad F_S = \frac{1}{N} + \left(1 - \frac{1}{N}\right)F'. \quad (8.7)$$

The relative dominance, F_S, within a sample is related to that in the parent stand, as, in the neutral model of a population where successive generations contain N adults apiece and each adult has equal chances of reproducing successfully, the relative dominance among alleles at a single locus among the adults of generation $t+1$ is related to that among adults of generation t. Thus F_S depends on sample size.

Table 8.2 Number of stems, N, number of species, S, and diversity, α, of stems on plots of different shapes at Barro Colorado and Pasoh (data from Condit et al. 1996b and R. Condit, S. P. Hubbell and R. B. Foster, p. c.

Plot dimensions (m)	Stems ≥ 1 cm dbh			Stems ≥ 10 cm dbh		
	N	S	α	N	S	α
Barro Colorado Island						
31.62 × 31.62	488	83	29.3	42	25	28.4
2 × 500	488	94	35.0	42	27	37.0
1 × 1000	488	98	37.3	42	28	40.0
100 × 100	4880	173	35.0	425	91	36.0
20 × 500	4880	181	37.1	425	96	38.8
10 × 1000	4880	189	39.2	425	100	41.5
Pasoh						
31.62 × 31.62	670	212	106.4	53	41	81.2
2 × 500	670	243	137.2	53	44	116.4
1 × 1000	670	253	147.8	53	44	127.6
100 × 100	6702	493	122.8	529	203	120.9
20 × 500	6702	513	129.3	529	217	137.2

N and S are rounded to nearest integer, but exact values were used to calculate α.

Table 8.3 Species diversity and parameters for the log series distribution of species abundances in selected tropical forest plots

Plot ID	Barro Colorado Island, Panama			Pakitza, Peru Old growth
	Ha 7.0	Ha 8,1	50-ha plot	
Plot area, ha	1.0	1.0	50.0	1.0
N	424	394	21233	655
S	92	88	229	161
α	36.17	35.18	35.86	70.29
x	0.9214	0.9180	0.9983	0.9057

N, number of trees ≥ 10 cm dbh on plot indicated; S, number of species among them. The parameter α is obtained from the equation $S = \alpha \ln(1 + N/\alpha)$; $x = N/(N + \alpha)$. Data for Barro Colorado from 1985 census of 50-ha plot (R. Condit, S. P. Hubbell, and R. B. Foster, p. c.); data for Pakitza from R. Foster (p. c.).

Where the log series applies,

$$F' = \frac{1}{\alpha + 1} \quad (8.8)$$

(Watterson 1974). F plays a crucial role in predicting the change in composition over time of a community's tree diversity in the null case where all trees are alike in the eyes of natural selection, regardless of their species (Leigh et al. 1993).

In 1-ha samples, tropical trees do seem to be distributed over species according to Fisher's log series (Tables 8.3, 8.4). In larger samples, this usually ceases to be true (Foster and Hubbell 1990, Leigh 1990b). Nonetheless, α varies little with the lower diameter limit of stems sampled (Table 8.1). Moreover, the number of species encountered in a plot

Table 8.4 Fisher's log series and the distribution of species abundances in selected tropical forest plots

Plot ID	$p(m)$, Barro Colorado Island, Panama						$p(m)$, Pakitza, Peru	
	Ha 7,0		Ha 8,1		50-ha plot		1-ha old growth	
m	P	O	P	O	P	O	P	O
1	33.33	37	32.30	33	35.80	18	63.67	70
2	15.36	14	14.83	14	17.87	17	28.83	30
3	9.43	12	9.07	10	11.89	7	17.41	18
4	6.52	8	6.25	8	8.91	5	11.83	6
5	4.81	0	4.59	2	7.11	3	8.57	3
6	3.69	5	3.51	2	5.92	7	6.47	2
7	2.91	1	2.76	3	5.06	4	5.02	4
8	2.35	0	2.22	2	4.42	4	3.98	4
9	1.92	3	1.81	2	3.93	1	3.20	3
10	1.60	2	1.50	0	3.53	6	2.61	2
11–15	4.95	4	4.60	7	13.66	18	7.74	5
16–20	2.35	2	2.15	3	9.73	11	3.38	0
21–30	1.88	2	1.67	2	13.65	29	2.39	1
31–40	0.59	2	0.50	0	9.58	12	0.63	0
41–50	0.20	0	0.17	0	7.33	8	0.18	1
51–70	0.10	0	0.08	0	10.82	15	0.07	3
71–100	0.02	0	0.01	0	11.02	16	0.01	0
101–140	0.00	0	0.00	0	9.83	8	0.00	0
141–200	0.00	0	0.00	0	9.60	9	0.00	0
201–280	0.00	0	0.00	0	8.06	11	0.00	0
281–400	0.00	0	0.00	0	7.25	8	0.00	0
401+	0.00	0	0.00	0	14.04	10	0.00	0

The number of species, $p(m)$ with m trees ≥ 10 cm dbh, predicted (P) according to Fisher's log series for different sample plots and observed (O) there. The predicted value of $p(m)$ is calculated from the formula $p(m) = \alpha x^m / m$: values of α and x are given in Table 8.3.

Data for Barro Colorado from R. Condit, S. P. Hubbell, and R. B. Foster (p. c.); data for Pakitza from R. Foster (p. c.).

sample of N contiguous individuals is much the same regardless of the lower diameter limit of stems sampled (Table 8.1). Thus 500 stems in a square plot will contain roughly the same number of species whether they be the 500 stems ≥ 10 cm dbh in a hectare or the 500 stems ≥ 1 cm dbh in 0.1 ha (Condit et al. 1996b). Small plots still lead to sampling error, but Fisher's α does allow comparison of relatively unbiased estimates of diversity among plots where different sizes of stems were sampled. Here, α serves as a measure of tree diversity, even though different tree species respond quite differently to environmental change, so that the neutral theory and its prediction of the log series distribution do not apply (Condit et al. 1992a, Leigh et al. 1993).

Empirical Correlates of Tree Diversity

The most obvious lesson from the data on tree diversity is that tree diversity is much greater in tropical forests than in the temperate zone (Table 8.5). Why this contrast is so striking is a question that will be explored at length.

Tree diversity also differs enormously from one tropical forest to another (Appendix 8.1), a fact that may lend clues to the causes of tropical tree diversity.

Tree diversity is much greater where total annual rainfall is higher and the dry season is weak or absent (Table 8.6), as Gentry (1988a), Wright (1992), and Clinebell et al. (1995) have emphasized. This is true even though the forest is taller and soils are often better where the dry season is more severe (Martin 1991: pp. 60–61; Clinebell et al. 1995).

Tree diversity also declines with altitude, once past a possible initial increase, which reflects the weakening with altitude of the dry season (Table 8.7). In the Andes, tree diversity declines sharply with altitude, but the decline only begins at 1500 m (Gentry 1988a), which is close to the lower limit of cloud forest (Terborgh 1971). On Costa Rica's Volcan Barva, tree diversity is declining by 750 m (Lieberman et al. 1996). Nevertheless, forest at 2600 m atop Volcan Barva, and even forest at 3280 m on Ecuador's Volcan Pasochoa, are clearly more diverse then temperate-zone forest of the eastern United States (Table 8.7).

As one passes from "modal" ultisols, however well weathered, to progressively impoverished white-sand soils, tree diversity declines. This decline is evident as one passes from tierra firme through caatinga to low bana at San Carlos de Rio Negro and from mixed forest to heath forests on successively poorer white-sand soils in Sarawak (Bruenig 1996: pp. 43–47).

What is the minimum area a rainforest requires to evolve substantial diversity? The question of how much area a forest needs to diversify is somewhat confounded by how much time is needed to diversify—as yet, we have no clear idea how rapidly trees speciate. Rainforest on Mauritius, which rose from the sea 7.8 million years ago (Lorence and Sussman 1988), is quite depauperate, but rainforest on Reunion, which is only 3 million years old (Strasberg 1995) is hardly less diverse than that of Mauritius (Table 8.8). Many of Hawaii's rainforests are dominated by single species, such as *Metrosideros*, as if Hawaii had too little land to evolve truly diverse rainforest. Puerto Rican rainforest is dominated by *Dacryodes excelsa*, as if Puerto Rico also lacks the "critical mass" for evolving diverse rainforest (Table 8.8). Sri Lanka could hardly be too small to support high diversity; the low tree diversities reported for South India and Sri Lanka probably reflect the drastic changes in climate to which the Deccan Plate was subject, either while drifting from Madagascar to Asia, or afterward, just as the low tree diversity in tropical Africa reflects the disastrous climate changes that afflicted it (Ashton and Gunatilleke 1987). Madagascar has been large enough to evolve diverse forest. On New Guinea, tree diversity on a 1-ha plot is quite high (Table 8.8), but during the last few million years, New Guinea has been readily accessible to plants colonizing from the Sunda Shelf (Pandolfi 1993).

On what scale does tree diversity vary? Mangrove is vastly more depauperate than adjoining rainforest. On the other hand, on Barro Colorado tree diversity of mature forest is much the same whether the plot is in a seasonal swamp, adjoining upland forest on the central plateau, Poacher's Peninsula, or a fragment of mature forest near

Table 8.5 Diversity of selected forests in the tropics and the temperate zone

Site	N	S	α
Old growth, deciduous, Coastal Plain, Western Triangle, Maryland	351	16	3.5
Old growth, deciduous, 700 m, Crabtree Woods, western Maryland	279	16	3.7
Oak-hickory forest, Missouri	334	20	4.7
Mixed conifer-broadleaf forest, Big Thicket, Texas	335	15	3.2
Las Tuxtlas, Mexico	359	88	37.2
Average 1-ha square, 50-ha plot, Barro Colorado Island, Panama	425	91	36.0
Llorona, Corcovado, Costa Rica	354	108	53.0
Cuyabeno, Ecuador (everwet forest)	693	307	211.0

N, number of trees ≥ 10 cm dbh; S, number of species, and Fisher's α for selected 1-ha plots.

Sources for the data are given in Appendix 8.1

Table 8.6 Effect of dry season on tropical tree diversity

Site	N	S	α
Chamela, Mexico (very dry forest, ≥ 3.18 cm dbh)	893	57	13.6
Guanacaste, C. Rica (dry forest, stems ≥ 2 cm)	15889	135	20.5
La Selva, C. Rica (wet forest, stems ≥ 10 cm)	1838	172	46.4
Barro Colorado, Panama (seasonal)	425	91	35.5
Nusugandi, Panama (very weak dry season)	559	191	102.3
Bajo Calima, Colombia (Choco, aseasonal)	664	252	148.0
Pakitza, Manu, Peru (seasonal)	655	166	70.3
Mishana, Peru (white sand, aseasonal)	842	275	142.1
Yanomano, Peru (modal soil, aseasonal)	580	283	218.2
Kirindy, Madagascar (dry forest, > 9.5 cm dbh)	778	45	10.4
Analamazaotra, 1000 m, Madagascar (rainforest)	539	125	51.1

N, number of trees ≥ 10 cm dbh; S, number of species; and Fisher's α for selected 1-ha plots (Costa Rican plots are of varied sizes). Sources of data are given in Appendix 8.1.

the laboratory clearing (Table 8.9). By and large, the controls on diversity of mature stands of terrestrial forest seem to be regional, not local.

How quickly does the diversity and species composition of tropical forest recover from major disturbance? There are several ways to judge. One is to examine the forest succession on accreting river banks. Near Cocha Cashu, in Peru's Parque Manu, Foster et al. (1986) established a sequence of 1-ha plots which, in their judgment, represent about 200 years of succession. The first two extended inward from accreting river bank, while the other three extended in from an oxbow lake. The youngest forest was dominated by *Cecropia membranacea* 20–30 m tall, the next three plots were dominated by *Ficus insipida* and *Cedrela odorata* up to 40 m tall, and in the oldest plot ("bosque transicional" of Cornejo), *Ficus* and *Cedrela* were being supplanted by other species. Stems ≥ 1 cm dbh on two 20 × 200 m strips were mapped and identified: the first strip ran through the youngest two forested hectares, the other through the two oldest hectares. There were 94 species among 897 stems ($\alpha = 26.46$) in the younger strip, whose average age was 60 years; there were 185 species among 1164 stems ($\alpha = 61.98$) in the older strip, whose average age was 180 years. Foster et al. (1986) also mapped and identified all trees ≥ 30 cm dbh on these hectares. Even at 200 years, the diversity of trees ≥ 30 cm dbh is far below that of mature floodplain forest (Table 8.10). Either periodic inundation of these plots depresses tree diversity, or tree diversity takes a long time to approach the levels characteristic of mature floodplain forest.

Another way to investigate recovery is to study a sequence of forests on abandoned farms of known age. Saldarriaga et al. (1988) examined a sequence of 23 stands, 4 in mature forest, the others on farms between 0.5 and 2 ha, abandoned between 9 and 80 years ago. These stands were all within 20 km of San Carlos de Rio Negro, in a region whose soils are very poor indeed—the nearby rivers are called "rivers of hunger." Tree diversity there is as low as on Barro Colorado Island. Saldarriaga et al. (1988) estimate that it takes 190 years after abandonment

Table 8.7 Relation of tree diversity to forest altitude

Site	A(m)	PS(ha)	N	S	α
Puerto Rico: Luquillos	500	4.05	3140	65	11.6
	1000	0.14	604	11	1.9
Costa Rica: Volcan Barva	100	1.00	458	108	44.6
	300	1.00	538	142	62.9
	750	1.00	561	123	48.7
	1250	1.00	610	82	25.5
	1750	1.00	479	64	19.9
	2600	1.00	650	29	6.2
Ecuador: Cuyabeno	260	1.00	693	307	211.0
Ecuador: Baeza	2000	1.00	482	45	12.2
Ecuador: Pasochoa volcano	3280	1.00	715	27	5.5
Java: Tjibodas	1400	1.00	283	59	22.7
	1550	1.00	427	57	17.7
	3000	0.04	152	9	2.1

A, altitude; PS, plot size; N, number of trees ≥ 10 cm dbh; S, number of species among them, and Fisher's α for selected plots. Sources of data are given in Appendix 8.1.

Table 8.8 Diversity of rainforest on different island sites

	Puerto Rico (400 m)	Mauritius (550 m)	Reunion (250 m)	Sri Lanka (50 m)	New Guinea (900 m)	Java (Tjibodas)		Madagascar	
						(1400 m)	(1550 m)	(1000 m)	(1550 M)
Plot size (ha)	4.0	1.0*	1.0	4.0	1.0	1.0	1.0	0.51	0.70
N	3140	1710	1079	2100	679	283	427	1387†	2808†
S	65	52	40	101	222	59	57	184	102
α	11.6	10.1	8.3	22.1	115	22.7	17.7	53.9	20.8

N, number of trees ≥ 10 cm dbh; S, number of species included among these trees, and Fisher's α, for selected plots on tropical islands. Sources of data are given in Appendix 8.1.
*In Mauritius, the trees are from ten disjunct 0.1 ha plots.
†For Madagascar, N is the number of non-liana stems ≥ 5 cm dbh.

for the forest's basal area and biomass to attain levels characteristic of mature stands. Yet diversity among stems ≥ 1 cm dbh attains levels characteristic of mature forest within 80 years (Table 8.11). Diversity recovers much more quickly in small clearings than it increases on newly formed land.

A third approach is to study forest regeneration after large-scale clearing. Piperno (1994) took a 4000-year core from Lake Wodehouse, in the Darien province of Panama. Judging by the core, pre-Columbian populations farmed this area extensively from 2600 to 300 years ago. The segment of the core representing this period is full of maize pollen, particulate carbon, and fragments of charred grass, while signs of mature forest are quite scarce during this period, much rarer than before or after. People clearly disrupted the integrity of this forest, reducing it to scattered fragments if not eliminating it altogether—a striking contrast to the slash-and-burn practices currently employed near San Carlos de Rio Negro (which form a few clearings in otherwise intact forest). Yet, just 300 years later, this lake is now surrounded by high-canopied, species-rich forest (Piperno 1994). Something does appear to "drive" tropical tree diversity rather forcibly.

NECESSARY CONDITIONS FOR TROPICAL DIVERSITY

What conditions had to be met for tropical diversity to evolve? How can managers preserve tree diversity when its causes are far from fully understood? A full answer, especially to the latter question, involves exploring all the interdependences that link tropical trees with each other and with the animals they support. This is the task of the next chapter. A few conclusions can be anticipated now.

Where there are many kinds of trees, many of these kinds must be rare. A rare plant species needs reliable means of conveying pollen to conspecifics and of dispersing its seeds. Flowering plants appear to have achieved dominance through their ability to diversify. They began to diversify in earnest when they attracted species-specific insect pollinators in the late Cretaceous (Crepet 1984). Flowering trees spread into mature forest when, in the early Tertiary, they were able to employ suitable animals to disperse the large seeds required to become established there (Regal 1977, Tiffney 1984, 1986, Tiffney and Mazer 1995). Flowering plants still need pollinators; many also need seed dispersers. Some of these pollinators and seed dis-

Table 8.9 Tree diversity in mature forest on various parts of Barro Colorado Island

	A (ha)	n	N	S	α
Trees ≥ 10 cm dbh					
Lutz, near laboratory (R. Werth p. c.)	0.22	1	120	52	34.9
Forest dynamics plot, avg.	0.25	200	106	45	30.4
Forest dynamics plot, avg.	1.00	50	425	91	36.0
Forest dynamics plot, swamp	0.84	1	277	75	33.8
Trees ≥ 20 cm dbh					
Forest dynamics plot, avg.	1.00	50	155	54	29.0
Forest dynamics plot, swamp	0.84	1	105	47	32.7
Poacher's Peninsula (S. J. Wright, p. c.)	1.00	4	145	55	32.3
Allee near laboratory (Thorington et al. 1982)	1.00	1	172	61	33.7

A, area of sample plots; n, number of plots sampled; N, the average number of trees; S, number of species per plot, and Fisher's α.
Forest dynamics plot data (1990 census) from R. Condit, S. P. Hubbell, and R. B. Foster (p. c.). Other data are from sources as noted or are unpublished.

Table 8.10 Successional increase of species diversity and basal area in the floodplain of Manu (trees ≥ 30 cm dbh; data from Foster 1990)

	Age (years)					
	40	80	120	160	200	Mature
Number of trees	66	63	83	83	86	110
Number of species	7	16	23	35	42	68
Fisher's α	1.70	6.92	10.5	22.8	32.4	75.9
Basal area	18	23	54	42	29	

Basal area (m²/ha) for stems ≥ 1 cm dbh, for 20 × 100 m subplots of 20-m wide strip running through transects, from Foster et al. (1986).

persers migrate seasonally, often between very different habitats. The needs of these animals must be understood and met if tree diversity is to be preserved (Leigh and de Alba 1992).

Trees also depend on animals for other services. Barro Colorado is surrounded by smaller islands, including islets of less than a hectare which cannot support *resident* mammals (Adler and Seamon 1991). Even on those islands that have been continuously forested (judging by aerial photographs and the abundance of drowned stumps surrounding their shores) since their isolation from the mainland in 1913 by the rising waters of Gatun Lake, tree diversity is far lower than on nearby mainland tracts of comparable area (Fig. 8.1, Table 8.12). If these islands initially had tree diversity similar to that today of nearby mainland tracts of similar size, tree diversity on these islands has dropped precipitously after isolation, far faster than successive random samplings of one generation's trees from the preceding generation's young could explain (Leigh et al. 1993).

Four species are spreading on these mammal-free islets: the clonal palm, *Oenocarpus mapora*, also common in gaps opened by windstorms on Barro Colorado; the stout, large-fronded palm *Scheelea zonensis*; the leguminous tree *Swartzia simplex*, which has an extraordinary capacity to resprout after its trunk is snapped; and, most common, *Protium panamensis*, which contributes over half the trees on two of the six islets studied (Table 8.12). All four of these species have fairly large seeds. Three have seeds that insects rarely or never attack. Larvae of bruchid beetles devastate *Scheelea* seeds, but they do not lay eggs on seeds falling after August (Wright 1990) because, in pristine settings, hungry mammals would eat all the seeds (and any larvae they contained) once the season of fruit shortage arrived in September or October (Forget et al. 1994). Thus late-falling *Scheelea* seeds, like all the seeds of the other three species of trees taking over these islands, do not need to be buried by mammals to escape insect attack.

Was this drop in diversity aggravated by the circumstance that many kinds of seed can only escape insect attack by being buried by agoutis, which these islands lack? In French Guiana, the seeds of *Moronobea coccinea* (Guttiferae), *Voucapoua americana* (Caesalpinoideae), and *Carapa procera* (Meliaceae) must be buried by agoutis to escape insect attack (Forget 1991a, 1994, 1996), whereas seeds of *Astrocaryum paramaca* (Palmae) and *Eperua grandiflora* (Caesalpinoideae) benefit by being buried by agoutis or hidden under the leaf litter by spiny rats (Forget 1991b, 1992). In chapter 2, I mentioned several trees of Barro Colorado that depend on agoutis burying their seeds. Are agoutis required to maintain the diversity of Neotropical trees?

To learn how to preserve tropical tree diversity will require intense study of the natural history of plant–animal interactions. Unfortunately, natural history is not popular: biologists, and their funding agencies, prefer more "important" topics.

EXPLAINING THE DIVERSITY OF TROPICAL TREES

Why are there so many kinds of tropical trees? This is really a two-part question. How species originate is a mystery all its own. Two elements are involved. Differences in habit or habitat must call forth "special adaptations, which are either irreconcilable or difficult to reconcile" (Fisher 1930: p. 126). Then mating preferences must evolve that prevent individuals adapted for different habits or habitats from crossing. This happens much more readily

Table 8.11 Successional increase of species diversity and basal area on abandoned farms near San Carlos de Rio Negro

No. of farms	Age (years)	≥ 1 cm dbh			≥ 10 cm dbh			Basal Area ≥ 1 cm
		Stems	Species	α	Stems	Species	α	
4	9–14	509 ± 65	46 ± 9	12.3 ± 3.0	31 ± 16	9 ± 4	4.4 ± 1.0	12.8 ± 2.6
4	20–23	640 ± 195	64 ± 9	17.8 ± 2.4	42 ± 11	17 ± 5	13.5 ± 7.7	16.9 ± 1.6
4	30–40	489 ± 88	73 ± 4	23.7 ± 0.7	49 ± 12	17 ± 2	9.2 ± 1.6	18.6 ± 4.2
3	55–60	258 ± 27	66 ± 3	28.8 ± 0.7	40 ± 5	20 ± 5	17.5 ± 7.0	24.5 ± 5.5
4	75–85	359 ± 69	71 ± 7	26.9 ± 3.4	59 ± 9	24 ± 5	15.5 ± 5.9	24.0 ± 1.5
4	Mature	365 ± 80	74 ± 13	28.4 ± 5.0	51 ± 2	24 ± 6	19.9 ± 9.3	34.8 ± 2.6

Data from Saldarriaga et al. (1988). For each farm, Saldarriaga et al. (1988) give the total number of stems, total number of species, and basal area (m²/ha) from three or more 10 × 30 m plots. Numbers given are means ± standard deviation for plots of each age class.

Table 8.12 Number, N, of stems ≥ 10 cm dbh, number, S, of species, Fisher's diversity, α, and Simpson's diversity, $1/F$, on selected small island and mainland plots near Barro Colorado Island, Panama (data from Leigh et al. 1993; island data are complete counts for those islands)

Site	N	S	α	$1/F$
Mainland quarter-plot	62	25	15.6	10.7
Vulture Island	59	10	3.5	4.1
Mainland half-plot	125	36	16.5	12.7
Camper Island	125	19	6.2	4.1
Aojeta Bay Island	128	26	9.9	6.6
Ormosia Island	135	32	13.3	8.5
Mainland full plot	250	49	18.2	14.3
NW of Juan Gallegos Island	340	25	6.6	2.4
Annie Island	399	37	10.0	2.4

if the individuals adapted to different habits or habitats initially live in separate populations, isolated from each other by some geographical barrier (Fisher 1930). Speciation can occur in the absence of any genetic incompatibility (Bradshaw et al. 1995, Grant and Grant 1997). Nevertheless, should the ranges of these populations spread and overlap, even a strong tendency for like to mate with like will not keep these populations distinct if hybrids survive and reproduce as readily as their parents (Grant and Grant 1997). Indeed, hybrids must be very poorly adjusted to their habitat for speciation to proceed (Jiggins et al. 1996). For speciation to occur, choice of mate must be aligned with ecological advantage: individuals must mate with others of similar habit. How this happens is less obvious than textbooks suggest.

How trees speciate is particularly mysterious. Biologists have formed their instincts about speciation from animals with short generation times, which are choosy about mates, yet somewhat adaptable in their choices. On occasion, when members of different plant species cross, offspring are produced with all the chromosomes of both parents. These polyploids can interbreed with each other, but not with their parent species: they have achieved instant speciation (Haldane 1932). The new species survives if its members occupy a "niche" that allows them to avoid competitive displacement. But many plant species could not have originated as polyploids. How did they evolve? How much time do trees need for their diversification to saturate their habitat with species? Indeed, is any habitat ever saturated with tree species? These questions are beyond my reach.

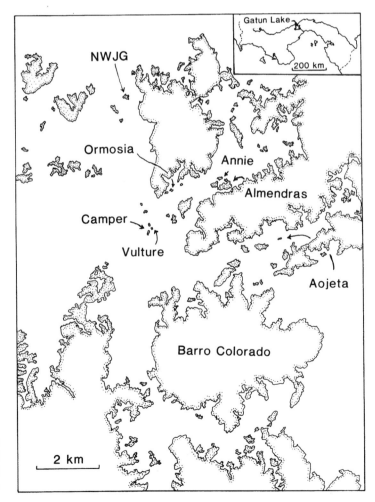

Figure 8.1. Locations of small islands with diversity listed in Table 8.12. Mainland plots were on both sides of the peninsula lying between Almendras and Aojeta.

Even though species of cichlid fish seem to be able to evolve in a few thousand years (Greenwood 1965), closely related but distinct species of rainforest vertebrates (Joseph et al. 1995), like closely related species of *Drosophila* (Coyne and Orr 1989) or populations of snapping shrimp sundered 3 million years ago by the Isthmus of Panama (Knowlton et al. 1993) usually require 1 or more million years of separation to become truly distinct species.

Second, how can so many species coexist in one place? Hubbell (1979) has emphasized the enormous lengths of time required for a tree species to disappear if it and its neighbors are "competitive equivalents." Hubbell and Foster (1986a) ascribed the enormous diversity of tropical trees to their competitive equivalence. Long ago, however, Fisher (1930, 1958: pp. 87–88) observed the other side of this coin: the enormous and improbable length of time required for a genotype (or a species) to become numerous if it is simply spreading by chance, aided by no competitive advantage. After all, what is declining must once have increased, and the time required to decline by chance to extinction is the time required to increase by chance to abundance. For Hubbell, speciation rate, and degree of disequilibrium, are the only variables available to explain diversity gradients. Although such explanations are not impossible in principle, they seem too implausible for serious attention. Moreover, chance, in the form of "demographic stochasticity"—the random variation in the number of surviving offspring per parent of the different members of a species—is altogether too slow a process to account for observed fluctuations in large populations (Leigh 1981; also see below).

Darwin already foresaw the ascription of plant diversity to chance as a temptation to be avoided: "When we look at the plants and bushes clothing an entangled bank, we are tempted to attribute their proportional numbers and kinds to what we call chance. But how false a view this is!" (Darwin 1859: p. 74). Darwin's reason for this opinion deserves examination

> Every one has heard that when an American forest is cut down, a very different vegetation springs up; but it has been observed that the trees now growing on the ancient Indian mounds, in the Southern United States, display the same beautiful diversity and proportion of kinds as in the surrounding virgin forests. (Darwin 1859: p. 74)

The preceding section documented the increase of tree diversity in the course of forest succession on land created as the Rio Manu wanders over its floodplain (Foster et al. 1986). Species composition and forest structure change radically during the course of forest succession on this new land. To see if succession at different floodplain sites tended to a common "climax," Terborgh et al. (1996) ranked the tree species on each of six plots of floodplain forest classified as mature by structural characteristics in order of their number of stems ≥ 10 cm dbh on that plot. Although these plots were scattered over 40 km and separated by stretches where the Rio Manu abuts directly against upland forest, which has a very different species composition, species compositions of these plots were much more similar to each other than any was to upland forest or to an earlier stage of succession. The more common species tended to rank similarly in abundance on all six plots. Terborgh et al. (1996) accordingly refused to ascribe the species compositions of these floodplain forests to chance.

How, then, can we make sense of the diversity of tropical forest? How can so many species coexist? Darwin (1859) provided a clue when he explained why natural selection favors diversity. Darwin (1859: pp. 75–76) began with a principle of competitive exclusion: competition is most intense, and the risk of competitive displacement of one population by another greatest, between those lineages that are most similar and closely related. This suggests that differences between species facilitate their coexistence, a phenomenon which provides one foundation for Darwin's (1859: pp. 111, 113ff) principle of the "divergence of character." Indeed

> Most of the animals and plants which live close round any piece of ground, could live on it . . . and may be said to be striving to the utmost to live there; but, it is seen, that where they come into the closest competition with each other, the advantages of diversification of structure, with the accompanying differences of habit and constitution, determine that the inhabitants, which thus jostle each other most closely, shall, as a general rule, belong to what we call different genera and orders. (Darwin 1859: p. 114)

More recently, MacArthur (1961, 1972; also see MacArthur and Levins 1964), following in Darwin's path, founded a program for analyzing biotic diversity upon the following principles:

1. Species limited by identical factors cannot coexist in the same habitat: the better competitor (taken in the wider sense of the one best able to multiply under the conditions of food availability, predator pressure, and prevailing weather patterns) will replace the other (Lotka 1925, Volterra 1931, Gause 1934, 1935, Kerner 1961).
2. The jack-of-all-trades is master of none (MacArthur 1961), or, to quote Darwin (1859: p. 116), "No physiologist doubts that a stomach, by being adapted to digest vegetable matter alone, or flesh alone, draws most nutriment from these substances."
3. Thus, to coexist, each species must be specialized to a different role or niche; that is to say, different species coexist only if they depend on different resources, live in different places, or are limited by different sets of natural enemies (serve as food for different sets of animals or pathogens). Perhaps it is more appropriate to say that a new tree species is likely to invade a community only if it possesses a decided advantage, at least when it is rare. I shall argue that in the tropics, such an advantage is most

likely to lie in a novel defense against pests. After all, as far as the means for coping with drought, shade, poor soil, and the like are concerned, there may be almost nothing new under the sun.

In short, we can understand why there are so many kinds of trees when we can discern what factors allow so many species to coexist. A more refined theory recognizes that diversity represents a balance between immigration and speciation, on the one hand, and extinction, on the other (MacArthur and Wilson 1967, Leigh 1981), but, as Darwin (1859) saw, the secret to survival for a species is to do something different from what its competitors are doing.

I will accordingly assume that coexistence and extinction are the outcome of selection. In particular, the extinction of large populations must be caused by changes in the physical or biotic environment (Leigh 1981). On Barro Colorado, environmental change elicits rapid responses from tree populations: the savage El Niño drought of 1982–1983, which devastated populations of some species, provided ideal conditions for the multiplication of others (Condit et al. 1996a).

Trade-offs, Habitat Heterogeneity, and Tree Diversity

Coexistence is possible because, in the biological world (MacArthur 1961), as in the business world (Smith 1776), the jack-of-all-trades is master of none. There are trade-offs between different capacities. For example, long life is inconsistent with early, copious reproduction: breeding for one compromises the other (Rose and Charlesworth 1981). Plant leaves face a trade-off between being able to photosynthesize rapidly in bright light and being able to turn a profit at low light: the protein required to support sustained, rapid photosynthesis entails maintenance costs which poorly lit leaves cannot afford (Givnish 1988). Similarly, plants of the tropical understory face a trade-off between expanding their leaves rapidly, in hopes of maturing them before herbivores find them, and keeping the nitrogen content of growing leaves low and their toxin content high, thus making them less attractive to herbivores (Kursar and Coley 1991, Coley and Kursar 1996). Such trade-offs render a degree of specialization advantageous to living things (Ghiselin 1974, Leigh 1990a), just as they render a division of labor advantageous to society and some degree of occupational specialization advantageous to its individuals (Smith 1776). Such trade-offs represent so many opportunities for different species to coexist. So we may rephrase our question: what trade-offs make possible the enormous diversity of tropical trees? In particular, why is tree diversity so much higher in the tropics than in the temperate zone?

Are different tropical trees specialized to colonize different sizes of forest light gaps? Sometimes a single tree falls, opening a light gap in the forest canopy no larger than the area of its crown, while another falling tree may bring many others down with it, opening a much larger gap. There is indeed a trade-off between the ability to survive in small gaps and the ability to grow fast in large ones. In large clearings on Barro Colorado, the tree *Trema micrantha*, which carries its narrow leaves along horizontal branches which radiate out from its straight white trunk to form a multilayer crown, can grow 14 m during its first 2 years of life; under the same circumstances, *Cecropia insignis*, with its large, deeply lobed leaves spiralled about erect branch-tips to form an open, domed, or umbrella crown, grows no more than 9.8 m, and *Miconia argentea*, with its dense, bushy, domed monolayer crown grows only 5 m. *Trema* survives to its ninth year only in gaps > 376 m^2 in area, whereas *Cecropia* can survive that long in gaps > 215 m^2, and *Miconia* (which grows more often at the forest edge than in the open) survives that long in gaps > 102 m^2 (Brokaw 1987).

Herbivores may be a driving force in this light–shade trade-off. In both sun and shade, seedlings of light-demanding species grow faster than seedlings of their shade-tolerant counterparts. Herbivores, however, do not allow light-demanding seedlings to survive long in the shade, while seedlings of shade-tolerant species have deeper roots and tougher stems and leaves, which slow their growth but allow them to resist herbivores and pathogens more effectively (Kitajima 1994). This is an aspect of a more general trade-off between escaping or outrunning herbivores by fast growth and deterring or resisting herbivores by being tough or poisonous, properties that require investments which slow growth (Coley et al. 1985). The species whose saplings grow fastest in large clearings devote most of their energy to stem growth (King 1994): they can survive only if their relatively few and ill-defended leaves are well enough lit to subsidize sufficient stem growth to permit their own replacement in the few months before they are eaten (cf. Coley 1983).

The trade-off between growing fast where light is abundant and low mortality when shaded is fundamental to the maintenance of diversity in northern hardwood forest (Pacala et al. 1996). As we have seen, this trade-off also contributes to the diversity of tropical trees. Yet the distributions of saplings of most species on Barro Colorado's 50-ha forest dynamics plot are not correlated with canopy height (Welden et al. 1991). They do not appear to specialize to different gap sizes. Lieberman et al. (1995) likewise concluded that, in the lowland rain forest of La Selva, Costa Rica, most tree species do not specialize to particular degrees of light availability and that segregation by size of light gap contributes little to tropical tree diversity.

Are different tropical trees specialized to colonize different microhabitats? There are striking trade-offs between the ability to survive in drier places and the ability to grow fast in wet ones. On Barro Colorado Island, leaves flushed for the dry season by the understory shrub *Psychotria marginata* have higher water use efficiency than their rainy season counterparts, but CO_2 levels are lower within dry season leaves, so when there is sufficient water these leaves are less efficient at low light levels, and they respond more slowly to sunflecks (Mulkey et al. 1992). Plants can grow in dry places by evading drought, either by having deep roots, and perhaps a high water storage capacity in their stems, like *Schefflera morototoni* (Tyree et al. 1991) or *Ochroma pyramidale* (Machado and Tyree 1994), or by dropping their leaves and ceasing to function during the dry season like *Pseudobombax septenatum* (Machado and Tyree 1994), and perhaps having to refill their cavitated vessels by root pressure after the onset of the rains, like Massachusetts grapevines in springtime (Sperry et al. 1987). Dry country plants that cannot evade drought must be able to tolerate drought by functioning when short of water (Tyree et al. 1994, Zotz et al. 1994). Shallow-rooted, evergreen plants must be drought tolerant. They must be able to withstand high xylem tensions; which is accomplished by using thick pit membranes pierced by narrow pores to separate adjoining vessel elements, but pit membranes of this type obstruct the rapid supply of water to the leaves under wet conditions (Tyree et al. 1994: p. 342). Their leaves must be adapted to conserve water and avoid wilting when the soil is dry (Wright et al. 1992). The adaptations to avoid or tolerate drought are either expensive, reduce productivity under wet conditions, or both. Yet Hubbell and Foster (1986b) found that most species on Barro Colorado's 50-ha plot, especially the most common ones, are distributed over this plot without regard to the distinction between the plateau, and the surrounding slopes and ravines, even though the latter are wetter, and their plants maintain higher predawn water potentials (Becker et al. 1988). Thus, despite the strong trade-offs involved, most of these tree species are apparently not habitat specialists.

It is true that habitat diversity enhances tree diversity (Ashton 1964, Gentry 1982, 1988a). Nonetheless, physical factors cannot tell us why there are so many more kinds of trees in a tropical hectare than in a hectare of temperate-zone forest. Are the trade-offs that should dictate specializations to aspects of the physical environment overriden by biotic factors? Are biotic factors more important than physical ones in the lives of tropical plants, as Dobzhansky (1950) suggested so long ago?

Temporal Sorting of Recruitment

Fluctuations in the abundance of pollinators, seed dispersers, seed-eaters, and seedling-browsers often have great impact on the reproductive success of tropical trees. Environmental conditions vary from year to year, affecting the mutualists, pests, and pathogens of different species in different ways. The conjunction of circumstances that favor abundant flowering, effective pollination, copious seed set, successful seed dispersal, and high survival of seeds and seedlings, all of which are needed for reproductive success on a truly grand scale, should occur for different species in different years. And, indeed, seed crops of different species vary quite differently from year to year (Table 8.13). Does this circumstance mean that one species or group of species secures the lion's share of gaps opened by the fall of mature trees in one year, while, in the next year, other species win these prizes, and, in the third, still others? Does this circumstance allow different tree species to coexist by enforcing a "temporal segregation" in their reproduction (Chesson and Warner 1981)?

To see how temporal "sorting" of recruitment allows different tree species to coexist, consider a forest with two species, A and B. Let $n_A(t)$ and $n_B(t)$ be the proportions of adult trees in this forest which belong to species A and B at year t, where $n_B(t) = 1 - n_A(t)$. Let a fraction, m, of the adults of each species die each year, and let each year's dead trees be replaced, one for one, by seedlings of that year which grow immediately into adults in their place. Let the proportion, $P_A(t)$, of year t's recruits which belong to species A be the total weight of seed produced by A adults that year, divided by the total weight of seeds produced that year by the forest as a whole. Then $P_A(t)$ can be expressed as

$$\frac{b_A(t)n_A(t)}{b_A(t)n_A(t) + b_B(t)n_B(t)} = \frac{n_A(t)R(t)}{1 + n_A(t)[R(t) - 1]}, \quad (8.9)$$

Table 8.13 Fruit fall of selected species on Barro Colorado in successive years

Species	1979	1980	1981	1982	1983	1984	1985
Dipteryx panamensis	12	3	19	29	13	6	
Virola surinamensis	2082	8579	3990	5612	2420		
Gustavia superba			6	2	10		
Quararibea asterolepis			44	85	0	16	22
Faramea occidentalis			<0.1	14.2	9.5	0.04	11.6

Dipteryx numbers are number of fallen fruit per square meter of crown (Giacalone et al. 1990); *Virola* numbers are median fruit production per sampled tree (Howe 1986); *Gustavia* numbers are median numbers of fruits per tree, measured in 1982, and estimated in other years from the ratio of seedling density that year to seedling density in 1982 (V. Sork, p. c.); *Quararibea* figures are numbers of seeds, ripe and unripe, falling into seed traps set around nine sample trees, and were provided from the Smithsonian's Environmental Monitoring program by D. Windsor (p. c.); *Faramea* numbers are fallen seeds per square meter (Schupp 1990).

where $b_A(t)$ and $b_B(t)$ are the total weight of seeds produced per tree by species A and B in year t, and $R(t) = b_A(t)/b_B(t)$ is the ratio of the seed weight produced by an A tree to the seed weight per B tree in year t. Then the proportion, $n_A(t + 1)$, of species A among the adult trees of year $t + 1$ is

$$n_A(t + 1) = n_A(t)(1 - m) + mP_A(t). \quad (8.10)$$

In other words, the proportion of species A in year $t + 1$ includes those surviving from year t, plus those recruited since.

Now let $n_A(t)$ be so much smaller than 1 that $|n_A(t)[R(t) - 1]| \ll 1$. Then

$$n_A(t + 1) = n_A(t)\{1 + m[R(t) - 1]\}. \quad (8.11)$$

If m is much smaller than 1, and if $R(t) = 3$ every third year, and 1/3 in other years, the numbers of A will increase when A is rare. In particular, $n_A(t + 3)/n_A(t) =$

$$(1 + 2m)\left(1 - \frac{2m}{3}\right)^2 \approx 1 + \frac{2m}{3} \quad (8.12)$$

Similarly, if B were rare, it would increase, indeed twice as fast, as the environment favors it twice as often. Thus temporal sorting in reproduction, even if incomplete, allows two species of different competitive ability to coexist (Chesson and Warner 1981).

Now let A represent an invading species, while B represents the ensemble of all other tree species in the community. A can invade only if, on the average, $\ln n_A(t + 1) > \ln n_A(t)$, that is to say, only if the average value of $\ln\{1 + m[R(t) - 1]\}$ is positive. Since $\ln(1 + x) \approx x$ when $x \ll 1$, then, if $m[R(t) - 1]$ is always far smaller than 1 in absolute value, A invades if the average value of $R(t) > 1$.

In this model, the factors permitting different species to coexist seem purely accidental. Nonetheless, in the tropics environmental variation impinges on reproduction of different tree species in more diverse ways than in the temperate zone because in the tropics, the biotic impact of environmental change is expressed through a far more complex array of mutualists and consumers. Thus, temporal sorting of recruitment could, in principle, maintain far higher diversity in the tropics than farther from the equator. Just how much temporal sorting does contribute to maintaining the diversity of tropical trees is far from clear. Insofar as plants can wait 20 years as saplings for a suitable gap to open overhead, temporal sorting of recruitment seems less likely to permit different species to coexist (R. Condit, p. c.). Even so, it is hard not to suspect that the seemingly senseless variety of ways plants arrange for the pollination of their flowers, the dispersal of their seeds, and the safety of their seedlings reflects the premium on finding a novel way to accomplish these ends that allows trees of the species in question to reproduce successfully when most others are having a bad year.

Pest Pressure and Tree Diversity: Is Pest Pressure Heavier in the Tropics?

The devastating effects on tropical crops of pests such as leaf-cutter ants suggest that pest pressure is an influence strong enough to override trade-offs imposed by physical factors. Can pest pressure maintain the extraordinary diversity of tropical trees?

Ridley (1930), Gillett (1962), Janzen (1970), and Connell (1971) inferred from the attractiveness of human monocultures to both disease and insect pests that species-specific pests enhance species diversity by keeping their host plants rare, thus making room for plants of other species. In the words of Ridley:

> Where too many plants of one species are grown together, they are very apt to be attacked by some pest, insect, or fungus. This is, of course, very well known to cultivators, whose plants in a limited area have to be carefully watched and guarded from disease. It is largely due to this also, in Nature, that one-plant associations are prevented and nullified by better means for dispersal of the seeds. When plants are too close together, disease can spread from one to the other, and become fatal to all. Where plants of one kind are separated by those of other kinds, the pest, even if present, cannot spread.... (p. xvi)

These authors implicitly assumed that trade-offs between abilities to parry different plant defenses would favor host-plant specialization, at least among the most destructive pests. Janzen (1970) and Connell (1971) thought that where a host plant's seedlings were most densely aggregated or most closely clustered about their parents, pest pressure would be disproportionately intense. Gillett (1962) expected the most common plant species would attract the greatest loads of pests. They all assumed that, in wet tropical climates, where neither winter nor seasonal drought reduce pest populations each year, pest pressure would be far more continuous and intense and promote far higher plant diversity than in the temperate zone.

Insect activity is distributed more evenly in the tropics than in the temperate zone (Wolda 1983: p. 96). Is pest pressure more intense in the tropics? Coley and Aide (1991) reported that tropical leaves are more heavily eaten, despite being tougher and more poisonous, than their angiospermous counterparts in the temperate zone. The differences involved, however, were less clear-cut than one would like. The case is now clearer. Tropical leaves suffer the majority of their losses to herbivores before they mature and toughen, while deciduous temperate-zone leaves suffer most of their losses after they mature (Coley and Barone 1996, Coley and Kursar 1996). Being longer lived, tropical leaves can afford to become inedibly tough when mature. Moreover, again unlike their temperate-zone counterparts, tropical leaves are more heavily stocked with antiherbivore toxins before they mature (Coley 1983, Coley and Barone 1996). Thus, if we focus on young leaves, tropical leaves are much more rapidly eaten, despite being much more poisonous, than their temperate-zone counterparts. The pressure of chewing insect folivores is thus much more intense in the tropics; I assume this is true for other classes of pests as well.

Specialization and Trade-offs in Parrying Different Plant Defenses

Only if other plant species are immune to a pest that keeps its host plant rare will that pest enhance plant diversity. In other words, the pest pressure hypothesis presupposes that pests, at least those that inflict the worst damage, are specialists. Basset (1992) questioned the prevalence of specialization among insect pests because, in a thorough sample of insect herbivores from an Australian rainforest tree, 89% of the herbivorous insect species he collected could feed on several kinds of plants, representing more than one plant family. Nonetheless, in tropical forest, the pests that inflict the most damage are usually specialists. On Barro Colorado, Barone tested the pest pressure hypothesis, not only by examining whether seedlings died faster when closer to their parent and/or to conspecific seedlings, but also by recording what herbivores were damaging these seedlings and assessing their degree of specialization in feeding trials. He found that most of the damage caused by chewing herbivores was inflicted by specialists (Coley and Barone 1996, Barone 1998). Similarly, in the dry forest of Costa Rica, Janzen (1988) found that major defoliations are almost always caused by a single generation of one or two species of caterpillar, consuming one or two species of host plant. If such results apply to all tropical forests and to pests consuming other items besides leaves, the diversity of nonspecialists on tropical trees is irrelevant.

Are damaging pests specialized because there are trade-offs among the abilities to parry different kinds of plant defenses? Long ago Fabre (1989, 1910) wondered why some insect herbivores were specialists and others were not. Few trade-offs between abilities to exploit different food plants have been demonstrated (Via 1991a). Bernays and Graham (1988) have questioned whether specialization benefits insect herbivores. Yet, in the dry forest of Costa Rica, sphingid caterpillars, which specialize to one or a few species, grow more rapidly, on the average, than saturniid caterpillars, most species of which have several species of host plant (Janzen 1984). What goes on here?

Trade-offs between the ability to use different food plants occur even in a species for which average individual performance (as measured by pupal weight and development time) does not depend on the type of food plant used. In Sampson County, North Carolina, the leaf miner *Liriomyza sativae* feeds on both tomato and cowpea plants, but those genotypes that do best on tomatoes tend to do poorly on cowpeas, and vice versa (Via 1984). Here, the "antagonistic pleiotropy," or genetically based trade-off between doing well on tomatoes and doing well on cowpeas, provides the potential for a population of specialists to evolve for each crop, but crop rotation has presumably prevented such specialist races from evolving (Via 1984). In Johnson County, Iowa, pea aphids, *Acyrtosiphon pisum*, feed on both alfalfa and red clover, but aphid clones tend to be most fit on the crop from which they are collected (Via 1991a), a feature which prolonged rearing on the alternative crop does not change (Via 1991b). Thus migration between crops is disadvantageous, and there is potential for the evolution of specialists on each crop. The evolution of host races has been documented for the soapberry bug *Jadera haematoloma*, which feeds on Sapindaceae. These bugs feed on seeds inside the fruits of these plants. Their beaks must accordingly be long enough to reach the seeds from the surface of the fruit, while overlong beaks are inconvenient in various ways. Each of three species of Sapindaceae recently introduced to the United States has led to the evolution of a "race" of soapberry bugs whose beak lengths are suited to exploit their fruits: the evolutionary change in soapberry lineages are documented by museum specimens collected at various times during the last two centuries (Carroll and Boyd 1992).

Trade-offs in detoxifying different compounds seem to explain the diversity and specialization of bruchid beetles, beakless weevils who lay their eggs on fruits inside whose seeds their larvae will feed and develop. The larvae of more than half the species of bruchid beetles studied by Janzen (1980) were found in the seeds of only one plant species. This specificity seems dictated by a bruchid larva's ability to circumvent the defenses of its host or turn them to its advantage (Janzen 1977). To what extent this is a trade-off between abilities to detoxify different poisons and to what extent the ability to detoxify a compound channels an adult bruchid's behavior in a manner which reduces its encounters with suitable alternate hosts is an open question. The bruchid *Caryedes brasiliensis* nourishes itself on canavanine, a compound which the arginyl transfer RNA's of many organisms mistake for arginine, with suitably disastrous consequences (Rosenthal et al. 1978). This bruchid breaks down canavanine into urea and canaline, which latter is mistaken for ornithine by the metabolic machinery of many insects, again with disastrous results, and it uses the nitrogen from urea (Rosenthal et al. 1976, 1977, 1978, 1982; Bleiler et al. 1988). To use the urea, this bruchid requires a suitable enzyme, which most insects lack, and which must tie up energy that could be used for other purposes, suggesting a metabolic trade-off. *C. brasiliensis* specializes on the seeds of the vine *Dioclea megacarpa*; 13% of the dry matter of these seeds is canavanine. Santa Rosa has two other species with canavanine-rich seeds big enough to permit development of this bruchid's larvae, but this beetle's adjustments to the habits of *Dioclea* apparently preclude using these other seeds (Janzen 1981): the trade-offs responsible for specialization may not always lie in the realm of metabolism.

Two questions remain. First, is plant diversity driven

by a coevolutionary race where plants are forever evolving novel defenses, each of which leads to the evolution of pests specialized to deal with it, as Ehrlich and Raven (1965) suggested? We know far too little to be able to answer this question. Second, is each species of plant kept rare enough by specialized pests to allow many plant species to coexist, as Gillett (1962), Janzen (1970), and Connell (1971) suggested? To this question we now turn.

Do Pests Kill the Young Nearest Their Parents?

In tropical settings with no dry season, insect activity continues more evenly through the year (Wolda 1983: p. 96; Wolda 1988: p. 5). Only in forests with a severe dry season can a tree escape insect pests by leafing out before the rains begin (Aide 1988, 1993, Murali and Sukumar 1993). The more nearly everwet the forest, the higher the proportion of tree species with animal-dispersed seeds (Table 8.14; also see Gentry 1982, Emmons 1989). It is as if, in everwet forest, the greater, less interrupted pressure from insect pests not only enhances tree diversity (Gentry 1988a, Wright 1992) but also places a greater premium on effective seed dispersal.

Most work on the pest pressure hypothesis has focused on whether pests prevent young plants near their parents from growing to maturity. Until recently, such studies focused on individual species, and their results have been as varied as the trees that were studied. Let us now turn to the results of studies on Barro Colorado.

Augspurger (1983, 1984) focused on seedling survival in wind-dispersed species, especially the canopy legume *Platypodium elegans*. The fruits of *P. elegans* are single-seeded pods that look like giant maple samaras ("winged seeds"), which are attached to the tree by their tips, rather than by the "head" where the seeds lie, as maple wings are. Most of the shaded *Platypodium* seedlings within 20 m of their parent die quickly from fungal attack, but this fungus does not attack seedlings in light gaps, however close to their parents these may be (Augspurger 1983). Augspurger (1984) also followed the fate of seedlings of one tree each of eight other wind-dispersed species. In five of these, no marked seedlings survived as long as a year in the shade. Shaded seedlings of another species, *Lonchophyllus pentaphyllus*, rarely survived unless they were > 20 m from their parent. In the remaining two species, seedlings suffered no detectable disadvantage from being near their parents, although one of these, *Aspidospermum cruenta*, dispersed its seeds so effectively that its seedlings were far more abundant 30–40 m from the parent than under the parent crown (Augspurger 1984).

Seeds of the canopy tree *Virola surinamensis* placed on the ground 40–49 m from a fruiting conspecific were 3.6 times more likely to become 6-week-old seedlings than were seeds placed less than 10 m away (Howe et al. 1985). Seedlings transplanted to the field 6 weeks after germination were 15 times more likely to survive the next 12 weeks if planted 30–40 m from a fruiting conspecific than if planted only 10 m away (Howe et al. 1985: Tables 3 and 4). In general, *Virola* seeds have vanishingly poor prospects if they fall near their parent or form part of a dense aggregation of conspecifics, even far from their parent (Howe 1989, 1990).

In contrast, 44% of the 3-month-old seedlings of the canopy tree *Tetragastris panamensis* < 20 m from their parent tree survive an additional year, 13 times the corresponding survival rate of *Virola* seedlings of comparable age (Howe 1990). Indeed, seedlings have a higher chance of surviving their first year under the parent crown than if dispersed farther away by monkeys, their primary disperser (Howe 1989, 1990). Nevertheless, recruitment of saplings ≥ 1 cm dbh is more abundant in this species some distance away from conspecific adults than near (a "partially repelled" distribution of recruits; see below), suggesting that pest pressure influences sapling distribution in this species (Condit et al. 1992a).

To analyze the factors responsible for the diversity of tropical trees, Hubbell and Foster (1983, 1986a,b, 1990a,b) established a 50-ha forest dynamics plot on Barro Colorado's central plateau. All nonliana stems on this plot were censused in 1980–1982, 1985, 1990, and 1995. I have already laid the forest dynamics plot under contribution for data on forest structure, dynamics, and diversity. Now I consider the testimony of this plot and its students concerning the effect on saplings of proximity to adult conspecifics.

The first advantage of this plot is that it allows an abundance of data to be brought to bear on individual species. Thus, in the most common species of canopy tree on the plot, *Trichilia tuberculata*, saplings 1–4 cm dbh grew more slowly and suffered higher mortality between 1982 and 1985 if their nearest neighbor ≥ 30 cm dbh was a conspecific. During this same period, half as many *Trichilia* per square meter attained 1 cm dbh within 5 m of a conspecific ≥ 30 cm dbh as elsewhere on the plot. Finally, the number of *Trichilia tuberculata* ≥ 1 cm dbh declined on those hectares with the most adult conspecifics and increased on those hectares with the fewest, as if the population were regulated by its own density, presumably thanks to its pests or pathogens (Hubbell et al. 1990).

Saplings of the canopy tree *Quararibea asterolepis* recruit more abundantly near conspecific adults (Condit et al. 1992a). Between 1982 and 1985, *Quararibea* saplings grew

Table 8.14 Percentage of animal-dispersed species among large trees in different tropical forests

Forest Type	Site	Percent
Dry	Santa Rosa, Costa Rica	64
Moist	Barro Colorado Island, Panama	78
Wet	Rio Palenque, Ecuador	94

faster the more conspecifics ≥ 16 cm dbh were between 10 and 50 m away (Hubbell et al. 1990). Nevertheless, *Quararibea* suffered severe defoliation from an outbreak of caterpillars of a noctuid moth (*Eulepidotis* sp.) in May and June of 1985. The caterpillars preferred young leaves. Saplings and adults with young leaves were far more likely to be attacked if near an infested adult, although proximity to uninfested conspecifics had no effect (Wong et al. 1990). Severely defoliated trees produced less fruit than usual that year, but the ill effects of defoliation were not apparent in later years. Would the defoliation by a specialist caterpillar, and/or the partial reproductive failure, make place for other species, either by allowing more light to reach the forest floor or by removing some competitors?

On the 50-ha plot, a canker disease is afflicting the canopy tree species *Ocotea whitei*. Between 1982 and 1985, sapling recruitment in this species was most abundant close to conspecific adults (Condit et al. 1992a). Now the disease is more prevalent where the density of its host species is greater, and saplings are more likely both to be infected and to die the closer they are to a conspecific adult. Thus this disease is killing juveniles closer to adults, spacing out the distribution of recruits, as the pest pressure hypothesis would suggest (Gilbert et al. 1994).

Condit et al. (1992a) have also used data from the 50-ha plot to undertake a comprehensive test of the pest pressure hypothesis. For each species A with enough members to permit a test, the distribution of distances of A recruits (A saplings attaining 1 cm dbh between 1982 and 1985) to the nearest A adult was compared with the distribution of distances of non-A recruits to the nearest A adult. If a significantly smaller fraction of A than of non-A recruits was within some distance $x \leq 20$ m of the nearest A adult, the distribution of A recruits was said to be "repelled." If the opposite were the case, and a significantly higher proportion of A than of non-A adults was within some distance $x \leq 20$ m of the nearest A adult, the distribution of A recruits was called "attracted." If the ratio of A to non-A recruits peaked at an intermediate distance from the nearest A adult, without leaving a significant shortage of A recruits close to their adults, the distribution of A recruits was called "partially repelled." All other distributions were said to show "no pattern." The first lesson from their test is that pest pressure is progressively less effective in spacing out recruitment as one descends from canopy trees to shrubs (Table 8.15). The second lesson is that species with clearly repelled distributions form a small minority even of canopy trees. On the other hand, in the absence of pests, all distributions should be attracted. It seems, therefore, that above the shrub layer, pests exert a profound influence on the distribution of recruits.

Wills et al. (1997) have provided more evidence for this conclusion. They found that over half the 84 most common species on the 50-ha plot recruit better (produce more saplings ≥ 1 cm dbh per conspecific adult) in 10 × 10 or 20 × 20 m plots where the basal area of conspecific adults

Table 8.15 Effect of nearness to a conspecific adult on sapling recruitment (data from Condit et al. 1992a)

No. of Species whose Recruit Distributions are	Canopy	Midstory	Understory	Shrub
Repelled	3	2	2*	0
Partially repelled	5	2	0	1
No pattern	10	15	11	2
Attracted	5	4	6	12
Total	23	23	19	15

*Counting *Faramea occidentalis* (Condit et al. 1994).

is lower. In particular, for over half these 84 species, the number of new stems ≥ 1 cm dbh appearing per conspecific adult between successive censuses, on 10 × 10 or 20 × 20 m plots, showed significant ($P < 0.05$) negative correlation with the total basal area of conspecifics on these plots. None showed significant positive correlation between these variates. For over half the significant negative correlations, $P < 0.002$. This effect was local. Less than 1 species in 8 showed significant negative correlation between recruitment per adult and basal area of conspecifics on 50 × 50 m plots. Some factor, presumably pests and/or pathogens, favors recruitment to neighborhoods where adult conspecifics are scarce, as the results of Condit et al. (1992a) suggest.

Moreover, the plot's most common canopy tree, *Trichilia tuberculata* (Hubbell et al. 1990), its most common midstory tree, *Hirtella triandra* (Condit et al. 1992a), and its most common understory treelet, *Faramea occidentalis* (Condit et al. 1994), all behave in accord with the pest pressure hypothesis. Do pests maintain tree diversity by limiting the populations of potential competitive dominants, as the starfish *Pisaster* maintains biotic diversity in the intertidal zones of rocky shores in the northeastern Pacific by preventing them from being overrun by a uniform carpet of mussels (Paine 1966, 1974)? It is true that in the 50-ha plot's most common species, *Hybanthus prunifolius*, with more than 41,000 stems ≥ 1 cm dbh, seeds are dispersed by a very inefficient mechanism. *Hybanthus* young recruit most abundantly near adults (Condit et al. 1992a), even though this species is subject to occasional massive defoliations (Wolda and Foster 1978). Nonetheless, recruitment of young *Hybanthus* (1–4 cm dbh) around the largest individuals (dbh ≥ 4 cm) is depressed, just as in *Trichilia* (R. Condit, p. c.). Are the largest *Hybanthus* old enough or big enough to permit buildup of species-specific pests or pathogens in their neighborhood?

Difficulties of Testing the Pest Pressure Hypothesis

I have explained in some detail why I think pest pressure plays an essential role in maintaining the diversity of tropical trees. Nonetheless, it is extraordinarily difficult to decide what constitutes a genuine test of this hypothesis. This difficulty illustrates some of the fundamental problems that confront those who would construct a true theoretical biology.

First, it is not clear whether the pest pressure hypothesis predicts a stable or a shifting species composition for the forest as a whole. Hubbell and Foster (1986a: p. 322) agree with Janzen (1970) that if pest pressure enhances species diversity, it does so by afflicting each species more heavily as its numbers increase, keeping all their populations low, thereby allowing many species to coexist. In this view, pests stabilize the species composition of the forest: in today's lingo, they treat the pest pressure hypothesis as an equilibrium theory of species diversity. Yet Gillett (1962) explicitly predicted that a new species with a novel defense would spread rapidly at first and then decline slowly to extinction as it accumulates an ever-growing load of pests, a process that would be reflected by continuous change in a forest's species composition.

Second, even those who treat the pest pressure hypothesis as an equilibrium theory of species diversity have disagreed on how to test it. To begin with, Janzen (1970) was unwary enough to assert that pest pressure would cause hyperdispersion among conspecific adults. Therefore, every tropical tree species whose adults have clumped distributions has been adduced as disproof of Janzen's theory (Hubbell 1979, Hubbell and Foster 1983). In the absence of herbivores, however, trees of a species *should* be clumped, for seeds move only a limited distance from their parents.

The logical next question is, how heavy must seedling mortality near the parent plant be, and how widely must this "radius of death" extend, to maintain a given level of tree diversity? This question bristles with mathematical difficulties; its solution presupposes a capacity to deal mathematically with the dynamics of spatial arrangements which today's theoreticians lack. Let us review some of the problems.

Hubbell (1980) constructed a theoretical lattice of hexagonal tree crowns where conspecific crowns never shared an edge, using only three species. He concluded that a consumption of seedlings near their parents heavy enough to prevent crowns of adult conspecifics from touching would not, of itself, permit many tree species to coexist. In this model, however, an invading plant, immune to the pests of these three species, could replace any newly dead tree within reach, while one of the three lattice species, excluded from hexagons bordered by conspecifics, can only replace a third of the trees that die (Leigh 1990, 1996). Hubbell grossly underestimated the potential of pest pressure for promoting tree diversity (Becker et al. 1985) because he did not see that his model implied a threefold advantage for invading trees with novel defenses.

Finding greater mortality among those seeds or seedlings of a species closest to their parents, however, does not necessarily imply that herbivores are controlling the numbers of that species. On Barro Colorado, Schupp (1992) examined the effects of distance from the nearest conspecific adult on the mortality of seeds of the treelet *Faramea occidentalis*, which averages 130 reproductive adults ≥ 4 cm dbh/ha on the 50-ha plot. On the average, seed survival during the first 28 weeks after seeds fell was higher 5 m away from the edge of a conspecific adult's crown than under it. However, the prospects of seeds 5 m from a conspecific adult's crown were independent of the density of adult conspecifics, while seed survival under conspecifics was higher where adults were more abundant. Where there were > 300 adult *Faramea* per hectare, seeds benefitted from being under adults. Thus seed-eaters were not capable of limiting the number of *Faramea* seeds which survived to germinate, even though, on the average, seeds survived better away from their parents.

To assess how pest pressure contributes to tree diversity, mortality patterns over all stages of growth of a plant must be considered, not just one. Thus, despite the fact that *Faramea* seeds survive best where adults are most abundant, between 1982 and 1990 fewer *Faramea* saplings attained 1 cm dbh where conspecific adults were most abundant (Condit et al. 1994). Moreover, between 1982 and 1990 *Faramea* saplings between 1 and 2 cm dbh grew significantly more slowly and suffered significantly higher mortality when they were within 2 m of conspecific adults. Finally, *Faramea* saplings grew most slowly and died most rapidly where the number of adult conspecifics within 30 or 60 m was highest. In short, even though *Faramea* seeds survive best where *Faramea* adults were most abundant, *Faramea* behaves in accord with the pest pressure hypothesis (Condit et al. 1994).

This last example leads us to the problem of scale. If we distinguish mortality and recruitment on the two 500 × 500 m halves of Barro Colorado's 50-ha plot, we will find several species that are surviving or recruiting more successfully on the half plot where they are less abundant (Table 8.16). Among them, *Capparis frondosa* has an attracted distribution of recruits (Condit et al. 1992a). A few species with an inverse relation between abundance and population increase is not proof that their populations are regulated by the concentration of pests on larger-scale aggregations of conspecifics, but it does suggest that the truth of the pest pressure hypothesis cannot be settled by studies on a single scale.

In the dry forest at Mudumalai, South India, on a rolling plateau full of elephants and tigers, just north of the massif of the Nilgiris, there is another 50-ha plot (Sukumar et al. 1992). There, the more common tree species suffer greater defoliation from insect herbivores (Murali and Sukumar 1993). This cannot be a universal truth: the single-dominant rainforests of Zaire, whose canopies consist almost entirely of *Gilbertiodendron dewevrei* (Hart et al. 1989), are not said to suffer disproportionately from insect herbivores. But if, on Barro Colorado, more common species were defoliated more heavily, would enough light reach the forest floor under the crowns of common trees

Table 8.16 Abundance, mortality, and recruitment of selected species on the two halves of BCI's 50-ha forest dynamics plot

Species	N (1982)	1982–1985			N (1985)	1985–1990			N (1990)
		D	R	U		D	R	U	
Trichilia tuberculata									
≥ 1 cm dbh, west half	4381	427 (63)	498	9	4380	579 (101)	652	9	4343
east half	8547	687 (205)	1141	10	8786	920 (282)	1385	22	8947
≥ 20 cm dbh, west half	643	82 (1)	34	3	591	100 (0)	43	2	532
east half	417	49 (3)	33	0	398	50 (0)	43	1	390
Capparis frondosa									
≥ 1 cm dbh, west half	1175	23 (30)	148	8	1262	39 (67)	116	6	1266
east half	2357	82 (117)	253	2	2409	113 (193)	180	6	2277
Beilschmeidea pendula									
≥ 1 cm dbh, west half	1430	187 (73)	307		1550	267 (123)	234		1517
east half	946	80 (34)	255		1121	130 (89)	242		1233
≥ 30 cm dbh, west half	71	11 (0)	2		62	6 (0)	4		60
east half	55	3 (0)	3		55	2 (0)	5		58

N (year), number of stems alive that year; D, the number of stems that died before the next census (in parentheses, the number crushed so that they fell below 1 cm dbh before the next census); R, the number of stems attaining 1 cm dbh before the next census; and U, the number of stems measured in this census that were not found during the following census. Data from R. Condit, S. P. Hubbell, and R. B. Foster (p. c.).

to permit seedlings of other species to become established? Did Murali and Sukumar (1993) conduct a valid test of the pest pressure hypothesis?

CONSEQUENCES OF TREE DIVERSITY

The factors that enhance plant diversity also allow extraordinary animal diversity. The H. J. Andrews Experimental Forest, a tract of over 50,000 ha in Western Oregon, has 60 species of true butterfly (excluding skippers, Hesperiidae): 37 species of Papilionidae, Pieridae and Nymphalidae, and 23 species of Lycaenidae (no Riodinidae) (Parsons et al. 1991). Forty-six species of true butterfly have been recorded from the 16,300 ha of Washington, D.C.: 31 species of Papilionidae, Pieridae, and Nymphalidae, and 14 species of Lycaenidae (no Riodinidae) (Clark 1932). A comparably urban site of 61,600 ha in the tropics, Singapore, has 142 species of Papilionidae, Pieridae, and Nymphalidae, and 155 species of Lycaenidae + Riodinidae (Fleming 1975).

Within the Neotropics, butterfly diversity appears to track tree diversity. On Barro Colorado Island, a visitor will see one species of large brown satyr, *Pierella luna*, flying up and down the trail, and one species of clearwinged satyr, *Cithaerias menander*, hovering over the forest floor, whose clear hind wings are stained with red in such a manner that when it flies it looks a bit like a noxious red bug. In Peru, where tree diversity is much greater, one sees several species of each, subtly but clearly distinguished by decorations on the margins of their hind wings. Numbers bear out this impression. In lowland Neotropical sites, species of Papilionidae, Pieridae, and Nymphalidae make up almost exactly half a site's total number of species of true butterfly (Robbins et al. 1996). Barro Colorado has 137 species of Papilionidae, Pieridae, and Nymphalidae (DeVries 1994), suggesting a total of about 274 species of true butterfly. La Selva, which has more species of tree per 4 ha than Barro Colorado, thanks to its wetter dry season, has 204 species of Papilionidae, Pieridae, and Nymphalidae (DeVries 1994), suggesting a total of about 408 species of true butterfly. A 4,000-ha lowland site at Pakitza, in Peru's Manu Park, has 425 species of Papilionidae, Pieridae, and Nymphalidae, and 427 of Riodinidae + Lycaenidae. Pakitza has three times the butterfly diversity of Barro Colorado; it has about twice Barro Colorado's tree diversity (Table 8.3).

Indeed, the diversity of tropical insects has become a legendary mystery. Erwin and Scott (1980) found at least 945 species among the 7712 beetles obtained by applying insecticidal fog to the crowns of 19 canopy trees of the species *Luehea seemannii* in central Panama. Here, Fisher's α exceeds 283. The insect species of the world are far from having been counted: we do not know whether the world contains 3 million, 30 million, or more insect species (May 1990).

Diversity among insects collected from a tropical tree's crown is markedly greater than diversity among insects from the crown of a British tree (Table 8.17). However, the diversity of beetles in the crown of a tropical tree is much the same in a dry forest of northern Venezuela, a moist forest in central Panama, and an everwet forest in Brunei (Table 8.17), even though tree diversity in these forests are strikingly different. In New Guinea, the diversity of beetles in the crown of an individual *Castanopsis acuminatissima* tree is as great at 2100 as at 500 m. Why is there so little congruence between tree diversity and beetle diversity per tree? We are far from being able to answer this question.

Making sense of the diversity of tropical insects would appear to be a hopeless task. It has seemed impossible to

Table 8.17 Diversity of all insects, and of beetles, from individual tree crowns at selected sites

Location and tree	N	S	α
All insects			
Brunei, lowland forest (Stork 1991)			
Shorea johorensis # 1	2649	637	266
Shorea johorensis # 4	2166	566	249
Pentaspadon motleyi # 1	1020	288	134
Pentaspadon motleyi # 2	1431	524	298
Castanopsis sp.	3573	684	251
Britain (Southwood et al. 1982)			
Betula pendula	19335	337	57
Quercus robur	8869	465	106
Salix cinerea	2094	322	110
Salix alba	1252	176	56
Beetles			
Central Panama, lowland moist forest (Erwin and Scott 1980)			
Luehea seemannii # 1	2085	335	113
Luehea seemannii # 2	1174	171	58
Luehea seemannii # 3	830	191	78
Luehea seemannii # 4	410	147	82
Luehea seemannii # 5	405	115	54
Northern Venezuela, lowland dry forest (Davies et al. 1997)			
Talisia sp.	975	292	141
Brownea grandiflora	393	168	111
Chrysophyllum lucentifolium	827	211	92
Brunei, lowland forest (Stork 1991)			
Shorea johorensis # 1	291	130	90
Shorea johorensis # 2	784	270	146
Shorea johorensis # 3	457	166	94
Shorea johorensis # 4	468	141	69
Castanopsis sp.	396	103	45
New Guinea, 500 m (Allison et al. 1993)			
Castanopsis acuminatissima 1	595	99	34
Castanopsis acuminatissima 2	496	125	54
New Guinea, 1200 m (Allison et al. 1993)			
Castanopsis acuminatissima 1	336	82	35
Castanopsis acuminatissima 2	299	86	40
New Guinea, 2100 m (Allison et al. 1993)			
Castanopsis acuminatissima 1	1829	234	71
Castanopsis acuminatissima 2	642	151	67

N, number of insects; S, number of species among them; and Fisher's α for insects collected by insectical fogging of the crowns of selected trees. Each row of numbers represents one tree's crown.

discern the principles governing the regulation of most insect populations (Wolda 1992), let alone the factors allowing these species to coexist. More careful study, however, may dispel this ignorance. Ray and Hastings (1996) found that a population's changes are much more likely to show detectable density dependence if > 10% of the population leave (or enter) the study area per generation. Use of a much larger study area sums the changes of a number of separate populations, each responding to its own local conditions, in a manner which averages out the density-dependent signal. Similarly, by accounting for the degree of aggregation in the distribution of numbers of different *Drosophila* species over patches of fruit on Barro Colorado's forest floor, Sevenster and van Alphen (1996) were able to account for the coexistence of the species involved.

They were also able to show that at Barro Colorado, the community of *Drosophila* species whose larvae grow in rotting fruit on the forest floor is nearly saturated with species. In all these cases, understanding hinges on an adequate reckoning with spatial heterogeneity.

The factors that allow the activity of tropical insects to last through the year and that permit fruit to be available in tropical forest year-round thereby allow a great increase in the diversity of birds and mammals (Table 8.18). The continuous supply of insects in tropical forest has provided openings for an extraordinary variety of insect-eating birds, bats, and other mammals; we have already discussed the integral role birds play in protecting tropical forest from insect pests. For both birds and mammals, however, the most spectacular contrast between temperate and tropical

Table 8.18 Numbers of bird and mammal species with different diets at selected tropical and temperate-zone sites (data from Emmons 1989)

Site	Dietary category	No. of Species
Mammals		
Gloucester Co., VA	Leaf-eaters	6
	Frugivores/granivores	6
	Insects and small vertebrates	15
	Carnivores	10
Ipassa, Gabon	Leaf-eaters	11
	Frugivores/granivores	56
	Insectivores	47
	Carnivores/omnivores	12
Birds		
Congaree, SC	Fruit and nectar	7
	Insectivores	27
	Carnivores	6
Manu, Peru	Fruit and nectar	84
	Insectivores	98
	Carnivores	25

forest is the vastly higher diversity of tropical fruit-eaters. Among these animals, population regulation and niche differentiation seems adequately understood and will not be discussed here. What is clear is that the need of tropical trees for animals to disperse their seeds has vastly multiplied the openings for fruit-eating vertebrates.

CONCLUDING REMARKS

Diversity of trees is far greater in the tropics than in the temperate zone. Tree diversity is greatest in the everwet forests of western Amazonia and Sarawak.

I assume that these trees coexist for a reason, not merely by chance. There are many environmental gradients in tree diversity. The latitudinal gradient is the most famous, but, within the tropics, tree diversity is lower the more severe the dry season. In everwet settings, tree diversity declines as one passes from "modal" oxisols and ultisols to bleached-white sand. Finally, climate and vegetation were very different just 12,000 years ago. How could such predictable diversity gradients arise so quickly by chance?

Diversity is nature's response to trade-offs, to circumstances whereby developing one capacity precludes another. Among tropical trees, pressure from pests and pathogens is a perennial threat. Different species of trees have different defenses. In natural forest, the most devastating pests are specialists on one or a few host species, as if the ability to penetrate one type of defense precludes the ability to parry others. Is each species of tropical tree kept rare by its pests, thereby allowing room for others?

Pest pressure is more intense and unrelenting in the tropics than in the temperate zone. Young tropical leaves are far more poisonous, yet eaten far more rapidly, than their counterparts in the temperate zone. In everwet forest, where tree diversity is greatest and pest pressure most continuous, nearly all trees have animal-dispersed seeds as if, in these forests, the premium on dispersing seeds beyond the range of their parents' pests is highest.

Although the enormous diversity of tropical trees appears to reflect the more intense pest pressure of tropical settings, a convincing test of this proposition requires an ability to reckon mathematically with changes in the spatial distribution of trees and saplings, which we lack.

The diversity of tropical trees supports a remarkable diversity of vertebrates and an extraordinary diversity of insects. Understanding how these insect populations are regulated and how so many species coexist requires appropriate attention to spatial heterogeneity.

Understanding diversity involves understanding how species compete and what circumstances allow them to coexist. To understand an ecological community, however, we must also understand the many ways different individuals and different species depend on each other. The next chapter is devoted to that task.

APPENDIX 8.1

Diversity in different forest plots

Site	Plot Size (ha)	No. of Trees	No. of Species	Fisher's α	Reference
Tropical sites					
Las Tuxtlas, Mexico					Bongers et al. (1988)
Plants ≥1 cm dbh	1.0	3344	185	42.2	
Plants ≥ 1 cm dbh	0.5	1584	136	35.6	
Nonvine plants ≥ 1 cm dbh	1.0	2976	142	31.0	
Trees/saplings ≥ 1 cm dbh	1.0	1202	119	32.8	
Nonvine plants ≥ 10 cm dbh	1.0	359	88	37.2	
Nonvine plants ≥ 10 cm dbh	0.5	176	58	30.2	
Trees ≥ 20 cm dbh	1.0	180	52	24.5	
Trees ≥ 40 cm dbh	1.0	75	31	19.8	
Chamela, Mexico, dry forest with 6–8 month dry season, ≥ 3.18 cm dbh					Martinez-Yrizar et al. (1996)
Upper plot	0.32	893	57	13.6	
Middle plot	0.32	1031	78	19.6	
Lower plot	0.32	673	88	27.1	
Finca Seacté, Guatemala, trees ≥ 10 cm dbh					
900 m	0.1	45	25	23.2	Kunkel-Westphal and Kunkel (1979)
1000 m	0.1	45	28	31.7	Kunkel-Westphal and Kunkel (1979)
La Selva, Costa Rica, trees ≥ 10 cm dbh					Lieberman et al. (1985)
Plateau	4.4	1838	172	46.4	
	1.0	418	93	37.1	
Hills	4.0	2114	165	41.9	
	1.0	529	102	37.6	
Guanacaste, Costa Rica, dry forest, stems ≥ 2 cm dbh					
	13.44	15889	135	20.3	Hubbell (1979)
Corcovado, Llorona, Costa Rica, stems ≥ 10 cm dbh					
	1.0	354	108	53.0	Hartshorn (1983)
Volcan Barva, Costa Rica, stems ≥ 10 cm dbh					Lieberman et al. (1996)
100 m	1.0	458	108	44.6	
300 m	1.0	538	142	62.9	
500 m	1.0	414	124	60.0	
1000 m	1.0	529	99	35.9	
1500 m	1.0	571	74	22.7	
2000 m	1.0	477	55	16.1	
2600 m	1.0	650	29	6.2	
Monteverde, Costa Rica 1480 m	1.0	559	76	23.7	Nadkarni et al. (1995)
Barro Colorado Island, Panama, 50-ha plot, 1982 census					
Stems ≥ 1 cm dbh (all)	50.0	235349	305	34.6	R. Condit and S. Loo de Lao (p. c.)
Trees ≥ 10 cm dbh (all)	50.0	20882	238	37.7	R. Condit and S. Loo de Lao (p. c.)
Trees ≥ 20 cm dbh (all)	50.0	7944	183	33.4	R. Condit and S. Loo de Lao (p. c.)
Trees ≥ 30 cm dbh (all)	50.0	4267	144	28.8	R. Condit and S. Loo de Lao (p. c.)
Stems ≥ 1 cm dbh, avg. ha	1.0	4615	173	35.5	R. Condit and S. Loo de Lao (p. c.)
Trees ≥ 10 cm dbh, avg. ha	1.0	409	91	36.2	R. Condit and S. Loo de Lao (p. c.)
Trees ≥ 20 cm dbh, avg. ha	1.0	156	54	29.1	R. Condit and S. Loo de Lao (p. c.)
Trees ≥ 30 cm dbh, avg. ha	1.0	84	35	23.1	R. Condit and S. Loo de Lao (p. c.)
Trees ≥ 10 cm dbh, ha 6.4	1.0	601	86	27.5	S. P. Hubbell and R. B. Foster (p. c.)
Trees ≥ 10 cm dbh, ha 6.4	0.3	153	48	24.0	S. P. Hubbell and R. B. Foster (p. c.)
Barro Colorado Island, lower Lutz Ravine					R. Wirth (p. c.)
Stems ≥ 5 cm dbh	0.22	198	68	36.6	
Stems ≥ 10 cm dbh	0.22	120	52	34.9	
Stems ≥ 20 cm dbh	0.22	58	32	29.3	
Barro Colorado Island, secondary forest surrounding laboratory clearing, trees ≥ 20 cm					Thorington et al. (1982)
Ha 1 (1975) Allee catchment	1.0	173	61	33.6	
Ha 2 (1975) Lutz, near weir	1.0	172	51	24.5	
Ha 3 (1975) TB 2–3	1.0	175	49	22.6	
Ha 4 (1975) TB 4–5	1.0	168	59	32.4	
Ha 5 (1975) Donato 5	1.0	168	44	19.4	

Site	Plot Size (ha)	No. of Trees	No. of Species	Fisher's α	Reference
Barro Colorado Island, secondary forest, Knight-Lang plot					Lang and Knight (1983)
Stems ≥ 2.5 cm dbh, 1968	1.5	4668	147	28.9	
Stems ≥ 2.5 cm dbh, 1978	1.5	4172	137	27.2	
Stems ≥ 5 cm dbh, 1968	1.5	2488	111	23.8	
Stems ≥ 5 cm dbh, 1978	1.5	2297	110	24.1	
Stems ≥ 10 cm dbh, 1968	1.5	768	79	22.1	
Stems ≥ 10 cm dbh, 1978	1.5	841	82	22.5	
Stems ≥ 20 cm dbh, 1968	1.5	286	46	15.5	
Stems ≥ 20 cm dbh, 1978	1.5	276	45	15.3	
Stems ≥ 30 cm dbh, 1968	1.5	164	26	8.7	
Stems ≥ 30 cm dbh, 1978	1.5	172	31	11.1	
Kuna Yala, Panama, everwet forest					
350 m, stems ≥ 10 cm dbh	1.0	559	191	102.3	R. Paredes (unpublished data)
Bajo Calima, Choco, Colombia, trees ≥ 10 cm dbh					Faber-Langendoen and Gentry (1991)
Plot A	1.0	664	252	148	
	0.1	62	46	80.7	
Plot B	0.5	352	154	104.3	
	0.1	60	41	57	
Cocha Cashu, Peru, mature floodplain forest (less seasonal than BCI)					Gentry (1988b)
Trees ≥ 10 cm dbh	1.0	650	189	89.5	
Trees ≥ 30 cm dbh	1.0	110	68	75.9	
Ecuador, everwet forest					
Cuyabeno, stems ≥ 5 cm dbh	1.0	1561	473	230.8	Valencia et al. (1994)
Cuyabeno, trees ≥ 10 cm dbh	1.0	693	307	211.0	Valencia et al. (1994)
Añangu, trees ≥ 10 cm dbh	1.0	734	153	58.8	Korning and Balslev (1994)
Jatun Sacha, trees ≥ 10 cm dbh	1.0	724	246	131.2	Phillips et al. (1994)
Yanamono, Peru (equatorial, aseasonal)					Gentry (1988b)
Trees ≥ 10 cm dbh	1.0	580	283	218.2	
Trees ≥ 30 cm dbh	1.0	110	81	138.7	
N. of Manaus, Brazil					
Trees ≥ 15 cm dbh	1.0	350	179	146.9	Prance et al. (1976)
Trees ≥ 20 cm dbh	1.0	214	126	128.6	Prance et al. (1976)
Trees ≥ 30 cm dbh	1.0	100	63	73.0	Prance et al. (1976)
Stems ≥ 1.5 m tall	0.2	1986	502	216.3	Fittkau and Klinge (1973)
Mishana, Peru (equatorial, aseasonal, white sand)					Gentry (1988b)
Trees ≥ 10 cm dbh	1.0	842	275	142.1	
Trees ≥ 30 cm dbh	1.0	83	54	67.0	
Maracá, Brazil, trees ≥ 10 cm dbh	0.25	105	40	23.6	Thompson et al. (1992)
Beni, Bolivia, seasonal forest, 1600 mm rain/year, trees ≥ 10 cm dbh					Boom (1986)
10 × 1000 m plot	1.0	649	94	30.2	
Neblina Base Camp, Brazil/Venezuela, bleached sand					Gentry (1988a)
Trees ≥ 10 cm dbh	1.0	493	89	31.7	
Trees ≥ 30 cm dbh	1.0	84	24	11.2	
San Carlos de Rio Negro, Venezuela					Uhl and Murphy (1981)
Trees ≥ 10 cm dbh	1.0	744	83	23.9	
ECEREX, Piste de St. Elie, Guyane Française					
Stems ≥ 5 cm dbh, parcelle P	0.25	335	120	66.9	Puig and Lescure (1981)
Stems ≥ 5 cm dbh, parcelle B	0.25	268	64	26.8	Puig and Lescure (1981)
Stems ≥ 5 cm dbh, parcelle C	0.25	253	74	35.2	Puig and Lescure (1981)
Stems ≥ 10 cm dbh	0.5	327*	130	79.8	Sabatier and Prévost (1988, p. 36)
Stems ≥ 10 cm dbh	1.0	654*	175	78.3	Sabatier and Prévost (1988, p. 36)
El Verde, Luquillo Mountains, Puerto Rico (450 m)					
Trees ≥ 4 cm dbh, 1951	0.72	1236	65	14.6	Crow (1980)
Trees ≥ 4 cm dbh, 1976	0.72	1019	50	11.0	Crow (1980)

(continued)

Appendix 8.1 (*continued*)

Site	Plot Size (ha)	No. of Trees	No. of Species	Fisher's α	Reference
Luquillo Mountains, Puerto Rico (virgin Tabonuco forest)					
Trees ≥ 10 cm dbh	4.05	3140	65	11.6	Odum (1970)
Trees ≥ 50 cm dbh	4.05	131	20	6.6	Odum (1970)
Elfin Forest, Luquillo Mountains, Puerto Rico, trees ≥ 10 cm dbh					Weaver et al. (1986)
Mt. Britton, 930 m	0.14	604	11	1.9	
Pichacos, 950 m	0.06	159	11	2.7	
Pico del Oeste, 1010 m	0.02	81	6	1.5	
Montane forest, Jamaica, trees ≥ 10 cm gbh					Tanner (1977)
Mor ridge, 1615 m, 5 m tall	0.08	388	16	3.36	
Mull ridge, 1605 m	0.1	515	35	8.5	
Wet slope, 1570 m	0.1	306	35	10.2	
Gap (saddle), 1590 m, 15 m	0.1	273	27	7.4	
Montane forest, Baeza, eastern Ecuador, 2000 m					Valencia (1995)
Stems ≥ 5 cm dbh	1.0	1622	55	11.0	
Stems ≥ 10 cm dbh	1.0	482	45	12.2	
Montane forest, Pasochoa volcano, Ecuador, 3280 m					Valencia and Jorgensen (1992)
Stems ≥ 5 cm dbh	1.0	1058	32	6.2	
Stems ≥ 10 cm dbh	1.0	715	27	5.5	
Cordillera Central, Colombia					Veneklaas (1991)
2550 m, stems ≥ 10 cm dbh	0.20	110	30	13.6	
Côte-d'Ivoire					Bernhard-Reversat et al. (1978)
Banco, stems ≥ 40 cm gbh	0.25	66	25	14.7	
Banco, stems ≥ 40 cm gbh	5.0	1325	104	26.4	
Yapo, stems ≥ 40 cm gbh	0.25	107	29	13.1	
Yapo, stems ≥ 40 cm gbh	5.0	2135	132	31.1	
Ipassa, Gabon					
Stems ≥ 4.5 cm dbh	0.4	297	89	43.1	Hladik (1982)
Stems ≥ 10 cm dbh	0.4	190	66	35.9	Hladik (1982)
Stems ≥ 20 cm dbh	0.4	89	35	21.3	Hladik (1982)
Stems ≥ 30 cm dbh	0.4	52	22	14.4	Hladik (1982)
Stems ≥ 30 cm dbh	0.9	93	41	28.0	Hladik (1982)
Douala-Edea, Cameroun					Newbery et al. (1986)
Stems ≥ 10 cm dbh	0.64	241	39	13.2	
Korup, Cameroun					Gartlan et al. (1986)
Stems ≥ 10 cm dbh	0.64	301	75	32.0	
Oveng, Gabon, trees ≥ 10 cm dbh	1.0	485	123	53.1	Reitsma (1988)
Stems ≥ 10 cm dbh	1.0	497	131	58.0	
Doussala, Gabon, trees ≥ 10 cm	1.0	413	100	42.0	Reitsma (1988)
Stems ≥ 10 cm	1.0	425	109	47.4	
Lopé, Gabon, trees ≥ 10 cm dbh	1.0	392	65	22.2	Reitsma (1988)
Stems ≥ 10 cm dbh	1.0	396	69	24.2	
Stems ≥ 10 cm dbh	1.0	386	100	43.8	White (1994)
Stems ≥ 10 cm dbh	1.25	511	116	46.8	White (1994)
Stems ≥ 70 cm dbh	11.25	175	48	21.8	White (1994)
Ekobakoba, Gabon, trees ≥ 10 cm	1.0	429	77	27.4	Reitsma (1988)
Stems ≥ 10 cm	1.0	438	85	31.4	
Bélinga, Gabon, stems ≥ 30 cm	1.0	119	50	32.5	Hladik (1978)
Analamazaotra, Madagascar, 1000 m					Abraham et al. (1996)
Stems ≥ 5 cm dbh	0.51	1423	189	58.5	
Stems ≥ 5 cm dbh, no lianes	0.51	1387	176	53.4	
Stems ≥ 10 cm dbh, no lianes	0.51	539	125	51.1	
Stems ≥ 20 cm dbh	0.51	110	55	43.8	

Site	Plot Size (ha)	No. of Trees	No. of Species	Fisher's α	Reference
Ranomafana, Madagascar, 1050 m, 20 × 500 m plot at Vatoharanana					Schatz and Malcomber (1994)
Stems ≥ 10 cm dbh	1.00	660	105–112	36.7	
Stems ≥ 50 cm dbh	1.00	47	19	11.9	
Ambohitantely, Madagascar, 1550 m					Abraham et al. (1996)
Stems ≥ 5 cm dbh	0.70	2808	102	20.8	
Stems ≥ 10 cm dbh	0.70	948	73	18.4	
Stems ≥ 20 cm dbh	0.70	134	25	9.1	
Kirindy, Morondava, Madagascar, dry forest					Abraham et al. (1996)
Stems ≥ 30 cm gbh	0.93	788	45	10.4	
Réunion, 250 m, stems ≥ 25 cm gbh	1.0	1270	43	8.6	Strasberg (1995)
Stems ≥ 10 cm dbh	1.0	1079	40	8.2	Strasberg (1995)
Upland forest 13 m tall (emergents to 17 m), 550 m, Mauritius, 20° S					Vaughan and Wiehe (1941)
Plot 1, stems ≥ 50 cm tall	0.1	1726	65	13.3	
Plot 1, stems ≥ 10 cm dbh	0.1	126	30	12.5	
Upland forest, Mauritius, 10 disjunct plots combined					Vaughan and Wiehe (1941)
Stems ≥ 10 cm dbh	1.0	1710	52	10.1	
Stems ≥ 20 cm dbh	1.0	371	35	9.5	
Stems ≥ 30 cm dbh	1.0	168	19	5.5	
Uppangala (12°30' N, 75°39' E), Western Ghats, India, 450 m (4500 mm rain/year, mostly between May and November)					
Stems ≥ 10 cm dbh	4.0	2607	98	20.1	Sinha and Davidar (1992)
Mudumalai, deciduous dry forest, South India (on northern flank of Nilgiris)					
Stems ≥ 1 cm dbh	50.0	25929	71	8.9	Sukumar et al. (1992)
Trees ≥ 10 cm dbh	50.0	15417	63	8.4	Sukumar et al. (1992)
Trees ≥ 20 cm dbh	50.0	8870	59	8.5	Sukumar et al. (1992)
Stems ≥ 1 cm dbh, avg. ha	1.0	514	26	5.8	Condit et al. (1996b)
Stems ≥ 10 cm dbh, avg. ha	1.0	301	22	5.5	Condit et al. (1996b)
Stems ≥ 20 cm dbh, avg. ha	1.0	177	18	5.0	Condit et al. (1996b)
Nilgiri Hills, India, 2100m					R. Sukumar and H. S. Suresh (p. c.)
Stems ≥ 3 cm dbh, Avalanchi	0.1	259	592	23.9	
Stems ≥ 3 cm dbh, Thaishola	0.1	260	28	8.0	
Kottawa, Sri Lanka, 50 m, everwet forest					
Stems ≥ 30 cm girth	4.0	2100	101	22.1	Gunatilleke and Ashton (1987)
Stems ≥ 20 cm dbh	4.0	970	64	15.4	Gunatilleke and Ashton (1987)
Pasoh, Malayan Peninsula					Condit et al. (1996b)
Stems ≥ 1 cm dbh	50.0	335100	817	100.7	
Stems ≥ 10 cm dbh	50.0	26452	683	127.9	
Stems ≥ 20 cm dbh	50.0	8403	525	124.1	
Stems ≥ 30 cm dbh	50.0	3770	377	104.3	
Stems ≥ 1 cm dbh	1.0	6702	495	123.3	
Stems ≥ 10 cm dbh	1.0	529	206	124.0	
Stems ≥ 20 cm dbh	1.0	168	92	83.3	
Stems ≥ 30 cm dbh	1.0	75	46	50.6	
Sungei Menyala, Malaya, stems ≥ 10 cm dbh					Manokaran and Kochummen (1987)
1947	2.0	1075	243	97.8	
1981	2.0	968	244	105.0	
Bukit Lagong, Malayan Peninsula Wyatt-Smith (1966)					
Stems ≥ 10 cm dbh	0.81	401	150	87.0	
Stems ≥ 20 cm dbh	0.81	204	103	83.1	
Stems ≥ 30 cm dbh	0.81	103	62	65.8	
Wanariset, E. Kalimantan, ≥ 10 cm	1.6	866	239	109.1	Newbery et al. (1992)
≥ 10 cm	1.0	540	180	94.5	Kochummen et al. (1990)
Lempake, E. Kalimantan, ≥ 10 cm	1.6	712	209	99.6	Newbery et al. (1992)

(continued)

Appendix 8.1 (continued)

Site	Plot Size (ha)	No. of Trees	No. of Species	Fisher's α	Reference
Danum Valley, Sabah					Newbery et al. (1992)
Trees ≥ 10 cm gbh, Plot 1	4.0	8975	388	82.6	
Trees ≥ 30 cm gbh, Plot 1	4.0	1936	247	75.1	
Trees ≥ 100 cm gbh, Plot 1	4.0	244	83	44.3	
Trees ≥ 10 cm gbh, Plot 2	4.0	8968	387	82.3	
Trees ≥ 30 cm gbh, Plot 2	4.0	1820	242	74.9	
Trees ≥ 100 cm gbh, Plot 2	4.0	262	75	35.1	
Gunung Mulu, Sarawak†					Proctor et al. (1983)
Alluvial forest, 50 m	1.0	615(613)	223	126.1	
Dipterocarp forest, 200–250 m	1.0	778(754)	214	99.6	
Heath forest, 170 m	1.0	708(702)	123	43.2	
Limestone forest, 300 m	1.0	644(640)	73	21.2	
Andulau Forest Reserve, Brunei, mixed dipterocarp forest					Becker (1996)
Trees ≥ 5 cm dbh	0.96	1341	393	187.3	
Badas Forest Reserve, Brunei, heath forest					Becker (1996)
Trees ≥ 5 cm dbh	0.96	1484	113	28.4	
Gunung Silam, Sabah, stems ≥ 10 cm dbh					Proctor et al. (1988)
280 m	0.4	203	83	52.4	
330 m	0.4	246	85	46.0	
420 m	0.4	231	85	48.6	
480 m	0.4	352	104	49.8	
540 m	0.4	330	103	51.4	
610 m	0.24	285	91	46.2	
700 m	0.24	251	91	51.3	
770 m	0.04	39	20	16.5	
790 m	0.04	34	23	31.2	
870 m	0.04	62	19	19.3	
Mt. Kinabalu, Sabah, stems ≥ 10 cm dbh along extended point-quarter line transects					
600 m, 50 m tall, 333 stems/ha		373	153	97.3	Kitayama (1992, 1995)
800 m, 45 m tall, 372 stems/ha		300	102	54.5	Kitayama (1992, 1995)
1000 m, 40 m tall, 369 stems/ha		208	93	64.6	Kitayama (1992, 1995)
1200 m, 30 m tall, 447 stems/ha		180	79	53.7	Kitayama (1992, 1995)
1400 m, 25 m tall, 759 stems/ha		188	70	40.5	Kitayama (1992, 1995)
1600 m, 30 m tall, 572 stems/ha		212	58	26.4	Kitayama (1992, 1995)
1800 m, 25 m tall, 593 stems/ha		204	41	15.5	Kitayama (1992, 1995)
2000 m, 30 m tall, 497 stems/ha		244	50	19.0	Kitayama (1992, 1995)
2350 m, 20 m tall, 778 stems/ha		112	26	10.6	Kitayama (1992, 1995)
2600 m, 20 m tall, 659 stems/ha		112	13	3.8	Kitayama (1992, 1995)
2800 m, 10 m tall, 1044 stems/ha		108	16	5.2	Kitayama (1992, 1995)
3000 m, 10 m tall, 1950 stems/ha		112	17	5.6	Kitayama (1992, 1995)
3200 m, 15 m tall, 1202 stems/ha		112	14	4.3	Kitayama (1992, 1995)
3400 m, 6 m tall, 1844 stems/ha		100	11	3.2	Kitayama (1992, 1995)
Tjibodas, Java					
1400 m, stems ≥ 10 cm dbh	1.0	283	59	22.7	Meijer (1959)
Stems ≥ 20 cm dbh	1.0	162	47	22.2	Meijer (1959)
Stems ≥ 30 cm dbh	1.0	118	33	15.2	Meijer (1959)
1550 m, stems ≥ 10 cm dbh	1.0	427	57	17.7	Yamada (1975)
3000 m, stems ≥ 4.5 cm dbh	0.04	386	10	1.9	Yamada (1976)
Stems ≥ 10 cm dbh	0.04	152	9	2.1	Yamada (1976)
New Guinea, trees ≥ 10 cm dbh					
900 m, plot 1	0.8	528	122	49.8	Paijmans (1970)
825 m, plot 2 (w/ Araucaria)	0.8	560	147	64.9	Paijmans (1970)
700 m, plot 3	0.8	426	145	77.5	Paijmans (1970)
600 m, plot 4	0.8	348	116	60.9	Paijmans (1970)
900 m, everwet forest, CMBRS	1.0	679	222	114.8	Wright et al. (1997)
Australia, Pin Gin Hill, 17°33' S, trees ≥ 10 cm dbh					
60 m	0.25	187	54	25.4	Spain (1984)
Australia, Gadgarra SFR, Atherton Tableland, trees ≥ 10 cm dbh					
680 m	0.25	216	62	29.1	Spain (1984)
700 m	0.25	212	59	27.1	Brasell et al. (1980)

Site	Plot Size (ha)	No. of Trees	No. of Species	Fisher's α	Reference
Australia, Davies Creek, 17°05' S, trees > 10 cm dbh					Connell et al. (1984)
840 m	1.68	1371	120	31.7	(basal area 61.5²/ha)
Mt. Glorious, Australia, 640 m, 17°20' S, stems > 10 cm tall					
Trees and shrubs ≥ 10 cm tall	1.0	10565	100	15.3	Hegarty (1991)
Lianes ≥ 10 cm tall	1.0	5771	42		Hegarty (1991)
Total ≥ 10 cm tall	1.0	16336	142	21.4	Hegarty (1991)
Trees and shrubs ≥ 2 m tall	1.0	4674	90	15.8	Hegarty (1988)
Trees and shrubs ≥ 6.4 cm dbh	1.0	1094	64	14.8	Hegarty (1988)
Temperate-zone sites					
Baber Woods, Westfield, IL					McClain and Ebinger (1968)
Stems ≥ 10 cm dbh	20.6	5449	31	4.3	
Gifford Woods, old-growth, Sherburne, VT					Bormann and Buell (1964)
Stems ≥ 10 cm dbh	0.4	156	8	1.8	
Stems ≥ 25 cm dbh	0.4	71	5	1.2	
Warren Woods, old-growth, MI					Cain (1935)
Stems ≥ 2.5 cm dbh	0.25	271	18	4.3	
Stems ≥ 10 cm dbh	0.25	73	10	3.1	
Heart's Content, virgin beech-hemlock, PA					Morey (1936)
Stems ≥ 10 cm dbh	5.06	1523	14	2.1	
Stems ≥ 25 cm dbh	5.06	773	12	2.0	
Cook Forest, virgin white pine/hemlock, PA					Morey (1936)
Stems ≥ 10 cm dbh	1.61	916	6	0.9	
Stems ≥ 25 cm dbh	1.61	699	5	0.7	
Old-growth forest, Western Triangle, Edgewater, MD					
Stems ≥ 2.5 cm dbh	1.0	1211	23	4.0	Burnham et al. (1992)
Stems ≥ 10 cm dbh	1.0	351	16	3.5	Burnham et al. (1992)
Stems ≥ 20 cm dbh	1.0	235	12	2.7	G. G. Parker (p. c.)
Stems ≥ 30 cm dbh	1.0	153	10	2.4	G. G. Parker (p. c.)
Liriodendron forest, Edgewater, MD					Parker et al. (1989)
Stems ≥ 2.5 cm dbh	0.5	597	20	4.0	
Cypress swamp, Battle Creek, MD					
Stems ≥ 2.5 cm dbh	1.0	731	22	4.3	Burnham et al. (1992)
Stems ≥ 10 cm dbh	1.0	404	9	1.9	Burnham et al. (1992)
Crabtree Woods, old growth, 700 m, western Maryland					McCarthy and Bailey (1996)
Stems ≥ 2.5 cm dbh	1.0	1296	21	3.6	
Stems ≥ 10 cm dbh	1.0	279	16	3.7	
Coweeta, North Carolina, WS2, 1989 census					J. A. Yeakley (p. c.)
Stems > 2.5 cm dbh	1.0	727	20	3.80	
Stems > 10 cm dbh	1.0	405	18	3.86	
Stems > 20 cm dbh	1.0	207	16	4.05	
Coweeta, North Carolina, WS18, 1989 census					J. A. Yeakley (p. c.)
Stems > 2.5 cm dbh	1.0	932	25	4.73	
Stems > 10 cm dbh	1.0	579	22	4.53	
Stems > 20 cm dbh	1.0	246	15	3.52	
Big Thicket, southeast Texas					P. Harcombe (p. c.)
Stems ≥ 10 cm dbh	1.0	335	15	3.2	
Stems ≥ 20 cm dbh	1.0	181	9	2.0	
Stems ≥ 30 cm dbh	1.0	99	7	1.7	
Indiana, old-growth oak forest, 1976 census					Parker et al. (1985)
Stems ≥ 10 cm dbh	8.5	2722	32	5.1	
Missouri, oak-hickory C. Hampe (p. c.)					
Stems ≥ 10 cm dbh	4.0	1362	29	5.1	
Stems ≥ 20 cm dbh	4.0	671	21	4.1	
Stems ≥ 10 cm dbh	1.0	334	20	4.7	
Stems ≥ 20 cm dbh	1.0	155	13	3.4	

*Inferred from average density of trees ≥ 10 cm dbh given by Puig and Lescure (1981).
†No. stems ≥ 10 cm followed by no. classified into spp.

APPENDIX 8.2

The Distribution of Tree Abundances

Consider a population of N adult trees, partitioned among species such that there are $\phi(1)$ species with one individual apiece, $\phi(2)$ species with two individuals apiece, and $\phi(j)$ species with j individuals apiece. Suppose that all species are alike in the eyes of natural selection, that is to say, a tree's chances of mortality or reproduction do not depend on what species it belongs to. To be specific, suppose that at each time-step, one of these N trees is selected at random to be duplicated and another (which might be the same) is chosen to die. Thus the dead tree is replaced by the new young, assumed immediately adult in its own right. With probability $1 - u$, this young is the same species as its parent; with probability u, it is an entirely new species. This is Moran's single locus, infinite alleles model, for a haploid population (Watterson 1974, Ewens 1979).

Thus, the probability $f(j, j + 1)$ that a species now with j trees has $j + 1$ after the next time-step is the probability $(N - j)/N$ that the tree dying then belongs to another species, times the probability j/N that the tree chosen to reproduce is of this species, times the probability $1 - u$ that the young is the same species as its parent. Thus $f(j, j + 1) = j(N - j)(1 - u)/N^2$. The probability $f(j, j - 1)$ that a species now with j trees has $j - 1$ after the next time-step is the probability $j(N - j)/N^2$ that the tree chosen to die is of this species and the tree chosen to reproduce is another species, *plus* the probability $j^2 u/N^2$ that both the tree which dies and the tree which reproduces are of this species, but that the young is of an entirely new species. Thus $f(j, j - 1) = [j(N - j) + uj^2]/N^2$.

At equilibrium, it is just as likely that some species now with j trees has $j + 1$ after the next time-step as that a tree now with $j + 1$ then has j. Thus, at equilibrium, $\phi(j)f(j, j + 1) = \phi(j + 1)f(j + 1, j)$ (Bartlett 1960, p. 23). Since the probability that the next time-step's young is a new species is u, and the number of species coming into existence should balance the number going extinct, $u = \phi(1)f(1, 0)$, so

$$u = \phi(1)(N - 1 + u)/N^2; \quad \phi(1) = N^2 u/(N - 1 + u).$$

Since

$$\phi(2)f(2, 1) = \phi(1)f(1, 2)$$
$$\phi(2)[2(N - 2 + 2u)]/N^2 = \phi(1)[(N - 1)(1 - u)]/N^2.$$

Thus $\phi(2) =$

$$\frac{N^2 u}{2(N - 2 + 2u)} \frac{(N - 1)(1 - u)}{N - 1 + u} \quad (8.14)$$
$$\approx \frac{N^2 u}{2(N - 2 + 2u)}\left[1 - \frac{uN}{N - 1}\right].$$

Similarly,

$$\phi(3) \approx \frac{N^2 u}{3(N - 3 + 3u)}\left(1 - \frac{Nu}{N - 1}\right)\left(1 - \frac{Nu}{N - 2}\right). \quad (8.15)$$

Generalizing, one obtains

$$\phi(j) = \frac{\theta N}{j(N - j)} \Pi_{i=1}^{j-1}\left(1 - \frac{\theta}{N - i}\right), \quad (8.16)$$

where $\theta = Nu$. The product is roughly $\exp[Nu \ln(1 - j/N)] \approx (1 - j/N)^{Nu}$. Thus

$$\phi(j) = \frac{Nu}{j}\left(1 - \frac{j}{N}\right)^{Nu-1} \approx \frac{Nu}{j} \exp -j(Nu - 1)/N. \quad (8.17)$$

If Nu is large enough that $\phi(j)$ is negligible for values of j much larger than 1, $\phi(j) \approx \alpha x^j/j$, a log series with $\alpha = Nu = \theta$, and $x = 1 - u + 1/N$. Representing the frequency distribution $\phi(j)$ by the probability density $P(p)$, where $p = j/N$ and the probability density $P(p)dp = P(j/N)/N = \phi(j)$, and setting $Nu = \theta$, we get

$$P(p) = \theta \frac{(1 - p)^{\theta - 1}}{p}. \quad (8.18)$$

Notice that the integral from 0 to 1 of $pP(p)dp$ is 1—the sum of the frequencies adds to 1, as they should.

A subsample from a log series distribution gives a log series with the same value of α but a smaller value of x. Condit (p. c.) has pointed out that the relation between Nu, the number r of individuals in a sample, and the number S of species in it can be derived as follows:

A species with frequency p is present in a sample of r individuals with probability $1 - (1 - p)^r$. Thus the expected number of species in a sample of r individuals is

$$\theta \int_0^1 [1 - (1 - p)^r] \frac{(1 - p)^{\theta - 1}}{p} dp. \quad (8.19)$$

Notice that, for $r = 1$,

$$(8.20)$$
$$\theta \int_0^1 [1 - (1 - p)] \frac{(1 - p)^{\theta - 1}}{p} dp = \theta \int_0^1 (1 - p)^{\theta - 1} dp = 1.$$

Moreover, the probability that individual r belongs to a species not present among the preceding $r - 1$ is

$$(8.21)$$
$$\int_0^1 [1 - (1 - p)^r - 1 + (1 - p)^{r-1}] \frac{(1 - p)^{\theta - 1}}{p} dp = \frac{1}{\theta + r - 1}.$$

By induction,

$$(8.22)$$
$$\int_0^1 [1 - (1 - p)^r] \frac{(1 - p)^{\theta - 1}}{p} dp = \sum_{i=0}^{r-1} \frac{1}{\theta + i}.$$

Since θ times this integral is the expected number of species S in a sample of r individuals,

$$(8.23)$$
$$S = \theta \sum_{i=0}^{r-1} \frac{1}{\theta + i} = \theta[\ln \theta + r - 1) - \ln \theta] \approx \theta \ln\left(1 + \frac{r}{\theta}\right).$$

Thus θ is simply Fisher's α for this sample.

References

Abraham, J.-P., R. Benja, M. Randrianasolo, J. Ganzhorn, V. Jeannoda, and E. G. Leigh Jr. 1996. Tree diversity on small plots in Madagascar: a preliminary review. *Revue d'Écologie (La Terre et la Vie)* 51: 93–116.

Adler, G. H., and J. O. Seamon. 1991. Distribution and abundance of a tropical rodent, the spiny rat, on islands in Panama. *Journal of Tropical Ecology* 7: 349–360.

Aide, T. M. 1988. Herbivory as a selective agent on the timing of leaf production in a tropical understory community. *Nature* 336: 574–575.

Aide, T. M. 1992. Dry season leaf production: an escape from herbivory. *Biotropica* 24: 532–537.

Aide, T. M. 1993. Patterns of leaf development and herbivory in a tropical understory community. *Ecology* 74: 455–466.

Allison, A., G. A. Samuelson, and S. E. Miller. 1993. Patterns of beetle species diversity in New Guinea rain forest as revealed by canopy fogging: preliminary findings. *Selbyana* 14: 16–20.

Ashton, P. S. 1964. *Ecological Studies in the Mixed Dipterocarp Forests of Brunei State*. Clarendon Press, Oxford.

Ashton, P. S., and C. V. S. Gunatilleke. 1987. New light on the plant geography of Ceylon. I. Historical plant geography. *Journal of Biogeography* 14: 249–285.

Augspurger, C. K. 1983. Seed dispersal of the tropical tree, *Platypodium elegans*, and the escape of its seedlings from fungal pathogens. *Journal of Ecology* 71: 759–771.

Augspurger, C. K. 1984. Seedling survival of tropical tree species: interactions of dispersal distance, light-gaps, and pathogens. *Ecology* 65: 1705–1712.

Barone, J. A. 1998. Host-specificity of folivorous insects in a moist tropical forest. *Journal of Animal Ecology* 67: 400–409.

Bartlett, M. S. 1960. *Stochastic Population Models*. Methuen, London.

Basset, Y. 1992. Host specificity of arboreal and free-living insect herbivores in rain forests. *Biological Journal of the Linnean Society* 47: 115–133.

Becker, P. 1996. Sap flow in Bornean heath and dipterocarp forest trees during wet and dry periods. *Tree Physiology* 16: 295–299.

Becker, P., L. W. Lee, E. D. Rothman, and W. D. Hamilton. 1985. Seed predation and the coexistence of tree species: Hubbell's models revisited. *Oikos* 44: 382–390.

Becker, P., P. E. Rabenold, J. R. Idol, and A. P. Smith. 1988. Water potential gradients for gaps and slopes in a Panamanian tropical moist forest's dry season. *Journal of Tropical Ecology* 4: 173–184.

Bernhard-Reversat, F., C. Huttel, and G. Lemée. 1978. Structure and functioning of evergreen rain forest ecosystems of the Ivory Coast, pp. 557–574. In *Tropical Forest Ecosystems*. UNESCO, Paris.

Bernays, E., and M. Graham. 1988. On the evolution of host specificity in phytophagous arthropods. *Ecology* 69: 886–892.

Bleiler, J. A., G. A. Rosenthal, and D. H. Janzen. 1988. Biochemical ecology of canavanine-eating seed predators. *Ecology* 69: 427–433.

Bongers, F., J. Popma, J. Meave del Castillo, and J. Carabias. 1988. Structure and floristic composition of the lowland rain forest of Los Tuxtlas, Mexico. *Vegetatio* 74: 55–80.

Boom, B. M. 1986. A forest inventory in Amazonian Bolivia. *Biotropica* 18: 287–294.

Bormann, F. H., and M. F. Buell. 1964. Old-age stand of hemlock-northern hardwood forest in central Vermont. *Bulletin of the Torrey Botanical Club* 91: 451–465.

Bradshaw, H. D. Jr., S. M. Wilbert, K. G. Otto, and D. W. Schemske. 1995. Genetic mapping of floral traits associated with reproductive isolation in monkeyflowers (*Mimulus*). *Nature* 376: 762–765.

Brasell, H. M., G. L. Unwin, and G. C. Stocker. 1980. The quantity, temporal distribution and mineral-element content of litterfall in two forest types at two sites in tropical Australia. *Journal of Ecology* 68: 123–139.

Brokaw, N. V. L. 1987. Gap-phase regeneration of three pioneer tree species in a tropical forest. *Journal of Ecology* 75: 9–19.

Bruenig, E. F. 1996. *Conservation and Management of Tropical Rainforests*. CAB International, Wallingford, UK.

Burnham, R. J., S. L. Wing, and G. G. Parker. 1992. The reflection of deciduous forest communities in leaf litter: implications for autochthonous litter assemblages from the fossil record. *Paleobiology* 18: 30–49.

Cain, S. A. 1935. Studies on virgin hardwood forest: III. Warren's Woods, a beech-maple climax forest in Berrien County, Michigan. *Ecology* 16: 500–513.

Carroll, S. P., and C. Boyd. 1992. Host race radiation in the soapberry bug: natural history with the history. *Evolution* 46: 1052–1069.

Chesson, P. L., and R. R. Warner. 1981. Environmental variability promotes coexistence in lottery competitive systems. *American Naturalist* 117: 923–943.

Clark, A. H. 1932. The butterflies of the District of Columbia and vicinity. *Bulletin of the U.S. National Museum* 157: 1–337.

Clinebell, R. R. II., O. L. Phillips, A. H. Gentry, N. Stark, and H. Zuuring. 1995. Prediction of neotropical tree and liana species richness from soil and climatic data. *Biodiversity and Conservation* 4: 56–90.

Coley, P. D. 1983. Herbivory and defensive characteristics of tree species in a lowland tropical forest. *Ecological Monographs* 53: 209–233.

Coley, P. D., and T. M. Aide. 1991. Comparison of herbivory and plant defenses in temperate and tropical broad-leaved forests, pp. 25–49. In P. W. Price, T. M. Lewinsohn, G. W. Fernandes, and W. W. Benson, eds., *Plant-Animal Interactions: Evolutionary Ecology in Tropical and Temperate Regions*. John Wiley and Sons, New York.

Coley, P. D., and J. A. Barone. 1996. Herbivory and plant defenses in tropical forests. *Annual Review of Ecology and Systematics* 27: 305–335.

Coley, P. D., J. P. Bryant, and F. S. Chapin III. 1985. Resource availability and plant herbivore defense. *Science* 230: 895–899.

Coley, P. D., and T. A. Kursar. 1996. Anti-herbivore defenses of young tropical leaves: physiological constraints and ecological tradeoffs, pp. 305–336. In S. S. Mulkey, R. L. Chazdon, and A. P. Smith, eds., *Tropical Forest Plant Ecophysiology*. Chapman and Hall, New York.

Condit, R., S. P. Hubbell, and R. B. Foster. 1992a. Recruitment near conspecific adults and the maintenance of tree and shrub diversity in a Neotropical forest. *American Naturalist* 140: 261–286.

Condit, R., S. P. Hubbell, and R. B. Foster. 1992b. Short-term dynamics of a neotropical forest. *BioScience* 42: 822–828.

Condit, R., S. P. Hubbell, and R. B. Foster. 1994. Density dependence in two understory tree species in a neotropical forest. *Ecology* 75: 671–680.

Condit, R., S. P. Hubbell, and R. B. Foster. 1996a. Changes in tree species abundance in a neotropical forest over eight years: impact of climate change. *Journal of Tropical Ecology* 12: 231–256.

Condit, R., S. P. Hubbell, J. V. LaFrankie, R. Sukumar, N. Manokaran, R. B. Foster, and P. S. Ashton. 1996b. Species-area and species-individual relationships for tropical trees: a comparison of three 50-ha plots. *Journal of Ecology* 84: 549–562.

Connell, J. H. 1971. On the role of natural enemies in preventing competitive exclusion in some marine animals and in rain forest trees, pp. 298–312. In P. J. den Boer and G. Gradwell, eds., *Dynamics of Numbers in Populations*. Center for Agricultural Publication and Documentation, Wageningen, The Netherlands.

Connell, J. H., J. G. Tracey, and L. J. Webb. 1984. Compensatory recruitment, growth and mortality as factors maintaining rain forest tree diversity. *Ecological Monographs* 54: 141–164.

Coyne, J. A., and H. A. Orr. 1989. Patterns of speciation in Drosophila. *Evolution* 43: 362–381.

Crepet, W. L. 1984. Advanced (constant) insect pollination mechanisms: pattern of evolution and implications vis-à-vis angiosperm diversity. *Annals of the Missouri Botanical Garden* 71: 607–630.

Crow, T. R. 1980. A rainforest chronicle: a 30-year record of change in structure and composition at El Verde, Puerto Rico. *Biotropica* 12: 42–55.

Darwin, C. R. 1859. *On the Origin of Species by Means of Natural Selection*. John Murray, London.

Davies, J. G., N. E. Stork, M. J. D. Brendell, and S. J. Hine. 1997. Beetle species diversity and faunal similarity in Venezuelan rainforest tree canopies, pp. 85–103. In N. E. Stork, J. Adis, and R. K. Didham, eds., *Canopy Anthropods*, Chapman and Hall, London.

DeVries, P. J. 1994. Patterns of butterfly diversity and promising topics in natural history and ecology, pp. 187–194. In L. A. McDade, K. S. Bawa, H. A. Hespenheide, and G. S. Hartshorn, eds., *La Selva: Ecology and Natural History of a Neotropical Rain Forest*. University of Chicago Press, Chicago.

Dobzhansky. T. 1950. Evolution in the tropics. *American Scientist* 38: 209–221.

Ehrlich, P. R., and P. H. Raven. 1965. Butterflies and plants: a study in coevolution. *Evolution* 18: 586–608.

Emmons, L. H. 1989. Tropical rain forests: why they have so many species and how we may lose this biodiversity without cutting a single tree. *Orion Nature Quarterly* 8(3): 8–14.

Erwin, T. L. 1982. Tropical forests: their richness in Coleoptera and other arthropod species. *The Coleopterists Bulletin* 36: 74–75.

Erwin, T. L., and J. C. Scott. 1980. Seasonal and size patterns, trophic structure, and richness of Coleoptera in the tropical arboreal ecosystem: the fauna of the tree *Luehea seemannii* Triana and Planch in the Canal Zone of Panama. *Coleopterists Bulletin* 34: 305–322.

Ewens, W. J. 1979. *Mathematical Population Genetics*. Springer, Berlin.

Faber-Langendoen, D., and A. H. Gentry. 1991. The structure and diversity of rain forests at Bajo Calima, Chocó region, western Colombia. *Biotropica* 23: 2–11.

Fabre, J.-H. 1989/1910. Les insectes végétariens, pp 965–975. In J.-H. Fabre, *Souvenirs Entomologiques*, vol. 2. Robert Laffont, Paris [rpt. of Souvenirs Entomolgiques, dixième série].

Findley, J. S. 1993. *Bats: A Community Perspective*. Cambridge University Press, Cambridge.

Fischer, A. G. 1960. Latitudinal variation in organic diversity. *Evolution* 14: 64–81.

Fisher, R. A. 1930. *The Genetical Theory of Natural Selection*. Clarendon Press, Oxford.

Fisher, R. A. 1958. *The Genetical Theory of Natural Selection* (2nd ed). Dover Press, New York.

Fisher, R. A., A. S. Corbet, and C. B. Williams. 1943. The relation between the number of species and the number of individuals in a random sample of an animal population. *Journal of Animal Ecology* 12: 42–57.

Fittkau, E. J., and H. Klinge. 1973. On biomass and trophic structure of the central Amazonian rain forest ecosystem. *Biotropica* 5: 2–14.

Fleming, W. A. 1975. *Butterflies of West Malaysia and Singapore* (2 vols). Longmans, Kuala Lumpur.

Forget, P.-M. 1991a. Comparative recruitment patterns of two non-pioneer canopy tree species in French Guiana. *Oecologia* 85: 434–439.

Forget, P.-M. 1991b. Scatterhoarding of *Astrocaryum paramaca* by *Proechimys* in French Guiana: comparison with *Myoprocta exilis*. *Tropical Ecology* 32: 155–167.

Forget, P.-M. 1992. Regeneration ecology of *Eperua grandiflora* (Caesalpiniaceae), a large-seeded tree in French Guiana. *Biotropica* 24: 146–156.

Forget, P.-M. 1994. Recruitment pattern of *Vouacapoua americana* (Caesalpiniaceae), a rodent-dispersed tree species in French Guiana. *Biotropica* 26: 408–419.

Forget, P.-M. 1996. Removal of seeds of *Carapa procera* (Meliaceae) by rodents and their fate in rainforest in French Guiana. *Journal of Tropical Ecology* 12: 751–761.

Forget, P.-M., E. Munoz, and E. G. Leigh Jr. 1994. Predation by rodents and bruchid beetles on seeds of *Scheelea* palms on Barro Colorado Island, Panama. *Biotropica* 26: 420–426.

Foster, R. B. 1990. Long-term change in the successional forest community of the Rio Manu floodplain, pp. 565–572. In A. H. Gentry, ed., *Four Neotropical Rainforests*. Yale University Press, New Haven, CT.

Foster, R. B., J. Arce B., and T. S. Wachter. 1986. Dispersal and the sequential plant communities in Amazonian Peru floodplain, pp. 357–370. In A. Estrada and T. H. Fleming, eds., *Frugivores and Seed Dispersal*. W. Junk, Dordrecht.

Foster, R. B., and S. P. Hubbell. 1990. Estructura de la vegetación y composición de especies de un lote de cincuenta hectáreas en la isla de Barro Colorado, pp. 141–151. In E. G. Leigh, Jr., A. S. Rand, and D. M. Windsor, eds., *Ecología de un Bosque Tropical*. Smithsonian Tropical Research Institute, Balboa, Panama.

Gartlan, J. S., D. McC. Newbery, D. W. Thomas, and P. G. Waterman. 1986. The influence of topography and soil phosphorus on the vegetation of Korup Forest Reserve, Cameroun. *Vegetatio* 65: 131–148.

Gause, G. F. 1934. *The Struggle for Existence*. Williams and Wilkins, Baltimore, MD.

Gause, G. F. 1935. *Vérifications Expérimentales de la Théorie Mathématique de la Lutte pour la Vie*. Hermann et Cie, Paris.

Gentry, A. H. 1982. Patterns of Neotropical plant species diversity. *Evolutionary Biology* 15: 1–84.

Gentry, A. H. 1988a. Changes in plant community diversity and floristic composition on environmental and geographical gradients. *Annals of the Missouri Botanical Garden* 75: 1–34.

Gentry, A. H. 1988b. Tree species richness of upper Amazonian forests. *Proceedings of the National Academy of Sciences, USA* 85: 156–159.

Ghiselin, M. T. 1974. *The Economy of Nature and the Evolution of Sex*. University of California Press, Berkeley.

Giacalone, J., W. E. Glanz, and E. G. Leigh Jr. 1990. Adición: fluctuaciones poblacionales a largo plazo de *Sciurus granatensis* en relación con la disponibilidad de frutos, pp. 331–335. In E. G. Leigh, Jr., A. S. Rand, and D. M. Windsor, eds., *Ecología de un Bosque Tropical: Ciclos Estacionales y Cambios a Largo Plazo*. Smithsonian Tropical Research Institute, Balboa, Panama.

Gilbert, G. S., S. P. Hubbell, and R. B. Foster. 1994. Density and distance-to-adult effects of a canker disease of trees in a moist tropical forest. *Oecologia* 98: 100–108.

Gillett, J. B. 1962. Pest pressure, an underestimated factor in evolution. *Systematics Association Publication* 4 (Taxonomy and Geography): 37–46.

Givnish, T. J. 1988. Adaptation to sun and shade: a whole-plant perspective. *Australian Journal of Plant Physiology* 15: 63–92.

Grant, P. R., and B. R. Grant. 1997. Genetics and the origin of bird species. *Proceedings of the National Academy of Sciences, USA* 94: 7768–7775.

Greenwood, P. H. 1965. The cichlid fishes of Lake Nabugabo, Uganda. *Bulletin of the British Museum of Natural History (Zoology)* 12: 315–357.

Gunatilleke, C. V. S., and P. S. Ashton. 1987. New light on the plant geography of Ceylon II. The ecological biogeography of the lowland endemic tree flora. *Journal of Biogeography* 14: 295–327.

Haldane, J. B. S. 1932. *The Causes of Evolution*, Longmans Green, London.

Hart, T. B., J. A. Hart, and P. G. Murphy. 1989. Monodominant and species-rich forests of the humid tropics: causes for their co-occurrence. *American Naturalist* 133: 613–633.

Hartshorn, G. S. 1983. Introduction (plants), pp. 118–157. In D. H. Janzen, ed., *Costa Rican Natural History*. University of Chicago Press, Chicago.

Hegarty, E. E. 1988. Canopy dynamics of lianes and trees in a subtropical rainforest. PhD thesis, Department of Botany, University of Queensland, Queensland, Australia.

Hegarty, E. E. 1991. Leaf litter production by lianes and trees in a sub-tropical Australian rain forest. *Journal of Tropical Ecology* 7: 201–214.

Hladik, A. 1978. Phenology of leaf production in rain forest of Gabon: distribution and composition of food for folivores, pp. 51–71. In G. G. Montgomery, ed., *The Ecology of Arboreal Folivores*. Smithsonian Institution Press, Washington, DC.

Hladik, A. 1982. Dynamique d'une forêt équatoriale africaine: mesures en temps réel et comparaison du potentiel de croissance des différentes espèces. *Acta Oecologia: Oecologia Generalis* 3: 373–392.

Howe, H. F. 1986. Consequences of seed dispersal by birds: a case study from Central America. *Journal of the Bombay Natural History Society* 83 (supplement): 19–42.

Howe, H. F. 1989. Scatter- and clump-dispersal and seedling demography: hypothesis and implications. *Oecologia* 79: 417–426.

Howe, H. F. 1990. Seed dispersal by birds and mammals: implications for seedling demography, pp. 191–218. In K. S. Bawa and M. Hadley, eds., *Reproductive Ecology of Tropical Forest Plants*. UNESCO, Paris, and Parthenon Publishing, Park Ridge, NJ.

Howe, H. F., E. W. Schupp,, and L. C. Westley. 1985. Early consequences of seed dispersal for a Neotropical tree (*Virola surinamensis*). *Ecology* 66: 781–791.

Hubbell, S. P. 1979. Tree dispersion, abundance and diversity in a tropical dry forest. *Science* 203: 1299–1309.

Hubbell, S. P. 1980. Seed predation and the coexistence of tree species in tropical forests. *Oikos* 35: 214–229.

Hubbell, S. P., R. Condit, and R. B. Foster. 1990. Presence and absence of density dependence in a neotropical tree community. *Philosophical Transactions of the Royal Society of London B* 330: 269–281.

Hubbell, S. P., and R. B. Foster. 1983. Diversity of canopy trees in a neotropical forest and implications for conservation, pp. 25–41. In S. L. Sutton, T. C. Whitmore, and A. C. Chadwick, eds., *Tropical Rain Forests: Ecology and Management*. Blackwell Scientific Publications, Oxford.

Hubbell, S. P., and R. B. Foster. 1986a. Biology, chance, and history and the structure of topical rain forest communities, pp. 314–329. In J. Diamond and T. J. Case, eds., *Community Ecology*. Harper and Row, New York.

Hubbell, S. P., and R. B. Foster. 1986b. Commonness and rarity in a Neotropical forest: implications for tropical tree conservation, pp. 205–231. In M. E. Soulé, ed., *Conservation Biology: The Science of Scarcity and Diversity*. Sinauer Associates, Sunderland, MA.

Hubbell, S. P., and R. B. Foster. 1990a. The fate of juvenile trees in a Neotropical forest: implications for the natural maintenance of tropical tree diversity, pp. 317–341. In K. S. Bawa and M. Hadley, eds., *Reproductive Ecology of Tropical Forest Plants*. UNESCO, Paris, and Parthenon Publishing, Park Ridge, NJ.

Hubbell, S. P., and R. B. Foster. 1990b. Structure, dynamics and equilibrium status of old-growth forest on Barro Colorado Island, pp. 522–541. In A. H. Gentry, ed., *Four Neotropical Rainforests*. Yale University Press, New Haven, CT.

Hutchinson, G. E. 1959. Homage to Santa Rosalia, or Why are there so many kinds of animals? *American Naturalist* 93: 145–159.

Janzen, D. H. 1970. Herbivores and the number of tree species in tropical forests. *American Naturalist* 104: 521–528.

Janzen, D. H. 1977. The interaction of seed predators and seed chemistry, pp. 415–428. In V. Labeyrie, ed., *Comportement des insectes et milieu trophique*. CNRS, Paris.

Janzen, D. H. 1980. Specificity of seed-attacking beetles in a Costa Rican deciduous forest. *Journal of Ecology* 68: 929–952.

Janzen, D. H. 1981. The defenses of legumes against herbivores, pp. 951–977. In R. M. Polhill and P. H. Raven, eds., *Advances in Legume Systematics*. Royal Botanic Gardens, Kew, England.

Janzen, D. H. 1984. Two ways to be a tropical big moth: Santa Rosa saturniids and sphingids. *Oxford Surveys in Evolutionary Biology* 1: 85–140.

Janzen, D. H. 1988. Ecological characterization of a Costa Rican dry forest caterpillar fauna. *Biotropica* 20: 120–135.

Jiggins, C. D., W. O. McMillan, W. Neukirchen, and J. Mallet. 1995. What can hybrid zones tell us about speciation? The case of *Heliconius erato* and *H. himera* (Lepidoptera: Nymphalidae). *Biological Journal of the Linnean Society* 59: 221–242.

Joseph, L., C. Moritz, and A. Hugall. 1995. Molecular support for vicariance as a source of diversity in rainforest. *Proceedings of the Royal Society of London B* 260: 177–182.

Karr, J. R. 1971. Structure of avian communities in selected Panama and Illinois habitats. *Ecological Monographs* 41: 207–233.

Kerner, E. H. 1961. On the Volterra-Lotka principle. *Bulletin of Mathematical Biophysics* 23: 141–157.

King, D. A. 1994. Influence of light level on the growth and morphology of saplings in a Panamanian forest. *American Journal of Botany* 81: 948–957.

Kitajima, K. 1994. Relative importance of photosynthetic traits and allocation patterns as correlates of seedling shade tolerance of 13 tropical trees. *Oecologia* 98: 419–428.

Kitayama, K. 1992. An altitudinal transect study of the vegetation on Mt. Kinabalu, Borneo. *Vegetatio* 102: 149–171.

Kitayama, K. 1995. Biophysical conditions of the montane cloud forests of Mount Kinabalu, Sabah, Malaysia, pp. 183–197. In L. S. Hamilton, J. O. Juvik, and F. N. Scatena, eds., *Tropical Montane Cloud Forests*. Springer-Verlag, New York.

Knowlton, N., L. A. Weigt, L. A. Solórzano, D. K. Mills, and E. Bermingham. 1993. Divergence in proteins, mitochondrial DNA, and reproductive compatibility across the Isthmus of Panama. *Science* 260: 1629–1632.

Kochummen, K. M., J. V. LaFrankie, Jr., and N. Manokaran. 1990. Floristic composition of Pasoh Forest Reserve, a lowland rain forest in peninsular Malaysia. *Journal of Tropical Forest Science* 3: 1–13.

Korning, J., and H. Balslev. 1994. Growth and mortality of trees in Amazonian tropical rain forest in Ecuador. *Journal of Vegetation Science* 4: 77–86.

Kunkel-Westphal, I., and P. Kunkel. 1979. Litter fall in a Guatemalan primary forest, with details of leaf-shedding by some common tree species. *Journal of Ecology* 67: 665–686.

Kursar, T. A., and P. D. Coley. 1991. Nitrogen content and expansion rate of young leaves of rain forest species: implications for herbivory. *Biotropica* 23: 141–150.

Lang, G. E., and D. H. Knight. 1983. Tree growth, mortality, recruitment, and canopy gap formation during a 10-year period in a tropical moist forest. *Ecology* 64: 1075–1080.

Leigh, E. G. Jr. 1981. The average lifetime of a population in a varying environment. *Journal of Theoretical Biology* 90: 213–239.

Leigh, E. G. Jr. 1990a. Community diversity and environmental stability: a re-examination. *Trends in Ecology and Evolution* 5: 340–344.

Leigh, E. G. Jr. 1990b. Introducción: ¿por qué hay tantos tipos de árboles tropicales? pp. 75–99. In E. G. Leigh, Jr., A. S. Rand, and D. M. Windsor, eds., *Ecología de un Bosque Tropical: Ciclos Estacionales y Cambios a Largo Plazo*. Smithsonian Tropical Research Institute, Balboa, Panama.

Leigh, E. G. Jr. 1996. Epilogue: research on Barro Colorado Island, 1980–94, pp. 469–503. In E. G. Leigh, Jr., A. S. Rand, and D. M. Windsor, eds., *Ecology of a Tropical Forest* (2d. ed.). Smithsonian Institution Press, Washington, D.C.

Leigh, E. G. Jr., and G. de Alba. 1992. Barro Colorado Island, Panama, basic research, and conservation. *The George Wright Forum* 9: 32–45.

Leigh, E. G. Jr., S. J. Wright, F. E. Putz, and E. A. Herre. 1993. The decline of tree diversity on newly isolated tropical islands: a test of a null hypothesis and some implications. *Evolutionary Ecology* 7: 76–102.

Lieberman, D., M. Lieberman, R. Peralta, and G. S. Hartshorn. 1996. Tropical forest structure and composition on a large-scale altitudinal gradient in Costa Rica. *Journal of Ecology* 84: 137–152.

Lieberman, M., D. Lieberman, G. S. Hartshorn, and R. Peralta. 1985. Small-scale altitudinal variation in lowland wet tropical forest vegetation. *Journal of Ecology* 73: 505–516.

Lieberman, M., D. Lieberman, R. Peralta, and G. S. Hartshorn. 1995. Canopy closure and the distribution of tropical forest tree species at La Selva, Costa Rica. *Journal of Tropical Ecology* 11: 161–178.

Lorence, D. H., and Sussman, R. W. 1988. Diversity, density and invasion in a Mauritian wet forest, pp. 187–204. In P. Goldblatt and P. P. Lowry, eds., *Modern Systematic Studies in African Botany*. Missouri Botanical Garden, St. Louis.

Lotka, A. J. 1925. *Elements of Physical Biology*. Williams and Wilkins, Baltimore, MD.

MacArthur, R. H. 1961. Population effects of natural selection. *American Naturalist* 95: 195–199.

MacArthur, R. H. 1972. *Geographical Ecology*. Harper and Row, New York.

MacArthur, R. H., and R. Levins. 1964. Competition, habitat selection and character displacement in a patchy environment. *Proceedings of the National Academy of Sciences, USA* 51: 1207–1210.

MacArthur, R. H., and E. O. Wilson. 1967. *The Theory of Island Biogeography*. Princeton University Press, Princeton, NJ.

Machado, J.-L., and M. T. Tyree. 1994. Patterns of hydraulic architecture and water relations of two tropical canopy trees with contrasting leaf phenologies: *Ochroma pyramidale* and *Pseudobombax septenatum*. *Tree Physiology* 14: 219–240.

Manokaran, N., and K. Kochummen. 1987. Recruitment, growth and mortality of tree species in a lowland dipterocarp forest in peninsular Malaysia. *Journal of Tropical Ecology* 3: 215–330.

Martin, C. 1991. *The Rainforests of West Africa*. Birkhäuser Verlag, Basel.

Martínez-Yrízar, A., J. M. Maass, L. A. Perez-Jimenez, and J. Sarukhán. 1996. Net primary productivity of a tropical deciduous forest ecosystem in western Mexico. *Journal of Tropical Ecology* 12: 169–175.

May, R. M. 1990. How many species? *Philosophical Transactions of the Royal Society of London* B 330: 293–304.

McCarthy, B. C., and D. R. Bailey. 1996. Composition, structure, and disturbance history of Crabtree Woods: an old-growth forest in western Maryland. *Bulletin of the Torrey Botanical Club* 123: 350–365.

McClain, W. E., and J. B. Ebinger. 1968. Woody vegetation of Baber Woods, Edgar County, Illinois. *American Midland Naturalist* 79: 419–428.

Meijer, W. 1959. Plantsociological analysis of montane rainforest near Tjibodas, West Java. *Acta Botanica Neerlandica* 8: 277–291.

Morey, H. F. 1936. A comparison of two virgin forests in northwestern Pennsylvania. *Ecology* 17: 43–55.

Mulkey, S. S., A. P. Smith, S. J. Wright, J. L. Machado, and R. Dudley. 1992. Contrasting leaf phenotypes control seasonal variation in water loss in a tropical forest shrub. *Proceedings of the National Academy of Sciences USA* 89: 9084–9088.

Murali, K. S., and R. Sukumar. 1993. Leaf flushing phenology and herbivory in a tropical dry deciduous forest, southern India. *Oecologia* 94: 114–119.

Nadkarni, N. M., T. J. Matelson, and W. A. Haber. 1995. Structural characteristics and floristic composition of a Neotropical cloud forest, Monteverde, Costa Rica. *Journal of Tropical Ecology* 11: 481–495.

Newbery, D. M., E. J. F. Campbell, Y. F. Lee, C. E. Ridsdale, and M. J. Still. 1992. Primary lowland dipterocarp forest at Danum Valley, Sabah, Malaysia: structure, relative abundance and family composition. *Philosophical Transactions of the Royal Society of London* B 335: 341–356.

Newbery, D. M., J. S. Gartlan, D. B. McKey, and P. G. Waterman. 1986. The influence of drainage and soil phosphorus on the vegetation of Douala-Edea Forest Reserve, Cameroun. *Vegetatio* 65: 149–162.

Odum, H. T. 1970. Summary: an emerging view of the ecological system at El Verde, pp. I-191–I-289. In H. T. Odum and R. Pigeon, eds., *A Tropical Rain Forest*. Division of Technical Information, U. S. Atomic Energy Commission, Washington, DC.

Pacala, S. W., C. D. Canham, J. Saponara, J. A. Silander, Jr., R. K. Kobe, and E. Ribbens. 1996. Forest models defined by field measurements: estimation, error analysis and dynamics. *Ecological Monographs* 66: 1–43.

Paijmans, K. 1970. An analysis of four tropical rain forest sites in New Guinea. *Journal of Ecology* 58: 77–101.

Paine, R. T. 1966. Food web complexity and species diversity. *American Naturalist* 100: 65–75.

Paine, R. T. 1974. Intertidal community structure: experimental studies on the relationship between a dominant competitor and its principal predator. *Oecologia* 15: 93–120.

Pandolfi, J. M. 1993. A review of the tectonic history of New Guinea and its significance for marine biogeography, pp. 718–728. In *Proceedings of the Seventh International Coral Reef Symposium, Guam, 1992*, vol. 2. University of Guam Press, Mangilao, Guam.

Parker, G. G., J. P. O'Neill, and D. Higman. 1989. Vertical profile and canopy organization in a mixed deciduous forest. *Vegetatio* 85: 1–11.

Parker, G. R., D. J. Leopold, and J. K. Eichenberger. 1985. Tree dynamics in an old-growth, deciduous forest. *Forest Ecology and Management* 11: 31–57.

Parsons, G. L., G. Gerasimos, A. Moldenke, J. D. Lattin, N. H. Anderson, J. C. Miller, P. Hammond, and T. D. Schowalter. 1991. *Invertebrates of the H. J. Andrews Experimental Forest, Western Cascade Range, Oregon. V. An Annotated List of Insects and Other Arthropods.* USDA Forest Service, Technical Report PNW-GTR-290.

Phillips, O. L., P. Hall, A. H. Gentry, S. A. Sawyer, and R. Vásquez. 1994. Dynamics and species richness of tropical rain forests. *Proceedings of the National Academy of Sciences, USA* 91: 2805–2809.

Piperno, D. R. 1994. Phytolith and charcoal evidence for prehistoric slash-and-burn agriculture in the Darien rain forest of Panama. *Holocene* 4: 321–325.

Prance, G. T., W. A. Rodrigues, and M. F. da Silva. 1976. Inventário florestal de um hectare de mata de terra firme km 30 da Estrada Manaus-Itacoatiara. *Acta Amazonica* 6: 9–35.

Proctor, J., J. M. Anderson, P. Chai, and H. W. Vallack. 1983. Ecological studies in four contrasting lowland forests in Gunung Mulu National Park, Sarawak. *Journal of Ecology* 71: 237–260.

Proctor, J., Y. F. Lee, A. M. Langley, W. R. C. Munro, and T. Nelson. 1988. Ecological studies on Gunung Silam, a small ultrabasic mountain in Sabah, Malaysia. I. Environment, forest structure and floristics. *Journal of Ecology* 76: 320–340.

Puig, H., and J. P. Lescure. 1981. Étude de la variabilité floristique dans la région de la piste de Saint-Elie. *Bulletin de liaison de groupe de travail sur l'écosystème forestier Guyanais (ECEREX)* 3: 26–29.

Ray, C., and A. Hastings. 1996. Density dependence: are we searching at the wrong spatial scale? *Journal of Ecology* 65: 556–566.

Regal, P. J. 1977. Ecology and evolution of flowering plant dominance. *Science* 196: 622–629.

Reitsma, J. M. 1988. *Forest Vegetation of Gabon.* Tropenbos, Ede, The Netherlands.

Ridley, H. N. 1930. *The Dispersal of Plants Throughout the World.* L. Reeve & Co., Ashford, Kent.

Robbins, R. K., G. Lamas, O. H. H. Mielke, D. J. Harvey, and M. M. Casagrande. 1996. Taxonomic composition and ecological structure of the species-rich butterfly community at Pakitza, Parque Nacional del Manu, Perú, pp. 217–252. In D. E. Wilson and A. Sandoval, eds., *Manu: The Biodiversity of Southeastern Peru.* Smithsonian Institution Press, Washington, D.C.

Rose, M. R., and B. Charlesworth. 1981. Genetics of life history in *Drosophila melanogaster* I. Sib analysis of adult females. *Genetics* 97: 173–186.

Rosenthal, G. A., D. L. Dahlman, and D. H. Janzen. 1976. A novel means for dealing with L-canavanine, a toxic metabolite. *Science* 192: 256–258.

Rosenthal, G. A., D. L. Dahlman, and D. H. Janzen. 1978. L-Canaline detoxification: a seed predator's biochemical mechanism. *Science* 202: 528–529.

Rosenthal, G. A., C. G. Hughes, and D. H. Janzen. 1982. L-Canavanine, a dietary nitrogen source for the seed predator *Caryedes brasiliensis* (Bruchidae). *Science* 217: 353–355.

Rosenthal, G. A., D. H. Janzen, and D. L. Dahlman. 1977. Degradation and detoxification of canavanine by a specialized seed predator. *Science* 196: 658–660.

Sabatier, D., and M.-F. Prévost. 1988. Quelques données sur la composition floristique et la diversité des peuplements forestiers de guyane française. *Bois et Forêts des Tropiques* 219: 31–52.

Saldarriaga, J. G., D. C. West, M. L. Tharp, and C. Uhl. 1988. Long-term chronosequence of forest succession in the upper Rio Negro of Colombia and Venezuela. *Journal of Ecology* 76: 938–958.

Schatz, G. E., and S. T. Malcomber. 1994. Botanical research at Ranomafana National Park: baseline data for long-term ecological monitoring, pp. 14–22. In *Ranomafana Parks Project Symposium.*

Schupp, E. W. 1990. Annual variation in seedfall, postdispersal predation, and recruitment in a neotropical tree. *Ecology* 71: 504–515.

Schupp, E. W. 1992. The Janzen-Connell model for tropical tree diversity: population implications and the importance of spatial scale. *American Naturalist* 140: 526–530.

Sevenster, J. G., and J. J. M. van Alphen. 1996. Aggregation and coexistence. II. A neotropical *Drosophila* community. *Journal of Animal Ecology* 65: 308–324.

Simpson, E. H. 1949. Measurement of diversity. *Nature* 163: 688.

Sinha, A., and P. Davidar. 1992. Seed dispersal ecology of a wind dispersed rain forest tree in the Western Ghats, India. *Biotropica* 24: 519–525.

Smith, A. 1776. *An Enquiry into the Nature and Causes of the Wealth of Nations.* W. Strahan and T. Cadell, London.

Southwood, T. R. E., V. C. Moran, and C. E. J. Kennedy. 1982. The richness, abundance and biomass of the arthropod communities on trees. *Journal of Animal Ecology* 51: 635–649.

Spain, A. V. 1984. Litterfall and the standing crop of litter in three tropical Australian rainforests. *Journal of Ecology* 72: 947–961.

Sperry, J. S., N. M. Holbrook, M. H. Zimmermann, and M. T. Tyree. 1987. Spring filling of xylem vessels in wild grapevine. *Plant Physiology* 83: 414–417.

Stork, N. E. 1991. The composition of the arthropod fauna of Bornean lowland rain forest trees. *Journal of Tropical Ecology* 7: 161–180.

Strasberg, D. 1995. Processus d'invasion par les plantes introduites à La Réunion et dynamique de la végétation sur les coulées volcaniques. *Écologie* 26: 169–180.

Sukumar, R., H. S. Dattaraja, H. S. Suresh, J. Radhakrishnan, R. Vasudeva, S. Nirmala, and N. V. Joshi. 1992. Long-term monitoring in a tropical deciduous forest in Mudumalai, southern India. *Current Science* 62: 608–616.

Tanner, E. V. J. 1977. Four montane rain forests of Jamaica: a quantitative characterization of the floristics, the soils and the foliar mineral levels, and a discussion of the interrelations. *Journal of Ecology* 65: 883–918.

Taylor, L. R., R. A. Kempton, and I. P. Woiwod. 1976. Diversity statistics and the log-series model. *Journal of Animal Ecology* 45: 255–272.

Terborgh, J. 1971. Distribution on environmental gradients: theory and a preliminary interpretation of distributional patterns in the avifauna of the Cordillera Vilcabamba, Peru. *Ecology* 52: 23–40.

Terborgh, J., R. B. Foster, and P. Nuñez V. 1996. Tropical tree communities: a test of the nonequilibrium hypothesis. *Ecology* 77: 561–567.

Thompson, J., J. Proctor, V. Viana, W. Milliken, J. A. Ratter, and D. A. Scott. 1992. Ecological studies on a lowland evergreen rain forest on Maracá Island, Roraima, Brazil. *Journal of Ecology* 80: 689–703.

Thorington, R. W. Jr., B. Tannenbaum, A. Tarak, and R. Rudran. 1982. Distribution of trees on Barro Colorado Island: a five-hectare sample, pp. 83–94. In E. G. Leigh, Jr., A. S. Rand, and D. M. Windsor, eds., *The Ecology of a Tropical Forest.* Smithsonian Institution Press, Washington, DC.

Tiffney, B. H. 1984. Seed size, dispersal syndromes, and the rise of the angiosperms: evidence and hypothesis. *Annals of the Missouri Botanical Garden* 71: 551–576.

Tiffney, B. H. 1986. Fruit and seed dispersal and the evolution of the Hamamelidae. *Annals of the Missouri Botanical Garden* 73: 394–416.

Tiffney, B. H., and S. J. Mazer. 1995. Angiosperm growth habit, dispersal and diversification reconsidered. *Evolutionary Ecology* 9: 93–117.

Tyree, M. T., S. D. Davis, and H. Cochard. 1994. Biophysical perspectives of xylem evolution: is there a tradeoff of hydraulic efficiency for vulnerability to dysfunction? *IAWA Journal* 15: 335–360.

Tyree, M. T., D. A. Snyderman, T. R. Wilmot, and J. L. Machado. 1991. Water relations and hydraulic architecture of a tropical tree (*Schefflera morototoni*). *Plant Physiology* 96: 1105–1113.

Uhl, C., and P. G. Murphy. 1981. Composition, structure and regeneration of a tierra firme forest in the Amazon basin of Venezuela. *Tropical Ecology* 22: 219–237.

Valencia R., R. 1995. Composition and structure of an Andean forest fragment in eastern Ecuador, pp. 239–249. In S. P. Churchill, H. Balslev, E. Forero, and J. L. Luteyn, eds., *Biodiversity and Conservation of Neotropical Montane Forests*. New York Botanical Garden, Bronx, NY.

Valencia, R., H. Balslev, and G. Paz y Miño C. 1994. High tree alpha-diversity in Amazonian Ecuador. *Biodiversity and Conservation* 3: 21–28.

Valencia, R., and P. M. Jørgensen. 1992. Composition and structure of a humid montane forest on the Pasachoa volcano, Ecuador. *Nordic Journal of Botany* 12: 239–247.

Vaughan, R. E., and P. O. Wiehe. 1941. Studies on the vegetation of Mauritius. III. The structure and development of the upland climax forest. *Journal of Ecology* 29: 127–160.

Veneklaas, E. J. 1991. Litterfall and nutrient fluxes in two montane tropical rain forests, Colombia. *Journal of Tropical Ecology* 7: 319–336.

Via, S. 1984. The quantitative genetics of polyphagy in an insect herbivore. I. Genotype-environment interaction in larval performance on different host plant species. *Evolution* 38: 881–895.

Via, S. 1991a. The genetic structure of host plant adaptation in a spatial patchwork: demographic variability among reciprocally transplanted pea aphid clones. *Evolution* 45: 827–852.

Via, S. 1991b. Specialized host plant performance of pea aphid clones is not altered by experience. *Ecology* 72: 1420–1427.

Volterra, V. 1931. *Leçons sur la Théorie Mathematique de la Lutte pour la Vie*. Gauthier-Villars, Paris.

Watterson, G. A. 1974. Models for the logarithmic species abundance distributions. *Theoretical Population Biology* 6: 217–250.

Weaver, P. L., E. Medina, D. Pool, K. Dugger, J. Gonzalez-Liboy, and E. Cuevas. 1986. Ecological observations in the dwarf cloud forest of the Luquillo Mountains in Puerto Rico. *Biotropica* 18: 79–85.

Welden, C. W., S. W. Hewett, S. P. Hubbell, and R. B. Foster. 1991. Sapling survival, growth and recruitment: relationship to canopy height in a neotropical forest. *Ecology* 72: 35–50.

White, L. J. T. 1994. The effects of commercial mechanized logging on a transect in lowland rainforest in the Lopé Reserve, Gabon. *Journal of Tropical Ecology* 10: 313–322.

Williams, C. B. 1964. *Patterns in the Balance of Nature*. Academic Press, London.

Wills, C., R. Condit, R. B. Foster, and S. P. Hubbell. 1997. Strong density- and diversity-related effects help to maintain tree species diversity in a neotropical forest. *Proceedings of the National Academy of Sciences, USA* 94: 1252–1257.

Wolda, H. 1983. Spatial and temporal variation in abundance in tropical animals, pp. 93–105. In S. L. Sutton, T. C. Whitmore, and A. C. Chadwick, eds., *Tropical Rain Forest: Ecology and Management*. Blackwell Scientific Publications, Oxford.

Wolda, H. 1987. Altitude, habitat and tropical insect diversity. *Biological Journal of the Linnean Society* 30: 313–323.

Wolda, H. 1988. Insect seasonality: why? *Annual Review of Ecology and Systematics* 19: 1–18.

Wolda, H. 1992. Trends in abundance of tropical forest insects. *Oecologia* 89: 47–52.

Wolda, H., and R. Foster. 1978. *Zunacetha annulata* (Lepidoptera: Dioptidae), an outbreak insect in a Neotropical forest. *Geo-Eco-Trop* 2: 443–454.

Wong, M., S. J. Wright, S. P. Hubbell, and R. B. Foster. 1990. The spatial pattern and reproductive consequences of outbreak defoliation in *Quararibea asterolepis*, a tropical tree. *Journal of Ecology* 78: 579–588.

Wright, D. D., J. H. Jessen, P. Burke, and H. G. de S. Garza. 1997. Tree and liana enumeration and diversity on a one-hectare plot in Papua New Guinea. *Biotropica* 29: 250–260.

Wright, S. J. 1990. Cumulative satiation of a seed predator over the fruiting season of its host. *Oikos* 58: 272–276.

Wright, S. J. 1992. Seasonal drought, soil fertility, and the species density of tropical forest plant communities. *Trends in Ecology and Evolution* 7: 260–263.

Wright, S. J., J. L. Machado, S. S. Mulkey, and A. P. Smith. 1992. Drought acclimation among tropical forest shrubs (*Psychotria*, Rubiaceae). *Oecologia* 89: 457–463.

Wyatt-Smith, J. 1966. Ecological studies on Malaysian forests. I. Composition of and dynamic studies in Lowland Evergreen Rainforest in two 5-acre Plots in Bukit Lagong and Sungei Menyala Forest Reserve and in two half-acre plots in Sungei Menyala Forest Reserve, 1947–59. Research Pamphlet 52: Forest Research Institute, Malaya.

Yamada, I. 1975. Forest ecological studies of the montane forest of Mt. Pangrango, West Java. I. Stratification and floristic composition of the montane rain forest near Cibodas. *South East Asian Studies* 13: 402–426.

Yamada, I. 1976. Forest ecological studies of the montane forest of Mt. Pangrango, West Java. II. Stratification and floristic composition of the forest vegetation of the higher part of Mt. Pangrango. *South East Asian Studies* 13: 513–534.

Zotz, G., M. T. Tyree, and H. Cochard. 1994. Hydraulic architecture, water relations and the vulnerability to cavitation of *Clusia uvitana* Pittier: a C_3-CAM tropical hemiepiphyte. *New Phytologist* 127: 287–295.

NINE

The Role of Mutualism in Tropical Forest

MUTUALISM IN A COMPETITIVE WORLD

The Interdependences of Tropical Forest

Tropical forest confronts us with a paradox. On the one hand, it is notorious for intense, unbridled competition. Plants struggle with each other for the available light, water, or nutrients. The timber content of tropical forest (Tilman 1988), the abundance of lianes (Putz 1984), hemiepiphytes (Todzia 1986), and epiphytes (Benzing 1990) taking advantage of other plants' wood, and the shadiness of the forest floor, all testify to the intensity of competition for light. In their struggle for water, trees of the evergreen forest in eastern Amazonia sink roots 18 m below the soil surface (Nepstad et al. 1994). To seek nutrients, some plants send roots into accumulations of detritus high above ground (Nadkarni 1981, Dressler 1985). Root production and respiration in the forest at Paragominas in eastern Amazonia account for nearly 50 tons of sugar (20 tons of carbon) per hectare per year, of which nearly 40 tons are consumed in the top meter of soil, where the struggle for nutrients is most intense (Davidson and Trumbore 1995, Trumbore et al. 1995). Moreover, animals struggle with each other for the available food. Predators seek ruthlessly for prey, and potential prey display the most amazing variety and originality in the artifices they employ to avoid being eaten. The *Passiflora* vine whose leaves have little yellow spots to persuade the approaching *Heliconius* butterfly that a predecessor has already laid eggs there (Williams and Gilbert 1981, Gilbert 1982), the enormous *Omphalea* liane with sugar-mimicking alkaloids in its leaves to derail the sensory organs by which insects detect food, the *Urania* caterpillars that ingest these compounds without harm and sequester them as defenses against their own predators (Lees and Smith 1991), and many another stratagem that insects employ to deter or hide from predators (Robinson 1969) are all part of the lore that makes tropical biology so fascinating for the naturalist.

Yet tropical forest is also a tissue of interdependences, an apex of mutualism (Jacobs 1988). It is a truism that tropical plants need mycorrhizae to take up nutrients from the soil (Janos 1980, 1983), animals to pollinate their flowers and disperse their seeds, and an enormous variety of organisms to recycle the nutrients in dead vegetation (Corner 1964). Many plants attract animals to defend them against herbivores or other plants (Janzen 1966, Schupp and Feener 1991). A plant may need two kinds of animals to assure the safety of its seeds. On Barro Colorado, the canopy tree *Dipteryx panamensis* needs bats, *Artibeus lituratus*, to remove its seeds from the parent and its pests, and agoutis to bury them out of the sight of other seed-eating mammals (Forget 1993). Similarly, the canopy tree *Virola surinamensis* needs toucans to remove seeds from the parent and agoutis to bury them, lest they be destroyed by weevils (Howe et al. 1985, Forget and Milleron 1991). These interdependences multiply. A tree that needs agoutis to bury its seeds also needs enough other species of tree nearby to keep the agoutis fed when it is not dropping fruit (Forget 1994). A *Dipteryx*, which needs the 70-g fruit bat *Artibeus lituratus* to disperse its seeds, also needs other forests within these bats' migrating range to feed them during its own forest's "off season" (Bonaccorso 1979). In the Malay peninsula, durian trees, which need the bat *Eonycteris spelea* to pollinate their flowers, also need caves within flying range where these bats can roost, and coastal mangroves, *Sonneratia alba*, within flying range to keep these bats in nectar during the month before the durians' main flowering season (Lee 1980).

Some interdependences are yet more intricate. In Southeast Asia, dipterocarps and many other tree species fruit gregariously every few years (Janzen 1974b), apparently in response to the nocturnal cooling during protracted spells of clear weather (Ashton et al. 1988). Such "gregarious fruitings" provide so huge a supply of fruit that frugivores leave much of it uneaten (Chan 1980). When these trees flower in preparation for a gregarious fruiting, the forest's demand for pollinators is enormous, far greater than usual (Appanah 1985). Groups of tree species (often, but not always, members of the same section of a genus) that share the same species of pollinator flower in sequence, each building up pollinator populations for the next while minimizing competition among species for pollinators (Appanah 1985), even though the species flowering in sequence fruit in synchrony to satiate seed-eaters (Chan 1980). A series of 10 unrelated species of trees and vines, all pollinated by carpenter bees (*Xylocopa* spp.), flower in sequence, attracting their pollinators from the edge habitats where they normally live, each increasing the number of *Xylocopa* in the forest for its successor. Six species of dipterocarp (*Shorea* spp., section *Mutica*) are pollinated by the same species of thrips. These *Shorea* flower in synchrony, each species passing on an ever larger population of thrips to its successor (Appanah 1985, 1993). In other groups of dipterocarps, members of which share in common one or more species of small, fast-growing pollinator such as planthoppers or beetles, member species also flower in sequence, each building up pollinator populations for its successor (Appanah 1985, 1993).

Mutualisms of pollination and dispersal have profoundly affected tropical forest. Other things being equal, dense single-species aggregations of plants are particularly susceptible to outbreaks of pests or disease (Dethier 1976, Burdon and Chilvers 1982, Burdon 1987, Gilbert and Hubbell 1995). To avoid being eaten, plants that live in dense aggregations must invest particularly heavily in antiherbivore defenses. Wind-pollinated conifers, which must be near conspecifics to get pollinated (Regal 1977), have long-lived needles whose antiherbivore defenses slow their decay once they fall and acidify the soil (Waring and Schlesinger 1985). Ferns, which are also common where found, can also slow decomposition. On the everwet flank of Sri Lanka, the fern *Dicranopteris linearis* dominates "fernlands" in disturbed settings near the rainforest. Both in the fernland and in the forest, fern frond litter decays more than twice as slowly as leaf litter from a dominant tree species of the nearby forest. Moreover, decay of litter of both species is markedly slower when placed in the fernland than when placed in the nearby forest (Maheswaran and Gunatilleke 1988). The fronds and stems of *Dicranopteris* in Hawaii are likewise slow to decompose, thanks in large part to their high ratio of lignin to nitrogen (Russell and Vitousek 1997). Species that can persist when rare presumably need to invest less in antiherbivore defenses to achieve a given level of protection, so members of such species can often outgrow individuals of social species that can only occur close to conspecifics. The ability to escape from aggregations of conspecifics, however, is the gift of reliable animal pollinators, and, for large-seeded plants, seed dispersers as well.

A glance at the fossil record suggests how the prevalence of efficient pollinators and seed dispersers can influence the productivity and diversity of a forest. At the end of the Jurassic and the beginning of the Cretaceous, land vegetation appears to have been packed with antiherbivore compounds, presumably because plants had to be near their conspecifics. The toxicity of the vegetation is suggested by the fact that the vertebrate herbivore fauna of the time was dominated by giant dinosaurs (Coe et al. 1987). Even today, reptilian herbivores such as giant tortoises can eat food whose tannin content is high enough to deter mammals: these late Jurassic dinosaurs must have been "walking compost heaps" (Coe et al. 1987).

In the early Cretaceous, however, pollen-eating insects were already pollinating plants (Crepet et al. 1991). In the middle or late Cretaceous, some angiosperms began using nectar to attract animals, primarily insects, to serve as relatively species-specific pollinators (Crepet et al. 1991, 1992, Nixon and Crepet 1993). This event triggered a cascade of diversification among flowering plants (Crepet 1984) which were restricted to open places and successional habitats (Wing et al. 1993). Flowering plants were able to dominate mature forest only when, in the early Cenozoic, they began using animals as seed dispersers (Tiffney 1984). This strategy, which allowed trees of shade-tolerant species to disperse large seeds far from their parents (Regal 1977, Tiffney and Mazer 1995), sparked adaptive radiations in many groups of birds and mammals (Sussman 1991). In turn, efficient pollinators and seed dispersers made possible the enormous diversity of flowering trees. This diversity is crucial to the defense of tropical forest against pathogens and pests (Gilbert and Hubbell 1995). The ability of rare plants to channel resources into fast growth rather than antiherbivore defense allows more efficient recycling of dead vegetable matter (Corner 1964, Regal 1977). Their pollinators and seed dispersers enable tropical plants to maintain remarkably high levels of genetic diversity, despite being so rare (Table 9.1).

Why Mutualism Is the Central Problem of Evolutionary Biology

We have seen the extraordinary degree of interdependence among tropical organisms, despite the intensity of competition among them. We can now appreciate why the impact of "civilization" on a tropical ecosystem is justly described as a disturbance or a disruption.

The more complex and highly adapted the characteristic affected by a gene, the less likely a random change in that gene will improve the characteristic in question (Fisher 1930, 1958a: pp. 42–44). Indeed, since Aristotle's time, the fact that "normal" organisms usually accomplish their goal of growth and reproduction (Aristotle 416b24 in Barnes 1984: p. 663) while abnormal organisms often cannot, has been considered evidence that organisms are adapted, that is to say, organized for the purpose of reproducing themselves (Aristotle 199a33–199b7 in Barnes 1984: p. 340; Pittendrigh 1958). Similarly, the fact that a "random"

Table 9.1 Genetic diversity in plants of the tropics and the temperate zone

Category	Reference	N	D		%P	%H
Nontropical trees and shrubs	Hamrick et al. (1992)	136			53.4	15.8
Common BCI trees and shrubs	Hamrick and Loveless (1989)	16	8	(0.3–1200)	60.9	21.1
Common BCI trees (no shrubs)	Hamrick and Loveless (1989)	9	2	(0.3–12.7)	60.1	21.4
Rare BCI trees and shrubs	Hamrick and Murawski (1991)	16	0.07	(0–0.3)	41.8	14.2
Rare BCI trees (no shrubs)	Hamrick and Murawski (1991)	13			46.3	15.8

N, number of species sampled; D, median density (number per hectare) of adults on Barro Colorado's 50-ha forest dynamics plot (data courtesy of R. B. Foster and S. P. Hubbell), with range of densities in parentheses; %P, percentage of studied loci per species that were polymorphic; and %H, proportion of studied loci per individual that were heterozygous, averaged over species.

change in the organization of a tropical ecosystem is usually disruptive suggests that organisms have a common stake in maintaining the integrity of their ecosystem. Despite the intensity of competition among its members, an ecosystem has some of the features of a gigantic mutualism.

Failure to recognize the importance of understanding how competition can bring forth mutualism led many to consider the idea of mutualism rather anti-Darwinian and thus to dismiss the phenomenon as unimportant, and many others, who considered mutualism all-important, to use it as a reason to dismiss natural selection and other forms of competition as fundamentally irrelevant (Sapp 1994). More recently, this division has been reflected by the split between "community ecologists," who study the phenomena of populations and communities in relation to natural selection, and "ecosystems ecologists" who view species and guilds in terms of how they serve the good of the ecosystem as a whole.

The theme of cooperation emerging from competition is, however, an old one. Adam Smith (1776) proposed competition as an effective force for the development and perfection of that most complex and delicate of all mutualisms, human civilization. Darwin argued that natural selection favors diversity because the resulting division of labor yields a more productive and luxuriant community:

> The advantage of diversification in the inhabitants of the same region is, in fact, the same as that of the physiological division of labour in the organs of the same individual body. . . . No physiologist doubts that a stomach by being adapted to digest vegetable matter alone, or flesh alone, draws most nutriment from these substances. So in the general economy of any land, the more widely and perfectly the animals and plants are diversified for different habits of life, so will a greater number of individuals be capable of there supporting themselves. (Darwin 1859: pp. 115–116)

The link between competition and mutualism is made particularly clear in Maynard Smith and Szathmáry's (1995) *The Major Transitions of Evolution*. They show how essential the role of mutualism was in the most crucial events of evolutionary history. In the packaging of genes and proteins into organisms, the evolution by endosymbiosis of eukaryotes, the coordination during sexual reproduction of the replication of different genes in an orderly and fair meiosis, and the evolution of complex multicellular organisms and social groups, simpler parts have joined in mutualism to form more complex and effective wholes. Eukaryotic cells, metazoans, and complex animal societies all preserve traces of the means by which natural selection shaped these aggregates into integrated wholes, thereby showing how evolutionary history can testify distinctively and decisively to evolutionary mechanism.

Indeed, a glance at current formulations of selection theory (Dawkins 1982, Williams 1992, Bourke and Franks 1995) shows how deeply competition and mutualism are intertwined. Genes are called the fundamental units of selection because they are replicated intact (Dawkins 1982), although not always accurately, so selection among genes is a meaningful process (Bourke and Franks 1995). Genes embody the information, the "codices" (Williams 1992), that program the development of their organisms: they are equally "the units of self-interest" whose differential replication drives evolution (Bourke and Franks 1995). There is no meaningful evolution without selection on genes, and evolutionary changes among individuals, social groups, etc., are caused only insofar as they serve the interest of some gene. And yet, genes of a genome are utterly interdependent. No single codex, no single gene, can be read, let alone replicated, unless it is a part of a genome in an appropriate organism. By itself, a codex is as impotent as a computer program without a suitable computer to execute it and an operator who knows how to use the computer and the program. Even this analogy understates the case: a gene contributes only a small fraction of a program, which is useless unless the other parts programmed by the rest of the genome provide a meaningful context for reading it out. Although ecosystems lack the individuality, coherence, and obvious purposiveness of organisms (Williams 1985) and offer an extraordinary collection of outrageously dysfunctional parasitisms and the like for proponents of selfish genes to scandalize us with (Dawkins 1982), an individual organism is likewise helpless outside an ecosystem that provides it with food, shelter, and even the air that it breathes.

Thus, the central problem, not only of tropical ecology, but of all evolutionary biology, is how, and under what circumstances, does competition lead to the evolution of mutualism and interdependence? It is one of the extraordinary ironies of our age that, in framing an argument to show why there are no grounds for believing that the good of group or community is relevant to evolution, Dawkins (1976) revealed, with a clarity few can match, just why mutualism is so central a question. The central importance of mutualism thus forces us to assess the nature and degree of the interdependences of tropical forest.

How Can Mutualisms Evolve?

Mutualisms are of crucial importance both in evolutionary history and in modern ecosystems (Leigh and Rowell 1995). Yet the only ordering principle of biology, natural selection, is a competitive one. What conditions allow natural selection to promote harmony within groups of interacting organisms? Members of a group depend on each other, but they are also each other's closest competitors. What circumstances could keep competition from overwhelming their common interest in their group's effectiveness and ruining the common good that their cooperation could create? This is the principal problem of ethology. We shall see that it is one of the grand unifying themes of biology and anthropology. An anthropological variant is Hardin's (1968) "Tragedy of the Commons." Villagers share a common interest in restricting grazing on their public pasture to a sustainable level. Yet a villager can derive immediate benefit by sneaking an extra animal onto the pasture. If too many yield to this temptation, the pasture will be ruined. Cancer reveals potential conflict between an individual and one of its cell lineages, showing that physiology, too, confronts this "standard problem of ethology." How can the common interest of members in the good of their habitat or group be rendered effective? Three primary mechanisms have been proposed for resolving this question: reciprocal altruism, selection among groups, and kin selection.

Selection among Groups and the Common Good

The oldest answer proposed for Hardin's dilemma [proposed, in fact, before Hardin (1968), but others had already outlined the problem (Frank 1995)] was that selection among groups favored those groups that exploited the environment in the most sustainable way (Wynne-Edwards 1963). In particular, Wynne-Edwards (1963) argued that individuals behave in accord with their group's good and refrain from overexploiting the group's resources because selection among groups or populations overrides selection within groups. Therefore, when Williams (1966) showed, in a devastating riposte, how rarely selection among groups can override selection within populations, studying the evolution of cooperation temporarily fell out of fashion among evolutionary biologists.

What circumstances would allow selection among groups to prevail? Here, I crudely sketch an argument outlined more carefully in Appendix 9.1 and presented in full by Leigh and Rowell (1995). The change per generation in a panmictic population's mean value of the magnitude, a, of a characteristic under selection is the population's heritable variation in a (the additive genetic variance in a; Fisher 1930, 1941, Falconer 1989) times the intensity of selection on a. This intensity is measured by the regression on this magnitude a on relative fitness v_a (the number of offspring per parent among parents with magnitude a, divided by the number of offspring per parent in the population as a whole). If the population is divided into distinct groups, then the influence exerted on the population's mean value of a by selection among groups is the variance among group means of a, times the intensity of "group selection" on these means (the regression of the relative fitness of groups on their mean value of a). Here, the influence of selection within groups is the average over all groups of the intensity of within-group selection on a, times heritable within-group variance in a (Price 1972, Crow and Aoki 1982).

Thus the ratio of variance among, relative to heritable variance within, groups is crucial to the effectiveness of selection among groups. The ratio of variance among to variance within groups is highest and selection among groups is most likely to be effective when groups exchange no migrants, each group descends from a single parent group, and there are many more groups than members per group (Leigh 1983).

Selection among groups, where each group consists of the mitochondria within a host cell, maintains the integrity of the endosymbiosis between mitochondria and their host cells (Margulis 1993) in modern eukaryotes and must have allowed it to evolve. Modern mitochondria are so dependent upon their host cells that they never migrate from cell to cell: they can only populate the progeny of their host cells. Moreover, when an egg and a sperm unite to form a zygote, it is almost invariably the rule that this zygote is populated by mitochondria from only one of the parents, usually the mother (Birky 1995); this circumstance minimizes genetic variation among mitochondria within a cell (Birky 1991). As a result, selection among cells on their mitochondria completely overrides selection within cells. Nevertheless, when the endosymbiosis was first evolving, there was scope for considerable disharmony between ancestral protomitochondria and their hosts (Blackstone 1995). How could this disharmony have been resolved? If a host cell's protomitochondria defended their cells against would-be immigrants (a "territoriality" suggested by the tendency of mitochondria carried in by the sperm to be eliminated by those in the egg, or vice versa; Eberhard 1980), the logical result would be the utter dependence of protomitochondrial reproduction on that of their hosts and a selection among cells on these mitochondria in their hosts' interest (Leigh 1983, Leigh and Rowell 1995), a selection which also served the common good of the mitochondria. Nonetheless, there would never have been a selection among cells on their mitochondria had not a complementation of functions created a common interest between mitochondria and their hosts (Blackstone 1995).

An analogous group selection maintains the symbiosis between advanced leaf-cutter ants, *Atta* spp., and the fungus they use to digest the leaf fragments they cut. *Atta* preferentially consume what would become the reproductive bodies of their fungus, and they spread chemical compounds about their nest that suppress the germination of fungal spores (Hölldobler and Wilson 1990). This fungus reproduces only when new leaf-cutter queens take a mouthful of fungus from the parental nest to start their own fungal culture (Hinkle et al. 1994). Thus the ants prevent exchange among nests of the fungus and assure that each nest's fungus descends from one parent nest, thereby creating a selection among nests on this fungus in the ants' interest. Subordinating fungus to ant evolution may have been crucial to the origin of abundant species of leaf-cutter ants with extraordinarily complex and tightly organized societies; after all, not just any fungus can digest freshly cut fragments of so great a variety of leaves (Hinkle et al. 1994). Subordinating fungus to ant evolution has ensured that the phylogenies of advanced leaf-cutter ants and their fungi coincide (Chapela et al. 1994, Hinkle et al. 1994).

On the other hand, the phylogenies of primitive attines, relatives of the leaf cutters, and the fungi they use to digest insect frass, corpses, or dead vegetable matter, do not coincide (Chapela et al. 1994, Hinkle et al. 1994). Just as leaf-cutter ants depend on their fungi for nourishment, so reef-building corals depend on symbiotic zooxanthellae (Trench 1987). Moreover, as in primitive attines, the phylogeny of corals is not coherent with that of their zooxanthellae (Rowan and Powers 1991). Apparently, some corals benefit by being able to choose their zooxanthellae and, if need be, to replace them by others (Rowan et al. 1997). Such corals "call" their zooxanthellae from the surrounding water, just as trees call their pollinators from the surrounding forest, rather than inheriting them from a parent (Trench 1987). Do primitive attines benefit from the ability to choose their fungus? This ability entails a possible cost. Where there is a potential conflict of interest between a host and its symbionts, the option of crossing to a new host allows symbionts to be less dependent on, and less careful of, their current host's welfare (Herre 1993). Under these conditions, selection among groups cannot enforce the good behavior of the symbionts. The harmony between corals and their zooxanthellae, or primitive attines and their fungi, appears to hinge on a common interest between the partners, the different and complementary gifts each partner brings to the mutualism, rather than on any particular style of enforcement.

Reciprocal Altruism: Mutually Enforced Common Interest

Axelrod (1984) caricatured the problem of evolving cooperation for the common good by a game called the "iterated Prisoner's Dilemma." There are two players: at each round, each player can cooperate or defect. If both cooperate, each earns three points. If one cooperates and the other defects, the defector earns five points, the cooperator, none. If both defect, each earns one point. At each round, defection is more profitable, whatever the partner's play. Yet each player can earn three points per round if they both cooperate every time, but only one point per round if they both keep defecting. Can they enforce the common interest in cooperating, when the immediate profit of defecting is greater and defecting is the safer strategy?

To answer this question, Axelrod (1984) imagined a tournament where the same pair of players confront each other for several rounds in succession. Let each player play by one of a large number, N, of possible rules or strategies. After m rounds, let there be a reassignment of strategies. Let w_i be the proportion of all the points during the preceding m rounds which accrue to strategy i: then a proportion, w_i, of the players are assigned strategy i for the next m rounds. If players tend to have partners that use the same strategy, then "tit for tat" (in round 1 cooperate, in round $n + 1$, play as the opponent did in round n) allows cooperation to spread through a world of defectors (Axelrod and Hamilton 1981, Axelrod 1984, Nowak and Sigmund 1992). When "tit for tat" prevails, "win-stay, lose-shift" (if one's play earned $\geq$ three points in round n, use it in round $n + 1$, otherwise shift) becomes advantageous, especially in an error-prone world where the opponent's play is sometimes misread, for "win-stay, lose-shift" is less seriously derailed than "tit for tat" when a cooperation is mistakenly read as a defection (Nowak and Sigmund 1993). In these games, the prospect of repeated interaction, with the opportunity it offers for retaliation for wrongs done, is essential for enforcing a common interest in cooperation.

Hamlets, *Hypoplectrus nigricans*, simultaneously hermaphroditic coral reef fish, use "tit for tat" to enforce a mutual reduction in competition for mates. Because a successful sperm contributes as many genes to future generations as a successful egg (Fisher 1930), selection normally favors those hermaphrodites that devote equal effort to male and female functions (Maynard Smith 1978). Because a sperm is usually much smaller than the egg it fertilizes, effort on male functions usually involves competition for mates, often by trying to produce the most sperm during a communal spawning bout. Indeed, this effort on "male functions" represents "the cost of sex," the decrease which sexual as opposed to asexual reproduction imposes on the net birth rate of a species. Hamlets reduce the cost of sex by pairing off and trading eggs with each other to fertilize (Fischer 1980). When two fish, A and B, pair off, in the first spawn A provides eggs for B to fertilize, and during the second spawn they reverse roles, with A fertilizing B's eggs, and so on. More eggs tend to be offered in later spawns, as if confidence were building up. If, however, one partner cheats by trying to fertilize eggs for two or three spawns in succession, cooperation ends, and the cheater must find and take the time and trouble to inspire confidence in another mate. This prospect assures fairly reliable cooperation. Because an egg-trading hamlet expends only the sperm required to fertilize its partner's eggs, these hamlets have testes that are much smaller relative to their ovaries than do related, less cooperative seabasses, such as *Serranus tortugarum* (Fischer 1981, 1987). Although this method of short-circuiting the cost of sex appears to be effective, it is surprisingly rare (Fischer 1988).

On an evolutionary time scale, a mutual policing analogous to reciprocal altruism appears responsible for the maintenance of honest meiosis. I explore this case in detail because it represents the most concrete illustration known of Adam Smith's (1776, 1790) idea that if members of a society can formulate and combine to enforce rules against "unfair" or "dysfunctional" competition, then competition among the members favors their group's good, in that it maximizes the average benefit per member. At most loci, heterozygotes have an equal chance of passing either allele to a given gamete. Indeed, meiosis is usually one of the fairest lotteries ever. Its honesty is surprising because biasing meiosis in one's own favor, as "segregation-distorters" do, seems such an easy way for an allele to spread itself. At a few loci, there are such segregation-distorter alleles (Crow 1979)—if there are a few, why aren't there more? Distorters causing a strong bias (so that 90% or more of the gametes produced by individuals heterozygous for the distorter carry it) can spread through a population amazingly rapidly, even if the distorter imposes a phenotypic defect whose spread causes the population's extinction (Lyttle 1977). Usually, a distorter's phenotypic defect is recessive, so the distorter's frequency rises until deaths among its homozygotes balances its spread in heterozygotes.

Why are such distorters so rare? Genes at unlinked loci cannot benefit from the distortion, but they do suffer from the distorter's phenotypic defect. Therefore, at every unlinked locus, selection favors mutants that restore the honesty of meiosis because such mutants spare a fraction of their bearers from a phenotypic defect the distorter would otherwise have inflicted on them. In this sense, the honesty of meiosis represents the common interest of the whole genome (Leigh 1991). Lyttle (1979) has demonstrated that selection of alleles at unlinked loci does actually diminish the bias caused by a distorter allele. Perhaps because the suppression of one allele's distortion effect eliminates the means by which a whole class of distorters could arise, the common interest of the genome in honest meiosis is effective at almost all loci, so an allele can spread only if it benefits the individuals that carry it. How effectively this common interest in honest meiosis is enforced is perhaps best illustrated by our utter inability to select a race of cows with female-biased sex ratio, despite the advantages such a race would have for farmers (Dawkins 1982: p. 43).

How was the stage set that allowed the evolution of honest meiosis? For the real mystery is not the common interest of modern genomes in honest meiosis, but how the circumstances evolved that allowed this common interest to develop and to take effect. It is difficult enough for constitutional lawyers to devise a legal framework where it is genuinely in each individual's interest to work for the good of the society. How can the inconscient mechanism of natural selection create an effective common interest among members in their group's good? As each gene depends on the expression of many others, it was presumably advantageous for them to be grouped into chromosomes to ensure that, during reproduction, each offspring received a copy of each gene (Maynard Smith and Szathmáry 1995). Chromosomes thus express the common interest of their genes in each other's presence. Segregation distortion is facilitated by sexual reproduction, which itself expresses the common interest of two parents in cooperating to produce more variable offspring (Williams 1975, Maynard Smith and Szathmáry 1995), for anisogamy (which ensures the integrity of the symbiosis between cells and their organelles; Cosmides and Tooby 1981) and the resulting overproduction of sperm provides maximum opportunity for segregation distortion. Sexual reproduction also sets the stage for eliminating distortion, for recombination between homologous chromosomes enhances the profit from sexual reproduction (Fisher 1930) and creates the genome's common interest in honest meiosis (Leigh 1971, Prout et al. 1973). If we trace the circumstances that led to the evolution of meiosis, we find that a common interest in pooling the products of different genes lies at the root of these developments: mechanisms of mutual enforcement would not evolve were there not a common good to defend.

Kin Selection and Social Behavior

An agent's genes can spread by behavior which benefits the agent's relatives (individuals that carry more of the agent's genes than the agent's "average competitor") if the benefit to the relative, as measured by the increase in its prospective reproduction, sufficiently exceeds the cost to the agent. The spread of an agent's genes by assistance to suitable relatives is called *kin selection* (Hamilton 1964a,b). The efficacy of kin selection shows that evolution is driven by selection on genes rather than selection on individuals (Dawkins 1982). Parental care reflects kin selection (Bourke and Franks 1995), and kin selection often influences the evolution of social behavior.

The closer the relative (the more genes it shares nonrandomly with the agent), the more the agent should sacrifice to secure a given increase in the relative's reproduction. In an outbred population, an individual should sacrifice its life to save two of its offspring (for half an individual's genes are copies of those of a given parent), two full siblings (for a quarter of the agent's genes are identical to those of a full sibling because they are copies of the same maternal genes, and another quarter are identical by descent from paternal genes), four half siblings (for a quarter of the agent's genes are identical by descent to the half sibling's by virtue of identity by descent from their common parent). In outbred populations, the coefficient of relationship, r, between parent and offspring, or between full siblings, the proportion of their genes that are "identical by descent," is 0.5. Between grandparent and grandchild, or between half siblings, $r = 0.25$, and so forth (Hamilton 1964a, West-Eberhard 1975). In a population aggregated into more or less endogamous groups, the average relatedness between members of the same group is their intragroup (intraclass) correlation coefficient (Fisher 1958b); indeed, selection among groups can be viewed as a form of kin selection (Crow and Aoki 1982). Bourke and Franks (1995) give a more general, and, alas, more complicated, discussion of coefficients of relationship.

Kin selection, however, cannot drive the evolution of social behavior unless the individuals involved derive some mutual benefit from associating in groups. Indeed, if individuals benefit greatly from living in groups, relatedness to that group's members does not affect an individual's decision to join that group, although relatedness can greatly influence the types of social organization that evolves once group living is established (Keller and Reeve 1994).

Coatis, *Nasua narica*, illustrate the role of kin selection in social behavior. Female coatis and their young form relatively tightly knit social bands. Daughters usually stay with their natal bands all their lives, whereas males leave their band when they mature and live alone (Kaufmann 1962, Russell 1982). The rules of band membership ensure that genetic drift normally causes a band's adult females to be closely related (Kaufmann 1962, Gompper et al. 1997).

Female coatis must live in bands. Only when banded together can female coatis prevent the larger males from chasing them from fruiting trees (Gompper 1996). Females in a band defend, and often nurse, each other's young (Russell 1982, Gompper 1994). There are more eyes to watch for predators, so each band member can spend more time foraging (Russell 1982). They groom each other for ectoparasites (Kaufmann 1962). Mutual care of the young allows adult females to search for fruiting trees and help each other find fruit.

Kin selection obviously enhances cooperation among band members. Nonetheless, if a band dwindles sufficiently, it will try to join another (Kaufmann 1962). Even though the newcomers are groomed less and suffer more aggression, the advantages of joining a larger band—greater safety from predators and competitors, more animals to look for fruiting trees, and the like—outweigh these annoyances. The driving force behind banding together is not kin selection, but the mutual benefit of each other's presence, that is to say, the common interest in being part of a group.

Kin selection plays a rather different role in the evolution of social behavior in Hymenoptera. If the disadvantage of reproducing alone is sufficiently severe, an insect will join a group of relatives even as a subordinate, for it is better to help related dominants reproduce, perhaps by performing tasks for which the dominants are less suited or lack time for, than not to reproduce at all (West-Eberhard 1981). For example, females defending nest cells and newly laid eggs have little time to forage, whereas subordinates whose newly laid eggs have just been replaced by those of a dominant may redirect their maternal instincts to care for the dominant's neglected brood. If dominants continue to prevent their subordinates from laying eggs, the latter's ovaries may degenerate, and their continued lack of mature eggs may prolong their tendency to care for their dominant's brood (West-Eberhard 1987). Thus, once group living is established, kin selection can bring forth a division of labor essential to group welfare from an inclination among dominants to social parasitism and an instinct among eggless subordinates to care for brood.

Truly complex insect societies evolve most readily when a single dominant, the queen, annihilates her subordinates' prospects of personal reproduction, thereby creating a common interest among subordinates in helping their queen reproduce. Once this happens, the queen no longer needs to force her subordinates to work, and the colony no longer needs to be small enough for her to "supervise every detail." Kin selection still plays an essential role, for colony organization is stable only if the subordinates (workers) spread their genes by helping their queen reproduce.

The roles of manipulation by dominants, kin selection, and mutual enforcement of rules of beneficial behavior in colony organization is illustrated by honeybees. A honeybee queen mates with 18–20 males and mixes the sperm thoroughly: thus, most of a worker's "colleagues" are half-sisters. Honeybee workers are capable of laying male eggs without mating. A worker benefits, however, by eating eggs laid by half-sisters, because a half-sister's son is less closely related to her than a brother, a son of the queen (Ratnieks and Visscher 1989). Indeed, most of the eggs workers do lay are eaten. Perhaps as a result, workers in hives with healthy queens lay few eggs (Ratnieks 1993). By eating each other's eggs, workers create a common interest in enhancing their queen's reproduction: the queen need do nothing further to force her workers to help (Ratnieks and Visscher 1989).

Honeybee colonies are marvels of self-organization (Seeley and Levien 1987, Seeley 1989a). Honeybee workers pass through a sequence of tasks as they age, dictated by their increase in titer of juvenile hormone (Robinson et al. 1989). The youngest workers assume the least risky tasks, such as nursing the brood, which are performed within the nest, while the oldest are foragers, the riskiest task of all (Robinson 1992), as if workers try to protect their reproductive options as much as possible, for as long as possible (West-Eberhard 1981). A worker's "cursus honorum," however, adjusts in response to her colony's needs. If most of the older workers are removed from the hive, some of the remaining workers become foragers at an earlier age than usual. If most of the hive's younger workers are removed, some of the remaining ones continue caring for the brood longer than usual (Robinson et al. 1989). Indeed, if all brood nurses are removed, some foragers will revert to nursing the brood (Robinson 1992). Thus honeybee workers adjust their division of labor to changes in their hive's needs. A honeybee colony also adjusts its foraging to the changing distribution of available pollen and nectar through the independent reactions of each forager to the quality of the food source it is visiting, the time required for food-storer bees to relieve the forager of the nectar it brings when it returns, and, perhaps, the protein content of the gland secretion which returning foragers receive from nurse bees (Seeley 1989b, 1994, Seeley et al. 1991, Camazine 1993).

Unenforced Mutualisms

Some animals live or forage in groups whose social organization is shaped only by their common interest. For these groups, relatedness, selection among groups, and mutual policing are irrelevant.

It pays cliff swallows, *Hirundo pyrrhonota*, to attract conspecifics to an insect swarm because when more swallows are present, the insect swarm is less likely to escape from them. Similarly, female evening bats, *Nycticeius humeralis*, which roost in groups of a few hundred, nurse the young of other mothers, perhaps because this will produce more bats to help search for insect swarms (Connor 1995).

When members of a group must "hang together or hang separately," this community of interest creates more tightly organized mutualisms. If a monkey group needs the help of all its members to coordinate its defense against predators, a monkey who causes the death of a fellow troop member by failing to cooperate properly endangers itself by weakening the group upon which its own safety depends. Its failure is yet more self-destructive if the predator returns to groups where it has already preyed successfully, like burglars returning to a house they have already robbed (Leigh and Rowell 1995).

A spectacular example of a mutualism enforced only by a binding common interest is provided by long-tailed manakins, *Chiroxiphia linearis*, at Monteverde, Costa Rica. There, male manakins form teams of 10 or more to attract and court females. Each team's alpha male garners all its team's matings, but he depends on close cooperation from his team's beta male in the energetic dancing and calling needed to attract females (McDonald and Potts 1994). The team's beta male is rarely related to its alpha, and the beta male is not rewarded for helping (nor is he punished for not helping). The beta tolerates this laborious servitude, which can last more than 5 years, because (1) females only mate with alpha males that are the heads of teams, and (2) when the alpha male dies, the beta inherits both the alpha's position and his female clientèle. In short, this mutualism is enforced by female choice. Female manakins prefer, and are loyal to, teams with the best coordinated dancing and singing. The more effectively a beta male cooperates, the greater its prospective future reproductive success (McDonald and Potts 1994).

Common interest can even lead to unenforced mutualism among members of different species. A famous example, characteristic of the wet tropics, is "mixed species bird flocks," in which several species of bird forage together. In many such flocks, each species is represented by an individual or a mated pair, perhaps with attendant young (Moynihan 1962: p. 8). Some flocks have a "core" consisting of pairs of one or more species, which defend the flock's "beat" as a joint territory. Birds with smaller territories join the flock when it passes over their territory; birds with larger territories may shift from flock to flock (Gradwohl and Greenberg 1980, Munn 1985).

Why do these birds flock together? Fruit-eating birds may discover fruit sources for each other. Insectivorous birds sometimes eat insects their neighbors have flushed and may no longer be able to reach (Moynihan 1979: p. 92). Nonetheless, in a mixed flock, each species searches for somewhat different foods, or searches for them in somewhat different ways or places, as if there were a premium on reducing competition (Jones 1977). The main advantage of flocking appears to be mutual protection from predators. Not only are there more eyes to watch for predators, birds in a flock can often "mob" smaller predators and drive them away. A bird may sacrifice little to join a flock. Even alarm calls may benefit the caller by coordinating the flock's responses to the predator thus signalled and also by informing the predator that, now the flock has been alerted, it no longer has any chance of preying successfully upon it (Smythe 1977). The community of interest among the members of bird flocks has sometimes had visible consequences. The social organization of some of these flocks is rather complex (Moynihan 1962), and coevolution among the species of some flocks has facilitated flocking behavior (Moynihan 1968).

How Mutualisms Evolve: Concluding Remarks

We have reviewed factors that contribute to the evolution of mutualisms requiring accommodation among their participants for their common good: selection among groups, reciprocal altruism, and kin selection. These factors can combine in many ways (Leigh and Rowell 1995). In many mutualisms, the fitness of the partner that provides the food depends on cooperating, so it need not be policed, whereas it must police the other partner (West and Herre 1994). For example, yucca moths, *Tegeticula* spp., lay eggs in flowers of yucca, *Yucca* spp. The moths pollinate the yucca flowers to provide their larvae with seeds to eat (Pellmyr and Thompson 1992). The yuccas do not reproduce unless the moths pollinate their flowers, but they restrict the proportion of seeds their moths eat by aborting flowers with too many moth eggs (Pellmyr and Huth 1994). Similarly, fig trees are pollinated by wasps whose larvae eat some of the fig tree's seeds (Corner 1940). To keep the relationship mutualistic, fig trees protect about half their seeds from wasp eggs by means as yet unknown (West and Herre 1994).

No mutualism, however, evolves unless there is a common interest among the organisms involved. The common interest underlying the origin of all mutualisms among conspecifics is one or another advantage of living in groups, such as defense against predators or competitors or improved ability to find or secure food or mates. The common interest underlying mutualisms involving different species, especially distantly related ones, normally derives from the combination of very different aptitudes (Douglas 1994). Thus heterotrophic corals provide nutrients and well-lit, well-defended sites to autotrophic zooxanthellae, which supply carbohydrates to the corals (Trench 1987). Autotrophic trees provide heterotrophic mycorrhizae with carbohydrates, while the mycorrhizae provide the trees with phosphorus and other mineral nutrients (Allen 1993). Leguminous trees provide heterotrophic rhizobial bacteria with carbohydrates, while the bacteria transform atmospheric nitrogen into nutrients the trees can use (Young and Johnston 1989). Trees provide mobile animals with nectar and some pollen, and these animals pollinate distant conspecifics of the tree. Trees provide other animals with seeds or fruit pulp to eat, and the animals carry other seeds away from the pests attracted to the parent (Janzen 1970) or toward better-lit, less crowded sites (Corner 1964). Some trees attract aggressive ants to defend them against herbivores that would eat them or vines that would smother them (Janzen 1966, McKey 1984, Schupp 1986). Epiphytic plants in nutrient-poor habitats provide ants with food and shelter in return for the nutrients derived from the ants' garbage dumps (Janzen 1974a, Treseder et al. 1995). The list goes on and on. Indeed, if the history of life can be viewed as the successive joining of smaller, already tested parts into larger wholes (Dyson 1985, Leigh and Rowell 1995), it can also be viewed as the history of parts with complementary functions joining to form wholes of enhanced effectiveness (Maynard Smith and Szathmáry 1995).

In mutualisms between different species, Darwin's principle of diversification has come full circle. Organisms escape competitive replacement by diversifying, as the principle of character displacement illustrates. Not only can organisms coexist more readily if they have different specialties than if they are engaged in more similar occupations, organisms with truly complementary specialties can join in fruitful mutualism. The greater diversity and more intense specialization of tropical organisms and the more intense predation pressure in tropical settings provide more, and more spectacular, opportunities for mutualism.

So far, however, none of these considerations have shed much light on the degree to which ecosystems might represent the common interest of their member organisms. The good of ecosystems does not "shell out" of the mutualisms that have been discussed so far, any more than the common good of all players shells out of the shifting coalitions predicted by von Neumann and Morgenstern's (1944) *Theory of Games and Economic Behavior*. The one significant difference is that the mutualisms of biology are built upon an enduring common interest rather than a temporary opportunity. The extent to which natural selection favors the transformation of ecosystems into "commonwealths" is the issue to which we next turn.

ECOSYSTEMS AS COMMONWEALTHS

Ecosystems and Economics

We have considered how the sensitivity of natural ecosystems to novel forms of disturbance or disruption suggests that organisms have a common stake in the integrity of their ecosystem. Polluting a river or lake may amount to nothing more than putting too much fertilizer into it, yet such pollution often diminishes the diversity of aquatic organisms and can render a body of water unfit as a source of human food (Odum 1971b). Opening a market for "bush meat" can lead forest peoples to hunt a tropical forest's animals (Feer 1993, Redford 1993), including essential dispersers of seeds, to near extinction. This circumstance could well cause a catastrophic decline in tree diversity, without a single tree being cut (Emmons 1989). To what extent are ecosystems organized for the common good of their members? How could they become so organized?

Ecosystems are not discrete individuals, as are genes or even species. In particular, ecosystems do not produce discrete daughter ecosystems. Therefore, there is no such process as selection among ecosystems. If ecological communities suit the common good of their inhabitants, this must reflect the outcome of selection on these inhabitants. Does selection on a community's individuals lead to a commonwealth, organized to benefit its members?

The simplest explanation of the apparent adaptedness of ecosystems is that organisms evolve in an ecological context. Disrupt or destroy that context, and the organisms suffer. To this extent, an ecosystem's members must share a common interest in their ecosystem's integrity. Is there any more significance to the adaptedness of ecosystems?

Let us begin with the "constants of tropical forest": lowland tropical moist and wet forests on "modal soils" (Baillie 1996) the world around let only 1% of the incident light reach the forest floor. They use about 1.4 m of water per year (as measured by actual evapotranspiration), have basal area near 30 m²/ha, a leaf area index near 7, drop about 7 tons dry weight of leaves/ha · year, and perhaps also have the same gross productivity, the equivalent of about 90 tons of sugar/ha · year. These constants of the forest suggest that there is something more to the adaptedness of ecosystems than the evolution of their members in a common context. It is often said that saints are splendidly diverse in their goodness, while evil has a certain dull sameness about it. In ecology, however, it is the common good that is predictable: what is arbitrary and unpredictable is the nature of monopolies and other forms of "ecological distortion" which impose losses on other species far greater than the gains of the perpetrator (Leigh 1991). Such distortion would be exemplified by ant-plants whose ants clear competing shoots from a wider space than their hosts need. Ant-mediated distortion would lead as readily to a uniform stand of slender poles as to a parkland of squat trees. The inefficient exploitation of sunlight, water, etc., would render this antplant monopoly inherently unstable. Either other species would penetrate the ants' defense and exploit the unused resources, or, were the ant defense truly impenetrable, antplants that use the available resources more efficiently would spread at the expense of their inefficient fellows. Put another way, unexploited energy represents a niche for a possible invading species, and ecological communities, which are forever being tested by potential competitors, presumably evolve until they are maximally resistant to the invaders they usually encounter (Leigh 1971).

In fact, most organic carbon produced in terrestrial ecosystems is respired (Lotka 1925, Odum 1971a: pp. 98–101): the "coal measures" are exceptional. Even today, when human disturbance is causing more organic detritus to wash into rivers, 99.3% of the carbon in the vegetable matter produced by terrestrial ecosystems is respired by some organism (Schlesinger 1991: p. 137). In other words, energy *is* a limiting factor. Nearly any temporary excess attracts a consumer. Moreover, there seems to be a premium on efficient exploitation of the available energy. A monopoly such as the one a group of Australian bellbirds establish when they defend a tract of eucalyptus forest against all other insectivorous birds to create an "artificial" abundance of insect prey (Loyn 1987) usually does not last long. Adam Smith (1776) would have rejoiced greatly if artificial monopolies and other such "market distortions" could have been so thoroughly banished from the Europe of his day.

The efficiency with which energy is exploited in the tropics, and the corresponding rarity of major "ecological distortion," simply reflects the intensity of competition among tropical organisms. The biota of small, long-isolated tropical islands have experienced a limited array of possible competitors. As a result, resources on these islands are not fully exploited, so these biotas are especially sensitive to disruption by organisms people introduce from abroad (Darwin 1859: pp. 105–106; Lorence and Sussman 1988, Macdonald et al. 1991, Strasberg 1995). It is on these isolated islands where competition has been least intense that one would expect to find the greatest divergences from the "constants of tropical forest." Alas, successive human migrations have made testing this prediction progressively less practical.

Harmonious Ecosystems

So far, a similitude has been established between ecosystems and free-market economies with respect to the efficiency of exploitation of available resources. Human societies, however, especially those where the forms of economic competition are least constrained, are more noted for the degree of interdependence among their members than for their harmony. Can competition among individuals or species in an ecosystem lead to an effective, harmonious common interest among the ecosystem's members?

Odum (1971a), following Lotka (1925), proposed that ecosystems are reasonably harmonious because any species whose behavior benefits another species derives benefit in return, thanks to selection favoring individuals in those other species whose behavior is more favorable to their benefactors. In Odum's view, herbivores benefit the grass they eat by fertilizing it and by speeding the recycling of dead plant matter. Odum expected suitable prey species to attract predators, but he presumed that predators evolve, insofar as possible, to benefit, or at least not to endanger, their prey.

Odum's idea fell on deaf ears, although Dawkins (1982: p. 240, 246) hinted that those species prosper best that interact harmoniously with other common species in their habitat. The mathematical ecology of the day suggested that a predator could not benefit by enhancing the growth of its prey (or of any other beneficial species), nor could a plant benefit by improving its soil, because the profit from so doing would accrue equally to all members of the agent's species (Wilson 1980). The existence of farmers and stockraisers suggests that our mathematics were limiting our vision. Later theory has suggested that there are circumstances under which an individual or its progeny benefit differentially by enhancing the growth of a different, beneficial species (Slatkin and Wilson 1979, Frank 1994a), or a plant by improving its soil (Wilson 1980). Let us consider one process that enhances harmony among species and enquire into the possible range of its effects.

Neighborhood Selection, Trees, and Mutualism among Species

Let us consider a single species whose population is aggregated into discrete trait groups or neighborhoods of N adults apiece, where an individual's reproduction is governed by other members of its trait group, but where each trait group's young join a common pool from which the next generation's trait groups are formed at random (Wilson 1975). Here, there are two layers of selection: selection among members of a trait group for characteristics increasing an individual's share of its trait group's reproduction, and selection among trait groups for characteristics increasing a trait group's total reproductive output (see Appendix 9.1). The smaller N, the stronger selection among, relative to selection within, trait groups. It makes no difference whether the population is divided into discrete trait groups of N adults apiece or whether the population is continuously distributed and each individual's reproduction is governed by interactions with its $N - 1$ (fixed) nearest neighbors, nor does it matter whether successive generations are distinct or if individuals die one by one, so long as young join a common pool from which individuals are drawn at random to replace dead adults (Wilson 1980, Leigh 1991, 1994).

The best-studied example of trait-group selection is provided by fig wasps (Herre 1989). Each species of fig tree has its own species of pollinating wasp (Corner 1940). The situation may be caricatured by assuming that a fixed number, N, characteristic of the species, of fertilized fig wasps enter a fig fruit—a head of flowers turned outside in to form a ball or *syconium* with flowers on the inside—and lay eggs in half its ovules. When the adult wasps emerge from this fruit's seeds, they mate at random, and the fertilized females fly off in search of new figs to pollinate (Janzen 1979, Herre 1989, West and Herre 1994). The N fertilized females entering a fig fruit can be viewed as forming a trait group of "triploids" whose young are the fertilized females leaving the fruit. Selection within the fig fruit favors producing nearly as many male as female young, for this sex ratio, which is under the control of the female wasp, maximizes the proportion of this female's genes among those in the fertilized females leaving the fruit (Appendix 9.1). Selection among fig fruits favors producing as many female wasps as possible and only enough males to fertilize these females (Colwell 1981). Theory predicts, and observation confirms, that in those species of fig wasp with the lowest N, sex ratio among the offspring is most biased in favor of females (Herre 1985).

Trait-group or neighborhood selection is thus a real process, not just a theoretician's dream. Moreover, trees with well-dispersed seeds are subject to selection among neighborhoods, insofar as a tree's reproductive output is governed by interactions with a few near neighbors, while its seedlings compete with those from many such neighborhoods. But what does this have to do with ecosystem harmony?

The answer is that it makes no difference whether the trees in our neighborhood are different genotypes of the same species or members of different species: in the latter case, the different species could be represented mathematically as different alleles at a haploid locus. Just as "neighborhood structure" in a population of conspecifics causes an allele's spread to be influenced by its contribution to the good of its bearers' neighborhoods, so does neighborhood structure in a diverse forest allow species composition to be influenced by the good of each individual's immediate neighbors. Put another way, if a seed cannot grow into a tree unless it is dispersed far beyond its parent's neighborhood, then a characteristic that enhances the quantity or quality of seeds a tree disperses benefits its bearers, even if it increases the reproduction of neighbors to a greater degree, because most of the seeds with which the bearer's seeds compete will be from other neighborhoods, beyond the reach of their parent's benefactions. Similarly, neighborhood selection prevents the spread of a "spiteful" trait that reduces its bearer's reproduction to annihilate the reproduction of its neighbors, for the young of such a tree will compete mostly with the young of trees beyond the reach of their parent's spite. Neighborhood selection only works, however, if seeds disperse farther than this spite propagates. In the "pine barrens" of New Jersey (Givnish 1981), fire-resistant pines produce flammable debris that fuels fires that kill competing oaks and ruin the soil. Since fires spread much farther than an individual tree's seeds, selection favors this form of ecological distortion (Bond and Midgley 1995).

Do Pests Promote Mutualism among Tropical Trees?

Resorting to rarity to escape insect pests creates a neighborhood selection among tropical trees, for such a tree can only endow its seeds with a reasonably promising future by dispersing them far beyond the ring of neighbors with which it competes for light, water, and nutrients (Leigh 1994).

How might neighborhood selection affect a forest? A tree's solution to a problem often affects neighbors as well as itself. Solutions of nearly equivalent value to the agent can affect neighbors differently. Some plants depend for their nutrient supply on vesicular-arbuscular mycorrhizae, which seedlings of most other species can acquire by root contact; others employ ectomycorrhizae, many of which only conspecific seedlings can share (Janos 1980). A plant may deploy leaves and roots in a manner that enhances its whole neighborhood's productivity or in a way that enhances its own competitive ability at the expense of its neighbors (Horn 1971, King 1993). Some plants form root grafts with neighbors that enable all the plants so grafted to resist windthrow by hurricanes (Basnet et al. 1993).

A plant's strategy perhaps affects its neighbors most readily through its effects on soil. Litter and root exudates serve as food and nutrients for soil organisms. Soil organisms, both microbes and invertebrates, maintain the crumb structure and porosity of the soil, which in turn affect the soil's aeration, its penetrability by roots, its water-holding capacity, and the ease with which rainwater infiltrates the soil (Bruenig 1996: pp. 16f). Soil microbes are involved in the formation of aggregates of humus with mineral soil, which serve as sink, store, and source of nutrients for plants

(Bruenig 1996: p. 20). Plants can affect the soil in many ways. Some plants lift water from the subsoil (Dawson 1993, 1996), part of which becomes available to neighboring plants (Dawson 1993, Emerman and Dawson 1996) and to earthworms and soil organisms which improve the soil's water-holding capacity and infiltrability, and improve its content of nutrients and organic matter (Joffre and Rambal 1988, 1993: p. 571). The leaves of some plants decay readily when they fall, improving the soil; the leaves of other plants are full of long-lived poisons which delay their decomposition when they fall and impair the quality of the soil (Hobbie 1992). Litter from some plants forms a raw humus, causing leaching of the soil below, forming a bleached sand (Bruenig 1996: p. 23). Does long-distance seed dispersal sometimes enable neighborhood selection to tip the balance in favor of solutions more favorable to neighbors?

One way to assess the impact of neighborhood selection is to compare a mixed rainforest whose trees have well-dispersed seeds with an adjacent "single-dominant" rainforest dominated by a species of tree whose seeds rarely disperse beyond their parents' crowns (Connell and Lowman 1989). In Zaire, forest whose canopy consists entirely of the caesalpinoid legume *Gilbertiodendron dewevrei* with large, heavy seeds that simply fall to the ground under their parent adjoins mixed rainforest whose trees have well-dispersed seeds (Hart et al. 1989, Hart 1990). As one might expect, *Gilbertiodendron dewevrei* employs ectomycorrhizae, presumably species specific, while vesicular-arbuscular mycorrhizae prevail in the surrounding mixed forest (Fassi and Moser 1991). *Gilbertiodendron* trees drop a thick decay-resistant litter which reduces herbaceous cover, and the even canopy of *Gilbertiodendron* forest lets through far less light than the uneven canopy of adjacent mixed forest (Hart 1990, Fassi and Moser 1991). Nevertheless, we need experimental studies comparing seedling and sapling regeneration in *Gilbertiodendron* forest and adjacent mixed forest and comprehensive contrasts of other monodominant forests with adjacent mixed forests to assess the role of neighborhood selection more reliably. Still, it appears that neighborhood selection favors a more dynamic forest, which is much richer in animals (Leigh 1994).

Fig Trees and a Replay of the Origin of Angiosperms

Fig trees have refined the art of pollination to an extraordinary degree (Nason et al. 1996). When a fig's flowers, crowded on the insides of their syconia, are ready for pollination, the tree "calls" pollinating wasps from an astonishingly wide area (Table 9.2). On Barro Colorado, one *Ficus obtusifolia* attracted wasps from at least 22 different pollen-bearing trees, which must have been scattered over 8000 ha of forest (Nason et al. 1996). These numbers are typical (Table 9.2). It is wonderful how a minute wasp, which grew within the confines of a single fig seed, can find a suitable syconium to pollinate within the few days

Table 9.2 Density of adults (D), and of wasp-supplying adults (DR), number of pollen parents per fruit crop (NP), supplying area for pollinators of a crop (SA), and genetic diversity in two species of strangler fig

Species	$1/D$ (ha)	$1/DR$ (ha)	NP	SA (ha)	%P	%H
Ficus obtusifolia	14	364	17	6198	84.2	25.4
Ficus popenoei	76	1976	6	11856	66.7	15.8

$1/D$, number of hectares per adult of reproductive size; $1/DR = 26(1/D)$, number of hectares per adult releasing fig wasps in time to pollinate sampled crop; %P, percentage of studied loci that were polymorphic, %H, percentage of studied loci per individual that are heterozygous. NP is the average of two trees per species. Data on genetic diversity are from Hamrick and Murawski (1991); other data are from Nason et al. (1996).

of its adult life. Thanks to their wasps, many figs maintain extraordinarily high levels of genetic diversity (Nason et al. 1996).

Dispersal of fig seeds is similarly effective. In the Neotropics, fig seeds are dispersed primarily by bats. When feeding, these bats all remove fruit well away from the parent tree before eating them (Morrison 1978a) because predators lurk in fruiting trees, waiting for unwary frugivores (Howe 1979, Kalko et al. 1996). Some of these bats fly several kilometers in search of trees bearing ripe figs (Morrison 1978a,b). Bats are effective dispersers of fig (and other) seeds to newly accreting riverbanks in Amazonia (see Foster et al. 1986). In the Old World, fig trees whose seeds were carried by bats and birds were prominent and important early colonists of the lifeless islands that remained after Krakatau exploded in 1883 (Thornton 1996: pp. 148–152).

Fig trees pay heavily for effective pollination. Because a fertilized female wasp must find figs to pollinate within the few days before she dies, most species of fig have some trees with figs ready to pollinate at any given time of year (Milton et al. 1982, Windsor et al. 1989). Consequently, there are figs ready to eat all through the year, whether or not the season is propitious for fig reproduction (Milton et al. 1982, Windsor et al. 1989). Moreover, figs with larger fruits apparently "pay extra" for more effective seed dispersal. Fig wasps are killed by relatively slight overheating: larger fig fruits must use precious water transpiring to keep their wasps cool enough (Patiño et al. 1994). Larger fig fruits are also pollinated by more wasps per fruit (Herre 1993). Where more wasps pollinate each fruit, the proportion of males among their offspring is higher, wasting more of the fig tree's resources on useless nonpollinators (Herre 1989), and the nematodes parasitizing these wasps are more virulent (Herre 1993). Apparently, the only advantage of larger fig fruits (Patiño et al. 1994) is that they attract larger bats, which fly farther (Kalko et al. 1996).

Fig trees appear to stand in much the same relation to other angiosperms that the early angiosperms did to gymnosperms. Fig trees grow quickly (Janzen 1979). *Ficus insipida* photosynthesizes as rapidly per unit area of leaf as crop plants, and much more rapidly than any other species of tree recorded in the literature (Zotz et al. 1995). As Corner (1967: p. 24) remarked, "By leaf, fruit and easily rotted wood fig-plants supply an abundance of surplus produce." Indeed, in many areas, figs are a resource of central importance. Since most species of fig tree must supply fruit all year long to maintain their pollinators, figs, especially in Asia and the Neotropics, are a "keystone resource" whose availability animals can rely on during a season of food shortage (Terborgh 1986, Lambert and Marshall 1991, Thornton 1996: pp. 142f). In many sites, figs are such a steady and reliable source of fruit that they support specialist frugivores. In Borneo, three species of green pigeon (*Treron* spp.) specialize on figs, and five species of barbet (*Megalaima* spp. and *Calorhamphus fuliginosus*) feed primarily on figs (Leighton and Leighton 1983: p. 191). On Barro Colorado, figs provide the staple food for a guild of bats specializing on canopy fruit (Bonaccorso 1979), which includes 10 of the island's 20 species of fruit-eating bat (Kalko et al. 1996), and fig fruit and nitrogen-rich fig leaves form an important part of the diet of howler monkeys (Milton 1980). In Panama, the rotting wood of *Ficus insipida* supports an extraordinarily diverse and busy community of insects and other arthropods (Zeh and Zeh 1991, 1992). Just as angiosperms grew faster, bore tastier and more abundant fruit and leaves, and rotted more readily once dead than gymnosperms, so figs grow faster, provide better food more reliably, and rot more readily once dead than most angiosperms. Thus has improvement in the pollinator mutualism enhanced the luxuriance, and the degree of mutualism, in tropical forest.

CONCLUDING REMARKS

Tropical rainforest illustrates competition at its most intense. It also represents an apex of mutualism. This should surprise no one. Competition and mutualism are interwoven all through the history of life. The most crucial events in evolutionary history are those in which smaller parts have joined in mutualism to form more effective wholes. These triumphs of mutualism leave traces in the mechanisms that suppress conflicts between parts and their wholes and maintain the community of interest among the parts in the good of their whole. These harmony-maintaining mechanisms are the most distinctive footprints of the decisive role of natural selection in evolution (Leigh 1995).

Mutualisms play a fundamental role in the function of both terrestrial and marine ecosystems. A variety of circumstances can favor the evolution of mutualism. The one thing *needful* is an effective community of interest among the potential mutualists. Among conspecifics, mutualism is originally triggered by some form of "safety in numbers," while different species are most likely to join in mutualism if they benefit by pooling different aptitudes for their common good.

To what extent is an ecological community a "commonwealth," in whose good its members share a common stake? Does this common stake influence the evolution of its member species? If so, how?

Selection will eliminate wasteful monopolies, for there are always individuals to benefit by exploiting poorly used resources. More particularly, the productivity of tropical forest hinges on the availability (and employment) of reliable long-distance pollinators and seed dispersers, which enable each species to be rare enough to avoid costly defenses against specialist herbivores, which interfere with the recycling of nutrients and diminish the quality of the soil.

Fig trees, with their wonderful pollinating wasps, have raised long-distance pollination and seed dispersal to new heights. They are a crucial source of fruit and foliage for many tropical animals. With their fast growth and rapid decay, they represent the ultimate in tropical dynamism.

Provided we choose to save it, the fate of tropical forest will depend on our willingness to understand the natural history of the various ways plants depend on animals and microorganisms, and vice versa. But we humans must first discern, and act on, our common interest in the preservation of rainforest. That topic is the subject of the final chapter.

APPENDIX 9.1 The Varieties of "Group Selection"

The Crucial Role of Variance Within and Among Subpopulations

Consider a characteristic of magnitude a in a population of haploids, where successive generations are distinct. Let this characteristic be governed by a single multiallelic locus; let a fraction, $q_i(t)$, of the individuals of generation t carry allele i, and let carriers of allele i have average magnitude, a_i, and relative fitness, $v_i = w_i/W$, where w_i is the average number of offspring per i-bearing parent, and W is the average number of offspring per parent in the population as a whole. If a_i does not change from one generation to the next, then the population's average value, $A(t)$, of a at generation t is $\Sigma_i q_i(t)a_i$, its average value $A(t+1)$ in generation $t+1$ is $\Sigma_i q_i(t+1)a_i = \Sigma_i w_i q_i(t)a_i/W = \Sigma_i v_i q_i(t)a_i$, and the change in this average value from generation t to generation $t+1$ is (Price 1970, Frank 1994b)

$$A(t+1) - A(t) = \Sigma_i q_i(v_i - 1) a_i = cov(v,a) = cV(a). \quad (9.1)$$

Here, cov (v, a) is the covariance between relative fitness, v (which is "normalized" so that its average is 1), and the average magnitude, a, for carriers of a given allele, evaluated over the ensemble of alleles at this locus. This covariance can in turn be expressed as $cV(a)$. Here, $V(a)$ is the heritable variance in this magnitude, $\Sigma_i q_i(a_i - A)^2$, and c is the intensity of selection on a, that is to say, the regression of the allelic average of relative fitness on the allelic average of a. If v depends linearly on a, $c = dv/da$.

Now suppose that this population is aggregated into distinct groups. Let A' be the average over all groups of this magnitude, A_g its average in groups of type g, and v_g the relative reproductive rate (rate of foundation of new groups of type g, less extinction rate of preexisting ones, of type g, minus the average of the same difference for all groups in the species). Thus, in analogy with selection within a population,

$$A'(t+1) - A'(t) = cov(v_g, A_g) + E[cov_g(v,a)] \quad (9.2)$$

$$A'(t+1) - A'(t) = b_g V_g(A_g) + E[cV(a)] \quad (9.3)$$

(Price 1972, Frank 1994b). Here, b_g is the intensity of selection among groups on A_g, that is to say, the regression of the relative reproductive rate of groups on their average value A_g of this characteristic, or dv_g/dA_g; $E[cov(v, a)] = E[cV(a)]$ is the average effect on A' of selection within groups. Here, we must remember that the mean value for a group of the magnitude a is not fixed, as we assumed it was for genotypes, but changes as a result of selection among members of that group. It follows from the equation of Price (1972) that variance among, relative to heritable variance within, groups of a, plays a decisive role in the effectiveness of selection among groups on the magnitude a.

Selection among Groups as a Form of Kin Selection

Can the condition for effectiveness of selection among groups be expressed in terms of the relatedness of members within a group relative to those of different groups? Suppose, with Slatkin and Wade (1978), that all subpopulations have N adults apiece, and that the heritable variance (the "additive genetic variance"; Fisher 1930, 1941, 1958a) in a, $V(a)$, is the same within each. The total variance, V_T, of the species in a is variance within groups plus variance among groups: $V_T(a) = V(a) + V_G(A_G)$. The average relatedness of two members of the same group, relative to that of two individuals chosen at random from the species as a whole, is the intragroup or "intraclass" correlation coefficient (Fisher 1958b):

$$r = \frac{V_G(A_G)}{V_T(a)} = \frac{1}{1 + V(a)/V_G(A_G)}. \quad (9.4)$$

If b_G is negative and c is positive, then decrease of a benefits a group's reproduction, but genotypes with lower a contribute less to the reproduction of their group. In this case, low a is called altruistic, because it benefits the group at the individual's expense. In this case, c represents the cost to an individual of a unit decrease in a, and $-b_G = b -$ c represents the net benefit to a group of a unit decrease in A_G. Selection will lower the species-wide mean, A', if

$$-b_G V_G = V_G(b-c) > cV; \quad V_G b > cV_T; \quad b\frac{V_G}{V_T} = br > c. \quad (9.5)$$

Equation 9.5 expresses Hamilton's (1964a) famous "decision-rule" which states that selection favors altruism if the benefit, b, to the recipient, times the relatedness, r, of recipient to donor exceeds the cost to the donor. This rule applies in many other contexts besides selection among groups. Bourke and Franks (1995) provide perhaps the most comprehensive discussion of different ways to define and calculate the coefficient of relatedness.

When Can Selection Among Groups Prevail over Within-Group Selection?

To learn when selection among groups can prevail, we must first enquire what governs the ratio of variance among groups to heritable variance within groups (Leigh 1983). Let us continue to assume a population of sexual haploids, where successive generations are distinct. Let the population be aggregated into many groups of N adults apiece. Let heritable variance within each group have the same variance, V. Genetic drift increases among-group variance, V_G, by V/N per generation (Lande 1976), this being the variance among means of samples of size N.

Now suppose that each generation, a fraction, m, of each group's members disperse at random as migrants to other groups. "On the average," this exchange of migrants replaces a proportion m of a group's members by migrants whose average deviation from the species-wide mean is 0. Thus, every generation, this exchange of migrants reduces the deviation of each group's mean value from the species-wide mean by the factor $1-m$, and consequently reduces V_G, the average squared deviation of group means from the species-wide mean, by the factor $(1-m)^2 \approx 1 - 2m$.

Suppose, moreover, that variance among $1/L$ of those groups is reduced from V_G to V_G/k, since each new group is the "average" of k parent groups. Thus, among-group variance in generation $t+1$, $V_G(t+1)$, is

$$V_G(t+1) = \frac{V}{N} + V_G(1-2m) - \frac{V_G}{L}\left(1 - \frac{1}{k}\right). \quad (9.6)$$

At equilibrium

$$\frac{V}{N} \approx 2mV_G + \frac{V_G}{L}\left(1 - \frac{1}{k}\right). \quad (9.7)$$

First, let $k = 1$, so that each group descends from a single parent group. Then, at equilibrium, $V = 2mNV_G$; variance among groups surpasses heritable variance within groups if less than one migrant is exchanged per population per two generations (Wright 1931). If the average group lifetime is L generations, the time scale of selection among groups is L times slower than that among individuals. Therefore, variance among groups must be L times that within groups if selection among groups is to meet selection within groups on equal terms. Meeting on equal terms means that the frequency of an allele A remains in balance if groups consisting only of A-bearers produce W times as many offspring groups per parent group as the "competition," whereas A-bearers in mixed groups produce $1/W$ times as many offspring (individuals) per parent as their competitors. Selection among and within groups therefore "meet as equals" if one migrant is exchanged per two groups per *group* lifetime.

Now, assume that no migrants are exchanged among populations, so $m = 0$, while $k > 1$. Then, at equilibrium,

$$\frac{V}{N} = \frac{V_G}{L}\left(1 - \frac{1}{k}\right); \quad V = \frac{V_G N}{L}\left(1 - \frac{1}{k}\right). \quad (9.8)$$

For selection among groups to meet selection within groups on equal terms, V_G must be equal to LV. This only happens if $N(1 - 1/k) < 1$, that is to say, if < 1 of every N groups is founded by colonists from more than one parent group. Thus groups must be very distinct, exchanging almost no migrants, and their reproduction must be nearly uniparental if selection among groups is to prevail over selection within groups. In other words, selection among groups only prevails if groups are "units of selection" (Lewontin 1970) in the same sense that genes are. This conclusion is argued more rigorously, with explicit attention to the factors governing variance within populations, by Leigh and Rowell (1995), following Lande (1992).

Trait-Group (Neighborhood) Selection

Now let our population of sexual haploids be aggregated into trait groups of N adults apiece. Let each adult's reproduction be governed by its interactions with fellow-members of its trait group, and let young from all trait groups join a common pool from which the next generation's trait groups are drawn at random (Wilson 1975, 1980). If the population's total heritable variation in the magnitude, a, of a quantitative characteristic is V_T, then the variance, V_G, among trait groups—the variance among means of samples of size N—is V_T/N, and the variance, V, within trait groups—the variance within samples of size N—is $V_T(N-1)/N$. If $-b_G$ is the regression of a group's reproductive output, divided by average reproduction per group, on its mean value of a, and c is the regression of the proportion an individual contributes to its group's reproduction on its value of a, then the change in the population's mean value of a between generations t and $t + 1$ is

$$A'(t+1) - A'(t) = \frac{-b_G V_T}{N} + \frac{cV_T(N-1)}{N} \quad (9.9)$$

(see Frank 1986). This equation ascribes the change in the population mean, A', to two layers of selection: selection among individuals of a group through the effect of an individual's value of a on its share of its group's reproduction, and selection among groups through the effect of a group's mean value of a on that group's total reproductive output.

If dispersal is limited and a trait group's young are competing only with the young of the M trait groups centered on its own, then selection among trait groups is driven by the variance among means in a sample of M trait groups, or $(M-1)V_T/MN$ (Leigh 1991, 1994), rather than V_T/N, the value for truly panmictic dispersal. Even if M is only 2, the effectiveness of selection among trait groups is only halved relative to the panmictic case.

The most familiar and best-studied example of trait-group selection is the influence of local mate competition (Hamilton 1967) on the sex ratio (Colwell 1981; Frank 1985, Herre 1985, 1989), mentioned in the text. This form of population organization makes the influence of an allele on the welfare of its bearers' trait groups at least marginally relevant to that allele's selective advantage (Wilson 1980).

Let us consider the evolution of sex ratio in a species of fig-pollinating wasp where, on average, N fertilized females pollinate a fig fruit. Suppose there are two types of foundresses, A and B: both produce the same numbers of offspring per mother, but a fraction, r_A, of the offspring of A foundresses, and a fraction, r_B, of B foundresses, are male. To calculate the change in the populationwide sex ratio, r', from one generation to the next according to equation 9.9, let $-b_G = d(\log W_g)/dr^*$ and $c = dv/dr$. W_g is the group fitness, the fig's total export of fertilized females, which is proportional to $1 - r^*$, the proportion of females among the offspring of that fig's foundresses. Thus $-b_G = -1/(1-r^*)$. To calculate c, let B_{ms} be the proportion of a mother's genes carried by a son, and B_{md} the proportion carried by a daughter. Now consider a fig pollinated by x A foundresses and $M - x$ B foundresses. In this fig, the average number of inseminations per son is $(1 - r^*)/r^*$, where $r^* = xr_A/M + (1 - x/M)r_B$, the overall proportion of males among this fig's young wasps. The ratio, v_A, of the proportion of a given A foundress's genes exported to the proportion $1/N$ it carried in is

$$V_A = \frac{r_A[(1 - r^*)/r^*]B_{ms} + (1 - r_A)B_{md}}{(1 - r^*)(B_{ms} + B_{md})}. \quad (9.10)$$

Here, $r_A[(1 - r^*)/r^*]B_{ms}$ is the number of genes per A foundress offspring contributed by her sons to fertilized females, $(1 - r_A)B_{md}$ is the number of genes per A foundress offspring in her daughters, and $(1 - r^*)(B_{ms} + B_{md})$ is the populationwide average of the number of genes exported per offspring. The corresponding ratio for B foundresses is

$$V_B = \frac{r_B[(1 - r^*)/r^*]B_{ms} + (1 - r_B)B_{md}}{(1 - r^*)(B_{ms} + B_{md})}. \quad (9.11)$$

Let $v_A - v_B = dv$, $r_A - r_B = dr$. Then dv/dr is

$$\frac{1}{B_{ms}+B_{md}}\left(\frac{B_{ms}}{r^*}-\frac{B_{md}}{1-r^*}\right)=\frac{1}{r^*(1-r^*)}\left(\frac{B_{ms}}{B_{ms}+B_{md}}-r^*\right). \quad (9.12)$$

The sex ratio favored by selection within figs is $B_{ms}/(B_{ms}+B_{md})$. The sex ratio r^* which is optimal for the foundress number M is $r^* = B_{ms}(M-1)/(B_{ms}+B_{md})M$.

To flesh out these results, we must calculate B_{ms} and B_{md}. As a male wasp is haploid, B_{ms} is necessarily 0.5. Finding B_{md} is harder. As foundresses aggregate at random from an infinite population, the relevant value of B_{md} is that for an average fig with N foundresses. Following Herre (1985: note 14), consider a representative locus. Let a_1 be the proportion of fertilized female foundresses where both alleles in the female, and the allele in its mate's sperm, are identical by descent (Ewens 1979: p. 225); call these type 1, or $AA\ A$. Let a_2 be the proportion of foundresses whose alleles are identical by descent, but differ from the allele at this locus in their mate's sperm; call these type 2, or $AA\ x$. Let a_3 be the proportion of fertilized females which are type 3, or $Ax\ A$, and let $a_4 = 1 - a_1 - a_2 - a_3$ be the proportion which are type 4, or $Ax\ y$, with no identity relations among the alleles she carries.

A type 1 ($AA\ A$) female contributes a proportion $1/N$ of A-bearers to the male young from its fig, and a proportion $1/N$ of AA to the fig's female young. If young mate at random with their fellows in their syconium, this foundress contributes a proportion $1/N$ type 1 and $(N-1)/N$ type 2 to its share of the fertilized females leaving this average fig. Similarly, a type 2 parent, $AA\ x$, contributes a proportion $1/N$ of A-bearers to this fig's male young and a proportion $1/N$ of Ax to this fig's female young. A type 2 foundress contributes a proportion $1/N$ of type 3 individuals to its share of the fertilized females leaving this fig. A type 3 female, $Ax\ A$, contributes a proportion $1/2N\ A$ and $1/2N\ x$ to its fig's males, and $1/2N\ AA$ and $1/2N\ Ax$ to its fig's females. Thus a type 3 foundress contributes a proportion $1/4N$ type 1 offspring (from matings of AA with A), $(2N-1)/4N$ type 2 offspring (from matings of AA with non-A), and $1/2N$ type 3 offspring (from matings of Ax with A or x) to its share of the fertilized females leaving the fig. Finally, a type 4 foundress, $Ax\ y$ contributes a proportion $1/2N$ of type 3 offspring (from matings of Ay with A and xy with x) to its share of the fertilized females leaving the fig.

From Table 9.3, the proportion of type 1 females among the fertilized females of generation $t+1$ is $a_1(t+1) = a_1(t)/N + a_3(t)/4N$, while $a_2(t+1) = (N-1)a_1(t)/N + (2N-1)a_3(t)/4N$, and $a_3(t+1) = a_2(t)/N + a_3(t)/2N + a_4(t)/2N$. At equilibrium, the values of a do not change from one generation to the next, so $a_3 = 4(N-1)a_1 = 2a_2$ and $a_4 = (2N-1)a_3$. Since $a_1 + a_2 + a_3 + a_4 = 1$, $a_1 = 1/(4N-3)(2N-1)$. The number of maternal genes per daughter is $1 + a_1 + a_2 = 1 + 1/(2N-1)$, so $B_{md} = N/(2N-1)$, $B_{ms}/(B_{ms}+B_{md}) = (2N-1)/(4N-1)$, and the optimal proportion, r^*, of males among the offspring of a fig with M foundresses in a species where the average foundress number is N is $(2N-1)(M-1)/(4N-1)M$ (Herre 1985). The smaller the foundress number, M, the more female-biased the sex ratio among the foundresses' offspring, but, for a given M, the sex ratio is more female-biased the lower the average foundress number (N) for the fig species in question. Both predictions have been confirmed (Herre 1985).

References

Allen, M. F. 1993. *The Ecology of Mycorrhizae*. Cambridge University Press, Cambridge.

Appanah, S. 1985. General flowering in the climax rain forests of South-east Asia. *Journal of Tropical Ecology* 1: 225–240.

Appanah, S. 1993. Mass flowering of dipterocarp forests in the aseasonal tropics. *Journal of Bioscience* 18: 457–474.

Ashton, P. S., T. J. Givnish, and S. Appanah. 1988. Staggered flowering in the Dipterocarpaceae: new insights into floral induction and the evolution of mast fruiting in the aseasonal tropics. *American Naturalist* 132: 44–66.

Axelrod, R. 1984. *The Evolution of Cooperation*. Basic Books, New York.

Axelrod, R., and W. D. Hamilton. 1981. The evolution of cooperation. *Science* 211: 1390–1396.

Table 9.3 Propagation of inbreeding from parent to offspring among fig wasps with average foundress number, N

Parent Foundress Type	Type 1 $AA\ A$		Type 2 $AA\ x$		Type 3 $Ax\ A$		Type 4 $Ax\ y$	
	male	female	male	female	male	female	male	female
Frequency of offspring genotypes among young of each sex	$1/N\ A$	$1/N\ AA$	$1/N\ A$	$1/N\ Ax$	$1/2N\ A$ $1/2N\ x$	$1/2N\ AA$ $1/2N\ Ax$	$1/2N\ A$ $1/2N\ x$	$1/2N\ Ay$ $1/2N\ xy$
No. of offspring foundresses of different type per parent								
Type 1		$1/N$		0		$1/4N$		0
Type 2		$(N-1)/N$		0		$(2N-1)/4N$		0
Type 3		0		$1/N$		$1/2N$		$1/2N$

Under each type of parent foundress are listed
(1) the fraction of offspring of different genotype one parent of type indicated contributes to the male, and the female, offspring produced by the fig's N foundress (evaluated before they mate),
(2) the expected proportions of different offspring types among the share of one parent of type indicated (assumed to be $1/N$) of the fig's total output of fertilized females.

Baillie, I. C. 1996. Soils of the humid tropics, pp. 256–286. In P. W. Richards, *The Tropical Rain Forest* (2nd ed). Cambridge University Press, Cambridge.

Barnes, J. (ed) 1984. *The Complete Works of Aristotle*, vol. 1. Princeton University Press, Princeton, NJ.

Basnet, K., F. N. Scatena, G. E. Likens, and A. E. Lugo. 1993. Ecological consequences of root grafting in Tabonuco (*Dacryodes excelsa*) trees in the Luquillo Experimental Forest, Puerto Rico. *Biotropica* 25: 28–35.

Benzing, D. H. 1990. *Vascular Epiphytes*. Cambridge University Press, Cambridge.

Birky, C. W. Jr. 1991. Evolution and population genetics of organelle genes: mechanisms and models, pp. 112–134. In R. K. Selander, A. G. Clark, and T. S. Whittam, eds., *Evolution at the Molecular Level*. Sinauer Associates, Sunderland, MA.

Birky, C. W. Jr. 1995. Uniparental inheritance of mitochondrial and chloroplast genes: mechanisms and evolution. *Proceedings of the National Academy of Sciences, USA* 92: 11331–11338.

Blackstone, N. W. 1995. A units-of-evolution perspective on the endosymbiont theory of the origin of the mitochondrion. *Evolution* 49: 785–796.

Bonaccorso, F. J. 1979. Foraging and reproductive ecology in a Panamanian bat community. *Bulletin of the Florida State Museum, Biological Sciences* 24: 359–408.

Bond, W. J., and J. J. Midgley. 1995. Kill thy neighbor: an individualistic argument for the evolution of flammability. *Oikos* 73: 79–85.

Bourke, A. F. G., and N. R. Franks. 1995. *Social Evolution in Ants*. Princeton University Press, Princeton, NJ.

Bruenig, E. F. 1996. *Conservation and Management of Tropical Rainforests*. CAB International, Wallingford, UK.

Burdon, J. J. 1987. *Diseases and Plant Population Biology*. Cambridge University Press, Cambridge.

Burdon, J. J., and G. A. Chilvers. 1982. Host density as a factor in plant disease ecology. *Annual Review of Phytopathology* 20: 143–166.

Camazine, S. 1993. The regulation of pollen foraging by honey bees: how foragers assess the colony's need for pollen. *Behavioral Ecology and Sociobiology* 32: 265–272.

Chan, H. T. 1980. Reproductive biology of some Malaysian dipterocarps II. Fruiting biology and seedling studies. *Malaysian Forester* 43: 438–451.

Chapela, I. H., S. A. Rehner, T. R. Schultz, and U. G. Mueller. 1994. Evolutionary history of the symbiosis between fungus-growing ants and their fungi. *Science* 266: 1691–1694.

Coe, M. J., D. L. Dilcher, J. O. Farlow, D. H. Jarzen, and D. A. Russell. 1987. Dinosaurs and land plants, pp. 225–258. In E. M. Friis, W. G. Chaloner, and P. R. Crane, eds., *The Origins of Angiosperms and Their Biological Consequences*. Cambridge University Press, Cambridge.

Colwell, R. K. 1981. Group selection is implicated in the evolution of female-biased sex ratios. *Nature* 290: 401–404.

Connell, J. H., and M. D. Lowman. 1989. Low-diversity tropical rain forests: some possible mechanisms for their existence. *American Naturalist* 134: 88–119.

Connor, R. C. 1995. Altruism among non-relatives: alternatives to the 'Prisoner's Dilemma.' *Trends in Ecology and Evolution* 10: 84–87.

Corner, E. J. H. 1940. *Wayside Trees of Malaya*. Government Printer, Singapore.

Corner, E. J. H. 1964. *The Life of Plants*. World Publishing Co., Cleveland, OH.

Corner, E. J. H. 1967. Ficus in the Solomon Islands and its bearing on the post-Jurassic history of Melanesia. *Philosophical Transactions of the Royal Society of London* B 253: 23–159.

Cosmides, L. M., and J. Tooby. 1981. Cytoplasmic inheritance and intragenomic conflict. *Journal of Theoretical Biology* 89: 83–129.

Crepet, W. L. 1984. Advanced (constant) insect pollination mechanisms: pattern of evolution and implications vis-à-vis angiosperm diversity. *Annals of the Missouri Botanical Garden* 71: 607–630.

Crepet, W. L., E. M. Friis, and K. C. Nixon. 1991. Fossil evidence for the evolution of biotic pollination. *Philosophical Transactions of the Royal Society of London* B 333: 187–195.

Crepet, W. L., K. C. Nixon, E. M. Friis, and J. V. Freudenstein. 1992. Oldest fossil flowers of hamamelidaceous affinity, from the Late Cretaceous of New Jersey. *Proceedings of the National Academy of Sciences, USA* 89: 8986–8989.

Crow, J. F. 1979. Genes that violate Mendel's rules. *Scientific American* 240(2): 134–146.

Crow, J. F., and K. Aoki. 1982. Group selection for a polygenic behavioral trait: a differential proliferation model. *Proceedings of the National Academy of Sciences, USA* 79: 2628–2631.

Darwin, C. R. 1859. *The Origin of Species*. London, John Murray.

Davidson, E. A., and S. E. Trumbore. 1995. Gas diffusivity and production of CO_2 in deep soils of the eastern Amazon. *Tellus* 47B: 550–565.

Dawkins, R. 1976. *The Selfish Gene*. Oxford University Press, Oxford.

Dawkins, R. 1982. *The Extended Phenotype*. Oxford University Press, Oxford.

Dawson, T. E. 1993. Hydraulic lift and water use by plants: implications for water balance, performance and plant-plant interactions. *Oecologia* 95: 565–574.

Dawson, T. E. 1996. Determining water use by trees and forests from isotopic, energy balance and transpiration analyses: the roles of tree size and hydraulic lift. *Tree Physiology* 16: 263–272.

Dethier, V. G. 1976. *Man's Plague?* Darwin Press, Princeton, NJ.

Douglas, A. E. 1994. *Symbiotic Interactions*. Oxford University Press, Oxford.

Dressler, R. L. 1985. Humus collecting shrubs in wet tropical forests, pp. 289–294. In *Ecology and Resource Management in Tropics*, vol. 1. Bhargava Book Depot, Varanasi, India.

Dyson, F. 1985. *Origins of Life*. Cambridge University Press, Cambridge.

Eberhard, W. G. 1980. Evolutionary consequences of intracellular organelle competition. *Quarterly Review of Biology* 55: 231–249.

Emerman, S. H., and T. E. Dawson. 1996. Hydraulic lift and its influence on the water content of the rhizosphere: an example from sugar maple, *Acer saccharum*. *Oecologia* 108: 273–278.

Emmons, L. H. 1989. Tropical rain forests: why they have so many species and how we may lose this biodiversity without cutting a single tree. *Orion Nature Quarterly* 8(3): 8–14.

Ewens, W. J. 1979. *Mathematical Theory of Population Genetics*. Springer, Berlin.

Falconer, D. S. 1989. *Introduction to Quantitative Genetics*. Wiley, New York.

Fassi, B., and M. Moser. 1991. Mycorrhizae in the natural forests of tropical Africa and the Neotropics, pp. 157–202. In *Funghi, Pianti e Suolo*. Centro di Studio sulla Micologia del Terreno, C. N. R., Torino.

Feer, F. 1993. The potential for sustainable hunting and rearing of game in tropical forests, pp. 691–708. In C. M. Hladik et al., eds., *Tropical Forests, People and Food*. UNESCO, Paris, and Parthenon Publishing, Park Ridge, NJ.

Fischer, E. A. 1980. The relationship between mating system and simultaneous hermaphroditism in the coral reef fish, *Hypo-*

plectrus nigricans (Serranidae). *Animal Behaviour* 28: 620–633.

Fischer, E. A. 1981. Sexual allocation in a simultaneously hermaphroditic coral reef fish. *American Naturalist* 117: 64–82.

Fischer, E. A. 1987. The evolution of sexual patterns in the seabasses. *BioScience* 37: 482–489.

Fischer, E. A. 1988. Simultaneous hermaphroditism, tit-for-tat, and the evolutionary stability of social systems. *Ethology and Sociobiology* 9: 119–136.

Fisher, R. A. 1930. *The Genetical Theory of Natural Selection*. Clarendon Press, Oxford.

Fisher, R. A. 1941. Average excess and average effect of a gene substitution. *Annals of Eugenics* (now *Annals of Human Genetics*) 11: 53–63.

Fisher, R. A. 1958a. *The Genetical Theory of Natural Selection* (2nd ed). Dover Press, New York.

Fisher, R. A. 1958b. *Statistical Methods for Research Workers*, Oliver and Boyd, Edinburgh.

Forget, P.-M. 1993. Post-dispersal predation and scatterhoarding of *Dipteryx panamensis* (Papilionaceae) seeds by rodents in Panama. *Oecologia* 94: 255–261.

Forget, P.-M. 1994. Recruitment pattern of *Vouacapoua americana* (Caesalpiniaceae), a rodent-dispersed tree species in French Guiana. *Biotropica* 25: 408–419.

Forget, P.-M., and T. Milleron. 1991. Evidence for secondary seed dispersal by rodents in Panama. *Oecologia* 87: 596–599.

Foster, R. B., J. Arce B., and T. S. Wachter. 1986. Dispersal and the sequential plant communities in Amazonian Peru floodplain, pp. 357–370. In A. Estrada and T. H. Fleming, eds., *Frugivores and Seed Dispersal*. W. Junk, Dordrecht.

Frank, S. A. 1985. Hierarchical selection theory and sex ratios. II. On applying the theory, and a test with fig wasps. *Evolution* 39: 949–964.

Frank, S. A. 1986. Hierarchical selection theory and sex ratios. I. General solutions for structured populations. *Theoretical Population Biology* 29: 312–342.

Frank, S. A. 1994a. Genetics of mutualism: the evolution of altruism between species. *Journal of Theoretical Biology* 170: 393–400.

Frank, S. A. 1994b. Kin selection and virulence in the evolution of protocells and parasites. *Proceedings of the Royal Society of London* B 258: 153–161.

Frank, S. A. 1995. Mutual policing and repression of competition in the evolution of cooperative groups. *Nature* 377: 520–522.

Gilbert, G. S., and S. P. Hubbell. 1995. Plant diseases and the conservation of tropical forests. *Bioscience* 46: 98–106.

Gilbert, L. E. 1982. The coevolution of a butterfly and a vine. *Scientific American* 247(2): 110–121.

Givnish, T. J. 1981. Serotiny, geography and fire in the pine barrens of New Jersey. *Evolution* 35: 101–123.

Gompper, M. 1994. *The Importance of Behavior, Ecology and Genetics to the Maintenance of White-nosed Coati (Nasua narica) Social Structure*. PhD dissertation, Department of Zoology, University of Tennessee, Knoxville.

Gompper, M. E. 1996. Sociality and asociality in white-nosed coatis (*Nasua narica*): foraging costs and benefits. *Behavioral Ecology* 7: 254–263.

Gompper, M. E., J. L. Gittleman, and R. K. Wayne. 1997. Genetic relatedness, coalitions and social behavior of white-nosed coatis, *Nasua narica*. *Animal Behaviour* 53: 781–797.

Gradwohl, J., and R. Greenberg. 1980. The formation of antwren flocks on Barro Colorado Island, Panama. *Auk* 97: 385–395.

Hamilton, W. D. 1964a. The genetical evolution of social behavior. I. *Journal of Theoretical Biology* 7: 1–16.

Hamilton, W. D. 1964b. The genetical evolution of social behavior. II. *Journal of Theoretical Biology* 7: 17–51.

Hamilton, W. D. 1967. Extraordinary sex ratios. *Science* 156: 477–488.

Hamrick, J. L., M. J. W. Godt, and S. L. Sherman-Broyles. 1992. Factors influencing levels of genetic diversity in woody plant species. *New Forests* 6: 95–124.

Hamrick, J. L., and M. D. Loveless. 1989. The genetic structure of tropical tree populations: associations with reproductive biology, pp. 129–146. In J. H. Bock and Y. B. Linhart, eds., *The Evolutionary Ecology of Plants*. Westview Press, Boulder, CO.

Hamrick, J. L., and D. A. Murawski. 1991. Levels of allozyme diversity in populations of uncommon Neotropical tree species. *Journal of Tropical Ecology* 7: 395–399.

Hardin, G. 1968. The tragedy of the commons. *Science* 162: 1243–1248.

Hart, T. B. 1990. Monospecific dominance in tropical rain forests. *Trends in Ecology and Evolution* 5: 6–11.

Hart, T. B., J. A. Hart, and P. G. Murphy. 1989. Monodominant and species-rich forests of the humid tropics: causes for their co-occurrence. *American Naturalist* 133: 613–633.

Herre, E. A. 1985. Sex ratio adjustment in fig wasps. *Science* 228: 896–898.

Herre, E. A. 1989. Coevolution of reproductive characteristics in 12 species of New World figs and their pollinator wasps. *Experientia* 45: 637–647.

Herre, E. A. 1993. Population structure and the evolution of virulence in nematode parasites of fig wasps. *Science* 259: 1442–1445.

Hinkle, G., J. K. Wetterer, T. R. Schultz, and M. L. Sogin. 1994. Phylogeny of the attine ant fungi based on analysis of small subunit ribosomal RNA gene sequences. *Science* 266: 1695–1697.

Hobbie, S. E. 1992. Effects of plant species on nutrient cycling. *Trends in Ecology and Evolution* 7: 336–339.

Hölldobler, B., and E. O. Wilson. 1990. *The Ants*. Harvard University Press, Cambridge, MA.

Horn, H. S. 1971. *The Adaptive Geometry of Trees*. Princeton University Press, Princeton, NJ.

Howe, H. F. 1979. Fear and frugivory. *American Naturalist* 114: 925–931.

Howe, H. F., E. W. Schupp, and L. C. Westley. 1985. Early consequences of seed dispersal for a Neotropical tree (*Virola surinamensis*). *Ecology* 66: 781–791.

Jacobs, M. 1988. *The Tropical Rain Forest*. Springer, Berlin.

Janos, D. P. 1980. Mycorrhizae influence tropical succession. *Biotropica* 12(2) (supplement): 56–64.

Janos, D. P. 1983. Tropical mycorrhizas, nutrient cycles and plant growth, pp. 327–345. In S. L. Sutton, T. C. Whitmore, and A. C. Chadwick, eds., *Tropical Rain Forest: Ecology and Management*. Blackwell Scientific, Oxford.

Janzen, D. H. 1966. Coevolution of mutualism between ants and acacias in Central America. *Evolution* 20: 249–275.

Janzen, D. H. 1970. Herbivores and the number of tree species in tropical forests. *American Naturalist* 104: 501–528.

Janzen, D. H. 1974a. Epiphytic myrmecophytes in Sarawak: mutualism through the feeding of plants by ants. *Biotropica* 6: 237–259.

Janzen, D. H. 1974b. Tropical blackwater rivers, animals, and mast fruiting by the Dipterocarpaceae. *Biotropica* 6: 69–103.

Janzen, D. H. 1979. How to be a fig. *Annual Review of Ecology and Systematics* 10: 13–51.

Joffre, R., and S. Rambal. 1988. Soil water improvement by trees in the rangelands of southern Spain. *Acta Oecologica/Oecologia Plantarum* 9: 405–422.

Joffre, R., and S. Rambal. 1993. How tree cover influences the water balance of Mediterranean rangelands. *Ecology* 74: 570–582.

Jones, S. E. 1977. Coexistence in mixed species antwren flocks. *Oikos* 29: 366–375.

Kalko, E. K. V., E. A. Herre, and C. O. Handley Jr. 1996. Relation of fig fruit characteristics to fruit-eating bats in the New and Old World tropics. *Journal of Biogeography* 23: 565–576.

Kaufmann, J. H. 1962. Ecology and social behavior of the coati, *Nasua narica* on Barro Colorado Island Panama. *University of California Publications in Zoology* 60: 95–222.

Keller, L., and H. K. Reeve. 1994. Partitioning of reproduction in animal societies. *Trends in Ecology and Evolution* 9: 98–102.

King, D. A. 1993. A model analysis of the influence of root and foliage allocation on forest production and competition between trees. *Tree Physiology* 12: 119–135.

Lambert, F. R., and A. G. Marshall. 1991. Keystone characteristics of bird-dispersed *Ficus* in a Malaysian lowland rain forest. *Journal of Ecology* 79: 793–809.

Lande, R. 1976. Natural selection and random genetic drift in phenotypic evolution. *Evolution* 30: 314–334.

Lande, R. 1992. Neutral theory of quantitative genetic variance in an island model with local extinction and colonization. *Evolution* 46: 381–389.

Lee, D. 1980. *The Sinking Ark: Environmental Problems in Malaysia and Southeast Asia*. Heinemann, Kuala Lumpur.

Lees, D. C., and N. G. Smith. 1991. Foodplant associations of the Uraniinae (Uraniidae) and their systematic, evolutionary, and ecological significance. *Journal of the Lepidopterists Society* 45: 296–347.

Leigh, E. G. Jr. 1971. *Adaptation and Diversity*. Freeman, Cooper & Company, San Francisco.

Leigh, E. G. Jr. 1983. When does the good of the group override the advantage of the individual? *Proceedings of the National Academy of Sciences, USA* 80: 2985–2989.

Leigh, E. G. Jr. 1991. Genes, bees and ecosystems: the evolution of a common interest among individuals. *Trends in Ecology and Evolution* 6: 257–262.

Leigh, E. G. Jr. 1994. Do insect pests promote mutualism among tropical trees? *Journal of Ecology* 82: 677–680.

Leigh, E. G. Jr. 1995. The major transitions of evolution (book review). *Evolution* 49: 1302–1306.

Leigh, E. G. Jr., and T. E. Rowell. 1995. The evolution of mutualism and other forms of harmony at various levels of biological organization. *Écologie* 26: 131–158.

Leighton, M., and D. R. Leighton. 1983. Vertebrate responses to fruiting seasonality within a Bornean rain forest, pp. 181–196. In S. L. Sutton, T. C. Whitmore, and A. C. Chadwick, eds., *Tropical Rain Forest: Ecology and Management*. Blackwell Scientific, Oxford.

Lewontin, R. C. 1970. The units of selection. *Annual Review of Ecology and Systematics* 1: 1–18.

Lorence, D. H., and R. W. Sussman. 1988. Diversity, density and invasion in a Mauritian wet forest, pp. 187–204. In P. Goldblatt and P. P. Lowry II, eds., *Modern Systematic Studies in African Botany*. Missouri Botanical Garden, St. Louis, MO.

Lotka, A. J. 1925. *Elements of Physical Biology*. Williams and Wilkins, Baltimore, MD.

Loyn, R. H. 1987. The bird that farms the dell. *Natural History* 87/6: 54–60.

Lyttle, T. W. 1977. Experimental population genetics of meiotic drive systems I. Pseudo-Y chromosomal drive as a means of eliminating cage populations of *Drosophila melanogaster*. *Genetics* 86: 413–445.

Lyttle, T. W. 1979. Experimental population genetics of meiotic drive systems II. Accumulation of genetic modifiers of segregation distorter (*SD*) in laboratory populations. *Genetics* 91: 339–357.

McDonald, D. B., and W. K. Potts. 1994. Cooperative display and relatedness among males in a lek-mating bird. *Science* 266: 1030–1032.

Macdonald, I. A. W., C. Thébaud, W. A. Strahm, and D. Strasberg. 1991. Effects of alien plant invasions on native vegetation remnants on La Réunion (Mascarene Islands, Indian Ocean). *Environmental Conservation* 18: 51–61.

McKey, D. 1984. Interaction of the ant-plant *Leonardoxa africana* (Caesalpiniaceae) with its obligate inhabitants in a rainforest in Cameroon. *Biotropica* 16: 81–99.

Maheswaran, J., and I. A. U. N. Gunatilleke. 1988. Litter decomposition in a lowland rain forest and a deforested area in Sri Lanka. *Biotropica* 20: 90–99.

Margulis, L. 1993. *Symbiosis in Cell Evolution*. W. H. Freeman, San Francisco.

Maynard Smith, J. 1978. *The Evolution of Sex*. Cambridge University Press, Cambridge.

Maynard Smith, J., and E. Szathmáry. 1995. *The Major Transitions in Evolution*. Freeman/Spektrum, Oxford.

Milton, K. 1980. *The Foraging Strategy of Howler Monkeys*. Columbia University Press, New York.

Milton, K., D. M. Windsor, D. W. Morrison, and M. A. Estribi. 1982. Fruiting phenologies of two Neotropical *Ficus* species. *Ecology* 63: 752–762.

Morrison, D. W. 1978a. Foraging behavior and energetics of the frugivorous bat *Artibeus jamaicensis*. *Ecology* 59: 716–723.

Morrison, D. W. 1978b. On the optimal searching strategy for refuging predators. *American Naturalist* 112: 925–934.

Moynihan, M. H. 1962. The organization and probable evolution of some mixed species flocks of Neotropical birds. *Smithsonian Miscellaneous Collections* 143 (7): 1–140.

Moynihan, M. H. 1968. Social mimicry: character convergence versus character displacement. *Evolution* 22: 315–331.

Moynihan, M. H. 1979. *Geographic Variation in Social Behavior and in Adaptations to Competition among Andean Birds*. Nuttall Ornithological Club, Cambridge, MA.

Munn, C. A. 1985. Permanent canopy and understory flocks in Amazonia: species composition and population density, pp. 683–712. In P. A. Buckley, M. S. Foster, E. S. Morton, R. S. Ridgely, and F. G. Buckley, eds., *Neotropical Ornithology*. Ornithological Monographs no. 36. American Ornithologists' Union, Washington, DC.

Nadkarni, N. M. 1981. Canopy roots: convergent evolution in rainforest nutrient cycles. *Science* 214: 1023–1024.

Nason, J. D., E. A. Herre, and J. L. Hamrick. 1996. Paternity analysis of the breeding structure of strangler fig populations: evidence for substantial long-distance wasp dispersal. *Journal of Biogeography* 23: 501–512.

Nepstad, D. C., C. R. de Carvalho, E. A. Davidson, P. H. Jipp, P. A. Lefebvre et al. 1994. The role of deep roots in the hydrological and carbon cycles of Amazonian forests and pastures. *Nature* 372: 666–669.

Nixon, K. C., and W. L. Crepet. 1993. Late Cretaceous fossil flowers of Ericalean affinity. *American Journal of Botany* 80: 616–623.

Nowak, M. A., and K. Sigmund. 1992. Tit for tat in heterogeneous populations. *Nature* 355: 250–253.

Nowak, M., and K. Sigmund. 1993. A strategy of win-stay, lose-shift that outperforms tit-for-tat in the Prisoner's Dilemma game. *Nature* 364: 56–58.

Odum, H. T. 1971a. *Environment, Power and Society*. Wiley, New York.

Odum, E. P. 1971b. *Fundamentals of Ecology*. W. B. Saunders, Philadelphia.

Patiño, S., E. A. Herre, and M. T. Tyree. 1994. Physiological determinants of *Ficus* fruit temperature and implications for survival of pollinator wasp species: comparative physi-

ology through an energy budget approach. *Oecologia* 100: 13–20.
Pellmyr, O., and C. J. Huth. 1994. Evolutionary stability of mutualism between yuccas and yucca moths. *Nature* 372: 257–260.
Pellmyr, O., and J. N. Thompson. 1992. Multiple occurrences of mutualism in the yucca moth lineage. *Proceedings of the National Academy of Sciences, USA* 89: 2927–2929.
Pittendrigh, C. S. 1958. Adaptation, natural selection and behavior, pp. 390–416. In A. Roe and G. G. Simpson, eds., *Behavior and Evolution*. Yale University Press, New Haven, CT.
Price, G. R. 1970. Selection and covariance. *Nature* 227: 520–521.
Price, G. R. 1972. Extension of covariance selection mathematics. *Annals of Human Genetics* 35: 485–490.
Prout, T., J. Bundgaard, and S. Bryant. 1973. Population genetics of modifiers of meiotic drive I. The solution of a special case and some general implications. *Theoretical Population Biology* 4: 446–465.
Putz, F. E. 1984. The natural history of lianas on Barro Colorado Island, Panama. *Ecology* 65: 1713–1724.
Ratnieks, F. L. W. 1993. Egg-laying, egg-removal and ovary development by workers in queenright honey bee colonies. *Behavioral Ecology and Sociobiology* 32: 191–198.
Ratnieks, F. L. W., and P. K. Visscher. 1989. Worker policing in the honeybee. *Nature* 342: 796–797.
Redford, K. H. 1993. Hunting in Neotropical forests: a subsidy from Nature, pp. 227–246. In C. M. Hladik et al., eds., *Tropical Forests, People and Food*. UNESCO, Paris, and Parthenon Publishing, Park Ridge, NJ.
Regal, P. J. 1977. Ecology and evolution of flowering plant dominance. *Science* 196: 622–629.
Robinson, G. E. 1992. Regulation of division of labor in insect societies. *Annual Reviews of Entomology* 37: 637–665.
Robinson, G. E., R. E. Page Jr., C. Strambi, and A. Strambi. 1989. Hormonal and genetic control of behavioral integration in honey bee colonies. *Science* 246: 119–112.
Robinson, M. H. 1969. Defenses against visually hunting predators. *Evolutionary Biology* 3: 225–259.
Rowan, R., N. Knowlton, A. Baker, and J. Jara. 1997. Landscape ecology of algal symbionts creates variation in episodes of coral bleaching. *Nature* 388: 265–279.
Rowan, R., and D. A. Powers. 1991. A molecular genetic classification of zooxanthellae and the evolution of animal-algal symbioses. *Science* 251: 1348–1351.
Russell, A. E., and P. M. Vitousek. 1997. Decomposition and potential nitrogen fixation in *Dicranopteris linearis* litter on Mauna Loa, Hawai'i. *Journal of Tropical Ecology* 13: 579–594.
Russell, J. K. 1982. Timing of reproduction by coatis (*Nasua narica*) in relation to fluctuations in food resources, pp. 413–431. In E. G. Leigh Jr., A. S. Rand, and D. M. Windsor, eds., *The Ecology of a Tropical Forest*, Smithsonian Institution Press, Washington, DC.
Sapp, J. 1994. *Evolution by Association*. Oxford University Press, New York.
Schlesinger, W. H. 1991. *Biogeochemistry: An Analysis of Global Change*. Academic Press, San Diego, CA.
Schupp, E. W. 1986. *Azteca* protection of *Cecropia*: ant occupation benefits juvenile trees. *Oecologia* 70: 379–385.
Schupp, E. W., and D. H. Feener Jr. 1991. Phylogeny, lifeform, and habitat dependence of ant-defended plants in a Panamanian forest, pp. 177–197. In C. R. Huxley and D. F. Cutler, eds., *Ant-Plant Interactions*. Oxford University Press, Oxford.
Seeley, T. D. 1989a. The honey bee colony as a superorganism. *American Scientist* 77: 546–553.
Seeley, T. D. 1989b. Social foraging in honey bees: how nectar foragers assess their colony's nutritional status. *Behavioral Ecology and Sociobiology* 24: 181–199.
Seeley, T. D. 1994. Honey bee foragers as sensory units of their colonies. *Behavioral Ecology and Sociobiology* 34: 51–62.
Seeley, T. D., S. Camazine, and J. Sneyd. 1991. Collective decision-making in honey bees: how colonies choose among nectar sources. *Behavioral Ecology and Sociobiology* 28: 277–290.
Seeley, T. D., and R. A. Levien. 1987. A colony of mind. *The Sciences* 27: 28–43.
Slatkin, M., and M. J. Wade. 1978. Group selection on a quantitative character. *Proceedings of the National Academy of Sciences, USA* 75: 3531–3534.
Slatkin, M., and D. S. Wilson. 1979. Coevolution in structured demes. *Proceedings of the National Academy of Sciences, USA* 76: 2084–2087.
Smith, A. 1776. *An Enquiry into the Nature and Causes of the Wealth of Nations*. W. Strahan and T. Cadell, London.
Smith, A. 1790. *The Theory of Moral Sentiments*. W. Strahan and T. Cadell, London.
Smythe, N. 1977. The function of mammalian alarm advertising: social signals or pursuit invitation? *American Naturalist* 111: 191–194.
Strasberg, D. 1995. Processus d'invasion par les plantes introduites à La Réunion et la dynamique de la végétation sur les coulées volcaniques. *Écologie* 26: 169–180.
Sussman, R. W. 1991. Primate origins and the evolution of angiosperms. *American Journal of Primatology* 23: 209–223.
Terborgh, J. 1986. Keystone plant resources in the tropical forest, pp. 330–344. In M. E. Soulé, ed., *Conservation Biology*. Sinauer Associates, Sunderland, MA.
Thornton, I. 1996. *Krakatau*. Harvard University Press, Cambridge, MA.
Tiffney, B. H. 1984. Seed size, dispersal syndromes, and the rise of the angiosperms: evidence and hypothesis. *Annals of the Missouri Botanical Garden* 71: 551–576.
Tiffney, B. H., and S. J. Mazer. 1995. Angiosperm growth habit, dispersal and diversification reconsidered. *Evolutionary Ecology* 9: 93–117.
Tilman, D. 1988. *Plant Strategies and the Dynamics and Structure of Plant Communities*. Princeton University Press, Princeton NJ.
Todzia, C. 1986. Growth habits, host tree species, and density of hemiepiphytes on Barro Colorado Island, Panama. *Biotropica* 18: 22–27.
Trench, R. K. 1987. Dinoflagellates in non-parasitic symbioses, pp. 530–570. In F. J. R. Taylor, ed., *The Biology of Dinoflagellates*. Blackwell Scientific, Oxford.
Treseder, K. K., D. W. Davidson, and J. R. Ehleringer. 1995. Absorption of ant-provided carbon dioxide and nitrogen by a tropical epiphyte. *Nature* 375: 137–139.
Trumbore, S. E., E. A. Davidson, P. B. de Camargo, D. C. Nepstad, and L. A. Martinelli. 1995. Belowground cycling of carbon in forests and pastures of eastern Amazonia. *Global Biogeochemical Cycles* 9: 515–528.
von Neumann, J., and O. Morgenstern. 1944. *Theory of Games and Economic Behavior*. Princeton University Press, Princeton, NJ.
Waring, R. H., and W. H. Schlesinger. 1985. *Forest Ecosystems: Concepts and Management*. Academic Press, Orlando, FL.
West, S. A., and E. A. Herre. 1994. The ecology of the New World fig-parasitizing wasps *Idarnes* and implications for the evolution of the fig-pollinator mutualism. *Proceedings of the Royal Society of London* B 258: 67–72.
West-Eberhard, M. J. 1975. The evolution of social behavior by kin selection. *Quarterly Review of Biology* 50: 1–33.

West-Eberhard, M. J. 1981. Intragroup selection and the evolution of insect societies, pp. 3–17. In R. D. Alexander and D. W. Tinkle, eds., *Natural Selection and Social Behavior: Recent Research and New Theory*. Chiron Press, New York.

West-Eberhard, M. J. 1987. Flexible strategy and social evolution, pp. 35–51. In Y. Itô, J. L. Brown, and J. Kikkawa, eds., *Animal Societies: Theories and Facts*. Japan Science Society Press, Tokyo.

Williams, G. C. 1966. *Adaptation and Natural Selection*. Princeton University Press, Princeton, NJ.

Williams, G. C. 1975. *Sex and Evolution*. Princeton University Press, Princeton, NJ.

Williams, G. C. 1985. A defense of reductionism in evolutionary biology. *Oxford Surveys in Evolutionary Biology* 2: 1–27.

Williams, G. C. 1992. *Natural Selection*. Oxford University Press, New York.

Williams, K. S., and L. E. Gilbert. 1981. Insects as selective agents on plant vegetative morphology: egg mimicry reduces egg laying by butterflies. *Science* 212: 467–469.

Wilson, D. S. 1975. A theory of group selection. *Proceedings of the National Academy of Sciences, USA* 72: 143–146.

Wilson, D. S. 1980. *The Natural Selection of Populations and Communities*. Benjamin Cummings, Menlo Park, CA.

Windsor, D. M., D. W. Morrison, M. A. Estribi, and B. de Leon. 1989. Phenology of fruit and leaf production by 'strangler' figs on Barro Colorado Island, Panamá. *Experientia* 45: 647–653.

Wing, S. L., L. J. Hickey, and C. C. Swisher. 1993. Implications of an exceptional fossil flora for late Cretaceous vegetation. *Nature* 363: 342–344.

Wright, S. 1931. Evolution in Mendelian populations. *Genetics* 16: 97–159.

Wynne-Edwards, V. C. 1963. Intergroup selection in the evolution of social systems. *Nature* 200: 623–626.

Young, J. P. W., and A. W. B. Johnston. 1989. The evolution of specificity in the legume-rhizobium symbiosis. *Trends in Ecology and Evolution* 4: 341–349.

Zeh, D. W., and J. A. Zeh. 1991. Dispersal-generated sexual selection in a beetle-riding pseudoscorpion. *Behavioral Ecology and Sociobiology* 30: 135–142.

Zeh, D. W., and J. A. Zeh. 1992. Emergence of a giant fly triggers phoretic dispersal in the neotropical pseudoscorpion, *Semeiochernes armiger* (Balzan) (Pseudoscorpionida: Chernetidae). *Bulletin of the British Arachnological Society* 9: 43–46.

Zotz, G., G. Harris, M. Königer, and K. Winter. 1995. High rates of photosynthesis in the tropical pioneer tree, *Ficus insipida* Willd. *Flora* 190: 265–272.

TEN

The Rainforest Endangered

This story began in Panama so that our knowledge of Barro Colorado Island could be fashioned into a lens through which to view the tropical rainforests, seasonal and aseasonal, of the world. It ends in Madagascar. Dry season is drawing to a close. A Malagasy graduate student, three Malagasy botanical technicians, a chauffeur, and I are in a bright new Toyota Land Cruiser, jolting over a rough laterite track from the forestry station where we sleep, toward the Forêt d'Ambohitantely, 8 km and 30 minutes away. This forest, with roughly the same acreage as Barro Colorado Island, is one of the largest tracts left of the "plateau forest" that used to cover perhaps a quarter of Madagascar.

We are crossing a rolling grassland, with a few scattered, ragged patches of pine and *Eucalyptus*, remnants of French attempts at reforestation, and rare fragments of the original forest, a gray-green, variegated canopy about 12 m tall, sheltering from dry season fires in draws and ravines. The land is still smoking from a fire which reddened the skyline above our quarters the preceding Saturday and Sunday nights. We were protected from this fire by the main road into the forest reserve, which fortunately proved an effective firebreak. The landscape we are crossing is lonely and barren at the best of times, but it is normally possessed of a certain spacious magnificence which inhabitants of eastern Wyoming would recognize at once. During this visit, I often went to admire the view from the edge of this forest, on the brow of a hill. A narrow path descended this hill to lose itself among its neighboring hills. Once I saw three people walking this path, leading a calf and carrying various small bundles, presumably going to a market 40 km farther on; who knows how far they had already walked through this loneliness? Now, however, the landscape we are crossing is black and smoking, dreary beyond words.

Suddenly, from the car's cassette player bursts forth the magnificent strains of the Kyrie eleison from Mozart's C minor mass: 20 minutes of pleading for the Lord to have mercy, the only Greek left in the Latin liturgy of Rome. Indeed, a great deal was crying out for forgiveness: the destroyed forests, which had been sources of water, wood, medicines, and meat to the people round about; this barren grassland which has replaced the forests, whose fires prohibit any other form of vegetation; a recent decree of foreign moneymen that the local currency must "float," which halved the value of my collaborator's salary in 6 short weeks; and the obscene facts that the car carrying us to the forest costs 10 times the pay of all three technicians combined, and that these technicians can only reach their forests if a foreigner brings money from over the sea.

The replacement of luxuriant forest by wasteland in a context of social injustice has counterparts all through the "third world." The agents of forest destruction are sometimes peasants who must clear land to feed their families, sometimes loggers or agribusinessmen, granted concessions or incentives by money-hungry governments struggling with an appalling debt crisis, none of whom have any time for the politically and economically unimportant aboriginals or tribals who live there. All too often, the contest is not conservation versus human well-being, but conservation versus an arrant waste, reflecting, at best, the victory of financial benefits for distant city-dwellers over the welfare of aboriginals and other country people.

Often, local governments are given little choice in the matter. Georges Bernanos (1946) warned the people of Brazil, in what many might now find prophetic words, about first-world advisors and moneymen who "confuse producing with creating":

> You create with time, and they neither respect nor love time: they consider it a merchandise like any other. They lack patience, that is to say, they will never understand the great achievements of life, because life is all patience. They do not know the patience of life, and they disdain yours, because in their grossness of heart they

mistake it for a sort of lazy resignation. They find you very slow to "exploit" your land, that is to say, slow to remake it in their image and resemblance. They find you slow to exploit your land, slow to open it up to them, slow to present it to them gaping, ready for use. They hear something stirring underground, they don't know what, and they dream only of possessing this thing right away, dead or alive. They measure the thickness of your mountains, the slope of your rivers, the depth of your valleys; they calculate what it will cost in marks, florins, yen, rubles, piasters, pounds or dollars to remove all obstacles and bring you happiness in the only form they know it: merchandise. They only forget that although mountains can be pierced, rivers rerouted, valleys blocked or filled, no force in the world can change a people's nature, its intimate, traditional, familial view of life, in sum, its idea of happiness. (Bernanos 1984: p. 9, my translation)

Indeed, tropical forest is a victim of a spiritual crisis in humanity. The destruction of tropical forest reflects an all-too-widespread willingness to submit to an economic determinism every bit as rigid as that of Marx. People pay homage to this economic determinism every time they insist that conservation be argued solely on the basis of the future economic benefits it will bring. We are assured that beauty has a place in the realm of economic determinism: we can decide how much we will pay for beauty, as we do for any other merchandise. This remark is hollow, meaningless. A rainforest tribe might place great store by its ability to occupy and preserve the forest where it has lived for generations, the framework of its life and culture. But if a government or large business firm wants the forest, for the sake either of the timber above ground or the gold or copper below, this tribe would be unable to keep their home however much they might love it and depend on it, for they have no money, and the use they make of the forest only generates money by endangering the tribe's balance with their environment.

Economic determinism is a facet of a spiritual crisis because the freedom to discern and choose the right that is now being sacrificed on the altar of economic determinism is an aspect of the human spirit. Although White (1968) declared that Genesis 1:28's notorious command to "fill the earth and subdue it, and have dominion . . . over every living thing that moves upon the earth" lies at the root of our ecological crisis, Genesis 2:15 proclaims that "the Lord God took the man and put him in the garden of Eden to till it and take care of it." In the eyes of the world, dominion and stewardship are very different things, and the commands of Genesis 1 and 2 are accordingly in blatant contradiction. Nevertheless, according to that faith which was the cradle of western civilization, Christ himself identified lordship, dominion, as a form of stewardship and service, not only in a saying or two, but in the very fabric of his life.

The idea that humanity is responsible for the rest of creation is a part of Christian faith. The Jew who composed Psalm 104 already recognized that plants and animals have intrinsic value as living praises of God, quite apart from their usefulness to ourselves (Lewis 1958: pp. 83–85). St. Francis of Assisi did not found a new religion: he brought Christianity back to itself, quite a different business. Similarly, Muslims believe that a primary duty of humanity is to intercede with God on behalf of the rest of creation (Nasr 1979: pp. 23–24). That Hindus find the Christians they meet shockingly careless of nature reflects a fault in Christian practice, not a flaw in the faith: love of the whole creation should be a bond in common among Jews, Christians, Muslims and Hindus.

Luther has been accused of reducing "the history of the cosmos to the drama of the individual soul's salvation" (Gilson 1991: p. 244). Yet the same tendency is manifest in the modern Catholic view that each new human soul is considered such a good in itself that restricting population growth—so essential to the health of all creation—borders on the sinful. The Biblical view that the whole of creation is God's work, and that humanity is responsible for all of it, is now in great peril.

The rise of economic determinism, conjoined with that of human self-centeredness, which so endangers the tropical forest and its peoples, matters to all of us, believers and unbelievers alike. Setting at nought the peoples of the rainforest because of their lack of economic clout hurts us all, morally and intellectually. There is no such thing as a noble savage, and forest tribes are no better at foreseeing the dangers of too many children than anyone else. Nonetheless, their feel for the plants and animals of the forest does allow them to live more nearly in harmony with their environment than outsiders ever could (Malhotra 1993, Morán 1993, Posey 1993), and there is much to be learned from them. The greatest lesson we can learn from them is that humans do not, indeed cannot, live by bread alone. Bernanos (1953: pp. 91, 100, 111) warned time and again that modern capitalist materialism is based on a false idea of what it is to be human. This fact becomes most painfully obvious when the world economy impinges on rainforest peoples (Grenand 1993, Posey 1993, Venkatesan 1993, Hladik et al. 1993).

Nonetheless, the rest of us are also at risk. Perhaps we can best comprehend the threat posed by economic determinism by considering its effects on science. Everyone recognizes that our society depends on science for solutions, not only to today's problems, but to future problems we may not even have imagined. Some, including myself, find this dependence dangerous, but this dependence is no less a fact for all that. What few stop to think about is how many depend on science to acquaint them with the wonders of nature. Television serials on tropical nature, like those on the vast wonders of the universe, are immensely popular. Science is a triumph of the human spirit. Einstein is our century's answer to Shakespeare, because Einstein, like Shakespeare, was willing to bet on beauty and truth. One wonders whether television serials on science are the nearest many people come to what Elizabethan Londoners were offered by the theater of their day.

Many have announced that science leaves no place for religious faith, Christian or otherwise (Monod 1970). Nonetheless, the health of science presupposes assumptions that were quite natural to medieval Christianity, but very much at risk in the current climate of economic determinism. After all, the science we know took root in medieval Europe from the triple conviction that nature is orderly, its order is accessible to human minds, and that, as God's work, nature is worth studying (Gilson 1936: 1991 pp. 242–243). Study of nature enables its practitioners to praise their Lord and Creator more intelligently and to serve their neighbor more effectively. What will become of science and society if belief in the fundamental presuppositions of scientific endeavor is swept away in the reaction against the Christian world view?

Make no mistake. These presuppositions *are* in peril. Many now find studying nature for its own or its Creator's sake too frivolous for serious attention. Even scientists have joined in the hue and cry against "curiosity-driven research." With so many human needs crying for attention, especially in the tropics, how could anyone be so brazen as to ask for money to study such topics as sex change in fish, or sex ratios among the offspring of fig-pollinating wasps?

Those who insist that science be directly relevant to "real human needs" implicitly assume, however, that we already know *all* the kinds of things we need to learn. This isn't true. In science, as in other realms, it often happens that "the stone that the builders rejected has become the chief cornerstone" (Ps 118:22). On Barro Colorado, a hectare of tropical forest may support 300 tons dry weight of vegetation because 300 g (fresh weight) of small birds eat the insects that would otherwise defoliate the trees on that hectare (see chapter 7). That forest may harbor so many kinds of trees because agoutis bury the seeds of many of them, protecting them from insect attack (see chapter 8). Comparing the nematodes parasitizing the wasps that pollinate different species of fig tree reveals factors that influence the virulence of parasites in general (Herre 1993). The rules of sex change in fish confirm in a convincing manner that evolution is driven by natural selection within populations (Leigh et al. 1985), which implies that cooperation only evolves when it serves the advantage of all parties involved (Leigh and Rowell 1995). The relation of the sex ratio among the offspring of wasps pollinating a fig fruit to the number of wasps laying eggs in that fruit shows how insect pests, by favoring the dispersal of seeds far from their parent trees, may favor the luxuriance of tropical forest (Leigh 1994). None of these phenomena claimed the attention of those who have time only for the relevant or the important. They were discovered by people who studied tropical forest because they loved it and were fascinated by it.

Indeed, the fact that what we know of the organizing principles of tropical forest hinges so largely on the studies of just such "trivia" is a source of great joy to the Taoist who has rejoiced with Chuang Tzu in the "usefulness of the useless," the Christian who rejoices in the radical idiosyncrasy of a universe which far surpasses anything that humans could ever imagine, and the atheistic biologist who has devoted her life to the study of tropical forest because it is quite the most gorgeous and wonderful thing she has ever seen. It is a strange business to consider how very many people's welfare depends on the communication of the wondrousness of tropical forest to a wider public, in tropical countries and in the world at large. For our future does depend on the welfare of the rest of creation, and the integrity of creation will be restored only if love for it becomes sufficiently widespread to motivate its restoration.

References

Bernanos, G. 1946. *Lettre aux Anglais*. Gallimard, Paris.
Bernanos, G. 1953. *La Liberté, Pour Quoi Faire?* Gallimard, Paris.
Bernanos, G. 1984. *Lettre aux Anglais*. Éditions du Seuil, Paris.
Gilson, E. 1936. *The Spirit of Medieval Philosophy*. Charles Scribner's Sons, New York.
Gilson, E. 1991. *The Spirit of Medieval Philosophy*. University of Notre Dame Press, Notre Dame, IN.
Grenand, F. 1993. Bitter manioc in the lowlands of tropical America: from myth to commercialization, pp. 447–462. In C. M. Hladik, A. Hladik, O. F. Linares, H. Pagezy, A. Semple, and M. Hadley, eds., *Tropical Forests, People and Food*. UNESCO, Paris, and Parthenon Publishing, Park Ridge, NJ.
Herre, E. A. 1993. Population structure and the evolution of virulence in nematode parasites of fig wasps. *Science* 259: 1442–1445.
Hladik, C. M., A. Hladik, O. F. Linares, H. Pagezy, A. Semple, and M. Hadley (eds). 1993. *Tropical Forests, People and Food*. UNESCO, Paris, and Parthenon Publishing, Park Ridge, NJ.
Leigh, E. G. Jr. 1994. Do insect pests promote mutualism among tropical trees? *Journal of Ecology* 82: 677–680.
Leigh, E. G. Jr., E. A. Herre, and E. A. Fischer. 1985. Sex allocation in animals. *Experientia* 41: 1265–1276.
Leigh, E. G. Jr., and T. E. Rowell. 1995. The evolution of mutualism and other forms of harmony at various levels of biological organization. *Écologie* 26: 131–158.
Lewis, C. S. 1958. *Reflections on the Psalms*. Harcourt Brace Jovanovich, New York.
Malhotra, K. C. 1993. People, biodiversity and regenerating tropical sal (*Shorea robusta*) forests in West Bengal, India, pp. 745–752. In C. M. Hladik, A. Hladik, O. F. Linares, H. Pagezy, A. Semple, and M. Hadley, eds., *Tropical Forests, People and Food*. UNESCO, Paris, and Parthenon Publishing, Park Ridge, NJ.
Monod, J. 1970. *Le Hasard et la Nécessité*. Éditions du Seuil, Paris.
Morán, E. F. 1993. Managing Amazonian variability with indigenous knowledge, pp. 753–766. In C. M. Hladik, A. Hladik, O. F. Linares, H. Pagezy, A. Semple, and M. Hadley, eds., *Tropical Forests, People and Food*. UNESCO, Paris, and Parthenon Publishing, Park Ridge, NJ.

Nasr, S. H. 1979. *Ideals and Realities of Islam.* Allen and Unwin, London.

Posey, D. A. 1993. The importance of semi-domesticated species in post-contact Amazonia: effects of the Kayapó Indians on the dispersal of flora and fauna, pp. 63–71. In C. M. Hladik, A. Hladik, O. F. Linares, H. Pagezy, A. Semple, and M. Hadley, eds., *Tropical Forests, People and Food.* UNESCO, Paris, and Parthenon Publishing, Park Ridge, NJ.

Venkatesan, D. 1993. Ecology, food and nutrition: the Onge foragers of the Andaman tropical forest, pp. 505–514. In C. M. Hladik, A. Hladik, O. F. Linares, H. Pagezy, A. Semple, and M. Hadley, eds., *Tropical Forests, People and Food.* UNESCO, Paris, and Parthenon Publishing, Park Ridge, NJ.

White, L. Jr. 1968. The historical roots of our ecologic crisis, pp. 75–94. In L. White, Jr., *Dynamo and Virgin Reconsidered.* MIT Press, Cambridge, MA.

Name Index

Abelard, Peter, 108
Abramsky, A., 161
Absy, M. L., 9
Adler, G. H., 28–29, 160
Aide, T. M., 22, 190
Appanah, S., 152, 212
Aristotle, 212
Augspurger, C. K., 192
Axelrod, R., 215

Ball, M. C., 110
Barat, C., 48, 49–50, 51
Barghoorn, E. S., 9
Bartlett, A. S., 9
Basset, Y., 191
Beatty, R. P., 28–29
Beck, J. W., 9
Bernanos, Georges, 232–233
Bernays, E., 191
Berner, R. A., 59
Blanc, P., 95
Bock, B. C., 33
Borchert, R., 153, 155
Bossel, H., 135
Bourke, A. F. G., 217
Brokaw, N. V. L., 188
Brown, M. J., 124
Bruenig, E. F., 105, 211–222
Bruijnzeel, L. A., 51, 63, 68–69
Brunig, E. F., 55, 106
Budd, A. F., 10
Bugbee, B. B., 130

Cavelier, J., 112
Chapin, F. S, 123
Charles-Dominique, P., 31, 160, 161
Chesson, P., 189–190
Chuang Tzu, 234
Clinebell, R. R., 182
Coley, P. D., 123, 164, 166, 167, 190

Colinvaux, P., 9
Condit, R., 10, 182, 190, 192, 193, 194
Connell, J. H., 161, 190, 192
Cook, M. T., 108
Corner, E. J. H., 26, 82, 84, 212, 219, 223
Crossley, D. A., Jr., 164

Darwin, C. R., 26, 187, 213, 219
Dawkins, R., 220
Dietrich, W. E., 55, 60, 71
Dobzhansky, T., 189

Eberhard, W. G., 35
Ehrlich, P. R., 192
Einstein, A., 233
Emmons, L. H., 159, 160, 161
Enders, R. K., 157
Erwin, T. L., 195

Fabre, J.-H., 191
Fan, S.-M., 128, 132
Fisher, R. A., 179, 181–182, 185–186, 187, 195, 212
Fittkau, E. J., 156, 164
Forget, P.-M., 16, 20, 28, 151, 152, 185, 211
Foster, R. B., 15, 149–152, 183, 187, 189, 192, 194
Franken, W., 48, 51
Franks, N. R., 217

Garwood, N. C., 151
Gautier, J. P., 160, 161
Gautier-Hion, A., 160, 161, 162
Gentry, A. H., 182
Germán-Heins, J., 76
Gillett, J. B., 190, 192, 194
Givnish, T. J., 100, 102–103, 108, 111, 136

Glanz, W. E., 31, 157
Gleason, H. A., 108
Gliwicz, J., 28
Goldstein, G., 112
Grace, J., 129
Graham, M., 191
Grier, C. C., 128, 129
Guilderson, T. P., 9

Hairston, N. G., 158
Hallé, F., 82, 84, 85
Hardin, G., 214
Harrington, R. A., 130
Hastings, A., 196
Herre, Edward Allen, 15, 26, 222, 225, 226
Herrera, R., 123
Hertzler, R. A., 67
Heuveldop, J., 51
Holbrook, N. M., 154
Hopkins, E. J., 63
Horn, H. S., 93–94
Howe, H. F., 21
Hubbell, S. P., 187, 189, 192, 194

Jackson, J. B. C., 10
Janson, C. H., 159
Janzen, D. H., 10, 15, 123, 190, 191, 192, 194
Jordan, C. F., 51

Karr, J. R., 167
Keyes, M. R., 129
King, D. A., 90–93, 100–101, 113, 131, 136, 156–157
Kira, T., 127, 128
Kitajima, K., 150, 158, 188
Klinge, H., 123, 156, 164
Krieger, H., 135
Kursar, T., 132, 164, 166

Leigh, E. G., Jr., 3, 85, 86, 111–112, 120, 165–166, 214, 225
Leopoldo, P. R., 48, 51
Lettau, H. H., 63
Lieberman, M., 188
Lloyd, C. R., 51
Lloyd, J., 129
Logan, R. S., 128
Long, J. N., 90
Lotka, A. J., 220
Loucks, O. L., 90, 113
Lovelock, J. E., 158–159
Lowman, M. D., 166–167
Lyttle, T. W., 216

MacArthur, R. H., 48, 187–188
Manokaran, N., 51
Marques Filho, A. de O., 51, 60, 61, 63
Martin, P. S., 10
Maynard Smith, J., 213
Medina, E., 132
Meinzer, F. C., 55, 106–107, 154
Michaloud, G., 160
Milton, K., 30, 160, 161
Monteith, J. L., 54
Montgomery, G. G., 31
Moran, V. C., 164
Morgenstern, O., 219
Mosley, M. P., 49
Moynihan, Martin, 11, 15, 35
Mulkey, S. S., 150
Murali, K. S., 195

Nagy, K. A., 30, 31, 32, 33, 156, 157, 159, 167
Nason, J., 26
Nepstad, D. C., 53
Nieuwolt, S., 63
Nortcliff, S., 72

Odum, H. T., 127–128, 220
Oldeman, A. A., 82, 84, 85
Oppenheimer, J. R., 30

Paine, R. T., 130
Parker, G. G., 124

Penman, H. L., 53–54, 55, 59, 60, 63
Piperno, D. R., 9, 10, 11, 184
Pockman, W. T., 154
Pócs, T., 112
Putz, F. E., 121, 125, 134–135

Rancy, A., 10
Rand, A. S., 34, 161
Raven, P. H., 192
Ray, C., 196
Reich, P. B., 153, 155
Reichle, D. E., 166
Retallack, G. J., 76
Ribeiro, M. de N. G., 60, 61, 63
Ridley, H. N., 190
Robinson, Douglas, 167
Robinson, Scott, 167
Rosenzweig, M. L., 161
Roubik, D. W., 37, 38
Rowell, T. E., 214, 225
Ryan, M. G., 135, 136
Ryan, M. J., 34

Saldarriaga, J. G., 183–184
Salisbury, F. B., 130
Schimper, A. F. W., 59, 77
Schoener, T. W., 161
Schowalter, T. D., 164, 165
Schupp, E. W., 194
Scott, J. C., 195
Seeley, T. M., 217
Sevenster, J. G., 196
Shuttleworth, W. J., 60, 61
Simpson, E. H., 180
Simpson, G. G., 8
Sinclair, A. R. E., 159
Slobodkin, L. B., 158
Smith, Adam, 213, 216, 220
Smith, F. E., 158
Smith, J. N. M., 161
Smythe, N., 8, 18, 27–28, 120, 149, 150, 151, 152, 160
Southwood, T. R. E., 164
Stallard, R. F., 70–71
Sternberg, L. S., 154

Sukumar, R., 195
Sunquist, M. E., 31
Szathmáry, E., 213

Terborgh, J., 94, 104, 158, 159, 162, 187, 223
Thornes, J. B., 72
Tilman, D., 123
Trumbore, S. E., 132
Tyree, M. T., 91, 154, 155, 189

Unsworth, M. H., 54

Van Alphen, J. J. M., 196
Van der Hammen, T., 9
Van Schaik, C. P., 162
Via, S., 191
Villa Nova, N. A., 60
Vis, M., 49, 50–51
Von Neumann, J., 219

Waring, R. H., 135
Webb, S. D., 8, 10
West Eberhard, M. J., 217, 218
White, L., Jr., 233
Williams, G. C., 214
Willis, C., 193
Willis, E. O., 35, 167
Windsor, D. M., 48, 61, 63, 166
Winter, K., 127, 133
Wirth, R., 36
Wofsy, S. C., 128
Wolda, H., 190, 192, 193, 196
Wolfe, J. A., 102, 152, 156
Wright, S. J., 3, 18, 150, 152, 153–154, 156–157, 166, 182
Wynne-Edwards, V. C., 214

Yamakura, T., 121

Zaret, T. M., 161
Zimmerman, J. K., 25
Zimmerman, U., 154
Zotz, G., 127, 133, 156

Subject Index

Abies spp.
   Abies amabilis, 134
   Abies balsamea, 158
Acalypha diversifolia, 10
Acid mor humus, 110
Acoustic emission, 155
Aerodynamic roughness of forest canopy, 55, 104–107
Africa, 8, 10
Agouti paca, 28
   creation of social pacas in one generation, 28
Agoutis. *See Dasyprocta punctata*
Agriculture, pre-Columbian, 11
Alfisols, 72, 73
Alouatta palliatta (Cebidae), 29–30
Alseis blackiana (Rubiaceae), 96
Amazonia
   animal biomass, 157
   convective thunderstorms, 47
   ectomycorrhizae, 131
   energy budget, 132, 133
   erosion rate, 51
   evapotranspiration, 53, 60
   forest respiration, 138, 139
   gross photosynthesis, 128, 129
   pan evaporation, 61, 63
   Pleistocene drought and vegetation change, 9
   rainfall, 48, 52–53
   rooting depth, 133–134, 211
   runoff, 48, 52–53
   soil respiration, 132, 211
   throughfall and stemflow, 51
Americas, the, 8
Anacardium excelsum (Anacardiaceae), 89, 150, 154
Animals
   agoutis, 8, 16, 18, 19, 20, 27–28, 185, 211
   bats, 8, 16, 31–32, 158, 211, 218, 222, 223
   blue duikers, 27
   coatis, 27, 29, 217
   dry season effects on, 47
   fiddler crabs, 34–35
   frogs, 34–35
   howler monkeys, 27, 28, 29–30
   iguanas, 33–34
   lizards, 34
   mammals of Barro Colorado, 8, 10, 11, 157, 172
   mammals of Manu, 157, 159, 164, 172
   pacas, 27, 28
   principal animals of Barro Colorado, 27–39
   regulation of animal populations, 156–168
   spiny rats, 27, 28–29
   squirrels, 8, 19
   three-toed sloths, 31
   unenforced mutualisms, 218
   white-faced monkeys, 30–31
   See also Birds; Insects
Anisogamy, 216
Anolis limifrons (Iguanidae), 34
Antarctica, 8
Antiherbivore defenses, 162–163, 188, 190–192, 212
Antirrhoea spp., 150
Antwren flocks, mixed, 32–33
Apis mellifera, 37
Artibeus spp. (Phyllostomidae)
   Artibeus jamaicensis, 31–32, 158
   Artibeus lituratus, 16
Aspidospermum cruenta, 192
Astrocaryum spp. (Palmae), 29, 31
   Astrocaryum paramaca, 185
   Astrocaryum standleyanum, 17–18
Atta spp. (Formicidae)
   Atta colombica, 36–37
   Atta sexdens, 37
Attines, 215
Australia, 8
   herbivory rates in *Eucalyptus* forests, 167
   leaf consumption by insect herbivores, 166–167
   marine herbivores, 158
   rainstorms and stemflow, 51–52

Balsam firs, 158
Bamboo, 155
Barro Colorado Island (BCI)
   climate, 3, 10, 11, 46–48, 52, 54–58, 61–63
   climate cycles of the Pleistocene, 8–11
   effects of human settlement, 10–12
   erosion, 70–74, 75
   forest dynamics, 121, 122, 136
   isthmus of Panama and the great biotic interchange, 5, 8
   origins of its biota, 5, 8
   principal animals of, 27–39
   principal plants of, 15–27
   seasonal rhythm of leaf flush, 150, 151, 152, 156, 163
   seasonal rhythm of fruit production and fruit fall, 150–154, 161, 163, 170–171
   the study area, 3, 4, 6–7
   tree diversity and its maintenance, 180, 181, 182, 184, 185, 186, 192–195
Barro Colorado Nature Monument, 3, 5, 90

Bats
- *Artibeus spp.*, feeding rate and population density, 31–32, 158
- as dispersers of fig seeds, 222

Biomass and productivity, 120–148
- biomass of the forest and its animals, 156, 162, 163, 164, 172
- carbon cycle, 131–133
- constants of tropical forest, 120–122, 123, 219–220
- energy budget of rainforest, 138–139
- gross production (total photosynthesis) of tropical forest, 125–130
- leaf area index and leaf biomass, 120–126, 134, 136, 167
- leaf fall and leaf production, 120–126, 133, 135, 139–142, 167
- leaves, 133
- partitioning of biomass in different forests, 123, 124, 125, 129, 134
- recovery of biomass after clearing, 123–125, 134
- roots, 133–134
- trade-off among stems, leaves, and roots, 130–131, 132
- what limits a forest's timber content, 135–138, 139
- wood, 134–135

Biotic adaptation, 75–76
Biphasic flow, 72
Birds
- of Barro Colorado, 8, 167–168
- food consumption by, 167, 168
- insect control by, 167–168
- mixed species bird flocks, 32–33, 218
- toucans as seed dispersers, 20

Black palm trees, 17–18
Boreal forest, 105
Borneo
- cost of tree growth, 136–137
- El Niño droughts in, 47
- Keystone fruit-bearing fig trees, 223

Bradypus variegatus (Bradypodidae), 31
Brazil (*See also* Amazonia)
- prolonged El Niño, 11–12

Butterfly diversity, 195

Cactus, epiphytic, 24–25
Calathea lutea (Marantaceae), 97
Calcium carbonate, 71, 72
California
- kelp's high production and frond area, 130
- redwood forests, 79, 108, 138

Calophyllum longifolium (Guttiferae), 89
Canavanine, 191
Capparis frondosa, 194
Carapa procera (Meliaceae), 185
Carboniferous period, late, and low CO_2, 59
Caribbean, during the Pleistocene, 9
Carica papaya, 84
Carnot "heat engines," 55
Castanopsis acuminatissima, 195
Casuarina nobilis, 76
Catasetum viridiflavum (Orchidaceae), 25–26
Cation exchange capacity, 76
Cebus capucinus (Cebidae), 30–31
Cecropia spp., 92, 94, 150
- *Cecropia insignis*, 188
- *Cecropia longipes*, 107
- *Cecropia membranacea*, 183
Cedrela odorata, 183
Ceiba pentandra, 154, 155–156
Cenozoic period, early, 212
Cephalophus monticola, 27
Choloepus hoffmanni, 31
Christianity and conservation, 233, 234
Chromosomes, 216
Chrysophyllum spp., 152
Chrysothemis friedrichsthaliana (Gesneriaceae), 96
Climate, tropical, 46–66
- Barro Colorado's, 3, 10, 11, 46–48, 52, 54–58, 61–63
- climate and vegetation type, 46
- evapotranspiration, 52–63
- hurricanes, 55–56, 74
- long-term climate cycles, 57
- Pleistocene changes in, 9–10, 56
- rainfall, 46–53

Cloud forests
- elfin forests and, 59, 107–112, 137
- evapotranspiration and, 59, 60, 111–112

Clusia uvitana, hydraulic architecture of, 155–156
Coatis, 27, 29, 217
Coendou rothschildii, 157
Cohesion theory of sap ascent, 154
Colombia
- during the Pleistocene, 9
- elfin forests, 108

Common good, common interest, and mutualism, 214–220, 223
Commonwealths, ecosystems as, 219–223
Community ecologists, 213
Conifers, 212
Convective thunderstorms, 47, 55
Cooperation. *See* Mutualism

Cordia bicolor (Boraginaceae), 89
Costa Rica
- animals and enforced mutualism, 218
- during the Pleistocene, 10
- elfin forests, 108
- flowering and fruiting, 152
- leaf arrangement, 100, 101, 104
- litter leaf size, 102, 103
- pests and plant defenses, 191
- tree architecture, 95
- tree diversity, 182, 188
- wood respiration, 136

Costa Rican Natural History (Janzen), 15
Costus spp. (Zingiberaceae)
- *Costus guanaiensis*, 98
- *Costus pulverulentus*, 98

Cretaceous period
- angiosperm evolution and herbivore escape, 212
- evolution of species-specific pollinators in, 184, 212
- plant geography and continental drift, 5, 8

Crop cultivation, humans and, 11
Cyclones, 56

Dacrydium comosum, 107
Dacryodes excelsa, 182
Dasyprocta punctata (Dasyproctidae), 27–28
- protecting seeds by burying them, 16, 18, 20, 27–28, 185, 186

Deforestation
- dangers to Panama Canal, 51
- rainfall and, 47–48
- runoff and, 68–70

Demography of
- *Anolis limifrons* (small lizard), 34
- *Artibeus jamaicensis* (bat), 32
- iguanas, 33–34

Dendroica pensylvanica, 32
Desmopsis panamensis, 100–101
Dichorisandra hexandra (Commelinaceae), 96
Dicranopteris linearis, 212
Didelphis marsupialis, 27
Didymopanax pittieri (Araliaceae), 92
Dimerandra emarginata, 25
Dinosaurs, 212
Dipterocarps, 211–212
Dipteryx panamensis (Leguminosae, Papilionoideae), 15–16, 18, 20, 29, 31, 90, 152, 211
Distorters, 216
Diversity
- recovery of tree diversity after disturbance, 183–185
- also see Tropical diversity

Diversity, maintenance of. *See* Species, coexistence of
"Doubly-labeled water" technique, 156
Douglas firs, 124, 127
Drip tips, 79
Drosophila species, coexistence of, 187, 196
Dry forest
  diversity of, 183
  production of, 129–130
Dwarfed forest, 107

Eciton burchelli (Formicidae), 35–36
Ecosystems
  and economics, 219–220
  ecosystems ecologists, 213
  harmonious, 220–223
Ectomycorrhizae, 131
Ecuador
  during the Pleistocene, 9
  tree diversity, 182
El Niño, 11–12, 47
  drought of 1982–1983, 10, 22, 34, 47, 121, 135, 152–153, 155, 188
  dry season of 1992, 53
  El Niño/Southern Oscillation event, 11, 47
Elfin forests, 59, 107–112, 120, 138
Energy budget
  of *Artibeus jamaicensis* (bat), 32
  of tropical forest (carbon cycle), 133, 138–139
  transport-limited, 72–73, 76
  weathering-limited, 71–72, 74–76
  See also Runoff, erosion, and soil formation
Eocene period, 5
Eperua grandiflora (Caesalpinoideae), 185
Epiphyllum phyllanthus (Cactaceae), 24–25
Epiphytes, 121, 211
Epiphytic cactus, 24–25
Epiphytic orchids, 25–26
Erosion
  of solutes, 71–75
  of suspended matter, 70–75
Erythrina spp.
  Erythrina costaricensis, 10
  Erythrina poeppigiana, 153
Eucalyptus forests, 232
  effect on runoff in Madagascar, 69
  herbivory rates in Australia, 167
Eugenia wrayi (Myrtaceae), 109
Eulaema cingulata, 26
Evapotranspiration, 52–53
  calculating, 59–60
  deforestation and, 69
  and evaporation, 61, 63
  factors affecting, 53–59

  in montane forests, 59
  similarity at different sites, 52–53, 120
  soil, vegetation and, 76
  water budgets, 62, 63
Evolutionary history, 212–214

Fagraea spp. (Loganiaceae), 112
Faramea occidentalis (Rubiaceae), 21–22, 31, 85, 152, 160, 193, 194
Ficus spp. (Moraceae), 26–27, 222–223
  efficient pollination of, 26–27, 222
  Ficus insipida, 26, 27, 32, 107, 183, 222–223
  Ficus obtusifolia, 26, 222
  Ficus yoponensis, 26
Fig trees, 26–27, 222–223
Fisher's alpha (diversity measure), 179–180, 204
Folivores, 29–30, 31, 33–34, 156, 157, 159, 162–163, 164–166
Food. *See* Seasonal rhythms of fruiting, leaf flush, and animal populations
Food consumption by
  army ants (*Eciton*), 36
  bats, 31–32, 158
  bees, 37–38
  birds, 167–168
  howler monkeys, 30
  leaf-cutter ants (*Atta*), 36–37
  leaf-eating insects, 164–168
  mammals (total), 156–158, 159, 164, 172
  primate communities, 162, 163
  sloths (*Bradypus*), 31
Forest
  biomass and production of. *See* Biomass and production
  diversity of. *See* Tropical diversity
Forest Dynamics Plot, BCI
  analyses of diversity, 180, 181, 194, 192–195, 198
  analyses of mortality and recruitment, 122, 136
Forest Dynamics Plots, Asia
  analyses of diversity, 180 194–195, 201
  analyses of mortality and recruitment, 122
Frugivores
  fig trees and, 31–32, 158, 223
  terrestrial, 27, 28
  vertebrate, 149, 152, 153, 156–157, 158
  Virola, attraction to, 19
Fruiting
  Faramea fruit, 21–22
  gregarious fruiting in Malaysia, 160, 211–212

  "Group Selection," 214–215, 223–225
  Hybanthus fruit, 22
  palm fruiting cycle, 18, 19
  Tetragastris fruit, 20–21
  Virola fruit, 19, 21
  See also Seasonal rhythms of fruiting, leaf flush, and animal populations

Gabon
  seasonal shortages of fruit, 162
  terrestrial frugivores, 160
  tree diversity, 179
Galapagos Islands
  during the Pleistocene, 9
  seasonal food shortages, 161
Garcinia, 85
Gatun Lake, 2, 11, 28, 51, 61, 185
Genes, 213, 216–217
Genesis, Book of, 233
Geophila repens (Rubiaceae), 98
Gilbertiodendron dewevrei, 194, 222
Glaciers, 8, 9
Gondwana, 5, 8
Greenhouse gases, 57, 59
Gross photosynthesis, 125–130
Guatemala, 9
Gustavia superba (Lecythidaceae), 87
Gymnosperms, 138
Gynacantha membranalis, 38

Hamlets, 215–216
Hawaii
  decomposition of fern fronds, 212
  gross photosynthesis, 129–130
  tree diversity, 182
"Heat engines," Carnot, 55
Herbivores, 149, 150, 153, 158–163, 188, 191
  antiherbivore defenses, 110, 162–163, 188, 191–192, 212
Hermaphrodites, 215–216
Hevea brasiliensis, 84
Hirtella triandra, 193
Holocene period, early, 10
Humidity, evapotranspiration and, 55
Hura crepitans (Euphorbiaceae), 87
Hurricanes, 55–56, 74
Hybanthus prunifolius (Violaceae), 22–23, 100–101, 153, 156, 193
Hydraulic architecture of plants, 107, 149, 154–156

Iguana iguana (Iguanidae), 33–34
India, 8
  pest pressure, 194
  tree diversity, 182
  vertebrate herbivores, 163

Industrial age, 11
Insects
    Africanized honeybees, 37
    aphids, 191
    army ants, 35–36
    Barro Colorado's, 8
    bee pollination, 16, 26, 212
    bruchid beetles, 18, 19, 191
    butterfly diversity, 195
    damselflies, 38–39
    euglossine bees, 26
    fig wasps, 26–27, 221
    herbivorous, 150, 167–168, 191
    honeybees, 37, 217–218
    how much foliage do insects eat?, 164–167
    and leaf flush, 132
    leaf-cutter ants, 36–37, 215
    numbers and biomass of arboreal insects, 164
    pest pressure and tree diversity, 190–195
    population limitation by birds, 167–168
    soapberry bugs, 191
    stingless bees, 5, 8, 37–38
    termites, 36
    of tree crowns, 163–168
    yucca moths, 219
Interated Prisoner's Dilemma (game), 215
Intertidal at Tatoosh Island, northeastern Pacific, 15, 130
Intertropical Convergence Zone, 47
Iridaea cornucopiae, 15
Isthmus of Panama
    and the great biotic interchange, 5, 8
    rainfall gradient across, 47

Jurassic period, 8, 212

Kaolinitic soils, 72
Kin selection, 216–218, 223, 224
    in coatis, 29, 217

Laminaria, 15
Landslides, 74–75
Laurasia, 5
Leaf area index and leaf fall, 120–125, 126, 128–129, 139
Leaf arrangement
    monolayers and multilayers, 93–95
    optimal divergence angle for spiral leaves, 80–82
    See also Tree shapes and leaf arrangements
Leaf flush, 150, 151, 152–156
Leaf size and shape
    altitudinal gradients in, 102, 103, 104
    relative uniformity in tropics, 79

use in inferring past climates, 79, 102
Leaves, 133
    herbivory and fallen, 165–166
    herbivory and marked, 166–167
Leonardoxa africana, 165
Leptospermum flavescens (Myrtaceae), 107
Lessoniopsis, 15
Lianes, 121, 134–135, 211
Light levels at forest floor, 124–126
Liriodendron tulipifera, 166
Lonchophyllus pentaphyllus, 192
Luehea spp., 150
    Luehea seemanii, 107
        insects in crowns of, 195
        water transport and hydraulic architecture in, 107

Mabea occidentalis (Euphorbiaceae), 89
Macropores, 72–73
Macrotermes carbonarius, 36
Madagascar, 9
    deforestation, 69, 232
    leaf arrangement in forests of, 100, 101, 104
    lemur populations and foliage quality, 162, 163
    litter leaf size, 102, 103
    rainfall, raindrop size and energy of impact, 49, 53
    seasonal shortages of fruit and new leaves, 162
    tree architecture, 95, 100
    tree diversity, 182, 183
Madden Lake, 51, 61
    evaporation from, 61
    sediment accumulation in, 51
Major Transitions of Evolution, The (Maynard Smith and Szathmáry), 213
Malaysia
    elfin forests, 107, 110
    evapotranspiration, 76
    forest dynamics and diversity at Pasoh, 122, 180
    gregarious fruiting of dipterocarps, 160, 211–212
    leaf area index, 121, 122
    leaf arrangement, 100, 101, 104
    litter leaf size, 102, 103
    pan evaporation, 61, 63
    termites, 36
    tropical forest canopy, 105
    wood respiration, 138
Mammals. *See* Animals
Mangroves, 103–104, 107, 121, 122, 182
Manu, see Peru (Amazonian)
Mauritius, tree diversity, 182, 184

Mazama americana, 157
Mecistogaster spp., 38–39
Megafauna, 10–11, 163
Megaloprepus coerulatus, 38–39
Meiosis, 216
Melipona panamica, 37
Meliponinae, 5, 8, 37–38
Miconia argentea, 188
Micropores, 73
Microrhapias quixensis, 32
Milankovitch cycles, 57
Mitochondria, 214–215
Mixed antwren flocks (Formicariidae), 32–33
Mixed bird flocks, 218
Montmorillonitic soils, 72
Moronobea coccinea (Guttiferae), 185
Mortality rates of trees. *See* Trees, death rates of
Mutualism in tropical forest, role of
    appendices, 223–226
    centrality of, 211–213
    do pests promote mutualism among tropical trees?, 221–222
    ecosystems and economics, 219–220
    fig trees, 222–223
    harmonious ecosystems, 220–223
    how can mutualisms evolve?, 214–218
    how mutualisms evolve, 218–219
    interdependences of tropical forest, 211–212
    kin selection and social behavior, 216–218
    neighborhood selection, trees, and mutualism among species, 220–221
    reciprocal altruism: mutually enforced common interest, 215–216
    relationship to competition, 213–214
    selection among groups and the common good, 214–215
    unenforced mutualisms, 218
    varieties of group selection, 223–226
    why mutualism is the central problem of evolutionary biology, 212–214
Mycorrhizae, 134
Myrmotherula spp., 33
    Mymotherula fulviventris, 32
    Myrmotherula axillaris, 32–33

Nasua narica (Procyonidae), 27, 29, 217
Natural selection and "selfish" genes, 213

Neea chlorantha (Nyctaginaceae), 94
Neotropics, 8, 10
New Guinea
   cost of tree growth, 136–137
   elfin forests, 108
   tree diversity, 182
New Zealand, 50
North Carolina
   insect biomass, 164
   pests and plant defenses, 191
Niche differentiation
   and coexistence, 187–188
   among giant damselflies, 37
   among terrestrial mammals, 187–188
   temporal sorting in tree recruitment, 189–190
Nutmeg family, *Virola* spp., 19–20
Nyssa aquatica, 108

Ochroma pyramidale, 189
Ocotea whitei, 10, 193
Oenocarpus mapora, 185
Oligocene Brandon lignite, 5
Olmedia aspera (Moraceae), 10, 88
Orchid Island, 28, 36
Orchids, epiphytic, 25–26
Oxisols, 72, 73, 76, 131

Pacas. *See Agonti paca*
Pacific, northeastern, kelp, 55, 130
Palm trees
   black, 17–18
   Scheelea, 18–19
Panama
   El Niño droughts in, 47
   pan evaporation, 63
Panama Canal, 3, 4, 11
   deforestation dangers to, 51
   Panama Canal treaties, 3
Panama Railroad, 11
Papaya tress, 84
Pentaclethra macroloba, 101
Permian period, early, 59
Permo-Triassic period, 5
Peru (Amazonian)
   birds: numbers, biomass and consumption, 157, 159, 160, 168
   butterfly diversity, 195
   elfin forests, 107, 108
   fruit production, 156
   mammal biomass and consumption, 157, 159, 160, 162, 164, 172
   tree diversity, 183, 195
   understory bird flocks, 33
Pesticides, 76, 110
Philydor, 33
Photosynthesis
   Michaelis-Menten equation for leaf photosynthesis, 126–127
   Calculation of whole-forest photosynthesis, 127
   Measurement of daily photosynthesis for sun-leaves, 127
   Measurement of whole-forest photosynthesis, 128–129
Phyllostomatidae, 8
Physalaemus spp. (Leptodactylidae)
   Physalaemus coloradorum, 34
   Physalaemus pustulosus, 34–35
Phytoliths, pollen and vegetation of the past, 9–10
Pinus contorta, 91
Piper spp., 10, 153
Pisaster, 15, 193
Pisolites, 72
Plants
   and animals, 8
   ferns, 212
   kelp, 15, 55, 130, 158
   principal plants of Barro Colorado, 15–27
   productivity, ultimate limits on, 130
   stomatal index, 59
   See also Seasonal rhythms of fruiting, leaf flush, and animal populations
Platypodium elegans (Leguminosae, Papilionoideae), 16–17, 25, 192
Pleistocene, cyclic revolutions of the, 8–10
Pollination
   and angiosperm diversification, 212
   by bees, 16, 26, 212
   of dipterocarps, 211–212
   of durians, 211
   of fig trees, 26, 222
   pollinators and seed dispersers, and tree diversity, 184–185, 212
Polypodium crassifolium, 25
Population densities
   in representative insects, 35, 37, 38
   in representative mammals, 28, 29, 30, 31, 32, 163, 164, 172
   in representative reptiles, 33, 34
Population limitation. *See* Population regulation
Population regulation
   by amount of food, 27, 28–29, 30, 31, 36, 38, 159–162
   by food quality, 162–164
   by predators, 158–159, 167–168
   in agoutis, 27, 159–160
   in *Anolis limifrons*, 34
   in giant damselflies, 38
   in howler monkeys, 30, 160
   in spiny rats (*Proechimys semispinosus*), 28–29
   in white-faced monkeys (*Cebus*), 34
   need to reckon with spatial heterogeneity, 195–196
Postelsia palmaeformis, 15, 130
Poulsenia armata, 10
Pourouma aspera, 101
Pressure-bomb measurements, 154
Prioria copaifera (Leguminosae), 99
Proechimys semispinosus (Echimyidae), 27, 28–29
Protium panamensis, 185
Pseudobombax septenatum, 155, 189
Pseudostigmatidae, 38–39
Pseudotsuga menziesii, 91
Psychotria spp. (Rubiaceae), 23–24, 153
   Psychotria acuminata, 23, 87, 91, 110
   Psychotria chagrensis, 23, 24
   Psychotria deflexa, 23, 87
   Psychotria furcata, 23, 24, 153–154
   Psychotria horizontalis, 23–24, 31, 153, 155
   Psychotria limonensis, 23, 24
   Psychotria marginata, 23, 24, 150, 153, 188–189
Puerto Rico
   cost of increasing tree height, 137
   elfin forests, 107, 108, 110
   leaf area index, 122
   rain forest gross production, 128
   tree diversity, 182

Quararibea spp. (Bombacaceae), 31
   Quararibea asterolepis, 90, 99, 152, 192–193
Quickflow, 68

Rainfall
   drop size and erosive power, 49–51
   duration of rainfall events, 48
   intensity, 48
   minus runoff: evapotranspiration?, 52–53
   perturbations of the rhythm, 47
   the seasonal rhythm, 46–47
   and transpiration, 47–48
   where does the rainwater go?, 51–52
Rainforest
   endangered, 232–234
   energy budget of, 138–139
Ramphastos swainsonii, 20

Reciprocal altruism, 215–216
Reiteration and tree architecture, 82, 84–85
Religion, biblical view of creation, 233
Republic of Panama, 5, 11
Rhipidocladum racemiflorum, 155
Rhizophora mangle (Rhizophoraceae), 105, 155
Rinorea viridifolia (Violaceae), 94
Rubber trees, 84
Runoff, erosion, and soil formation, 67–78
   erosion and soil formation on Barro Colorado, 51, 70–74, 75
   implications for other sites, 74–75
   modes of runoff, 67–70
   runoff and weathering, 70

St. Francis of Assisi, 233
Sapindaceae, 191
Savanna habitats, 8–9, 10
Scheelea zonensis (Palmae), 18–19, 31, 151, 185
Schefflera spp., 82, 94
   Schefflera morototoni, 90, 91, 154–155, 156, 189
   Schefflera pittieri, 91, 110
Sea surface temperatures, 9
Seasonal rhythms of fruiting and leaf flush, 149–163, 170–171
   gregarious fruiting of dipterocarps (Malaysia), 160, 211–212
   how timing of leaf flush, flowering, and fruiting is controlled, 152–156
   impact on vertebrate folivores and frugivores, 156–162
   regulation of populations of vertebrate folivores and frugivores, 156–157
   seasonality of flowering and fruiting, 150–153, 160, 161, 163, 170–171
   seasonality of leaf flush, 150, 151, 163
Seed dispersers
   Faramea seed dispersal, 22
   fruit-eating bats as, 16, 222
   impact on angiosperm diversification, 212
   megafauna and, 10
   role in maintaining tree diversity, 184–185
   seed dispersal, 8, 16–17
   two-stage dispersal, 16, 20, 185, 211
   Virola seed dispersal, 19–20
Sex allocation in hamlets, 215–216
Sex ratio in fig wasps, 221, 225–226

Sexual selection
   in frogs, *Physalaemus*, 34–35
   in giant damselflies, *Megaloprepus*, 37–38
   and speciation, 35
Shrubs, 22–24
Silver fir trees, 134
Smectites, 72
Smithsonian Institute, 3
Smithsonian Tropical Research Institute (STRI), 3, 15
Soil and vegetation, 75–76
Soil formation. *See* Runoff, erosion, and soil formation
Soil organisms, 221–222
Solar radiation
   Similar levels at floors of different forests, 124, 126
   Similarity for different tropical canopies, 58, 128–129
South America, 8
   during the Pleistocene, 9
   megafauna of, 10–11
Southern Oscillation, 11, 47
Speciation
   mechanisms of, 185–187
   time required for, 35, 187
Species, coexistence of
   and seasonal shortage of food, 161
   general considerations, 187–189
   imposed by pest pressure, 190–195
   imposed by temporal sorting of recruitment, 189–190
Species diversity. *See* Tropical diversity
Spondias mombin, 107
Sri Lanka
   decomposition of plant matter in fernland and forest, 212
   tree diversity, 182
Starfish, 15, 193
Stomatal resistance, 106–107
Strangler figs, 26
Suboscines, 8
Sumatra, 131
Swartzia simplex, 185

Tabebuia spp. (Bignoniaceae), 156
   Tabebuia guayacan, 25, 153
   Tabebuia neochrysantha, 153
   Tabebuia rosea, 85
Tanzania, elfin forests, 108, 112
Tatoosh Island (northeastern Pacific), 15
   kelp productivity on, 130
Taxodium distichum, 108
Tayassu tayacu, 27
Tectaria incisa (Polypodiaceae), 97
Television, tropical nature serials on, 233–234

Temperature, evapotranspiration and, 56–59
Temperature regulation in sloths, 31
Terminalia spp., 82, 85
   Terminalia catappa, 82
Tertiary period, early, 184
Tethys, 5, 8
Tetragastris panamensis (Burseraceae), 20–21, 192
Thamnomanes spp., 33
Thamnophilus punctata, 32
Theory of Games and Economic Behavior (von Neumann and Morgenstern), 219
Thevetia ahouvai (Apocynaceae), 88
Thorn scrub, savanna and, 8–9, 10
Throughfall, impact energy of, 50
Thunderstorms, convective, 47, 55
Tit for Tat, see Iterated Prisoner's Dilemma
Trade-offs and species diversity, 187–191
"Tragedy of the Commons" (Hardin), 214
Transpiration
   rainfall and, 47–48
   tree crown resistance and, 106–107
   See also Evapotranspiration
Tree shapes and leaf arrangement, 79–119
   canopy roughness, 104–107
   designing trees: goals and constraints, 85–94
   empirical measures of tree-performance, 91–93, 94
   herbs of the forest floor, 95
   leaf size, leaf physiognomy, and their correlates, 102–104
   limits on tree height, 85–86, 90, 107–113, 136–138
   monolayers and multilayers, 93–94
   optimal leaf arrangement, 80–82
   tree architecture and leaf arrangement, 95, 100–101
   tree architectures: classification, 82–85
   trunk and branch design, 90–91
   why are elfin forests stunted?, 107–112, 120, 137–138
Trees
   basal area of, 121, 122, 123, 124, 132, 134, 138, 139–142
   on Barro Colorado, 15–22, 26–27
   death rate of, 55, 121, 122, 136, 195
   diversity of, 179–195
   leaf biomass, 121, 123, 124, 125, 134, 136, 137

leaf fall, 120, 121, 122, 124, 134, 135, 138, 139–142, 167
leaf production, 133
leaves, 133
mutualism among tropical, 220–223
pest pressure and diversity of, 190–195
root production, 131, 133–134
rooting depth, 133-134, 211
sapwood area, 91, 93, 107, 122, 123, 136
wood production, 134–138
wood respiration, 135–136
See also Seasonal rhythms of fruiting, leaf flush, and animal populations; Tropical diversity
Trema spp., 94
Trema micrantha, 188
Trichilia spp.
Trichilia cipo, 152
Trichilia tuberculata, 192, 193
Trigona, 5, 8
Tropical diversity, 179, 197
appendices, 198–204
consequences of tree diversity, 195–197
difficulties of testing the pest pressure hypothesis, 193–195
do pests kill the young nearest their parents?, 192–193
empirical correlates of tree diversity, 182–184
explaining the diversity of tropical trees, 185–195
how can we measure tree diversity?, 179–182
necessary conditions for, 184–185
pest pressure and tree diversity, 190–191
specialization and trade-offs in plant defenses, 191–192
temporal sorting of recruitment, 189–190
trade-offs, habitat heterogeneity, and tree diversity, 188–189
Tulip poplars, 166
Typhoons, 56

Uca beebei, 34–35
Ultisols, 76, 182
Unonopsis pittieri (Annonaceae), 88
Urea, 191
Urera spp., 150

Vegetation, soil and, 75–76
Venezuela
during the Pleistocene, 9
elfin forests, 107
leaf fall and leaf area index, 122, 123
rainfall, 53
tree diversity, 195
tree leaves, 103
Vertebrate folivores, 156–157
Vertebrate frugivores, 149, 152, 153, 156–157, 158
Vertebrate herbivores, 149, 150, 153, 158–163, 188, 191
antiherbivore defenses, 212
Virola spp. (Myristicaceae), 19–20, 82, 95
Virola "bozo", 19
Virola nobilis, 19–20, 21
Virola sebifera, 19–20, 83, 96
Virola surinamensis, 10, 19, 192, 211
Voucapoua americana (Caesalpinoideae), 185

Weinmannia spp. (Cunoniaceae), 111
Western Hemisphere Convention on Nature Protection and Wildlife Preservation, 3
Wind, evapotranspiration and, 55–56

Xanthosoma spp. (Araceae)
Xanthosoma helleboricum, 97
Xanthosoma pilosum, 97
Xylem tension, plant, 154, 155, 189
Xylopia frutescens, 94

Zaire
pest pressure, 194
trees and neighborhood selection, 222
Zuelania guidonia (Flacourtiaceae), 88